THE ROUTLEDGE HANDBOOK OF DEVELOPMENT AND ENVIRONMENT

The handbook seeks to illuminate the key concepts in the study of development-environment through showcasing some of the Majoritarian (formerly "Developing") world's emerging scholars in order to explore theoretical connections through critical/radical theory, "small" theory, various conceptual frameworks, and non-Western and subaltern viewpoints.

The volume examines the themes around the study of the relationship between economic and social development and the environment. Part 1 covers theoretical and conceptual approaches to the study of development and environment by examining the diverse ways in which people perceive, understand, and act upon the world around them. Cross-scalar topics such as neo-liberalism and globalization, human rights, climate change, sustainability, and technology are covered in Part 2. The book shifts to examinations of resources and production in Part 3, where authors with a focus on one or more environmental resources or types of economic production are presented. Topics range from water, agriculture, and food, to energy, bioeconomy, and mining. The fourth section presents chapters where people are at the center of the development-environment nexus through topics such as gender relations, children, health, and cities. Finally, policy and governance of development and environment are explored in Part 5. The section includes both academics and practitioners who have worked with policy makers and are policy makers themselves.

The book is primarily intended for scholars and graduate students in geography, environmental studies, and development studies for whom it will provide an invaluable and up-to-date guide to current thinking across the range of disciplines, which converge in the study of development and environment.

Brent McCusker is Professor of Geography and Department Chairperson in the Department of Geology and Geography at West Virginia University, USA. He has published extensively on land use and livelihoods systems in sub-Saharan Africa. He also works with USAID on livelihood vulnerability analysis and mapping across a range of countries in sub-Saharan Africa and south Asia.

Waquar Ahmed is an assistant professor of geography at the University of North Texas, USA. He has been Chair of the Development Geographies Specialty Group of the Association of American Geographers (2015–2017). His research interests are in development-underdevelopment, state theory, foreign direct investments, and social movements.

Maano Ramutsindela is Professor in the Department of Environmental and Geographical Science at the University of Cape Town, South Africa. He has conducted research in Botswana, Namibia, Lesotho, South Africa, and Tanzania focusing on transfrontier conservation, land, and regions.

Patricia Solís is Associate Research Professor in the School of Geographical Sciences and Urban Planning at Arizona State University, USA. She is the executive director of the Knowledge Exchange for Resilience at Arizona State University, an interdisciplinary effort to link multi-sector community needs with research innovations around resilience.

THE ROUTLEDGE INTERNATIONAL HANDBOOK SERIES

THE ROUTLEDGE HANDBOOK OF DIGITAL LITERACIES IN EARLY CHILDHOOD
Edited by Ola Erstad, Rosie Flewitt, Bettina Kümmerling-Meibauer and Íris Susana Pires Pereira

ROUTLEDGE INTERNATIONAL HANDBOOK OF UNIVERSITIES, SECURITY AND INTELLIGENCE STUDIES
Edited by Liam Gearon

THE ROUTLEDGE INTERNATIONAL HANDBOOK OF RESEARCH ON DIALOGIC EDUCATION
Edited by Neil Mercer, Rupert Wegerif and Louis Major

THE ROUTLEDGE INTERNATIONAL HANDBOOK OF YOUNG CHILDREN'S RIGHTS
Edited by Jane Murray, Beth Blue Swadener and Kylie Smith

ROUTLEDGE INTERNATIONAL HANDBOOK OF WOMEN'S SEXUAL AND REPRODUCTIVE HEALTH
Edited by Jane M. Ussher, Joan C. Chrisler and Janette Perz

THE ROUTLEDGE INTERNATIONAL HANDBOOK OF STUDENT-CENTRED LEARNING AND TEACHING IN HIGHER EDUCATION
Edited by Sabine Hoidn and Manja Klemenčič

THE ROUTLEDGE HANDBOOK OF ENGLISH LANGUAGE EDUCATION IN BANGLADESH
Edited by Shaila Sultana, M. Moninoor Roshid, Md. Zulfeqar Haider, Mia Md. Naushaad Kabir, and Mahmud Hasan khan

For more information about this series, please visit: www.routledge.com/Routledge-International-Handbooks-of-Education/book-series/HBKSOFED

THE ROUTLEDGE HANDBOOK OF DEVELOPMENT AND ENVIRONMENT

Edited by Brent McCusker, Waquar Ahmed, Maano Ramutsindela, and Patricia Solís

Routledge
Taylor & Francis Group
LONDON AND NEW YORK

First published 2022
by Routledge
2 Park Square, Milton Park, Abingdon, Oxon OX14 4RN

and by Routledge
605 Third Avenue, New York, NY 10158

Routledge is an imprint of the Taylor & Francis Group, an informa business

British Library Cataloguing-in-Publication Data
A catalogue record for this book is available from the British Library

Library of Congress Cataloging-in-Publication Data
Names: McCusker, Brent, editor. | Ahmed, Waquar, editor. | Ramutsindela, Maano, editor.
Title: The Routledge handbook of development and environment / edited by Brent McCusker, Waquar Ahmed, Maano Ramutsindela, and Patricia Solís.
Description: 1 Edition. | New York, NY : Routledge, 2022. | Series: Routledge international handbooks | Includes bibliographical references and index.
Identifiers: LCCN 2021020692 (print) | LCCN 2021020693 (ebook) | ISBN 9781138325661 (hardback) | ISBN 9780367710231 (paperback) | ISBN 9780429450310 (ebook)
Subjects: LCSH: Economic development—Developing countries. | Environmental policy—Developing countries. | Sustainable development—Developing countries.
Classification: LCC HC59.7 .R668 2022 (print) | LCC HC59.7 (ebook) | DDC 338.9009172/4—dc23
LC record available at https://lccn.loc.gov/2021020692
LC ebook record available at https://lccn.loc.gov/2021020693

ISBN: 978-1-138-32566-1 (hbk)
ISBN: 978-0-367-71023-1 (pbk)
ISBN: 978-0-429-45031-0 (ebk)

DOI: 10.4324/9780429450310

Typeset in Bembo
by Apex CoVantage, LLC

CONTENTS

FIGURES

TABLES

CONTRIBUTORS

Ana Sofia Solano Acuña is a doctoral candidate in history and humanistic studies at the Pablo de Olavide University in Spain. She holds a position as a research associate in the Colegio de América at the same university and is a member of the teaching and research faculty in the Social Sciences Department at the National University of Costa Rica. She has been a senior researcher for ten years in the Population Studies for Equity with a Gender and Cultural Diversity Perspective program. Her research expertise focuses on indigenous forms of resistance, indigenous political thought, dialogue between the State and subaltern sectors, historical memory, ethnic interpretation of history, and collaborative processes for a more inclusive hertoriography.

C. Nancy Aguirre is Leader of the Geography Commission Research Committee of the Pan American Institute of Geography and History, an international, scientific, and technical organization of the Organization of American States. Her expertise centers on human-environment relations, spatial analysis, land use Science, Latin America, and the Caribbean. She was Vice-Chair for South America, of the Association for Global Spatial Data Infrastructure Outreach and Membership Committee, and co-lead of the Group on Earth Observation Biodiversity Observation Network (GEOBON), Working Group 3: Terrestrial ecosystem change, deliverable #7 on Capacity Building. She earned her PhD in Geography and the Environment from the University of Texas at Austin.

Waquar Ahmed is an assistant professor of geography at the University of North Texas, USA. He has been Chair of the Development Geographies Specialty Group of the Association of American Geographers (2015–2017). His research interests are in development-underdevelopment, state theory, foreign direct investments, and social movements.

Alexandra Areiza-Tapias has a bachelor's degree in environmental administration and a master's in geography from the University of the Andes, Colombia. Her experience includes territorial planning processes with an emphasis on biodiversity conservation, especially in aspects related to the integration of protected areas, as well as complementary biodiversity conservation strategies in territorial plans. Alexandra is interested in analyzing the power relations around conservation processes, especially those that involve high restrictions on land use, land transactions, and livelihoods of local communities in Colombia.

Carlos Morera Beita is Professor at the National University of Costa Rica, where since 1993 he has been Head of the Geography School and President for Research. He has published more than 50 articles and five books related to environment and geography. He is a geographer with a bachelor's degree from the National University of Costa Rica, a master's degree from the University of Arizona, and a PhD from the University of Barcelona, Spain.

B.S. Butola completed his Masters and MPhil/PhD from the Center for the Study of Regional Development, Jawaharlal Nehru University. He taught for 16 years in the Department of Geography, North Eastern Hill University Shillong, Meghalaya. He joined Jawaharlal Nehru University as Associate Professor in 1998 and taught there for 22 years. He has supervised 35 PhD and 50 MPhil scholars and contributed in Geography Text Book writings for IX, X, XI, and XII standards for National Council for Educational Research and Training. He also contributed in writing an Encyclopedia for the Ministry for Human Resource Development, Government of India. He had been a visiting Professor at King's College London. He is currently a visiting Professor at the School of Planning and Architecture, New Delhi.

Ipsita Chatterjee is Associate Professor in the Department of Geography and the Environment, University of North Texas. She completed her PhD in geography at Clark University, Massachusetts and was an Assistant Professor at Pennsylvania State University, and later, at the University of Texas at Austin. Chatterjee focuses on globalization, urban change, class, identity, labor, Marxism, and feminism. Chatterjee's first book titled *Displacement, Revolution, and the New Urban Condition*, published by Sage, deals with globalization in an Indian city, gentrification, and associated displacement of urban poor, particularly, the Muslim poor. Her second book titled *Spectacular Cities: Religion, Landscape and the Dialectics of Globalization,* published by Oxford, deals with the Disneyization of urban landscapes. In this book, Chatterjee explores migrant temple landscapes in Dallas and Atlanta in the US and their counterparts in India to reveal how landscapes are narratives that ground globalization, citizenship, patriotism, gender, and sexism. Her most recent book is titled *Alt-Right Movement: Dissecting Racism, Patriarchy and Anti-Immigrant Xenophobia* and has been published by Sage in May 2021. Chatterjee has also authored numerous articles in academic journals like *Urban Studies*, *Transactions of the Institute of British Geographers*, *Geoforum*, *Gender, Place and Culture* among others.

Bikramaditya K. Choudhary teaches at the Centre for the Study of Regional Development, Jawaharlal Nehru University (JNU). Prior to joining JNU, he taught at Banaras Hindu University. He has over two dozen publications in different Journals and Books on issues of political ecology, development, informal economy, waste-pickers, history of tuberculosis, ecology of disease, regional disparity, and other related issues of development. He has edited a book titled *City, Space and the Politics in the Global South* (Manohar 2019/Routledge 2020). His current area of research includes urban services, urban identity formation, and spatial segregation. Currently, he is engaged with three externally funded projects on the issue of water security, access, and transformative pathways towards urban sustainability.

Jamison Conley is Associate Professor of Geography at West Virginia University. His research is on spatial statistics and their applications. Recent projects include analysis of access to public health services and analysis of demographics to support international development.

Alicia Danze is a doctoral student and teaching assistant at the University of Texas at Austin in the Department of Geography and the Environment. She studies navigation of immigration institutions, focusing on immigration court proceedings in Texas. Her research includes discourse analysis surrounding the development of detention centers in suburban and rural communities across the state. Previously, she served as immigrant and refugee rights paralegal at the Washington Lawyers Committee for Civil Rights and Urban Affairs in Washington, DC, USA.

Ana Gómez Donoso leads a group to integrate gender studies on the research committee of the geography commission of the PanAmerican Institute for Geography and History. Her research expertise spans topics in rural development, indigenous communities, feminist geography, ecofeminist philosophy, and gender violence. Her previous positions include work with the World Bank, and as a humanitarian advocate with Doctors Without Borders. She holds an engineering degree in geography and sustainable development from the Catholic Pontific University of Quito, Ecuador and a master's in socio-environmental studies from the Latin-American School of Social Sciences, Mexico. Her PhD is in gender studies from the University of Granada, Granada.

Samantha Espinoza-Villejo is critical geographer from San Antonio, Texas. Specializing in disaster and geography, Samantha is most interested in ways that governments can reimagine themselves to serve vulnerable environments and communities. The work she does centers on philosophies that rejoin the concepts of human and environment into a singular complex web.

Tim Essam is Senior Data Scientist of the Office of HIV/AIDs at the United States Agency for International Development (USAID). He holds a PhD in Agricultural and Resource Economics from the University of Maryland.

Caroline Faria is Assistant Professor in the Department of Geography and the Environment at the University of Texas at Austin. She is a feminist political and cultural geographer whose research seeks to understand nationalism and neoliberal globalization, with a focus on the Gulf-East African regions. Her work has focused on the US-based South Sudanese diaspora and contemporary processes of gendered development and nation-building. Through a feminist and critical race lens, she examines how the nation is bounded, reproduced, and contested in the politics and performances of gendered bodies. She earned a PhD from the University of Washington, USA.

Robert Melchior Figueroa is Associate Professor of environmental justice and philosophy in the School of History, Philosophy, and Religion at Oregon State University. He has advanced recognition and restorative justice in environmental justice studies, introduced philosophy of environmental identity and heritage with communities, peoples, and spaces around the globe.

Alistair Fraser is lecturer in the Department of Geography at Maynooth University, Ireland. His research, which cuts across political, economic, and cultural geography, engages the struggle for land in South Africa, rural change generally, and foodscapes. He is the author of *Global Foodscapes: Oppression and Resistance in the Life of Food* (Routledge, 2017) and co-author with Rob Kitchin of *Slow Computing: Why We Need Balanced Digital Lives* (Bristol University Press, 2020).

Jayson J. Funke formerly worked in investment banking before earning his PhD from Clark University. He has taught at Brandeis University, Mount Holyoke College, and Framingham

State College. He is currently privileged to teach geography at Medgar Evers College, CUNY and serves as Editorial Associate for the radical journal *Human Geography*.

Richard Furtick is a geographer, musician, and poet interested in the intersections of music and place, and the recuperation of radical music into mainstream society. He studied at the University of North Texas and still lives in the Denton area. He spends most of his day staring at sentences and trying to write his own.

Margaret Hinrichs is a postdoctoral research associate in the School for the Future of Innovation in Society at Arizona State University. She also serves as Program Manager in the Decision Theater Network, and supports research in the Urban Resilience to Extremes Sustainability Research Network at ASU. Her expertise focuses on participatory agenda setting, governance informatics, transdisciplinary organizing and decision-making. She earned her PhD in organizational communication from Arizona State University, a master's of arts in organizational communication from Saint Louis University and a bachelor's degree from the University of Missouri School of Journalism.

Daniel Avendaño Leadem is an academic professor and member of the faculty of the School of Geographic Sciences at the National University of Costa Rica. He specializes in sustainable development, and has published on integration of ecosystem services in land use planning. He belongs to the Program for Territorial Tourism Studies where he coordinates research and social extension projects related with tourism, sustainability, and economic geography. He holds a master's of science in urban and regional planning from the University of Greenwich, and in applied resource efficiency in architecture and planning from HafenCity University, Germany.

C. Melinda Levin, Professor, University of North Texas, is a documentary film director focusing on the human-environmental interface. Her films and research have taken her to fragile landscapes and waterways, including the American Southwest, Argentina, Chile, China, Cuba, Europe, Mongolia, Tajikistan, and Turkey. She is currently developing a web application for geo-storytelling, using the design tools of biomimicry.

Carlos Jorge Guilbe López is Professor in the Department of Geography at the University of Puerto Rico, Recinto de Río Piedras. His expertise centers on territorial planning, geographic education, and historical geography. He coordinates the First Time Use Survey Project in Puerto Rico. He is a member of the Puerto Rico Geographical Alliance and the American Association of Geographers. He also collaborates as a member of the Editorial Board of *The Professional Geographer*. He earned his PhD in geography and urban planning from the University of Wisconsin-Milwaukee, USA.

Salvatore Engel-Di Mauro is Professor at the Geography Department of SUNY New Paltz. His research encompasses critical physical geography, dialectical and historical materialism, gender-environment processes, socialism and environment, socialist histories, and soil acidification and contamination. He is chief editor for the journal *Capitalism Nature Socialism*.

Brent McCusker is Professor of Geography and Department Chairperson in the Department of Geology and Geography at West Virginia University. He has published extensively on land use and livelihoods systems in sub-Saharan Africa. His current research focuses on the implications of climate change on rural livelihoods and broader economic development in Malawi. He also works with USAID on livelihood vulnerability analysis and mapping across a range of countries in sub-Saharan Africa and south Asia.

Anand Prasad Mishra is Professor in the Department of Geography, Banaras Hindu University. He studies geography with the help of Marxist methodology, especially social geography through tools of geographical historical materialism.

Park Muhonda holds a PhD in geography from West Virginia University. He is currently working with Dartmouth College, Department of Geography as Research Associate. Park is interested in multidisciplinary research, employing a mixed methods approach. Recently, his research has focused on understanding rural livelihoods, vulnerability, and adaptation.

Luis Fernando Sandoval Murillo is a geographer with undergraduate degrees from the National University of Costa Rica, and a master's degree in GIS and remote sensing from the University of Costa Rica. He has participated in projects concerning landscape ecology and spatial planning and teaches courses related to GIS and remote sensing in the School of Geographic Sciences at the National University of Costa Rica.

Asubisye Mwamfupe is lecturer in the Department of Geography at the University of Dar es Salaam in Tanzania. He is also coordinating the Population Studies and Research Centre of the same University. His PhD training at the University of Dar es Salaam examined the potentials of local institutions to enhance the adaptability of smallholder farmers to climate change in rural Tanzania. Mwamfupe specializes in natural resource governance, climate and society, rural development, and the geography of health. He is currently involved in research projects on New Partnerships for Sustainability (NEPSUS) and Climate-Smart Coffee (PACSMAC) both in collaboration with Copenhagen Business School in Denmark. He is a co-author of *Public and Private Authority Interactions in Conservation and Development* (2020 with Stefano Ponte and Christine Noe).

Sagie Narsiah is Geographer by training. He received a PhD from Clark University, Massachusetts, USA. Presently, he is Associate Professor in the School of Social Sciences at the University of KwaZulu-Natal, Howard College Campus, Durban, South Africa, where he teaches and coordinates the Social Policy Program. While his primary focus is on the geography of development exploring, in the main, the impact of macro-economic development policies on local communities in particular, more recently, his research has included governance, public participation, sustainable development, and the social impacts of environmental change.

Christine Noe is Associate Professor of Human Geography at the University of Dar es Salaam. Her research is on conservation and development politics, and she seeks to explain how local and international development politics influence rural livelihoods, land tenure, and security dynamics in Tanzania. Her specific research foci in the recent years have been on conservation partnerships, transfrontier conservation areas, and rural livelihood changes. Noe's recent publications include *Regionalising Tourism Using Transfrontier Conservation Areas in Southern Africa* (2020); *Berlin Curse in Tanzania's Selous World Heritage Site* (2019); and *Graduated Sovereignty in Tanzania's Mineral Sector* (2019). She is co-editor (with Dan Brockington) of *Prosperity in Rural Africa? Insights from Longitudinal Studies in Tanzania* (2021). Noe is actively involved in providing leadership in research collaborations and mentorship of young African scholars.

Rhoda Nyirenda is a geography PhD student at West Virginia University. She received her master's in geography at West Virginia University and her bachelor's degree at the University of Malawi, Chancellor College. Born and raised in Malawi where livelihoods directly depend on nature and, having experienced the impact of climate change on agriculture, she is interested in rural livelihoods and climate change adaptation strategies.

Kwame N. Owusu-Daaku is Assistant Professor in the Department of Earth and Environmental Sciences at the University of West Florida. Kwame is a qualitative researcher with interests in human geography, urban planning, international development, sustainability, community engagement, and civic scholarship. His current research interests include human-environment interactions, climate change adaptation and resilience, contaminated land revitalization, science education, and the social aspects of biophysical science.

Maano Ramutsindela is Professor in the Department of Environmental and Geographical Science at the University of Cape Town, South Africa. He has conducted research in Botswana, Namibia, Lesotho, South Africa, and Tanzania focusing on transfrontier conservation, land, and regions.

Luis Sánchez-Ayala obtained his PhD in geography from Florida State University, Tallahassee. Currently, he is an associate professor of geography at the Department of History at the Universidad de los Andes, Bogotá, Colombia. His research interests focus on the spatiality of political and cultural issues, specifically questions related to human mobility; conflict and territory; land grabbing; conservation of nature and people's livelihoods; and borders and frontiers. He earned his master's in geography from the University of Akron, and bachelor's in geography from the University of Puerto Rico.

James-Eric Simon, a lifelong student of philosophy, discovered in the discipline of geography a lens through which to examine the intersection of knowledge, action, and space. Researching in the critical tradition, he studies the co-constitution of language and knowledge, as they relate to the urbanization of water and the production of hydro-social space. He is currently serving as a Research Engineering Science Associate with the University of Texas at Arlington, providing analysis for the Texas Commission on Environmental Quality.

Patricia Solís is Executive Director of the Knowledge Exchange for Resilience at Arizona State University, an interdisciplinary effort to link multi-sector community needs with research innovations around resilience. Her expertise centers on the application of participatory geospatial technologies for humanitarian, environment, and development needs. She is Associate Research Professor in the School of Geographical Sciences and Urban Planning at ASU. She co-founded and directs YouthMappers, a consortium of student-led chapters which have grown to 223 university campuses in 51 countries, working to create spatial data in collaboration with USAID GeoCenter. She is President of the Geography Commission of the PanAmerican Institute for Geography and History, of the Organization of American States. She earned her PhD in geography from the University of Iowa, USA.

Porter Swentzell is enrolled member of Santa Clara Pueblo, New Mexico. He grew up participating in the traditional life of his community and is passionate about Indigenous language and cultural vitality. Porter lives in his community with his wife and three kids. He is Chair and Associate Professor of Indigenous Liberal Studies at the Institute of American Indian Arts in Santa Fe, New Mexico.

Jean-Marie Théodat was born in Port-au-Prince in 1961. Following secondary studies in Haiti, he studied geography at the University of Paris Sorbonne up to the agrégation. He subsequently undertook PhD studies on the theme of the double insularity caused by the border between Haiti and the Dominican Republic. He was recruited as a lecturer at the University Paris 1 Panthéon Sorbonne, where he lectured in political, economic, and cultural geography from 1998 to 2010. He decided to return to Haiti in April 2010 post-earthquake to help with rebuilding efforts, in particular, the restructuring of higher education. Currently, he is the Director of the Caribbean Delegation of the Agence Universitaire de la Francophonie. He has been working as an International Technical Expert at the French Ministry of Foreign Affairs since 2012.

Rebecca Maria Torres is Associate Professor in the Department of Geography and the Environment, associate of the Teresa Lozano Long Institute of Latin American Studies, associate of the Population Research Center, and affiliate of the Center for Women's and Gender Studies at the University of Texas at Austin. Her areas of interest include (im)migration, children/youth geographies, gender, feminist geography, and activist/engaged scholarship. Since 2015, she has collaborated with a bi-national, transdisciplinary team of scholars focusing on refugee/migrant children and youth from Mexico and Central America. She holds a bachelor's degree in history, Ibero-American studies and Spanish from the University of Wisconsin-Madison, and a master's of science in international agricultural development. She earned her PhD in geography from the University of California at Davis.

Johannes Tsheola is Professor of Development Planning & Management and researcher at the Institute of Public Affairs & Governance in South Africa. He holds a doctoral degree in Development Geography and has been in academia and management since 1993, culminating into the position of Executive Dean which he held until April 2, 2019, at the University of Limpopo, South Africa. Since 2016, he has served as Deputy Editor for both the Journal and International Conference on Public Administration & Development Alternatives. During 2015–2020, the National Research Foundation (NRF) awarded him a C3 Rating. Tsheola's research specializations are in governance, state capitalism, and development, approached from the contested philosophical foundations of globality of coloniality vis-à-vis coloniality of globality.

Paula Soto Villagrán is a tenured research professor in the Department of Sociology in the Social Sciences and Humanities Division of the Autonomous Metropolitan University, Iztapalapa in Mexico. She has published more than 30 papers and chapters, and four books. Her expertise centers on sociology, urban studies, gender, cultural geography, and qualitative methodology, and explores the relationships among the city, mobility, urban habitat, fear of violence from a spatial and gender dimension. She holds a master's degree and PhD in anthropological sciences and is a member of the National System of Researchers of the National Council of Science and Technology of Mexico.

Steve Wolverton is Professor in the Department of Geography and the Environment at the University of North Texas. His research focuses on the historical ecology of Holocene South and North America – particularly in western Mendoza, Argentina and southwestern Colorado – with an interest in human-animal interactions.

Introduction

1

INTRODUCTION

Development and environment in the 2020s

Brent McCusker, Waquar Ahmed, Maano Ramutsindela, and Patricia Solís

As scholars in a field of study that examines development and environment equally vigorously, we need to move mainstream thinking past the idea that somehow, someway our current economic system of the (over)production of goods and service can be brought into alignment with long-term environmental sustainability. At the dawn of the 2020s, development and environment as both a field of study and as a domain of practice is fraught with doubt and uncertainty. Notions of linear paths of increasing levels of development have long since been dismissed. Nations that long dominated the global economy through manufacturing have declined dramatically, while others that suffered through famines and cultural chaos a mere 50 years ago have recently shown impressive growth. Environmental contradictions loom larger than ever across scales, from the global coronavirus pandemic reaching into households to accelerated global climate change melting ice caps, amplifying dramatic and devastating wildfires, and flooding coasts.

In the face of these challenges and contradictions, is the study of "development and environment" really all that valuable any longer? If we are truly "post-development" as some authors have suggested, and the political consensus on action to counter global climate change has been stalled by at least a few of the major actors, is pursuing twined studies of development and environment not little more than just documenting the damage? What is the benefit of continued studies of development and environment?

In this text, we explore themes around the study of the relationship between economic and social development and the environment. The co-editors of this text do not view these as separate fields of study that are "interrelated", but rather view them in dialectical relation with each other, inseparable as discrete objects of study. Far too often, development and environment are conceived as separate, somehow to be balanced or brought into greater harmony to secure an increase in well-being over the long term. While we review work on sustainable development, eco-Marxism, deep ecology, and other fields, we remain critical about the possibilities for any balance or sustainability of the current development-environment relationship. The remarkable ability of the interests of capital to escape any limits placed on it continues unabated. With the recent withdrawal (and then re-entry) of the United States from the Paris Accords on Climate Change, economic growth has once again been shown to be perhaps *the* single overriding objective of the global system. Not that one country's withdrawal would collapse the entire agreement, but when that country is one of the world's largest contributors to global carbon

 DOI: 10.4324/9780429450310-2

emissions, the impact of such actions on current and future sustainability cannot be underestimated. Climate scientists often warn of increasing variability in the climate system. We agree and submit that we also need to consider this in the context of increasing variability of many of the world's *political systems*.

The approach of this handbook is slightly different from others covering such a broad field of study. Rather than compiling the past accomplishments of well-established authors into a rarified compendium of the field, we chose a different tack. We sought to summarize thinking on some of the key topics in development-environment studies by showcasing some of the developing world's emerging scholars. Our preferences were to elevate voices that often get drowned, if heard at all, by more established, better-resourced, and generally more privileged authors from the global north, the US and UK in particular. We asked our authors to compile new, original research or significantly update past work that corresponded to some of the critical sub-areas of study in the topic, but to do so in their own voice, free of any heavy hand from the editors. We realize some of the shortcomings of such an approach. We know, for instance, that we couldn't possibly cover every topic and that there would inevitably be overlap with closely aligned fields of study. The chapters in this book were written before the recent novel coronavirus (COVID-19) outbreak and do not reflect the nexus of health, environment, and development. The contradictions that arise between development and environment around human health would take many a book to cover adequately.

We decided to accept these limitations if it meant this book would showcase voices (largely) from the global south. We chose a handbook of *immersion*, where the people live experiences with development and environment daily, rather than one that objectifies the topic from the rarified view of academia's ivory towers.

Snapshot of the field

While the benefit of such approaches to development and environment may be contested, the cost of inaction is clear. Continued exploitation of the environment as a cost of development has appeared to run up against hard limits. Continued tinkering with development to make it more "sustainable" also appear to have met their limits. In this text, we examine the fraught and contested relationship between development and environment. The topic is one that engages many related fields of study – environmental economics, sustainable development, political ecology, and green Marxism, to name only a few. We do not claim that "development and environment" is an exclusive field of study, somehow conceptually distinct from these other fields. In fact, there is great overlap. Environmental economics treats the environment, often conceptualized as "natural capital" as a commodity to be "priced" and factored into cost-benefit or profit equations. Much of the literature focused, therefore, on technical, methodological, or measurement issues around accurately pricing environmental goods and/or services. The market, many authors argue, will (eventually) correctly price these resources and the market can be relied upon to manage the development-environment nexus. As Redclift and Sage (1994) noted, such valuation is often based on exchange-values and not use values. The result is a wholesale undervaluation of the environment. We raise the prospect in this chapter that a dialectical understanding of development and environment will illuminate critical obstacles that have blocked any path toward sustainability.

Sustainable development

If eco- and green Marxism reject the notion that capitalism and nature can be held in dialectical tension, sustainable development adherents maintain that such a "dancing on the head of the pin" balance is not only possible, but a desirable state. The notion that capitalist development can be somehow managed or regulated into a less destructive relationship with the environment is a hallmark of approaches that slowly coalesced into what is now commonly referred to as "sustainable development". As Sachs and George (2015) noted,

> "sustainable development", although it can mean many things to many people, nevertheless contains a core message: keep the volume of human extraction/emission in balance with regenerative capacities of nature. That sounds reasonable enough, but it conceals a conflict that has yet to win public attention . . . sustainability, yes, but at what level.
>
> *(42)*

Sustainable development had been evolving as a concept after critical interventions questioning human impact on the environment, such as the publication of Rachel Carson's et al. *Silent Spring* (Carson, et. al, 2002) and at the UN Conference on the Human Environment in 1972. At the time, sustainable development was viewed by some as a challenge to modernization theories and approaches (Baker, 2006) that had largely set aside any environmental concerns and squarely promoted economic growth at any cost.

The idea of balance and sustainability crystalized in the *Brundtland Report* in 1987, where the driving notion of development and environment was utilizing the earth's natural resources for the greatest number of people in current and future generations. As has become commonplace, the issue of sustainability was raised and justified through the use of a crisis narrative. The report described the situation this way:

> Until recently, the planet was a large world in which human activities and their effects were neatly compartmentalized within nations, within sectors (energy, agriculture, trade), and within broad areas of concern (environment, economics, social). These compartments have begun to dissolve. This applies in particular to the various global "crises" that have seized public concern, particularly over the past decade. These are not separate crises: an environmental crisis, a development crisis, an energy crisis. They are all one.
>
> *(World Commission on Environment and Development, 1987)*

What is fascinating about the resort to the crisis narrative in the beginning of the report, is that the solutions proposed focused on either better management and/or governance of the *existing relationship* between development and environment. A core theme of many strands of sustainable development was that the worst effects of development on the environment could be "managed" (Redclift and Sage, 1994) to reduce the consumption of the environment and preserve some of it for the future. Management regimes, from corporate environmental policies to international carbon emissions agreements, all sought the same outcome – to somehow reduce the rapacious consumption of the environment that had long been a hallmark of economic development under capitalism (and, indeed, Soviet-style socialism!) but which continued with increased intensity throughout the latter half of the 20th century.

Management, as a technical approach, could only achieve so much progress toward sustainability without the twin concept of governance. Managing the development-environment relationship required rules and regulations, laws and treaties, and institutional arrangements that are the hallmarks of "good" governance. As Baker (2006) noted,

> the term "new practices of environmental governance" refers to the participation of non-state actors, alongside state and international organization, as well as the utilization

> of a wide range of policy instruments (including legal, voluntary and market instruments) and normative and governance principles to promote sustainable development.
>
> *(Baker, 2006, 9)*

What was not particularly clear in either managerial or governance approaches to sustainable development was the role of democratic representation. Bureaucrats, scientists, and economic elites often displaced local knowledge and democratic participation (Redclift and Sage, 1994).

While it could be argued that focusing on governance and regulation prioritizes the development side of the development-environment dialectic, approaches that foreground environmental management, conservation, and resource use were and are in no short supply. The direction of the previously mentioned approaches are from the development side of the dialectic. The approaches accept environmental constraints and ask, generally, how development should adjust. Environmental approaches may be considered ones that flip the binary from development-environment to environment-development. Sayer and Campbell (2004) summarized several dominant environment-focused approaches that flourished within sustainable development thinking. They refer to these approaches as, collectively, "integrated natural resource management" (ibid. 4). Variations include:

- Integrated catchment management
- Integrated water resource management
- Community-based natural resource management
- Integrated rural development
- Integrated conservation and development
- Ecosystem approaches
- Landscape management
- Adaptive collaborative management
- Multifunctional agriculture or forestry

(ibid. 4–5)

A strong theme of management runs through such approaches. A foundational assumption is that better management of environmental resources will prolong their availability. Relatively few approaches question the economic system that maintains the necessity of continuous growth. Issues such as who determines the level of economic growth that impinges on long-term environmental sustainability are less often driving concerns of these and related approaches.

Political ecology

Political ecologists contest foldings of development and environment together and often directly address who defines and controls notions of "growth". However, they often see the environment through a scaled, politicized lens, rather than in relation to "development". To many political ecologists, development has largely been left behind as a hallmark of modernist thinking. The post-development turn in the social sciences countered and delegitimized the concept of progress through economic growth for many fields. In political ecology, the post-structuralism that led to post-developmentalism also undermined political ecology's Marxist underpinnings. Notions of progress through growth, be it based in neo-classical or historical materialist thinking, quickly evaporated by the end of the 1990s.

Rather than emphasize macro-economic approaches to development, political ecologists examine issues such as the politics of scale (Peet and Watts, 2004, 4) from local to global and how such scaled power relations shape nature-society interactions. Watts and Peet further argue that political ecology has four general themes of study including refining political economy to make the relationship between capitalist growth and environmental outcomes explicit, redressing a perceived absence or underrepresentation of politics in examining human-environment interaction, highlighting the role of civil society in political ecologies, and emphasizing the importance of discursive approaches in political ecological studies.

Robbins (2020) positions political ecology as a direct challenge to apolitical ecological approaches and suggested "five theses" including degradation and marginalization, conservation and control, environmental conflict and exclusion, environmental subjects and identity, and political objects and actors (Robbins, 2020, 18). Such positions emerged after years of debate in part due to the field's opposition to the notion in sustainable development that capitalist political economy was, in fact, sustainable at all (Bryant and Bailey, 1997, 4).

We define development and environment studies here as those that examine the inseparable nature of economic growth and nature/environment/ecology. We do not consider "development" and "environment" as fields that "interact" or are "interrelated" but as fields that are one and the same – where one could not possibly exist without the other. We avoid examining the "impacts of" one on the other, rather preferring a *dialectical* view, where development and environment are bound together as one – moving through contradiction and sublation. This dialectical view need not be politically Marxist, although some authors hold more or less to that approach.

Eco/Green Marxism

Eco-Marxist, green Marxist, and other related schools of thought take a different *ontological* approach to the relationship between people and their environment. The dialectical method that Marx used to analyze capitalism is applied to the nature-society divide, or in this book, the development-environment divide. Before moving to describe a dialectical approach, a bit of ontological explanation is needed.

Smith and O'Keefe (1980) challenged three dominant ideologies concerning the relationship between nature and society that were rife in the social sciences at that time. The first asserted that "natural scientists study nature and social science does not" (30) while the second asserted that social and natural sciences study nature in different ways. The final approach "dissolves human nature into external nature" (30), a neo-Malthusian approach. Against these "common-sense" understandings of the relationship between humans and nature, Smith and O'Keefe used a reading of Marx to offer the idea that nature is produced as humans transition from first to second nature. This transition, which roughly corresponds to production for use to production for exchange, occurs through the labor process. Citing Marx, Smith and O'Keefe positioned labor at the center of the process: "by 'his industry', the producer 'changes the forms of the materials furnished by nature, in such a way as to make them useful to him'" (1980, 34). The dialectical nature of the relationship becomes obvious:

> but labor effects more than just a simple change in the form of matter; it produces a simultaneous change in the laborer, and it is from this abstract definition of the relation with nature that a unified Marxist theory of nature can be constructed.
>
> *(1980, 34)*

But how do we recognize or measure this "simultaneous change"? Empirically, Smith and O'Keefe only hinted at possibilities.

With a reconfigured relationship between humans and nature established and under the capitalist mode of production, second nature becomes the dominant state of nature. First nature "is produced by and within the confines of the second" (Smith and O'Keefe, 1980, 35). Under capitalism, the production of second nature shifts as labor is alienated by capital from nature through the specific arrangement of the factors of production. For example, workers move into factories and out of direct procurement of the products of first nature. This creates a nature fetishism in which workers' knowledge of nature is alienated from them; and indeed, alienation occurs where capitalist relations permeate.

The first-second nature divide is necessary under capitalism to get workers to accept their continued, alienated production of nature. This is precisely how capitalism exploits nature for profit, although socialist states parroted this model. Once nature is alienated from the worker, peasant, or other producer, it is "freely" available to capital as a no- or low-cost input into production and profit. In late Keynesian capitalism, the state attempted to shift the real cost of the consumption of nature onto firms, common in managerial approaches to sustainable development; under neoliberal regimes of accumulation, firms have worked tirelessly to shift this burden off themselves and to accelerate the privatization of communal natural resources as a source of profit, or what Harvey (2006) referred to as accumulation by dispossession. Privatization locks out non-market claims to use of resources or monetizes them, giving capital the upper hand in the allocation of such resources. This leaves those producers who actively use nature to reproduce themselves "stuck" between a lack of natural resources (that have been consumed by, reserved for, or otherwise alienated by capital) for reproduction and a weak labor market. The alienation of nature from producers in rural areas results in "excess" laborers who have few employment options. This is the dilemma in which many developing nations still find themselves. The "ecological rift" between development and environment is a product of the way dominant discourses view development on one side and ecology on the other (Foster et al., 2010, Kindle file).

This ontological position leads to fundamentally different observations about the relationship between development and environment. Eco/green Marxists such as Martin O'Connor (1994) argue that "for capitalism, sustainability is taken to mean sustained growth; technology is the vehicle of an irreversible ascension; and accumulation is asserted to be a destiny for all individuals and nations who work hard at it" (O'Connor, 1994, 1). The unrelenting tether of society to capitalism, in this approach, means that the idea of a sustainable capitalism – and by extension, development – is an unattainable state and to posit a "sustainable development" approach is fallacy. For Foster et al. (2010)

> the notion of "sustainable development," though an essential concept in the context of growing ecological crisis insofar as it emphasizes the need for ecological sustainability, has often been used as a category to reinforce the need for sustaining economic growth.
>
> *(Foster et al., 2010, Kindle file)*

As discussed earlier, "greening" development often reduces environment to "merely a question of management and markets" (ibid.).

Adopting a full Marxist position, however, is quite problematic. Marx himself had contributed to the optimistic view that these [ecological] problems were resolvable "technologically" wrote O'Connor (ibid. 3). When environmental limits press up against continued capital accumulation, technology is often posited as a solution for the contradiction. The problem, O'Connor and others wrote, is that such contradictions can not always be solved technologically and we are witnessing contradictions (global climate change, ocean acidification, pollution, etc.) that appear insurmountable. This leads to what James O'Connor has called the "second contradiction of capitalism" (O'Connor, 1994). While Marx's original, or first, contradiction of capital occurs due to contradictions between the production and realization of value, J. O'Connor argued that a second contradiction also undermines capitalism's sustainability. This second contradiction is "between capitalist production relations and productive forces and conditions of production" (O'Connor, 1994, 205). The "conditions of production" include the environment, which capital tends to treat as an externality and therefore easily exploited. Continued and intensified capitalist production undermines its own environmental conditions that allow it to produce cheap goods derived in part from nature and in part from the exploitation of human labor. The ecological crises facing the world, therefore, directly contradict capital's use of them in expanded production and increased exploitation of labor. Put simply, if natural resources are over-consumed in the drive for profit, and the conductions of life that sustain human labor is so degraded that human health suffers, then the capitalist mode of production is actively undermining its ability to reproduce itself.

David Harvey, a long standing critic of capitalism and close reader of Marx's works, casts doubt on notions that capital is necessarily undermined by such environmental contradictions. He noted that "capital has a long history of resolving its ecological difficulties", that the "nature we are supposedly exploiting and exhausting and which then supposedly limits or even 'takes revenge' on us is actually internalized within the circulation of capital" (see earlier), that "capital has turned environmental issues into big business" that could defer contradictions forward, and that "it may be perfectly possible for capital to continue to circulate and accumulate in the midst of environmental catastrophes" (Harvey, 2014, 246–249). For Harvey then, it is not at all a foregone conclusion that capital cannot overcome the environmental disasters of its own making. More problematic is how capital alienates humans from nature. If the ongoing environmental catastrophes do lead to a breakdown of capital, in this view, it is not due to a mechanistic contradiction between capital and nature, but rather active resistance and revolt of living human beings who can no longer tolerate the alienation imposed upon them by capital, both their alienation from nature and from each other.

Dialectical challenges to current thought

The competing approaches to development and environment, both epistemologically and ontologically, are, unhappily, not resolved by a single handbook. There is no ultimate answer for the scholar seeking a resolution to the contradiction. What we do offer in this introduction, however, is an alternative *method* for understanding the relationship. We have alluded to dialectics thus far in this chapter and now turn to a more nuanced explanation. It must be noted that dialectics, while heavily influenced by Marxist thought, need not necessitate one's alignment with all of the foibles and cul-de-sacs therein. Later, we offer a few key methodological principles and concepts for examining contradictions that do not necessarily require a Marxist approach to be useful.

Dialectics is a contentious subject. To its detractors it is teleological, closed, and rigid; to its adherents, it is a limitless fount of explanation. We engage what has been termed "new" dialectics to sublate the contradictions between development and environment. In this transcendence, neither object completely vanishes, rather a political economy of nature results from the process of resolving the contradictions between the two.

We engage dialectics at a meso-theoretical scale to avoid what Bhaskar (1993, 4) calls epistemic and ontic fallacies, defined as the "reduction of ontology to epistemology" and "the ideology of the compulsive determination of knowledge by being", respectively. In other words, we do not attempt to determine whether or not the world is inherently dialectical or if dialectics is just a useful way of understanding the world around us. Within the discipline of geography, perhaps the most well known proponent of a dialectics is David Harvey. His approach to dialectics is strongly indebted to Bertell Ollman (1976, 2003), who has spent his career explaining, clarifying, and expounding Marx's dialectics. Harvey, through Ollman, engages a dialectic that is more open and less deterministic, teleological, or immanent than other variations. His dialectics are relational and flexible. His 11 points explaining his dialectics left considerable space both for openness and for engagement with other epistemologies (Harvey, 1996, 48–57). On this point, Harvey went so far as to say that "other systems of thought" can be engaged alongside his dialectics, a point we will return to later. He wrote, "it would be wrong to view Cartesian and dialectical conceptions as fundamentally incompatible in all senses" (Harvey, 1996, 61).

As dialectics can be a language unto itself, we briefly review three dialectical concepts that are particularly useful for transcending binary views of development and environment: contradiction, sublation, and temporary stabilization. We note here that we draw upon only a partial vocabulary, not to indicate that these concepts are more or less useful, but because a full dialectical exposition of the development-environment relation is outside of the scope of this introductory chapter.

Contradiction is a pillar of dialectical approaches. For Hegel, "contradiction is the root of all movement and life, and it is only in so far as it contains a contradiction that anything moves and has impulse and activity" (Hegel et al., 1929, 67). Where much of science is constructed around hypotheses, contradiction fuels dialectical inquiry. In a strict Hegelian sense, an idea meets an opposed idea and presents a contradiction, which is negated (as a thesis-antithesis relationship). As Sheppard noted, "Hegel's account of the dialectic is one of a progressive movement toward enlightenment, proceeding through sequences of self-reflection as it moves seemingly inexorably via 'aufhebung' (sublation), towards 'absolute Wissen' (absolute knowing)" (Sheppard, 2008, 2605). The strict Hegelian reading of the concept has, however, been thoroughly critiqued as teleological, closed, and deterministic. As much as Marxist philosophers inducted Hegelian dialectics into their thinking, they too inherited such characteristics.

Ollman (2003) explained his Hegelian view of contradiction in its relation to his overarching philosophy of internal relations. A philosophy of internal relations insists that everything is related to everything else, in distinction to a philosophy of external relations, where objects are held discrete and unique, which is the prevalent view in mainstream science. This clarification is important, as without it, contradiction becomes nothing more than "driving force":

> contradiction is understood here as the incompatible development of different elements within the same relation . . . the commonsense notion of contradiction is that it applies to ideas about things and not to things themselves, that it is a logical relation between propositions . . . and not a real existing relation in the world. Whereas non-dialectical thinkers in every discipline are involved in a nonstop search for the "outside agitator", for something or someone that comes from outside the problem . . . and is the cause for whatever change that occurs, dialectical thinkers attribute the main responsibility for all change to the inner connections of the system or systems in which it occurs.
>
> *(2003, 18)*

The problem with Ollman's view of contradiction is that it is nearly as closed as Hegel's, as suggested in reference to "all change" is attributed to internal relations. A more open reading of contradiction is found in Bhaskar. He challenged Hegel's absolutist reading of negation, insisting that "his failure to sustain certain crucial distinctions and categories [of negation] (including in the end that of absence itself) – drive[s] the dialectic past and beyond him" (Bhaskar, 1993, 5). For Bhaskar, negation is not a final closing act nor is the only expression of negation contradiction. The point here is not to enter the debate over open versus closed dialectics read on the basis of negation/contradiction, rather to show that dialectical thinkers have considered these criticisms and responded with more openness and possibility.

In dialectics a contradiction is sublated, producing a synthesis. In Hegel's terminology, sublation (*aufhebung*) refers to the process through which the component parts of the contradiction are both retained, yet changed into a new form. This is the moment where the contradiction is resolved but also a moment of potential openness in dialectics, such as described previously at the stage of contradiction/negation. The charge of teleology is leveled at Hegelian and Marxist dialectics as the synthesis must emerge from the sublation of the contradiction. It is often assumed that the synthesis is singular, as in socialism's emergence from the sublation of the contradictions of capitalism in many Marxist writings. However, following Bhaskar, there can be many possible syntheses of many contradictions where negations are partial.

This concept is critical to the argument in this book: that a new relationship between development and environment is possible and can emerge in the process of sublating the development-environment divide precisely by retaining elements of both and applying new insight. Applying a dialectical approach to the problem at hand, we would ask "what are the elements of development and environment that have shown usefulness in addressing the problem and how can they be mobilized to both continue providing useful explanation through sublation, either partially or fully, of their contradictory elements?" A dialectical resolution to the issue (sublation → synthesis) precludes either development or environment simply subsuming the other into its epistemology. Because new dialectics is considerably more flexible and open (see Bhaskar, 1993; Sheppard, 2008), we suggest just one of the possibilities: renewed examination of the production of nature under capitalism that results in a new development-environment conceptualization.

To make such synthesis possible, we draw heavily upon the concept of temporary stabilization or moment (after Ollman, 2003; Levins and Lewontin, 1985). This allows for an element of a system to be held constant in order to be examined. Dialectical approaches focus on systems and wholes, showing the relationship between parts, but not fetishizing the part as somehow unique or isolated, a common charge level by dialectics against contemporary science. Ollman succinctly explains Marx's use of the concept:

> to consider the past and likely future development of anything as integral to what it is, to grasp this whole as a single process, does not keep Marx from abstracting out some part or instant of this process for a particular purpose and from treating it as relatively autonomous. Aware that the units into which he subdivided reality are the results of his abstractions, Marx is able to reabstract [sic] this reality, restricting the area brought into focus in line with the requirements of his current study. But when he does this, he often underscores its character as a temporarily stable part of a larger and ongoing process by referring to it as a "moment".
>
> *(Ollman, 2003, 66)*

The idea of the moment/temporary stabilization allows for integration of parts of both development and environment as separate objects of study into one post-binary, post-positivist reconstructed form. Notice in the quote that this process differs from treating objects as discrete units, as in a philosophy of external relations. Parts of the whole are only held as "relatively autonomous" so as to examine their constitution; they are immediately re-abstracted (placed back) into their relations. Their characteristics are not a result of being a separate entity, rather dependent on their relations with other parts of the whole. When combined with Harvey's admonition that seemingly incompatible systems of thought can provide rich insight to each other, a sublation of the contradictions between "development" and "environment" biases appears quite possible. Of course, the question remains as to whether or not either field would consider a dialectical approach viable. Our optimism, therefore, only exists in as much as both fields get transformed (sublated) into something(s) new.

Emergent themes in development-environment studies

We group chapters in this text in broad themes that not only link to the latest thinking on the topic at hand, but also foreshadow where the field is going. Our choice to enhance the voices of emergent and established scholars from the majoritarian world ensures that this is not only an anthology of knowledge but will push the field forward with perspectives that infrequently get such a platform. We open Part 1, which is focused on theoretical approaches to the study of development-environment, with two synthetic overview chapters from the book's co-editors that summarize the state-of-the-field in their respective areas of expertise. Margaret Hinrichs and Patricia Solís explore the concept of boundary objects and demonstrate how several of the contributors to the text attempt to overcome arbitrary boundaries between development-environment binaries. They note that such boundary erasure can "have dramatic influence over the effectiveness of the object as a tool for collaboration". The note of collaboration is warmly welcomed in this handbook. Far too often, the divisions and tight communities that form around isolated concepts arise in academia and practice precisely because it is more comfortable to retreat to intellectually comfortable spaces where dominant ideas are reinforced rather than challenged. Boundary erasure is challenging and, perhaps, even uncomfortable, yet necessary to confront the challenges of an economy (mode of production) that so clearly undermines its environment (conditions of production).

Maano Ramutsindela deconstructs both "development" and "environment" as fields of study and shows how constellations of ideas came to get represented with those signifiers. He challenges value neutral definitions of "environmentalism" and demonstrates a visceral connection to development though an examination of the practices of the poor. "Development" is as much a loaded term for Ramutsindela as "environment". Development as interventions, he notes, has been the main operational definition. The convergence of these two fields is adeptly recounted as the chapter concludes.

From these synthetic overviews, we move into more focused theoretical issues. Jayson J. Funke provides a compelling argument that nature has been financialized, by which he means nature has become a target of wild finance schemes, either to generate capital or as a sink for excess capital. After describing the financialization of capital, he adeptly pins late-stage capitalist "investment" as a cure for "the crises and risks of over-accumulation by managing the reproduction of land/nature, labor, and money as key capitalist commodities". Johannes Tsheola provides much needed insight into how colonialism continues to trouble the development-environment nexus. He shows that colonizers and the colonized saw the productive use of nature alongside its spiritual dimension. Colonialism, in his view, was an attempt to "rationalize" the use of natural resources. Because post-colonialism has failed to transform understandings of the state, it remains an active force in determining the development-environment nexus.

Eco-socialism as a potent force for transformation of mainstream thinking around development-environment is the theme of Salvatore Engel-Di Mauro's essay. Engel-Di Mauro highlights eco-socialist perspectives, their historical origins, and the relation to current struggles. "These struggles", he writes, "present occasions for building political affinities and connecting with existing movements. Eco-socialists should and often already do have a role to play in such political projects". Creating organizations around eco-socialist activism moves us from theoretical debates to practical, dialectical actions in his view (and we couldn't agree more!).

The chapters in Part 2 of the handbook cover topics that touch upon key resources, such as food, land, and water. In the expression of securing these needs, a development-environment dialectic is negotiated in daily life. Alistair Fraser examines the digitalization of the global food system and assumption that this will lead to "development" and "modernization". Digitalization as the cure for all that ails us has not exempted the global food system from its promises of miracles. Fraser reviews emergent food technologies and critically poses them against various development-environment scenarios and narratives. Importantly, he leaves open the possibility that digital technologies may enhance development-environment futures.

Anand Prasad Mishra illustrates the development-environment dialectic through exploration of the position of land in the relationship. Mishra traces the historical situation of land, soil, and humans vis-á-vis conceptualizations of economic growth, mode of production, and "development". This relationship has meant displacement, productivism, and domination at various moments and in various places. Mishra illustrates the dialectical inseparability of our key terms and warns that the "blind race of development is bent on making this planet a vast dump yard". Ana Sofia Solano Acuña writes on the struggles of indigenous western Panamanians, who faced constant attempts to strip them of their land. From struggles against local elites to mining and hydroelectric development, indigenous people have stood against both legal frameworks and extra-judicial attempts to remove, harass, and dispossess them. Against this backdrop, local indigenous movements have recently joined international networks to ensure their long term survival and maintenance of their lands. Imaginaries of land are rife throughout history and society and none more so than imaginaries captured in song. Furtick and Ahmed capture this notion in their discussion of Pete Seeger's *This Land Is Your Land*. They explore the corporate hold over cultural creativity and demonstrate how more radical themes are "whitewashed" out to appeal to mainstream consumers in a process they call "music colonization". The depoliticization of music tied to the land is yet another strategy of accumulation by dispossession and commercialization. If land can be depoliticized and dispossessed in the imaginary of music, it becomes easier to justify when doing so in the material.

When it comes to nature and humanity, development and environment, water is critical. Given climate change, water issues will become even more important in the near future. Park Muhonda describes changes in rainfall patterns in Malawi and juxtaposes climatological data with farmer perceptions of change. To investigate climate change, one must not only be able to quantify such changes in the physical environment, but they must also register with people who are vulnerable to such changes. Muhonda shows that farmer perceptions of precipitation change in Malawi is largely accurate. People so closely connected to water ought to be believed when they maintain their environment has changed. Muhonda's paper validates that normative statement with climate science. Part 2 features a discussion by James-Eric Simon and Waquar Ahmed that examines water infrastructure across the Dallas-Fort Worth metropolitan area in Texas, USA. They argue that "covert colonialism" was persistent in the creation of the Grapevine Reservoir. Narratives of modernization and "progress" permeated the creation of the reservoir. The domination of the concept of "development" silenced other voices and other water regime possibilities. "Development", indeed.

Sage Narsiah reviews the concept of "sustainable development" and explores its application to governance and public participation in South Africa. Narsiah describes corruption in the post-apartheid era and its pervasive effect on public confidence in the government. Corruption led to the widespread incapacitation of many state institutions.

Part 3 of the handbook contains chapters that intersect people and their communities in some aspect with several of the more widespread themes in development-environment studies. The first four chapters examine communities in intersection with sustainable development, climate change, environmental (in)justice, and neoliberalism through illuminating case studies of how such ideas are experienced in the daily lives of people across the world.

C. Nancy Aguirre poses an intriguing juxtaposition of climate change knowledges in Colombia. Her main contention is that more information from local scales is needed to balance narratives coming from above. Concepts such as "climate change" and "sustainability" may seem abstract and remote from an academic perspective, but as she shows in her case study, there are real-world gaps that need to be addressed from local people. B.S. Butola conceptualizes environmental (in)justice and accumulation by dispossession in India through examples across several categories of development-environment. Butola provides examples of these issues in various contexts in India and how such struggles relate to broader concepts of justice. In a cogent analysis of the intersection of disasters and neoliberalism, Samantha Epinoza-Villejo details the costs of "neoliberal success" in Houston, Texas, USA. She notes the class-based spatial configurations that resulted in poor and marginalized communities feeling "interlaced frustrations and terrors" of being "stuck in place". The immobility that results from neoliberal, market-based decision was on full display in Houston during Hurricane Harvey in 2007.

Conservation has certainly been a highly contested issue within development-environment studies and received continuous (and justified) scrutiny, particularly when practice has been posed in contrast to "people" or "communities". Carlos Morera Beita's case study in Costa Rica shows that even in a country often viewed as progressive on issues of conservation, prickly questions arise when taking the concept out of protected regions and extending it into new geographical and social landscapes. As contested as the concept of conservation is when considering natural resources, it becomes even more problematic when examining issues of heritage of indigenous people. The displacement of their narratives has been the norm in "conservation" of the past. Steve Wolverton, Robert Melchior Figueroa, C. Melinda Levin and Porter Swentzell probe such issues as archaeologists, or in their words, "tourists of the past". Protest and resistance by indigenous people should replace sanitized narratives designed to comfort contemporary tourists to archaeological sites. Their intersection of environmental justice and archaeology demonstrates the possibilities of development-environment work where boundaries are challenged, as Hinrichs and Solís challenge in Part 1. Similarly, Luis Sánchez-Ayala and Alexandra Areiza-Tapias challenge "official" or government conservation strategies in Colombia with ones that originate from community needs. Their case studies demonstrate how community-based action better conserved fish and cattle. Community development initiatives were better managed and had better outcomes. This lesson is relevant regardless of the geographic region of the world. Christine Noe and Asubisye Mwamfupe draw on work in Tanzania to examine the impact of conservation on local livelihoods, finding that recent "partnerships reflect conspicuously the power of neoliberalism, its reshaping of landscapes and profound alterations of the lives of rural people in areas targeted for conservation". Even under the rubric of "participation", they found, neoliberal impulses prevail.

Gender and development-environment is a literature unto itself; however, we highlight three contributions from emerging scholars that illustrate the complexities, contests, and emerging thinking on the topic. Using a case from Gareno, Ecuador, Ana Gómez Donoso and Paula Soto

Villagrán converge gender, space, and environment to understand how women have resisted and challenged marginalization and disempowerment. Through a participatory methodology they show how territory has been reconfigured into urban and rural spaces and how such spaces have been gendered. Extending the concept of "environment" into hurricanes, Carlos Jorge Guilbe López examined women's resistance in Puerto Rico following hurricanes Irma and Maria. López reviews Naomi Klein's shock doctrine and extends it to women's resistance to Washington's policies from their situation as community organizers. Resistance sprang from the poor treatment of communities by privatized utility providers and the neoliberal state. The importance of local organization and resistance became very clear in the aftermath of these two disasters. Women's access to environmental resources is a key part of livelihood systems around the world. Rhoda Nyirenda's chapter on women's access to flood land agriculture in Malawi demonstrates that women still have significant obstacles accessing quality land in the absence of male relatives to substantiate their claims. Nyirenda artfully combines qualitative and quantitative research to conclusively argue that women's livelihood systems are made more vulnerable by the obstacles placed in front of them by patriarchal land tenure arrangements. In Part 3, Jean-Marie Théodat reflects in a personal narrative on his sense of urbanity and memory as a "dialectic of place" in Port-au-Prince, Haiti. He reflects on the contrasts of his city and the attempts to "modernize" it over the recent past. Ipsita Chatterjee explores the concept of eco-cities and deconstructs the notion that "ecological sustainability" and "economic sustainability" can "strengthen each other". She shows that green movements are as often inseparable from the interests of capital in pursuit of profit. Using Kolkata, India as an example, she shows how "green" language is used for continued dispossession and alienating of people from their land and livelihoods.

We conclude the handbook with Part 4, which draws together themes of governance and policy-making. The regulation of concepts through their implementation is often *how* much development-environment work gets transformed into material space. If the community of scholars and practice wishes to influence such actions, a new vocabulary of the dialectic must encourage uptake by those in governance and policy-making. Bikramaditya K. Choudhary approaches territoriality through relationship with nationality and justice in indigenous communities in India. Choudhary examines the contested role of development, particularly as it resulted in dispossession and disempowerment of indigenous groups in post-independence India. Through "modernizing" policies of the Indian state, large groups of people suffered land dispossession, relocation, and forced resettlement. Choudhary argues that these forces outweighed counter-narratives and counter-voices and left the powerless with few avenues of recourse. Governmentality over development-environment poses opportunities and challenges. Kwame N. Owusu-Daaku employs a Foucauldian approach to governmentality to show how "good-intentioned" efforts at promoting climate change has resulted in maladaptation in Ghana. His case study on coastal protection illustrates how sea defense systems end up causing unintended negative impacts, such as coastal erosion and saltwater intrusion. Owusu-Daaku shows how the triangulation of authorities in Ghana have realized their initial goals but not how to undo the damage their actions created. The penultimate chapter in the text is a reproduction of a chapter by Brent McCusker from *The Routledge Handbook of Political Ecology* and, as such, is the only chapter in the handbook that is not wholly new for this handbook. McCusker describes the act of engaging a political ecology approach to international development at the US Agency for International Development (USAID) in the US. In his chapter, he writes of navigating the often discordant worlds of academia and policy-making. He notes that translation of concepts and expectations of the time scale of outcomes requires careful attention for any attempt at engagement to be successful. Concluding our handbook is a chapter that extends McCusker's previous chapter by explaining exactly how geographic analysis of a development analysis was implemented at USAID. Tim Essam, Brent McCusker, and Jamison Conley describe how they illustrated the geographic distribution of shocks to livelihoods in Niger. Their research was undertaken to directly answer questions about strengthening development projects and planning in regions of the world facing shortages of data for development planning. They describe their geo-spatial method and challenges in communicating such analyses to decision makers. They found that careful attention is needed to communicate the results of such research to decision makers.

Our goal with this handbook is to illuminate some of the key concepts in the study of development-environment by promoting voices that are often overlooked in such texts. A final "emergent theme" that we submit to add to the list mentioned earlier is that of providing space to voices from the global south and globally marginalized to be heard in this field of study. With that in mind, we now turn to our collection of authors and their illustrative and innovative chapters.

**Note: Parts of the section* Dialectical challenges to current thought *first appeared in McCusker (2013).*

References cited

Baker, S. 2006. *Sustainable development*. London: Routledge.

Bhaskar, R. 1993. *Dialectic: the pulse of freedom*. London: Verson.

Foster, J. B., et al. 2010. *The ecological rift: capitalism's war on the earth*. New York: Monthly Review Press. Kindle file.

Harvey, D. 1996. *Justice, nature and the geography of difference*. Malden, MA: Blackwell.

Harvey, D. 2006. *The limits to capital*. Third Edition. London: Verso.

Harvey, D. 2014. *Seventeen contradictions and the end of capitalism*. Oxford: Oxford University Press.

Hegel, G. W. F., et al. 1929. *Hegel's science of logic*. New York: Macmillan Co.

Levins, R. and R. C. Lewontin 1985. *The dialectical biologist*. Cambridge, MA: Harvard University Press.

McCusker, B. 2013. The Political Economy of Nature. *Human Geography* 6(3): 12–24.

O'Connor, M. 1994. *Is capitalism sustainable? Political economy and the politics of ecology*. New York: Guilford Press.

Ollman, B. 1976. *Alienation: Marx's conception of man in capitalist society*. Second Edition. Cambridge: Cambridge University Press.

Ollman, B. 2003. *Dance of the dialectic: steps in Marx's method*. Urbana, IL: University of Illinois Press.

Peet, R. and M. Watts 2004. *Liberation ecologies: environment, development, social movements*. London: Routledge.

Redclift, M. R. and C. Sage 1994. *Strategies for sustainable development: local agendas for the Southern Hemisphere*. Chichester: J. Wiley & Sons.

Robbins, Paul. 2020. *Political ecology: a critical introduction*. http://www.vlebooks.com/vleweb/product/openreader?id=none&isbn=9781119167464.

Sachs, W. and S. George 2015. *Planet dialectics: explorations in environment and development*. London: Zed Books.

Sayer, J. and B. M. Campbell 2004. *The science of sustainable development: local livelihoods and the global environment*. Cambridge, UK: Cambridge University Press.

Sheppard, E. 2008. Geographic Dialectics? *Environment and Planning A* 40: 2603–2612.

Smith, N. and P. O'Keefe 1980. Geography, Marx and the Concept of Nature. *Antipode* 12(2): 30–39.

World Commission on, E. and Development 1987. *Our common future*. Oxford: Oxford University Press.

PART 1

Theoretical approaches and syntheses

2

DEFINING AND TRANSGRESSING BOUNDARIES IN DEVELOPMENT AND ENVIRONMENT CONTEXTS

Margaret M. Hinrichs and Patricia Solís

In the hyphen between development-environment, we might awkwardly trip over the contrast that this boundary makes clear. Or we might easily slip along the dash as if it were a smooth lane carrying us across the ideas each side evokes. Either way, as we consider the mutually constitutive meanings of development and environment, we also take a moment to notice the act of placing that dash – of simultaneously defining and transgressing these notions. This is an act of boundary making. It is an important element of producing knowledge, especially in collaborative contexts. As we ponder this dialectic, "where development and environment are bound together as one – moving through contradiction and sublation" (McCusker et al., this volume, Chapter 1), we also consider the many ways in which our efforts towards synthesis, towards producing new insights and knowledge to solve actual, real problems of the world, involves drawing certain lines in certain places – sometimes materially, and always discursively.

This chapter explores that act, contextualizing and pointing out some of the ways that the authors of the chapters in this book discover, identify, name, define, blur, critique, and transgress boundaries in their experiences of development and environment. We first tackle a rationale for understanding this process as integral to knowledge coproduction. Then, we lay out boundary objects as particular tools by which the creative tension and synthesis of knowledge about development-environment can be performed. An example of a boundary object then helps to illustrate the process in a concrete way, drawing on our experiences working with geospatial technologies in the humanitarian sector. With that context in place, we review some of the ways in which we believe the authors of this *Handbook* make boundaries and use boundary objects in their diverse works. As we do so, we underscore that this act may range from deliberate to accidental, from a subtle subtheme to a main idea. We caveat our recounting of these instances by saying our intent is to illustrate our interpretation of the appearance of a dialectic, rather than to (mis)characterize the approaches of each author.

 DOI: 10.4324/9780429450310-4

The need for knowledge coproduction amidst shifting development-environment boundaries

Coproduction is the process of active involvement and engagement of relevant actors in efforts to create the necessary arrangements where policy-relevant and socially robust knowledge can be produced collaboratively, leading to knowledge that can be shared and understood across sectors (Frantzeskaki & Kabisch, 2016; van Kerkhoff & Lebel, 2015; Muñoz-Erickson, 2014). Coproduction includes processes that "iteratively unite ways of knowing and acting – including ideas, norms, practices, and discourses – leading to mutual reinforcement and reciprocal transformation of societal outcomes" (Wyborn et al., 2019, p. 3). When successful, a collaborative coproduction approach can lead to increased government accountability, greater civic engagement, consistent downstream implementation, and most importantly, higher levels of process and program success (Freeman, 1997; Fung & Wright, 2001; Hicks et al., 2008; Lasker & Weiss, 2003). The common theoretical assumption is that a collaborative coproduction process can transform combative stakeholders into cooperative partners that contribute to resolving challenges (Ostrom, 2000).

However, while knowledge coproduction is necessary for more inclusive development-environment challenges, it is not a sufficient condition for success. The manner in which diverse communities define and construct their perspective of a development-environment challenge often features identification of transgressed boundaries, the history of which can make collaborative coproduction of knowledge difficult. Understanding various approaches to managing such conflict can be insightful and shed light on respective problems and the tension between development and environment itself. In this *Handbook*, we observe and point out how the respective authors define and frame particular development-environment struggles through making these boundaries visible, as well as demonstrating how actors in these discursive and material struggles engage in blurring them.

Sometimes struggles unfold along the jurisdictional fault lines of modern development in the creases of the actual boundaries of the nation-state. For instance, Butola's study (Chapter 19) reveals how despite fights assuming the contrary, boundaries of citizenship do not necessarily extend or coincide with bestowing environmental justice. Instead, he argues that "rights in the developed countries have been possible at the cost of justice to the marginalised sections within the developed countries and a large mass of population in the colonies". Aligned with this same logic, Choudhary (Chapter 27) examines the "diaspora within" where the national border is the jurisdiction platform within which a significant development-induced displacement of people, primarily indigenous communities, occurs within India. Also aligned, Tsheola (Chapter 5) grapples with the oxymoron of the boundary between colonialism and post-colonialism by charting where new and old ideologies create "continuities and disjunctures, underpinning 'complex historical episodes' of relations of 'material phenomena and ideological controversies'". In the post-colonial global space marked by the near universal adoption of the Sustainable Development Goals (SDGs), Narsiah (Chapter 14) distinguishes boundaries between governance and public participation, "a movement from emancipation and freedom to domination and control . . . towards an environmentally destructive, unsustainable future".

Other times the challenge can be articulated along familiar environmental divides between conservation versus development. Morera Beita, Avendaño, and Sandoval (Chapter 18) explore how separate protected areas are considered an indispensable strategy for biodiversity conservation, but how increasingly such spaces in the era of the Anthropocene extend their focus to establish connectivity spaces, such as biological corridors. Their study reveals blurred boundaries exemplified by tourist lodging companies have developed their own natural attractions, in turn triggering a fiscal crisis in protected areas due to loss of visitor fees. Noe and Mwamfupe (Chapter 21) explore how conservation unfolds in communal lands and impacts local livelihoods. Their work maps out the boundaries which circumscribe protected areas into partnerships that indirectly aggravate local power struggles.

Finally, shifting boundaries among gender, poverty, and natural systems is fundamental to understanding struggles in modern development and environment. As Gómez and Soto (Chapter 22) explain, "it has shown how the domestic organization of work, consumption patterns, food insecurity, pollution and environmental destruction, as dynamic processes, specifically affect the lives of women; but at the same time places them as principal actors for environmental change". In the face of increasingly frequent and stronger disasters that climate change will bring, female resilience and resistance will factor significantly in the definition of knowledge boundaries. As Guilbe's study illustrates (Chapter 23), in the wake of two hurricanes in 20 days that hit Puerto Rico, the extended blackout literally defined boundaries of disaster and resilience, where images of darkness in women's lives "were effective propaganda to convert the island into a disaster zone and re-orient the redevelopment with a neoliberal compass".

Boundary objects are a critical tool for managing conflict in boundary-spanning processes

Research on collaborative work has refined the definition and use of various tools and approaches to support coproduction of knowledge, cross-sectoral collaboration and maintain a productive tension between science and other forms of life in modern society (Guston, 2001; Miller, 2001). Perhaps the most widely adopted mechanism for interdisciplinary and cross-sectoral work in this space is the boundary object, a term coined by Star and Griesemer (1989; Star, 1998) to identify objects which are plastic enough to adapt to immediate, local needs yet robust enough to maintain a common identity across time and space. The four different types of boundary objects identified are (Starr & Griesemer, 1989; Star, 1998; Star, 2010):

1. *Repositories* which store and index objects with varying units of analysis, such as a library, database, or museum.
2. *Ideal type or platonic object* such as a map, model, or atlas, which is sufficiently vague as to be adaptable across diverse circumstances.
3. *Terrain with coincident boundaries*, such as the outline of the state of Arizona or building blueprint, which allows collaborators to use the same referent but for different purposes.
4. *Forms* which serve to standardize communication of information across dispersed collaborators.

Since Star's seminal piece, the use and adaptation of boundary objects to understand interdisciplinary, cross-sectoral work has exploded across a range of fields and areas of research, including ecosystem services (Abson et al., 2014), education policy (Emad & Roth, 2009), software development (Barrett & Oborn, 2010), collaboration (Nicolini et al., 2012), resilience (Brand & Jax, 2007), healthcare (Allen, 2009; Kimble et al., 2010), new product development (Carlile, 2002), conflict resolution (Iorio & Taylor, 2014), negotiation (Koskinen & Mäkinen, 2009), water resource management (White et al., 2015), and environmental decision-making (Zurba & Berkes, 2014).

Other research has explored alternative but related concepts to boundary objects, such as prototypes (Subrahmanian et al., 2003) and boundary negotiating artifacts. Lee (2005, 2007) builds on the concept of boundary objects and argues that perhaps boundary negotiating, in her case through the use of museum artifacts, is itself part of a process by which methods of collaboration are developed. In reviewing the research outlined earlier, it is clear that boundary objects and boundary negotiating artifacts are useful in supporting interdisciplinary, cross-sectoral work. The proliferation of these terms and their application to studying the wide range of collaborative scenarios listed demonstrates the salience and relevance of these terms to the burgeoning interdisciplinary research sphere. This can be illustrated through the growing presence of technologies in development and environment contexts – from communication and IT, to navigating and mapping with geospatial technologies. Next, we explore the latter as an example of how these processes unfold within development-environment contexts.

The case of geospatial technologies: boundary objects for knowledge coproduction in digital humanitarianism

When geospatial technologies went online through the introduction of open mapping on the web, the nature of access to these tools changed dramatically (Elwood et al., 2010), representing "a paradigmatic shift in how geographic information is created and shared and by whom, as well as its content and characteristics". The subsequent emergence of volunteered geographic information (VGI) engendered a body of scholarship that typically constructs "the spatial data production activities of states in opposition to bottom-up, amateur, asserted" mapping (Cinnamon, 2015). Such technologies represented a melding of geographic data and technology with geospatial and communication technologies which helped actors define and frame their struggles with respect to others. Understanding which boundary objects are of most use to different actors and the ways in which they get defined and constructed is insightful and can help shed light on what communities are trying to learn or say about a particular development-environment problem or challenge.

Open geospatial technologies, and internet mapping in particular, play an increasingly important role in enabling digital humanitarianism and represent a novel medium for citizen participation in a digital era. Harvey & Chrisman (1998) were among early scholars recognizing that geospatial technologies represent a special instance of a "boundary object" connecting different actors in the construction of new social arrangements, separating them at the same time that they delineate important points of reference between them, and around which a negotiation of differences happens. Studies of how society, politics, and culture affect scientific research and technological innovation, and how these, in turn, affect society, politics, and culture comprise a body of literature on visibility and representation that seeks to uncover and expose silences in methods and data that marginalize particular groups in order to empower these populations in new and creative ways (Kwan, 2004, 2007; McLafferty, 2002), even while such empowerment is often recognized as contested, shifting, and context-dependent (Kwaku-Kyem, 2004).

However, this calls into question the knowledge politics that frame such interactions which are responding to the timely challenge of understanding citizenship in a digital age, asking critical questions such as how to define "good participation" in the context of new opportunities for political efficacy through digital media to people who might otherwise not be easily heard (Light & Allen, 2015). The special case of citizen participation known as "digital humanitarianism" speaks to the means by which these media are used for crowd-based engagement in disaster or development efforts. Burns (2014) describes this as a "set of social and institutional networks, technologies, and practices that enable large numbers of remote and on-the-ground individuals to collaborate on humanitarian projects" in which he notes "struggles for legitimacy and means of representation are fluid and contested, yet become more stable when implemented through technology".

Digital humanitarianism efforts to make advances in research on resilience and vulnerability are hampered by access to reliable data (Barrett & Headey, 2014), with decision makers bemoaning lack of data as one of the biggest obstacles to progress towards development goals (UN, 2014; Stuart et al., 2015). Even with a robust conceptual framework using the highest quality, complete data, the potential of leveraging knowledge coproduction, collaborative use of boundary objects and integration of boundary-spanning technologies towards better decision-making is not automatically guaranteed. Despite increasing amounts of readily available data, or perhaps because of it, a myth has emerged that having data, particularly large data sets, offers better or more advanced knowledge than was previously possible, and this leads uncritically to novel insights based on accuracy and objectivity (boyd & Crawford, 2012). This persistent misconception is likely due in part to the prevailing idea that an uninterrupted chain exists leading from data to information to knowledge to wisdom (Cooley, 1980; Cleveland, 1982; Ackoff, 1989; Sharma, 2008). Now taken for granted, the hierarchy has been abbreviated by proponents of the Data Revolution who give short shift to the process, or conflate the chain altogether, simply as Data to Decisions (UN, 2014; Stuart et al., 2015). This model also has the propensity to gloss over missing data, to imply filtering (Weinberger, 2010); to assume simple inductivism (Frické, 2009); and to lead to misinformation (Burns, 2008). As Cowell and Lennon's meta-analysis shows (2014: 263), "actively cultivating wide stakeholder buy-in" to new approaches "may secure wider support, but not necessarily translate into major influence on decisions". In the end, this understanding should be grounded in context, uncertainty, and reality (Kates & Burton, 1986; Gober et al., 2010; Hermann & Neumeier, 2008).

Boundary object creation processes affect boundary-spanning outcomes

Development literature provides ample guidance on strategies to adopt a critical subjectivity and aim toward explicit awareness of potential bias, which is conceived as participative inquiry (Reason, 1994; Reason & Bradbury, 2008). We recognize knowledge as partial and situated (Haraway, 1988) and the research process as a social one (Sayer, 1992), relying on corroboration, which is integral to a participatory methodology. When models act as boundary objects which engage participation and inclusion of multiple parties, actors on both sides of the boundary benefit (Cash, 2001). Indeed, when actors on both sides of a boundary are part of the boundary object creation process, they are more likely to use and advocate for the results of the research being produced (Crona & Parker, 2011). The context for the design of information technology is increasingly an interconnected mosaic of responsive adaptive systems (MoRAS) including people's ideas, organizations, communities, markets, and culture (Furnas, 2000).

As researchers continue to expand boundary objects' areas of application, it is becoming clear that their creation process can have dramatic influence over the effectiveness of the object as a tool for collaboration. Henderson's (1999) work exploring the use of boundary objects in the engineering design process contains a substantiated warning that the creation of a boundary object is imbued with "power issues" (1999, p. 53) and details how the inclusion or exclusion of certain groups during the creation process can affect its eventual success as a boundary object. Likewise, Monteiro's (2010) explanation of digital modeling within the scientific community elaborates on the creation process as the determining factor of a model's perceived reliability and ultimately its usefulness. He draws attention to how, in addition to the quality of the data feeding the model, people are fundamentally involved in building the model, actively interpreting and manipulating the data and making decisions on how to visualize the information, ultimately shaping how the model behaves and is understood by others. Henderson articulates similar concern, noting the observed desire to limit access "to those whose input and motives they trust" (1999, p. 115). Confidence in the boundary object is inherently linked to trust of the creation process and those who were involved. Concerns regarding power issues and trust are dynamics of all collaborative efforts, and while they are therefore integral to boundary objects as well, the process of boundary object creation has the potential to anticipate and address these concerns. As such, we need a better understanding of ways in which boundary object creation processes can facilitate healthier, more collaborative, and more productive teams.

While Lee (2007) demonstrates how individuals can make an object that comes to be used in a boundary capacity, there is another realm of capabilities left to be explored when the object is designed intentionally and collaboratively from the beginning. Young and Lutters (2017) give a solid example of what that intentional accommodation of multiple disciplines can look like in the creation of a boundary object, but overlook the need for discussion of collaboration within the creation process. Henderson (1999), meanwhile, delves deep into the collaborative process of object creation and how the participation of multiple perspectives not only enhances the final product but becomes itself a form of boundary object. However, she falls short of exemplifying intentional coordination of the group effort.

As collaboration becomes increasingly crucial to addressing the complex development-environment problems of society, more exploration into the conflict-ridden but innovative process of knowledge coproduction is needed to establish ways it can be done most effectively. Boundary objects are a key tool in that endeavor, but their effectiveness can easily be dulled through subpar creation that doesn't take into account interpersonal dynamics including interpretation, power, and trust. The clearest way to assuage these hindrances is to create the boundary object through intentional collaborative effort; however a process that includes contrasting viewpoints from the beginning will inherently be more complex than that of an individual creating an object that evolves to function at the boundary, and little research has taken up the challenge. What is not well understood is how actors with seemingly opposing goals overcome their conflict and integrate ideas while sustaining divergent interests (O'Mahony & Bechky, 2008). Even less considered within the boundary community are ways in which these objects can be intentionally created.

In practical terms, this means we should, in advance of mapping, thoughtfully consider the explicit analytical processes by which open spatial data can influence development decisions, taking advantage of cross-sectoral and international perspectives. Moreover, one of the most critical current challenges of harnessing the data revolution related to decision-making has to do with the timescale and spatial scale for decision-making and governance (Juhola & Westerholf, 2011; Measham et al., 2011). Similarly, short-term opportunities and threats are linked to longer-term dimensions; what decisions are implemented now will impact and be affected by a later phenomenon (Betsill & Bulkeley, 2006; Cash & Moser, 2000). Solís et al. (2016) articulated this as a formal methodological problem that delineates the spatial incongruence between decision-making and accountability. This work recognizes that the scale and spatial unit of the jurisdiction of decisions that are made may be incongruent with the data that influence those decisions. It furthermore recognizes that the impact of such decisions may affect and be affected recursively by behavior and outcomes in yet different spatial areas of different scales, and, ultimately, the spatial unit and scale to which decision makers are held accountable for such decisions may furthermore be incongruent with either data, decisions, or impact.

Examples of development-environment boundary objects in this *Handbook*

In the broader context of boundary negotiation in development-environment, the use of boundary objects in knowledge coproduction processes acknowledges various disciplinary perspectives, methods of analysis, and languages used to describe complex problems. When diverse collaborators understand the lens through which each team member understands a complex system, they can create stronger common ground with greater potential for creative innovation moving forward. There is a need for critical self-reflection of the value-laden processes of creating boundary objects, the role(s) of researchers in those processes, and the impact of such processes on decision-making and social arrangements. However, despite the increased attention and research into mechanisms of support for Triple Helix arrangements and interdisciplinary research, the vast majority examine before and aftereffects of the use of collaboration support tools, rather than the process by which collaboration happens (Hinrichs et al., 2017).

Among the chapters of this *Handbook*, a number of striking examples of boundary objects can be detected within the various approaches, as subtle indicators of negotiation mechanisms, even while not serving as the main focus of the authors' analyses. We point out here, where we see these objects embedded within the scholarly narrative, to illustrate our own theoretical observations, rather than to (mis)characterize the intentions or perspectives of each author.

Most notably, within modern development and environment projects, the role of digital data figures heavily. In light with the typology of boundary objects noted previously, such data can be thought of as repositories which store and structure both meaning and material reality in ways that make negotiation of the tensions of development and environment across time and space possible. Fraser (Chapter 7) calls attention to new connections between digital technologies and food that can be found all across the world. He traces how in "majority world" or wealthier contexts, new arrays of digital technologies and the *data and databases* they create can transform systems from agricultural production to consumer food purchases. The "smart farming" processes have overturned prior assumptions of development-environment dynamics, which represent a powerful array of forces encoded into the data objects that threaten to reduce food sovereignty. Still, Fraser sees within this data (a repository boundary object) the potential as a tool of resistance to oppose dominant interpretations and algorithms, and build alternative digital meanings and thus food system realities that could lend themselves to negotiating for social or environmental justice.

Among the platonic objects that Star and Griesemer describe in their second category, an increasingly critical ideal item that development and environment action entails is *water*. Unique is its ability to be contained but fluid, and to be at once a fundamental physical necessity for life as well as a mystical article of cultural heritage, water may also ultimately become an actual physical embodiment signifying other constructed boundaries. In Chapter 13, Simon and Ahmed narrate the urbanization of water in Texas, and we notice how water itself seems to serve as a boundary object in this case. They examine broad implications of the creation of an artificial lake, which significantly changed surrounding communities' relationships with water. Water as boundary object had been transformed by government intervention into a commodified tool to reconfigure a "cultural-institutional worldview" of urban-rural space. The mechanisms wielding this object through what the authors suggest may even amount to a "benign, non-exploitative form of colonization" embodied within the physical presence of the artificial lake, resulted in the transformation of a city, from being considered as periphery to serving as a new urban core.

Boundary objects that themselves are literal boundary sites, like the third category of terrain, are familiar mechanisms within development and environment research. Danze, Torres, and Faria (Chapter 11) lend insights that bring this understanding to a new level as they trace a 20-year trajectory of *detention infrastructure* in Texas. Their scholarly examination of Conroe lays out the tension and the resolution to hide from view local sites holding first prisoners, then immigrant detainees. They convincingly argue that rendering these boundary objects invisible simultaneously offers the opportunity to assuage Conroe's undesirable association with the (increasingly privatized) industries of incarceration, while it pushes asylum seekers' advocacy out of sight. Danze and her co-authors take a turn in the end, however, whereby instead of concluding for detention infrastructure to serve as terrain boundary object that allows for renegotiation of meanings of development, they recast the object of interest to center on the people detained.

With respect to the fourth category, Aguirre's study (Chapter 15) offers a realm to consider how standardization of communication and ideas can be embodied into boundary objects as *planning documents*. These particular tangible items structure the practices deployed upon actual land parcels in her study of the municipality of Ricaurte, Department of Cundinamarca, Colombia. She speaks to how the political and institutional apparatus constructed a National Plan for Adaptation to Climate Change, focused on ecological and socioeconomic aspects of climate change, but one that "consistently blurs when reaching more detailed scales of policy and action from global discourse to local level actions". This form-type of boundary object has "cascaded from global structures, particularly the United Nations Framework for Climate Change Convention", with the result of uncovering internal contradictions and flaws when articulated as a planning document, leaving open space for local adaptation.

Other authors in this *Handbook* offer the possibility to add a fifth category of boundary object to Star and Griesemer's topology, which is in effect a hybrid of the first (repository) and third (terrain) embodied in the tangible spoken memories of place. To wit, Solano Acuña (Chapter 9) evokes memories of past struggles over indigenous places as resistance against present mining within the Ngäbe-Buglé communities of Panama. In Haiti, Théodat (Chapter 25) examines memories in place and memories of place in a dialectic of "an almost physical memory of which people, who everywhere have eyes, ears and a memory, keep a diffuse memory that resurfaces on occasion". Wolverton, Melchior Figueroa, Levin, and Swentzell (Chapter 19) dig up the human past within Mesa Verde National Park to make palpable the stories of environmental justice as a tangible heritage, at once terrain and at once repository of memories of the marginalized.

Towards a synthesis

The value of raising attention to the contrasts, contradictions, conflations, and connections that a dialectical approach to development-environment affords us lies within the opportunity to generate insights, new knowledge, a synthesis of ideas – and ultimately towards transformative change. The chapters in this *Handbook* have inspired us to seek out that opportunity as we listen to the voices of diverse scholars grappling with challenges to equitable and sustainable socioeconomic development from all parts of planet Earth.

References

Abson, D. J., von Wehrden, H., Baumgärtner, S., Fischer, J., Hanspach, J., Härdtle, W., & Walmsley, D. (2014). Ecosystem services as a boundary object for sustainability. *Ecological Economics*, *103*, 29–37.

Ackoff, R. L. (1989). From data to wisdom. *Journal of Applied Systems Analysis*, *15*, 3–9.
Allen, D. (2009). From boundary concept to boundary object: the practice and politics of care pathway development. *Social Science & Medicine*, *69*(3), 354–361.
Barrett, C. B., & Headey, D. D. (2014). *Measuring resilience in a risky world: why, where, how, and who?* 2020 Conference Brief 1. May 17–19, Addis Ababa, Ethiopia. Washington, DC: International Food Policy Research Institute.
Barrett, M., & Oborn, E. (2010). Boundary object use in cross-cultural software development teams. *Human Relations*, *63*(8), 1199–1221.
Betsill, M. M., & Bulkeley, H. (2006). Cities and the multilevel governance of global climate change. *Global Governance: A Review of Multilateralism and International Organizations*, *12*(2), 141–159.
Boyd, D., & Crawford, K. (2012). Critical questions for big data: provocations for a cultural technological and scholarly phenomenon. *Information, Communication & Society*, *15*(5), 662–679.
Brand, F., & Jax, K. (2007). Focusing the meaning (s) of resilience: resilience as a descriptive concept and a boundary object. *Ecology and Society*, *12*(1).
Burns, C. (2008). *Deadly decisions: how false knowledge sank the Titanic, blew up the shuttle and led America into war*. Amherst, NY: Prometheus Books.
Burns, R. (2014). Moments of closure in the knowledge politics of digital humanitarianism. *Geoforum*, *53*, 51–62.
Carlile, P. R. (2002). A pragmatic view of knowledge and boundaries: boundary objects in new product development. *Organization Science*, *13*(4), 442–455.
Cash, D. W. (2001). "In order to aid in diffusing useful and practical information": agricultural extension and boundary organizations. *Science, Technology, Human Values*, *26*(4), 431–453.
Cash, D. W., & Moser, S. C. (2000). Linking global and local scales: designing dynamic assessment and management processes. *Global Environmental Change*, *10*(2), 109–120.
Cinnamon, J. (2015). Deconstructing the binaries of spatial data production: towards hybridity. *Canadian Geographer*, *59*, 35–51.
Cleveland, H. (1982). Information as a resource. *Futurist*, *16*(6), 34–39.
Cooley, M. (1980). *Architect or Bee? The human/technology relationship*. Monroe, ME: South End Press.
Cowell, R., & Lennon, M. (2014). The utilization of environmental knowledge in land-use planning: drawing lessons for an ecosystem services approach. *Environment and Planning C: Government and Policy*, *32*, 263–282.
Crona, B. I., & Parker, J. N. (2011). Network determinants of knowledge utilization: preliminary lessons from a boundary organization. *Science Communication*, *33*(4), 448–471.
Elwood, S., Goodchild, M. F., & Sui, D. Z. (2010). Researching volunteered geographic information: spatial data, geographic research, and new social practice. *Annals of the Association of American Geographers*, *102*(3), 571–590.
Emad, G., & Roth, W. M. (2009). Policy as boundary object: a new way to look at educational policy design and implementation. *Vocations and Learning*, *2*(1), 19.
Frantzeskaki, N., & Kabisch, N. (2016). Designing a knowledge co-production operating space for urban environmental governance – lessons from Rotterdam, Netherlands and Berlin, Germany. *Environmental Science & Policy*, *62*, 90–98. https://doi.org/10.1016/j.envsci.2016.01.010
Freeman, J. (1997). Collaborative governance in the administrative state. *UCLA Law Review*, *45*, 1.
Frické, M. (2009). The knowledge pyramid: a critique of the DIKW hierarchy. *Journal of Information Science*, *35*, 131–142.
Fung, A., & Wright, E. O. (2001). Deepening democracy: innovations in empowered participatory governance. *Politics & Society*, *29*(1), 5–41.
Furnas, G. W. (2000). Future design mindful of the MoRAS. *Human – Computer Interaction*, *15*(2–3), 205–261.
Gober, P., Kirkwood, C. W., Balling, Jr., R. C., Ellis, A. W., & Deitrick, S. (2010). Water planning under climatic uncertainty in Phoenix: why we need a new paradigm. *Annals of the Association of American Geographers*, *100*(2), 357–372.
Guston, D. H. (2001). Boundary organizations in environmental policy and science: an introduction. *Science, Technology, Human Values*, *26*(4), 399–408.
Haraway, D. (1988). Situated knowledges: the question of science in feminism and the privileges of partial perspective. *Feminist Studies*, *14*(3), 575–99.
Harvey, F., & Chrisman, N. (1998). Boundary objects and the social construction of GIS technology. *Environment and Planning*, *A30*, 1683–1694.

Henderson, K. (1999). *On line and on paper: visual representations, visual culture, and computer graphics in design engineering*. Cambridge, MA: The MIT Press.
Hermann, S., & Neumeier, S. (2008). Social implications of developing a web GIS. In A. Lovett & K. Appleton (Eds.), *GIS for environmental decision-making*. Boca Raton, FL: Taylor & Francis, pp. 241–251.
Hicks, D., Larson, C., Nelson, C., Olds, D. L., & Johnston, E. (2008). The influence of collaboration on program outcomes: the Colorado nurse – family partnership. *Evaluation Review, 32*(5), 453–477.
Hinrichs, M. M., Seager, T. P., Tracy, S. J., & Hannah, M. A. (2017). Innovation in the knowledge age: implications for collaborative science. *Environment Systems and Decisions, 37*(2), 144–155.
Iorio, J., & Taylor, J. E. (2014). Boundary object efficacy: the mediating role of boundary objects on task conflict in global virtual project networks. *International Journal of Project Management, 32*(1), 7–17.
Juhola, S., & Westerholf, L. (2011). Challenges of adaptation to climate change across multiple scales: a case study of network governance in two European cities. *Environmental Science & Policy, 14*, 239–247.
Kates, R. W., & Burton, I. (Eds.). (1986). *Geography, resources and environment, volume 2: themes from the work of Gilbert F. White*. Chicago, IL: University of Chicago Press.
Kimble, C., Grenier, C., & Goglio-Primard, K. (2010). Innovation and knowledge sharing across professional boundaries: political interplay between boundary objects and brokers. *International Journal of Information Management, 30*(5), 437–444.
Koskinen, K. U., & Mäkinen, S. (2009). Role of boundary objects in negotiations of project contracts. *International Journal of Project Management, 27*(1), 31–38.
Kwaku-Kyem, P. A. (2004). Of intractable conflicts and participatory GIS applications: the search for consensus amidst competing claims and institutional demands. *Annals of the Association of American Geographers, 94*(1), 37–57.
Kwan, M.-P. (2004). Beyond difference: from canonical geography to hybrid geographies. *Annals of the Association of American Geographers, 94*(4), 756–763.
Kwan, M.-P. (2007). Affecting geospatial technologies: toward a feminist politics of emotion. *The Professional Geographer, 59*, 22–34.
Lasker, R. D., & Weiss, E. S. (2003). Broadening participation in community problem solving: a multidisciplinary model to support collaborative practice and research. *Journal of Urban Health, 80*(1), 14–47.
Lee, C. P. (2005). Between chaos and routine: boundary negotiating artifacts in collaboration. In *ECSCW 2005*. Dordrecht: Springer, pp. 387–406.
Lee, C. P. (2007). Boundary negotiating artifacts: unbinding the routine of boundary objects and embracing chaos in collaborative work. *Computer Supported Cooperative Work (CSCW), 16*(3), 307–339.
Light, J., & Allen, D. (2015). *From voice to influence: understanding citizenship in a digital age*. Chicago: University of Chicago Press, p. 376.
McLafferty, S. L. (2002). Mapping women's worlds: knowledge, power and the bounds of GIS. *Gender, Place and Culture: A Journal of Feminist Geography, 9*(3), 263–269.
Measham, T. B., Preston, T., Smith, C., Brooke, R., Gorddard, G., Withycombe, G., & Morison, C. (2011). Adapting to climate change through local municipal planning barriers and challenges. *Mitigation & Adaptation Strategies for Global Change, 16*, 889–909.
Miller, C. (2001). Hybrid management: boundary organizations, science policy, and environmental governance in the climate regime. *Science, Technology, Human Values, 26*(4), 478–500.
Monteiro, M. (2010). Reconfiguring evidence: interacting with digital objects in scientific practice. *Computer Supported Cooperative Work (CSCW), 19*, 335–354.
Muñoz-Erickson, T. A. (2014). Co-production of knowledge – action systems in urban sustainable governance: the KASA approach. *Environmental Science & Policy, 37*, 182–191. https://doi.org/10.1016/j.envsci.2013.09.014
Nicolini, D., Mengis, J., & Swan, J. (2012). Understanding the role of objects in cross-disciplinary collaboration. *Organization Science, 23*(3), 612–629.
O'Mahony, S., & Bechky, B. A. (2008). Boundary organizations: enabling collaboration among unexpected allies. *Administrative Science Quarterly, 53*(3), 422–459.
Ostrom, E. (2000). Collective action and the evolution of social norms. *Journal of Economic Perspectives, 14*(3), 137–158.
Reason, P. (1994). Three approaches to participative inquiry. In N. K. Denzin & Y. S. Lincoln (Eds.), *Handbook of qualitative research*. Thousand Oaks: Sage, pp. 324–339.
Reason, P., & Bradbury, H. (Eds.). (2008). *Sage handbook of action research: participative inquiry and practice*. 2nd Edition. London: Sage Publications.
Sayer, A. (1992). *Method in social science: a realist approach*. 2nd Edition. New York: Routledge.

Sharma, N. (2008). The origin of the "data information knowledge wisdom" hierarchy. *Online Blog*, 4 February.

Solís, P., Vanos, J., & Forbis, R. (2016). The decision-making/accountability spatial incongruence problem for research linking science and policy. *The Geographical Review*. DOI: 10.1111/gere.12240.

Star, S. L. (1998). The structure of ill-structured solutions: boundary objects and heterogeneous distributed problem solving. In *Distributed artificial intelligence*. pp. 37–54.

Star, S. L. (2010). This is not a boundary object: reflections on the origin of a concept. *Science, Technology, & Human Values*, *35*(5), 601–617.

Star, S. L., & Griesemer, J. R. (1989). Institutional ecology, 'translations' and boundary objects: amateurs and professionals in Berkeley's museum of vertebrate zoology, 1907–39. *Social Studies of Science*, *19*(3), 387–420.

Stuart, E., Samman, E., Avis, W., & Berliner, T. (2015). *The data revolution: finding the missing millions*. ODI Research Report 03. London: Overseas Development Institute, p. 51.

Subrahmanian, E., Monarch, I., Konda, S., Granger, H., Milliken, R., & Westerberg, A. (2003). Boundary objects and prototypes at the interfaces of engineering design. *Computer Supported Cooperative Work (CSCW)*, *12*(2), 185–203.

United Nations Independent Expert Advisory Group (UN). (2014). *A world that counts: mobilising the data revolution for sustainable development*. A Report to the UN Secretary General. New York, NY: United Nations, p. 28.

van Kerkhoff, L. E., & Lebel, L. (2015). Coproductive capacities: rethinking science-governance relations in a diverse world. *Ecology and Society*, *20*(1), art14. https://doi.org/10.5751/ES-07188-200114

Weinberger, D. (2010). The problem with the data-information-knowledge-wisdom hierarchy. *Harvard Business Review*, 2 February.

White, D. D., Wutich, A. Y., Larson, K. L., & Lant, T. (2015). Water management decision makers' evaluations of uncertainty in a decision support system: the case of WaterSim in the Decision Theater. *Journal of Environmental Planning and Management*, *58*(4), 616–630.

Wyborn, C., Datta, A., Montana, J., Ryan, M., Leith, P., Chaffin, B., Miller, C., & Van Kerkhoff, L. (2019). Co-producing sustainability: Reordering the governance of science, policy, and practice. *Annual Review of Environment and Resources*, *44*, 319–346.

Young, A. L., & Lutters, W. G. (2017). Infrastructuring for cross-disciplinary synthetic science: meta-study research in land system science. *Computer Supported Cooperative Work (CSCW)*, *26*(1–2), 165–203.

Zurba, M., & Berkes, F. (2014). Caring for country through participatory art: creating a boundary object for communicating indigenous knowledge and values. *Local Environment*, *19*(8), 821–836.

3

FRAMING DEVELOPMENT THROUGH ENVIRONMENTALISM

Maano Ramutsindela

Introduction

Though environmentalism and development – as concepts, processes, and practices – have developed in distinct pathways, they have partially converged under different ideological milieus and changing socio-political, economic, and environmental conditions. Each of them carries multiple situated meanings. There are several varieties of environmentalism, some of which are associated with saving nature, technological innovation, and the struggles for social and environmental justice (Martinez-Alier, 2012; Domènech, 2013; Yao et al., 2021). Regarding development, it has been associated with economic growth, human wellbeing, human freedom, and so on (Sen, 2001). The historical trajectories of environmentalism and development raise two points for consideration. The first is their ideological roots, while the second is the way in which these two processes are involved in dynamic and unfolding relations and the outcomes they engender. The outcomes from development processes have triggered mainstream environmentalism which in turn framed development. Environmentalism considers persistent and deepening poverty in many parts of the Global South as a reason for using nature conservation for poverty alleviation (Ramutsindela, 2004). However, the preoccupation with environmental protection is seen as contributing to the very poverty environmentalism seeks to alleviate. This chapter focuses on this intersection and the framing of development through nature conservation. To shed light on this process, the chapter first presents an overview of environmentalism and development in their own terms. Thereafter it discusses how the two converge before it dwells on how various measures used to protect the environment shape the idea of development.

Overview of environmentalism and development

Analyses of environmentalism reveal the association between environmentalism and political ideology (Corrigan and Rawn, 2016), and its relations with popular politics (Bosworth, 2020), national identities (Milfont et al., 2019), and compliance often with little or no regard to social consequences (Hu, 2020). Our purpose here is to conceptualize environmentalism within the ambit of the movements that seek to protect nature. There are three main currents of environmentalism that are important for understanding the relationship between environmentalism and development (Martinez-Alier, 2002, 2012). The first current relates to the preoccupation

DOI: 10.4324/9780429450310-5

with saving nature from human destruction and the protection and (re)creation of the wilderness. Saving nature is central to the activities of many environmentalists and environmental movements. For example, environmentalists are described as courageous women and men who stepped out to preserve remnants of the wilderness with their own resources (Butler et al., 2010). They are devoted to the protection of nature and can also participate in or sponsor environmental movements that mobilize social action and support for environmental protection. However, the universal solutions they promote are often, but not always, national interests disguised as global goals (Shiva, 2005). The relationship between environmental movements and the state is informed by goals and strategies, that is the movements cooperate with the state in order to influence policies necessary for the achievement of environmental goals and "to access bureaucratic and informal opportunities offered by government institutions and agencies" (Ramutsindela, 2020: 259). Environmental movements have also emerged to challenge the state's preoccupation with materialist concerns and the consequent destruction of nature.

The second current relates to technological innovation and other means necessary for healing the planet even as the same technology has been used to plunder it. Technology has been deployed to mediate the historical tension between environmental protection and development, and to realign environmental objectives with business interests. More broadly, technological solutions to environmental problems chime with anti-Malthusian views rooted in the belief that humans have the capacity to oversee their destiny and that of nature. This ability is demonstrated by scientific and technological innovations that make it possible to overcome the constraints imposed by nature on human life and aspirations. However, pandemics such as COVID-19 and climate-related disasters have laid bare the limits of human capacity to control and manage nature and natural processes.

The third current, termed the environmentalism of the poor, refers to the ways in which ordinary people protect their environment and to their struggles for environmental justice (broadly understood as the justice for the environment and for the people's relation with their environment). It is predicated on the clash between the economy and the environment and between culture and environmental protection. It also challenges the Eurocentric conception of the environment and the associated moral discourses of people-environment relations. However, the environmentalism of the poor does not imply that "poor people are always and everywhere environmentalists"; rather it means that "in ecological distribution conflicts, the poor are often on the side of resource conservation and a clean environment, even when they themselves do not claim to be environmentalists" (Martinez-Alier, 2002: viii). Thus, they mobilize their actions to protect the environment as a source of their livelihood and to defend the evolving living economies and life on earth. Non-governmental organizations such as Survival International have adopted a human rights approach to defend indigenous peoples' rights to live on their land and to determine their futures.

What about development? The purpose of reflecting on development as a concept and process here is to locate meanings of development that chime with or are deployed by environmentalism. This is important because conceptions of development have discursively shifted over time to reflect the thinking of major players such as the World Bank (WB), the International Monetary Fund (IMF) and the United Nations (UN). Development has variously been associated with the processes of growth and change, the contradictions and unevenness of capitalism, a set of interventions, human freedom, human rights, and so on (Sen, 2001; Sengupta, 2004; Bård and Stephen, 2006; Harvey, 2011; Werner, 2019). It is a process that is also culturally situated in two senses. The first is that it is a product of Enlightenment with its European values and meanings of progress that place societies in a hierarchical order. In that order, the "gaps" between societies constitute obstacles that need to be removed to enable them to move up the developmental ladder. As literature has shown, development gaps are hard to close because of the structure of the capitalist world economy and the inequalities it perpetuates (Smith, 2010; Harvey, 2011). The second sense is that individuals and societies pursue development goals that are aligned with what is valued in cultures outside the confines of Eurocentrism (Sen, 2001). These two senses are important for understanding how one type of environmentalism pursues development goals aligned with European values and economic outcomes while the other seeks to maintain and defend cultural values against Eurocentric development planning.

It is worth noting that the development enterprise is largely pursued as a set of interventions (Hart, 2001). Jolly (2005) has argued that the UN led many of these interventions, which the IMF and WB promoted as development policies and approaches. The past decades witnessed various forms of development interventions globally. For example, in the 1950s development intervention took the form of concessional finance for poorer countries. Subsequently the 1960s was marked as a development decade. The failure of economic growth to translate into the reduction in poverty prompted the search for interventions geared towards meeting the basic needs of people in the Global South (Streeten and Burki, 1978; Leipziger, 2019). Institutional reforms in the form of structural adjustments were also considered necessary for development and a precondition for the transition to economic growth. The reforms were underpinned by high-conditionality lending as a strategy for the management of the debt crisis. Sachs (1989: 255) views high-conditionality lending as "the process in which the international institutions make loans based on the promise of the borrowing countries to pursue a specified set of policies". Ironically, these reforms led to the increase in poverty as many jobs were lost in the process of creating a lean and hopefully efficient state. As we shall see later, conditionalities were also implemented in the environmental domain.

There has been a shift away from measuring development by economic growth alone to encompass people and their capabilities. This shift resulted in the development of the UN Human Development Index (HDI) which focuses on indicators such as living standards, health, and education. The HDI opened debates on national policy choices that undermine or enhance human development. The discussion in the next section will demonstrate that the environmental projects are yet to center human wellbeing as part of development thinking and planning. Instead, these projects narrowly focus on economic benefits – that hardly accrue to local people – at the expense of cultural, social, and political development.

The convergence of environmentalism and development

Given the dynamism of environmentalism and development, there is a need to explain how these two converge in space and time. Such an intersection is not only a useful entry point into the analysis of the intersection of environmentalism and development but it also enables us to better grasp the framing of what I call the environmentalism of development: the ways in which ideas of the environment and the mechanisms to protect and manage it frame and universalize the developmental pathways, needs, and aspirations of different nations and diverse cultures. The convergence of environmentalism and development takes place in various forms and involves many activities. In this chapter I refer to some of the key moments in which this convergence unfolded. The first is the resolution of the antithesis between environmentalism and development while the second involves the ideology of conditionalities. The third moment is the global environmental crisis and the final refers to the impact of the coronavirus pandemic on the thinking about the environment and development.

Resolving the antithesis between environmentalism and development

Coincidentally, the outcomes of economic development that kept environmentalists and developmentalists apart have become the glue that bind them! Historically, the two were related as an antithesis of each other. For example, development projects spurred by the rapid pace of urbanization and industrialization in Europe and North America were considered inimical to nature. In response, environmentalists stepped out to defend nature against the juggernaut of industrial capitalism with their own resources (Butler et al., 2010). The Club of Rome emerged in 1968 to grapple with the questions of the interdependence of economic, political, natural, and social components of the world system and the equilibrium between available natural resources and the growing human population (Meadows et al., 1972). Fifty years later, the Club is still concerned with the integration of economic growth and human wellbeing in what is called living economies (Korten, 2015). The clearest meeting point between environmentalism and development is articulated through the popular notion of sustainable development, which took root in the 1970s as a global mandate. Sustainable development conceptualizes the environment and development in neoliberal terms. The ideals of sustainable development are elusive mainly because of the global capitalist economy and its consequent inequalities in societies and among nations (Narsiah, this volume). However, these ideals remain on the agenda and underpinned both the Millennial Development Goals and the subsequent Sustainable Development Goals.

Environmental conditionalities

Environmentalism adopted and readjusted the conditionalities of development referred to previously and applied them in the form of debt-swap for conservation. Thus, both environmentalism and development responded to the global debt crisis of the 1980s in a similar fashion. Environmental NGOs viewed the crisis as an opportunity to force the state to behave in a certain way, namely to rearrange their national debts in return for support for nature conservation projects (Gray, 1991; Nelson, 2003; Macekura, 2016). As a result of the debt-for-nature swaps, the state gave away land at the expense of the livelihood of ordinary people. These swaps laid the foundation for the current wave of green grabbing, a process by which the state enables and licenses the appropriation and alienation of land and other resources for environmental purposes (Apostolopoulo and Adams, 2015). Whereas foreign multi-national companies and local elites are involved in wider land grabbing processes for various purposes, environmental lobby groups and donors target pieces of land and forests for conservation.

The global environmental crisis and its developmentalism

In *Development and the Environmental Crisis*, Michael Redclift (2010) sought to place the environmental crisis into the political dialogue about development. Since then, the threats and anxieties related to the global environmental crisis have generated debates on the notions of crisis and their discursive outcomes. Of relevance to the theme of this chapter and the volume as whole is the implication of the global environmental crisis for development. Many environmentalists believe that the environmental crisis demands a global approach, including the establishment of planetary boundaries – a safe space within which human development should take place in order to avoid "major human-induced environmental change on a global scale" (Rockström et al., 2009). Proposals have been made to put aside part of the planet – ranging from 30% to half the earth – for creating such a space (Wilson, 2016). Approaches and responses to the global environmental crisis constrain but also direct development planning at the local level. Accordingly, integrated development planning is promoted in areas designated for conservation.

The coronavirus pandemic and environment-development thinking

The COVID-19 pandemic has refocused greater attention to, as well as concerns with, the interaction between humans and animals. Questions have been raised about the causes of the pandemic and how outbreaks associated with the human-animal interactions could be prevented in future. The conservation lobby has used the pandemic to achieve several goals, including firming up existing conservation narratives and policies. Their pandemic narrative is clear: nature is better off without human presence and animals benefit from this condition. Examples of this include the absence of cruise ships that restored freedom to the whales, and the silence of motor cars in national parks that allowed animals to roam freely. Corlett et al. (2020) went on to say that "declines in visitor numbers caused by travel restrictions and park closures have reduced stresses on sensitive animals and trampling pressure on popular trails". It is said that the pandemic benefited the environment more broadly by limiting air pollution as a result of the grounding of planes and the shutdown of factories (Lenzen et al., 2020). This narrative of the human absence hacks back to the idea of the wilderness and its socio-political trappings.

The COVID-19 pandemic has invigorated two long-standing debates on wildlife, namely the regulation of wildlife trade and poaching. It has been used as an opportunity to relaunch the call for the ban on wildlife trade and for increased investment in wildlife protection, especially poaching (Roe et al., 2020). Accordingly, Lindsey et al. (2020) and Marshall (2020) commented that the lockdown increased the opportunity for poaching and wildlife trade. There is also a call for investment in forensic science (Halbwax, 2020). As in the economic, social, and political spheres, the pandemic has exposed the fragility of conservation, especially in the Global South. What do all these mean for environment-development thinking? While it is too early for a definite answer, we can tease out what the responses tell us.

The COVID-19 pandemic and the responses it elicited from conservationists and economists challenge some of the areas of convergence between environmental protection and development needs. Two competing propositions have emerged from the pandemic. The first seeks to rebuild the economy as a necessary condition for resolving the high unemployment rate caused by, and in some cases, worsened by the pandemic. This proposition is silent about environmental imperatives. The second proposition seeks to consolidate the apparent environmental gains from the pandemic and calls for more investment in the protection of the environment. This call does little, if any, to locate conservation as an avenue for development. There is clearly a division of labor between professionals and specialists but the pandemic appears to reopen the gap between environmentalism and development that had been closing after many years of debates and changes in conservation and development thinking.

Not all is lost in the environment-development nexus. The pandemic draws attention to the importance of health systems for human wellbeing and survival. It helps refocus development as human wellbeing, thereby placing the lives of ordinary people at the center of the development enterprise. In this context, it offers an opportunity to fast track the implementation of SDGs. There is however a growing concern that the pandemic delays the realization of SDGs, exacerbates lack of access to health care, and further marginalizes displaced populations (Lambert et al., 2020; Leal Filho et al., 2020).

Development through the wild

The environmentalist notion of development and its emphasis on economic outcomes permeates the conservation of different kinds of natural resources (Martinez-Reyes, 2014). This is evident in national parks and community-based conservation as well as private game reserves. The dominant perspective in each of these conservation initiatives is that the protection of wildlife contributes to local development through income generated mainly from tourism. In national parks, the environmentalist notion of development has recently been promoted through phrases such as "parks and people" and "beyond boundaries". Thus, governments are encouraged to develop schemes for poverty alleviation in protected areas more as a strategy to reduce the cost of poverty to protected areas than to promote local development per se. This rationale is reflected in the Durban Accord of the 5th World Parks Congress held in Durban (South Africa) in September 2003, and the 6th World Parks Congress held in Sydney (Australia) in 2014. While the creation of national parks was inspired by the need to protect wild animals against human use and the expansion of settlements, a broader category of protected areas has developed over time to allow access to resources for livelihood in these areas under certain conditions. These conditions are clear in the IUCN categories of protected areas that distinguish areas in which people can harvest natural resources for their livelihoods from those which prioritize the conservation of nature. UNESCO introduced biosphere reserves (BRs) in the 1970s to promote forms of environmental protection that are sensitive to human needs and cultures. The sites for BRs grew from 24 in 1974 to 709 in 2019 (Read and Price, 2020). However, in practice BRs are biased towards the environment (Sundberg, 1998).

The concept of development in community-based conservation (CBC) initiatives seeks to achieve at least three main goals, namely to conserve biodiversity outside formally established protected areas, to include local people in the management of their resources, and to respond to what environmentalists see as the unsustainable use of natural resources. In this context, CBC links together development and conservation imperatives at the local level. The developmental outcomes of CBC initiatives are mixed depending on the resources in question, their structure, and the conditions under which they operate. The general criticism of CBC initiatives is that they are often imposed on communities by governments, donors, and the private sector (Wright, 2017). Noe and Mwamfupe (this volume) observed that many top-down and externally driven CBC initiatives in the wildlife and forest sectors have increased the livelihood insecurities and undermine local development initiatives. These initiatives constrain agriculture, which is one of the main sources of development for local communities. However, alternative grassroots community development strategies can simultaneously benefit communities and the conservation of the environment. This is so because local people have the agency to protect fisheries resources and to realign traditional methods of production with the conservation of nature.

Private nature/game reserves and conservancies are gaining momentum in many parts of the world and are presented as part of the solution to environmental problems. These areas are set aside by landowners to conserve natural resources in private property (Nature Conservancy, 2001). Private game reserves have developed in different contexts and are largely motivated by the desire to make profit from wildlife through, say, hunting, ecotourism, and the sale of wildlife products (Wels, 2003; Jones et al., 2005; Van Hoven, 2015). This begs the question how development is conceptualized in these privately protected areas. These areas are linked to development in many ways, including the establishment of settlements inside game reserves as a form of rural development. Private nature reserves can also enact development through their philanthropic foundations that serve as a developmental arm. The foundations support local projects that promise to meet the developmental needs of communities living in or around game reserves. Taken together, these efforts are biased towards the conservation of nature and promote the notion of economic development even though development is secondary to, and a vehicle for, conservation.

Economic development permeates various types of environmental projects mainly because it is seen as necessary for encouraging local people, especially in rural areas, to develop positive attitudes towards protected areas. This assumption is problematic in that it undermines the agency of local and indigenous people and the various methods they use to protect nature. The current mantra of sustainable development is underpinned by discourses that promote the use of the environment for monetary benefits to local communities. This is clearly articulated in the Mexican notion of *aprovechamiento*, which NGOs link with conservation for economic benefits while the Tres Reyes Maya view it as the preservation of the forest for livelihood and future generations (Martinez-Reyes, 2014). Western-oriented NGOs deployed *aprovechamiento* to achieve two interrelated goals, namely to create a new income stream for local people and to regulate their relations with the environment. Thus, conservation imposes certain notions of development. More, importantly the emphasis on generating income from conservation places nature conservation in the hands of a capitalist class and facilitates the neoliberalization of conservation.

The penetration of capitalism into nature has resulted in the neoliberalization of nature and the creation of capital markets believed to be a solution to socio-cultural problems created by capitalism in the first place. Here capitalism thrives because it is admired for its capacity to fix environmental problems, to intervene to save nature and itself, and to lighten the burden of the state (Castree, 2011). The rationales for the neoliberalization of nature are that the private sector is best placed to protect nature, governments cannot be trusted with the stewardship of nature and have other pressing priorities to meet, and the increasing demands by tourists require services that governments on their own cannot provide (Büscher et al., 2012; Corson et al., 2013; Ramutsindela and Shabangu, 2013). These rationales are captured in the notion of green capitalism: a form of environmentalism centered on the multiple ways in which markets operate in the environmental domain (Scales, 2017). Funke (this volume) discusses the financialization of markets to underscore the role of markets in creating, allocating, and governing socio-natural assets. Environmental finance confirms the centrality of economic rationalities in the protection and governance of the environment.

Conclusion

Economic rationalities have not only profoundly changed the relations of people and their biophysical environment but have permanently changed how we think about the link between development and environmentalism. At the national and global levels, economic growth remains an important indicator of development. Government policies on natural resources underscore the use of natural resources to increase the country's gross domestic product. At the local level, communities have become entrepreneurs on their own or coerced into business ventures by their own governments. The outcomes of these policies are that local communities and indigenous people have entered business ventures in CBC initiatives and are increasingly shifting towards an economically driven form of development as the only alternative for their livelihood in a capitalist economy. However, there are still pockets of indigenous peoples who continue to embed development within a culture which appreciates and embraces the entanglement of people and their environment.

The pursuit of environmentalism does not end with the protection of the environment but goes a long way to shape ideas of development. The environmental framing of development takes place through conservation ideas and practices, and this process is facilitated by the natural resource being conserved. The meanings of environment and development are tied together by a Eurocentric worldview. There are three broad categories of environment-related conceptions of development. The first and dominant one focuses on economic outcomes. The second refers to culturally embedded forms of community development that reflects the environmentalism of the poor. These alternative forms of development have remained peripheral to national and global policy formulations. The third one is development as access to natural resources in protected areas. This is achieved by rezoning protected areas and by community conservation in which nature can be harvested for livelihood as well as profit-making. Governments have embraced environmentalism because of the advantages it offers them. These include passing on the cost of development to external actors and local communities, decentralizing development imperatives, and creating a false sense of resource ownership and localized democracy.

References

Apostolopoulou, E., and Adams, W.M. 2015. Neoliberal capitalism and conservation in the post-crisis era: the dialectics of "green" and "un-green" grabbing in Greece and the UK. *Antipode* 47(1): 15–35.

Bård, A.A., and Stephen, P.M. 2006 *Development as a Human Right: Legal, Political and Economic Dimensions*. Cambridge, MA: Harvard University Press.

Bosworth, K. 2020. The people's climate march: environmental populism as political genre. *Political Geography* 83: 102281.

Büscher, B., Sullivan, S., Neves, K., Igoe, J., and Brockington, D. 2012. Towards a synthesized critique of neoliberal biodiversity conservation. *Capitalism Nature Socialism* 23(2): 4–30.

Butler, T., Vizcaíno, A., and Brokaw, T. 2010. *Wildlands Philanthropy: The Great American Tradition*. San Rafael, CA: Earth Aware.

Castree, N. 2011. Neoliberalism and the biophysical environment 3: putting theory into practice. *Geography Compass* 5(1): 35–49.

Corlett, R.T., Primack, R.B., Devictor, V., Maas, B., Goswami, V.R., Bates, A.E., Koh, L.P., Regan, T.J., Loyola, R., Pakeman, R.J. and Cumming, G.S. 2020. Impacts of the coronavirus pandemic on biodiversity conservation. *Biological Conservation*, 246: 108571.

Corrigan, L.M., and Rawn, M. 2016. Evangelical environmentalism: An analysis of gender and ideology. In Godfrey, P. and Torres, D. (eds.), *Systemic Crises of Global Climate Change: Intersections of Race, Class and Gender*. London: Routledge, pp. 300–313.

Corson, C., MacDonald, K.I., and Neimark, B. 2013. Grabbing 'green': markets, environmental governance and the materialisation of natural capital. *Human Geography* 6(1): 1–15.

Domènech, L., March, H., and Saurí, D. 2013. Contesting large-scale water supply projects at both ends of the pipe in Kathmandu and Melamchi Valleys, Nepal. *Geoforum* 47: 22–31.

Gray, A. 1991. The impact of biodiversity conservation on indigenous peoples. In Shiva, V., Anderson, P., Schücking, H., Lohmann, L. and Cooper, D. (eds.), *Biodiversity: Social and Ecological Perspectives*. London: Zed Books, pp. 59–76.

Halbwax, M. 2020. Addressing the illegal wildlife trade in the European Union as a public health issue to draw decision makers attention. *Biological Conservation* 251: 108798.

Hart, G. 2001. 'Development critique in the 1990s'. *Progress in Human Geography* 25: 649–658.

Harvey, D. 2011. Roepke lecture in economic geography – crises, geographic disruptions and the uneven development of political responses. *Economic Geography* 87(1): 1–22.

Hu, Z. 2020. When energy justice encounters authoritarian environmentalism: the case of clean heating energy transitions in rural China. *Energy Research & Social Science* 70: 101771.

Jolly, R. 2005. The UN and development thinking and practice. *Forum for Development Studies*, 32(1): 49–73.

Jones, B.T., Stolton, S.U.E., and Dudley, N. 2005. Private protected areas in east and southern Africa: contributing to biodiversity conservation and rural development. *Parks* 15(2): 67–76.

Korten, D.C. 2015. *Change the Story, Change the Future: A Living Economy for a Living Earth*. Oakland, CA: Berrett-Koehler Publishers.
Lambert, H., Gupte, J., Fletcher, H., Hammond, L., Lowe, N., Pelling, M., Raina, N., Shahid, T., and Shanks, K. 2020. COVID-19 as a global challenge: towards an inclusive and sustainable future. *The Lancet Planetary Health* 4(8): e312–e314.
Leal Filho, W., Brandli, L.L., Lange Salvia, A., Rayman-Bacchus, L., and Platje, J. 2020. COVID-19 and the UN sustainable development goals: threat to solidarity or an opportunity? *Sustainability* 12(13): 5343.
Leipziger, D.M. 2019. The basic human needs approach and north-south relations. In *The Challenge of the New International Economic Order*. London: Routledge, pp. 255–279.
Lenzen, M., Li, M., Malik, A., Pomponi, F., Sun, Y.Y., Wiedmann, T., Faturay, F., Fry, J., Gallego, B., Geschke, A., and Gómez-Paredes, J. 2020. Global socio-economic losses and environmental gains from the coronavirus pandemic. *PloS One* 15(7): e0235654.
Lindsey, P., Allan, J., Brehony, P., Dickman, A., Robson, A., Begg, C., Bhammar, H., Blanken, L., Breuer, T., Fitzgerald, K., and Flyman, M. 2020. Conserving Africa's wildlife and wildlands through the COVID-19 crisis and beyond. *Nature Ecology & Evolution* 4(10): 1300–1310.
Macekura, S. 2016. Crisis and opportunity: environmental NGOs, debt-for-nature swaps, and the rise of 'people-centred' conservation. *Environment and History* 22(1): 49–73.
Marshall, M. 2020. Conservation in crisis. *New Scientist* (30 May 2020), p. 8.
Martinez-Alier, J. 2002. *The Environmentalism of the Poor: A Study of Ecological Conflicts and Evaluation*. Cheltenham: Edward Edgar.
Martinez-Alier, J. 2012. The environmentalism of the poor: its origins and spread. In McNeill, J.R. and Mauldin, E.S. (eds.), *A Companion to Global Environmental History*. Chichester: Wiley & Sons, pp. 513–529.
Martinez-Reyes, J.E. 2014. Beyond nature appropriation: towards post-development conservation in the Maya Forest. *Conservation and Society* 12(2): 162–174.
Meadows, D.H., Meadows, D.L., Randers, J., and Berherens, W.W. 1972. *The Limits to Growth*. New York: Universe Books.
Milfont, T.L., Davies, C.L., and Wilson, M.S. 2019. The moral foundations of environmentalism. *Social Psychological Bulletin* 14(2): 1–25.
Nelson, R. 2003. Environmental colonialism: 'saving' Africa from Africans. *Independent Review* 8(1): 65–86.
Ramutsindela, M. 2004. *Parks and People in Postcolonial Societies: Experiences in Southern Africa*. Dordrecht: Kluwer.
Ramutsindela, M. 2020. Social movements. In Kobayashi, A. (ed.), *Encyclopedia of Human Geography*, 2nd edn, Volume 1. Amsterdam: Elsevier, pp. 257–262.
Ramutsindela, M., and Shabangu, M. 2013. Conditioned by neoliberalism: a reassessment of land claim resolutions in the Kruger National Park. *Journal of Contemporary African Studies* 31(3): 441–456.
Read, M.G., and Price, M.F. 2020. Introducing UNESCO biosphere reserves. In Read, M.G. and Price, M.F. (eds.), *UNESCO Biosphere Reserves: Supporting Biocultural Diversity, Sustainability and Society*. London: Routledge, pp. 1–10.
Redclift, M. 2010. *Development and the Environmental Crisis: Red or Green Alternatives*. London: Routledge.
Rockström, J., Steffen, W., Noone, K., Persson, Å., Chapin III, F.S., Lambin, E., Lenton, T.M., Scheffer, M., Folke, C., Schellnhuber, H.J., and Nykvist, B. 2009. Planetary boundaries: exploring the safe operating space for humanity. *Ecology and Society* 14(2).
Roe, D., Dickman, A., Kock, R., Milner-Gulland, E.J., and Rihoy, E. 2020. Beyond banning wildlife trade: COVID-19, conservation and development. *World Development* 136: 105121.
Sachs, J.D. 1989. Conditionality, debt relief, and developing country debt crisis. In Sachs, J.D. (ed.), *Developing Country Debt and Economic Performance, Volume 1*. Chicago: University of Chicago Press, pp. 255–296.
Scales, I.R. 2017. Green capitalism. In Richardson, D., Castree, N., Goodchild, M.F., Kobayashi, A.L., Liu, W. and Marston, R. (eds.), *International Encyclopedia of Geography: People, the Earth, Environment, and Technology*. Chichester: Wiley, pp. 1–8. doi: 10.1002/9781118786352.wbieg0488
Sen, A. 2001. *Development as Freedom*. Oxford: Oxford University Press.
Sengupta, A. 2004. The human right to development. *Oxford Development Studies* 32(2): 179–203.
Shiva, V. 2005. *Earth Democracy: Justice, Sustainability, and Peace*. London: Zed Books.
Smith, N. 2010. *Uneven Development: Nature, Capital, and the Production of Space*. Athens: University of Georgia Press.
Streeten, P., and Burki, S.J. 1978. Basic needs: some issues. *World Development* 6(3): 411–421.

Sundberg, J. 1998. NGO landscapes in the Maya biosphere reserve, Guatemala. *Geographical Review* 88(3): 388–412.
The Nature Conservancy. 2001. *Private Land Management and Conservation: A Guide to Organizations*. Arlington: The Nature Conservancy.
van Hoven, W. 2015. Private game reserves in southern Africa. In *Institutional Arrangements for Conservation, Development and Tourism in Eastern and Southern Africa*. Dordrecht: Springer, pp. 101–118.
Wels, H. 2003. *Private Wildlife Conservation in Zimbabwe. Joint Ventures and Reciprocity*. Leiden: Brill.
Werner, M. 2019. Geographies of production I: global production and uneven development. *Progress in Human Geography* 43(5): 948–958.
Wilson, E.O. 2016. *Half-Earth: Our Planet's Fight for Life*. London: Liveright Publishing.
Wright, V.C. 2017. Turbulent terrains: the contradictions and politics of decentralised conservation. *Conservation and Society* 15(2): 157–167.
Yao, Z., Zhong, Q., Xiao, J., Ye, S., Tang, L., and Zhang, Z. 2021. An environmental-friendly process for dissociating toxic substances and recovering valuable components from spent carbon cathode. *Journal of Hazardous Materials* 404: 124120.

4
THE FINANCIALIZATION OF NATURE

Jayson J. Funke

Scientists and environmentalists claim we live in the Age of the Anthropocene (Waters et al., 2016). A new geologic epoch reflecting the scale, extent, and totality of human modification of the environment. The human ability to modify and "produce nature" has coincided with the historic growth and geographic expansion of capitalism. The globalization of capitalism through liberalization and neoliberal deregulation is said to have inaugurated an era of predatory and parasitic "disaster capitalism" that ties value to risks associated with socio-natural hazards and catastrophes, infrastructure destruction, environmental degradation, and social violence, displacement, and conflict (Klein, 2008). Managing socio-natural risks is traditionally the responsibility of the capitalist state, which ensures the reproduction of land/nature, labor, and money as key, albeit "fictitious" commodity inputs for the continuous production of value and exchange. Under the neoliberal state, socio-natural risks and the production and reproduction of these fictitious commodities have increasingly come under the control of financial markets, agents, and institutions.

The idea that nature can influence financial markets may be more intuitive than the idea that financial markets can in turn influence or produce nature. Commodity derivatives have long been used to manage the financial risks associated with farming and natural hazards (hurricanes, floods, etc.). The development of carbon markets and emissions trading schemes, however, are attempts to construct financial markets that promise to produce a particular, lower-carbon atmosphere, as a form of nature. Environmental risks and market externalities, like carbon dioxide emissions, are treated as commodities deemed best managed as financial assets and subject to the discipline of market price movements. Both examples can be said to offer evidence of the "financialization of nature" involving the quantification and monetization of nature and the non-human labor involved in biophysical processes and geologic systems. Environmental Finance[1] has emerged as an ideological and institutional response to the failure of capitalist markets to "properly value" and mediate socio-natural relations, and to effectively redress environmental concerns through "sustainable consumption". Financial technologies and Social and Environmental Management Systems (SEMS) have been essential in this endeavor. They allow environmental risks, non-human labor, and socio-natural assets to be quantified, commoditized, and monetized as investment returns, and subject to the management and discipline of markets.

The first part of this chapter introduces key concepts essential to understanding the financialization of capitalism by exploring the foundations of modern money, capitalist finance, and

DOI: 10.4324/9780429450310-6

neoliberal governance. The explicitly critical concepts of rentierism, finance capital, and neoliberal debtfare states are used herein to help frame financialization as a particular regime of capital accumulation that attempts to manage the persistent capitalist crises of over-accumulation and class struggles through consumption and the production of debt-based financial rents. The financialization of nature is then introduced through exploration of Environmental Finance, its governance technologies, risk management practices, and investment products. The conclusion offers a radical critique that argues that neoliberal debtfare states and Environmental Finance seek to conceal and displace the crises effects of capitalist over-accumulation by displacing risks, normalizing the expansion of credit/debt, and by diverting public attention to market exchange and consumption, and away from capitalism's tendency to over-produce money at the expense of labor and nature. The discourse of Environmental Finance is essential to mystifying the outcomes of financialization, and the value-capturing practices of speculation and rentierism that displace the market risks tied to environmental crises, and that foster the market bubbles that manifest in financial crises. Crises reinforce patterns of uneven development by consolidating wealth, power, and, increasingly, green environments, among an elite few.

The financialization of capitalism

Capitalist finance has evolved in conjunction with the territorial jurisdictions of sovereign states with the purpose of circulating money and credit to sustain economic development. Capitalist money has historically functioned as a universal measure of market value, means of exchange, and unit of account, and the form of money itself has often assumed commodity form, such as gold or silver. Historically, sovereign states claimed the exclusive right to create money. However, since the end of the Bretton Woods System in the mid-1970s, most money is created through credit/debt instruments. Either through the monetization of state/sovereign debt as reserves in central banks and the bond market, or, more significantly, as a financial commodity through credit-based instruments in the private banking and financial system (Funke, 2017; McLeay et al., 2014). Financial profits and growth are thus tied to expanding credit/debt and rents (e.g. fees, investment returns).

The rise of monopoly capitalism in the early 20th century began with the creation of the modern corporation, and its shareholder, value-based governance system, which in turn spurred securities markets for corporate equities, thus inaugurating the era of finance capitalism. The growth of rentierism has been described as constituent to the financialization of capitalism, a term deployed by critical scholars to refer to, most broadly, the processes and outcomes associated with neoliberal globalization and the increasing rule and power of financial prerogatives, financial markets, and financial actors and institutions over domestic and international economies, politics, and societies (Epstein, 2006; Lapavitsas, 2011; Martin, 2002). Financialization infers a regime change in the political economy of modern capitalism away from Keynesian, social democratic welfare states and the closely managed Bretton Woods international monetary system, to the rise and institutionalization of free market neoliberal globalization under the management of liberalized financial capital. The liberalization of finance de-territorialized money and inaugurated the geographic race to further free capital from regulatory boundaries. The deregulation of finance combined with advancements in information technologies to increase competition through disintermediation[2] and enabled the creation of new markets and spaces of financial production. Offshore holding companies and special purpose entities in lightly regulated offshore financial centers and markets stretched the boundaries of traditional financial production, increasing the volume and types of financial players to include other service providers, retailers, and manufacturers, as well as nonprofit nongovernmental organizations

(NGOs), individuals and households, and governments. Financialization thus blurs the conceptual distinction between the "real" and "financial" sectors of the economy, by blending industrial and merchant capital together under the control of finance capital.

The international financial system has subsequently transitioned from highly regulated national and regional systems of capital circulation and financial intermediation under Bretton Woods to a more liberalized global geography of tiered capital flows among a hierarchy and network of international centers of financial production. The regulatory architecture of international finance consists of overlapping jurisdictions of territorially based monetary and financial laws (typically national or regional), and international soft law systems of industry "best practice" standards, benchmarks, and norms, which lack a strong international legal and institutional framework (Lastra, 2015). Discipline is primarily achieved, we are told, through technocratic institutional standards and best practices governed through a geo-financial, panoptic form of market governmentality ultimately linked to asset price movements (Nadvi, 2008; O'Tuathail, 1997; Sheppard & Leitner, 2010). Regulatory reforms following the 2007–2009 Financial Crisis and Economic Recession have done little to end the liberalized system of volatile global capital flows or constrain the power of panoptic finance capital or the reach of its market-based governance system.[3] The focus of reforms has been on financial consumption typically carried out at the national and regional scales, thus preserving the world's institutionally variegated and multi-scalar regulatory landscape.

Development, policy, and power

Neoliberal globalization has further increased the concentration of capital atop a global hierarchy of international financial centers, located in global cities and controlled by giant economic entities like institutional investors, multinational corporations, and financial conglomerates. Great pools of financial assets and capital are concentrated in firms housing elite crews of financial technocrats and policy elites institutionally clustered together and connected via markets, telecommunication systems, and social networks. These centers of financial power have accumulated the capacity for widespread political-economic control by serving as concentrated command points in the organization of the world economy (Sassen, 1999). This geofinancial complex exercises power over development policies through a combination of control over money, expertise, and access to the complex policy networks of institutional elites. The power of financial capital over global development and policy has further increased *in lieu* of declining rates of official development aid, and the emergence of new forms of private development finance (e.g. securitization, micro-finance) (D'Arista, 2005; Peet, 2007).

Investment banks, institutional investors, and rating agencies in particular exercise a range of interconnected powers within the institutional complex[4] of the development policy network as they control access to capital and direct large flows of credit/debt, FDI and portfolio capital for governments, firms, and households. Credit ratings are used to asses a country's overall development policies, especially fiscal and monetary policies, and a low rating can signal possible risks to investors and increase the cost of borrowing or spur capital flight. These large investors are also important players in generating ideas within the policy regime power complex. They house both financial elites and technocratic experts who translate ideas into policy, and who often hold revolving-door positions and consultancies in government and regulatory institutions. Internationally, they play similar roles within the broader context of international institutions involved in global governance, including the Organization for Economic Cooperation and Development (OECD), the International Monetary Fund, World Bank, and so forth.

The foundations of modern finance: rentierism, finance capital, and neoliberal debtfare states

Critical debates on financialization hinge on the concept of rent as key to the growth of credit/debt and class struggles over the social distribution of surplus value. Rent in neoclassical economics is most simply the amount that an owner can charge someone for the use of an asset. Since money is treated as a commodity under capitalism, money extended as credit acquires additional value derived from interest or fees (e.g. intermediation or rent). Rentierism is the capturing of surplus value through absolute, differential, and especially monopoly rents from financial circuits, including credit/debt instruments and fees, forms of regulatory and market arbitrage (e.g. bouts of "capital switching"), and by speculating on swinging market prices of assets and activities in production and exchange. This is similar to Marx's concept of "absolute rent" involving class struggle between landed and financial capital over the distribution of surplus value valorized through control of property and land rents. "Differential rents" refers to the variation in the values generated from the market exchange of similar socio-natural assets produced from different parcels of land. Differential rents can arise from differences in land fertility or from differences in capital investments in production. Monopoly rents involve exclusive control over both the market value of a socio-natural asset as well as exclusive control over its means or conditions of production and reproduction. Intangible assets like data and information, and natural phenomenon like weather, carbon emissions, and even biophysical processes like "eco-system services", have become important sources of financial rents (Andreucci et al., 2017; Birch, 2016; Bryant, 2018; Felli, 2014; Harvey, 2006: 349–353; Kallis & Swyngedouw, 2017; Kenney-Lazar & Kay, 2017; Walker, 2017).

Finance capital: the crises and fixes of over-accumulation

Crises and uneven development are endemic to financialization as a regime of accumulation. The capitalist pursuit of constantly expanding value leads to monopoly tendencies, and provides incentives to over-produce value. Increasing production *in lieu* of increasing demand leads to persistent crises, with over-accumulation occurring in some economic spaces or sectors simultaneously with under-production in other spaces or sectors. One way to manage accumulation crises is to switch capital into new markets ("capital switching"), or invest it in some type of "spatial-temporal fix". Under financialization these fixes increasingly involve rent-based financial instruments. The rents derived from credit, arbitrage, and speculation hinge on market volatility and the unevenness of the world's regulatory spaces and market territories, and are short-term, risk management fixes that only temporarily offset the crises of over-accumulation.

Hyman Minsky (1993) identified three types of financial risk management: hedged, speculative, and Ponzi. While capitalist finance has an inbuilt proclivity toward instability, Ponzi (or monopoly) finance fosters crises. Modern fractional reserve banking has been compared to Ponzi finance, but one where bankers and their money pyramid scheme is protected by law and backed by government debt. The basic principle of commercial banking technically involves borrowing short (accepting demand deposits as liabilities) and lending long (issuing loans that are worth more than the underlying deposits) as the basis of profit/rents. The "fractional" part of banking refers to this "money-multiplier" mirage, wherein banks monetize client deposits as interest-bearing credit. Bank profits thus expand in lieu with the growth of credit/debt, and thus require liquidity, an ever-increasing flow of new money from depositors (including official reserves) to keep it going or face the prospects of a bank run. In theory, by holding reserves in a central bank, banks hedge their liability risks through the liquidity of state credit/debt, which allows banks to leverage their money and satisfy all demand deposits on request. Bank reserve requirements (e.g. leverage ratios) are one of the few banking regulations existing under international law (Lastra, 2015: 46–47).

The liberalization of capital and the neoliberal deregulation of financial intermediation extended the pyramid scheme-like structure underlying modern fractional reserve banking to the speculative structure of finance capital. The hierarchy of finance capital rests on the universalization of asset/liability-based contracts that underlie all financial instruments as money/value equivalents (Mehrling, 2000). This financial pyramid is mirrored in the hierarchical ranking of financial assets, including foreign exchange and the settlement of international trade, and is explicit among financial market agents and institutions, including their risk management practices, like credit scoring, fair value accounting practices, and so on. Not all financial instruments are thus equal, but are ranked according to value at risk and rate of return, and subsequently organized into a hierarchy of asset classes and tiers. The top tiers are comprised of the highest quality, lowest-risk assets like sovereign debt, then tiers of good quality, slightly higher-risk assets like bank credit, followed by middle and lower tiers of more high-risk, lower-quality commercial and consumer assets. Legal theories of finance point out that the hierarchy of financial systems and their agents and institutions are legally bound and subject to law (Pistor, 2013). Financial systems are comprised of complex, interdependent webs of contractual legal obligations that link investors one to another. Law defines a contract's enforceability, and can influence the distribution of gains and losses. Law tends to be more binding on low-quality, high-risk assets toward the bottom of the hierarchy, and relatively more elastic for high-quality assets at the pyramid's apex. As with banking, financial rents expand in lieu with growing credit/debt relations and thus require liquidity, an ever-increasing flow of new credit (including high-quality sovereign debt) to keep it going.

The spatial and temporal displacement of risks is important to capturing rents and reinforcing the hierarchy of capitalist finance. Derivatives are the largest, most liquid, volatile, and risky assets and comprise the base of the pyramid. Derivatives allow investors to speculate and hedge risks and thus, like banks holding reserves, displace many non-systemic risks.[5] Financial risk management involves the commodification of risks themselves, and market devices play a key role in this process. Breaking down an asset's values in terms of its future stream of revenue and value at risk assessments allows financial engineers to develop a range of commercial financial assets based on those revenue streams. Through quantification, valuation, and securitization these tranched bundles of risk are marketed to the public, in effect displacing the original asset's risks to markets and investors. The goal of this financial engineering is to find a high-quality asset and income stream to use as collateral. For example, low interest rates in the US have fueled the buying and borrowing of US government debt directly for the purpose of derivative speculation. Most international commercial derivative instruments are priced in US dollars, and high-quality US treasuries are the most common asset used as collateral for derivatives. Thus, sovereign debt structures the pyramid as if functions as universal money of account at the apex of the international hierarchy of finance, while also supporting debt-based speculation at the bottom of the pyramid (Bryan & Rafferty, 2007; Green, 2000; Tickell, 2000; Wainwright, 2011).

High rents from financial assets encourage investors to increase their financial holdings. For many firms this means directing investment returns into riskier financial instruments rather than reinvestment back into production. Speculation drives market bubbles and the value expressed in the money pyramid becomes uncoupled from valorization in production and exchange. During a crisis, credit becomes scarce and market makers seek to shift their holding of assets upward in the hierarchy of finance (for example, a lower-quality unsecured asset is replaced by higher-quality secured asset). Credit ratings continuously assess and rank assets in real time, influencing how financial rents and asset devaluations are distributed upward or downward to higher- or lower-quality assets in the hierarchy (Marandola & Sinclair, 2017; Mehrling, 2000). Crises effects are transmitted throughout the money pyramid and across the global economy via capital flows, exchange rate regimes, and monetary policies (Vasudevan, 2008: 11). High rents, asset devaluations, and increased instability typically come at a steep cost to low-income groups, leading to higher default risks, lower credit scores, and greater economic stress rising from the high costs of financial exclusion (e.g. credit red-lining) or predatory inclusion (e.g. payday lenders). Faced with inflation, unemployment, and a lack of affordable housing, healthcare and so on, low-income groups face the discipline of both markets and neoliberal debtfare states.

Neoliberal debtfare states

As a regime of accumulation, financialization requires an endless capture of value as rents cannot be sustained without continuous valorization in production, which requires state intervention and governance for stability and legitimacy (Jessop, 1990). The terms "debtfare" and "debtfare state" have been used to describe the neoliberal state's role and technologies for governing the risks associated with the capitalist crisis of over-accumulation and its class struggles surrounding the social distribution of surplus value. Debtfare emerged in the 1980s in tandem with neoliberal Washington Consensus policies that pushed structural adjustments on indebted developing economies, compelling them to pursue "good investment climates" in order to attract capital and in turn maximize the ability to fully repay all debts and obligations. Neoliberal debtfare governance involves the institutionalization of technocratic practices and market-based governance technologies (market devices) for centralizing institutions and power through standards and performance benchmarks. Market devices involve particular quantitative and calculative practices and are useful for monetizing non-market things and making them subject to market discipline. Debtfare also involves a rhetorical developmental anti-poverty strategy that seeks to institutionalize the consumption of credit/debt to offset the loss of living wages and Keynesian welfare benefits. Expanding credit/debt in turn helps offset over-production through rents resulting from new forms of enclosure, and credit/debt becomes the basis for accumulation by dispossession, and increasingly a requirement for the social reproduction of more and more of the population (Hager, 2016; Jessop, 1990; Soederberg, 2014; Streeck, 2017).

States are fundamental to securing property rights, markets, and rents, and are not only mediators between flows of foreign capital and domestic assets, they are themselves often investors in and owners of land, minerals, and other socio-natural assets (Emel & Huber, 2008: 1395; Parenti, 2015). The neoliberal debtfare state thus plays an important role in externalizing/socializing socio-ecological risks and their costs by "de-risking" investments and socially normalizing financial risk management, and by advocating consumption as the way to achieve sustainability. The net effect is to shift the public gaze, activism, and policy toward the seemingly apolitical consumer realm of exchange/reproduction (creditor/debtor and services/consumer relations) rather than direct them at the more politically volatile realm of production (capital/labor relations). Neoliberal debtfare states and finance capital work together to concentrate and centralize money and power unevenly across classes, space, and time, producing a form of "monopoly finance capital" that in turn wields hegemonic power over socio-natural relations (Foster & Holleman, 2010).

Environmental Finance and the financialization of nature

Environmental Finance refers to the complex of market-based financial instruments, institutions and agents and their practices and activities directed at monetizing, trading, and managing non-human labor and socio-natural assets. This includes state-led and publicly funded institutions and investments in sustainability and infrastructure projects (e.g. carbon markets, hydro-electric plants, etc.) as well as Sovereign Wealth Funds (e.g. resource and land grabs), and public private partnerships, like the Nordic Environment Finance Corporation. Various investments and services provided by NGOs, unilateral (e.g. USAID) and multilateral financial institutions (e.g. World Bank) and development banks are also included as forms of climate finance (Green Climate Fund) and initiatives that fund climate-related projects and environmentally oriented public policies. Private sector agents, multinational corporations, and institutions provide various intermediary services and instruments for portfolio investors, including various green investment funds and a variety of nature-based derivative instruments (e.g. commodity futures), as well as for FDI in mineral extraction, timber production, energy generation, and agribusinesses. This includes private equity firms that target sustainable investments in nature-based industries, as well as venture capital in clean technologies and green infrastructure (Knox-Hayes, 2017a; Sandor, 2016; Sandor et al., 2014).

Environmental Finance rests on environmental economics and the new institutional economics, and involves the use of market devices, like SEMS, as market-based governance technologies. Market devices are "performative technologies of power" in that they institutionalize a standardized and centralized framework (e.g. institutional culture) that shapes how a group acts and makes decisions. SEMS are commonly used by groups and institutions for managing socio-natural risks, and for aligning a group's operations and investments with particular environmental objectives. Market devices frame distant fields, quantify, value, monetize, and thus price non-market, abstract things or intangible processes, rendering them visible and calculable and thus subject to surveillance and market discipline (Callon et al., 2007). Non-human labor is framed as an economic "service", and the human-related risks and costs associated with the harmful aspects of non-human labor (e.g. weather, droughts) are essentialized into natural hazard "events" that are spatially and temporally bound, quantified, and priced. SEMS are used to price market externalities like environmental pollution, and the potential environmental impacts, costs, or benefits associated with producing socio-natural assets, or for pricing non-human labor, like biophysical processes and "ecosystem services". Pricing and data is assessed against sustainability performance standards or benchmarks (e.g. Equator Principles), which allow for calculations of potential environmental risks, costs or benefits that may result from the group's activities and investments. SEMS include specialized management systems comprised of computerized databases and operating platforms that can be used by organizations or groups, as well as multilateral, transnational systems, like cap-and-trade-based carbon markets (Bracking, 2012; International Finance Corporation, 2007).

The market-based governance of nature is part of the sustainable development policy discourse, strategy, and practice. Carbon markets were established under the Kyoto Protocol (1997) and the 2015 Conference of Parties to the United Nations Framework Convention on Climate Change introduced a new era for climate finance. The institutional mechanisms of carbon markets are emission trading schemes, which are predicated on cap-and-trade. In essence, the amount of CO_2 emissions that a specified area and its emitters can produce during a set period of time is "capped", and property rights are assigned through corresponding permits that are allocated and exchanged among emitters. As the emissions cap tightens and permit prices rise over time, the costs of extra permits becomes an incentive for emitters to reduce their carbon footprint. Carbon offsets, like REDD+ projects, ostensibly encourage sustainable development by tying technical and financial support to the production of "carbon sinks" (conservation projects or sustainable technology transfers) in countries of the Global South. Offsetting allows for the spatial displacement of excess emissions, while sequestered carbon is monetized as credit and marketed as a green financial asset. REDD+ projects have demonstrated difficulties meeting the performance measures that verify their value as carbon sinks, which in turn impacts the market pricing of their services and thus the "green grabbing" value of their rents, and many projects have been criticized for displacing resident indigenous groups (Bond, 2012; Bumpus & Liverman, 2008; Clark & Hermele, 2013: 34–35; Ervine, 2013; Fletcher et al., 2016; Knox-Hayes, 2017b; Lansing, 2012; Toni, 2011).

The financial production of nature

In the 1980s, debt-for-nature swaps emerged to help highly indebted poor governments (e.g. Bolivia, 1987) fund conservations efforts. The idea being that the revenue generated from monetizing sovereign debt would be written off, in turn freeing funds for conservation efforts. Likewise, payment for ecosystem services projects have emerged as a form of market-based financing and management of conservation, using financial incentives to compel efficient ecosystem maintenance. Conservation finance has even emerged as a financial assets class (e.g. green bonds), solidly reflecting the shift in funding conservation management. Value-capturing via conservation finance typically involves investing in revenue-generating activities centered around transforming non-human labor, and often common land and resources, into financial commodities. For example, value created through the production of sustainable forest products, or "encounter value" projects involving spectacle-based value-capturing of non-human labor, like wildlife safaris and ecotourism (Barua, 2017; Corson et al., 2013; Huff & Brock, 2017; Sullivan, 2013).

Financial profits and rents accrue through direct and portfolio investments. Direct investment is more long term and with the intent to purchase a controlling equity interest in a firm or asset in order to directly manage it. Most direct investment is undertaken by institutional investors, multinational firms, fund managers (including sovereign wealth funds), private equity, REITs, and other large-scale investors. Direct investment is often paralleled with portfolio investing. Portfolio investing is speculative by nature and typically involves managing financial risks by grouping relatively short-term investments across a pool of instruments and asset classes comprised of a variety of risk/return thresholds. Portfolio investing is widely practiced with a broad range of assets including equities, bonds, index and mutual funds, and derivative instruments (forwards, futures, options, swaps). This includes securitized products and various green investment funds. Green funds include index and mutual funds that employ SEMS to screen and select environmentally or socially responsible assets (commercial equities and bonds) for investment, or to screen and exclude those assets not meeting such standards.

Direct investments in land and infrastructure are attractive because they involve long-term value streams (e.g. tax revenues) or public assets that can be monetized into investments returns, and associated risks can be hedged. Historically most infrastructure (e.g. systems for transportation, communication, sanitation etc.) have been public goods financed by governments in various ways, depending on the scale and requirements of the project. Increased private investment in infrastructure has resulted in increasingly conjoined public-private infrastructure partnerships (O'Brien & Pike, 2017). Investments in land surged across Asia, Latin America, Eastern Europe, and especially Africa following the crisis of 2008–2009, expanding in tandem with global commodity prices, especially food. These so-called "land grab" deals have grown from approximately one million hectares per annum to over ten million hectares per annum, and include investments from Sovereign Wealth Funds (SWFs), Real Estate Investment Trusts (REITs), and investors seeking land, primarily for the production of crops. SWF have command over some of the world's largest pools of investment capital, and SWF investments in foreign land and infrastructure are more long-term investments that are designed to help improve the investing state's access to foreign technology, raw materials, or markets. REITs, both publicly traded and private, own, operate, and fund income-producing real estate or mortgages and mortgage-backed securities. REITs have been used for owning or financing land development and the construction of commercial property, as well as for managing land for production, such as timberland. Millions of acres of productive rural land have been brought under the ownership and control of investors, and rents are generated through rising land prices and profits derived from commodities. While agro-fuels, food, and forestry account for nearly 90% of the crops on verified land grabs, they play minor roles in value-grabbing, which is dominated by derivatives instruments tied to minerals (especially hydrocarbons) (Chen, 2013; Fairbairn, 2014; Gunnoe, 2014; Kay, 2017; Kenney-Lazar, 2018; Ouma, 2014).

While long-term investments in land and infrastructure are exposed to higher-risk premiums, parallel portfolio investments in derivatives can hedge risks and even provide investors speculative returns. Derivatives markets exist for hedging and speculating against the costs of environmental risks (e.g. weather derivatives and disaster bonds), as well as for hedging and speculating in raw material commodities (e.g. crops), extractive industries (e.g. mining, timber), and hydrocarbons (e.g. oil, gas). Growth in commodity derivatives ignited in the early 1990s with new commodity index funds and commodities futures. Futures are derivatives tied to individual commodities, while index funds group commodities together. Speculators have in turn developed futures that can be used to speculate against various price movements in indices and other derivatives. As a result, commodity derivatives and index funds themselves have become subject to futures speculation. Between 2002 and 2008 major grain traders (Archer Daniels Midlands, Cargill, etc.) and intermediaries had pandered agricultural futures for hedging risks and profiting from speculation. Processing companies, like cereal manufacturers, competed against index funds and speculators in bidding up the price of commodity futures, pushing global food prices to near-record highs (McMichael, 2013; Ouma et al., 2018: 501).

The financial value and production of nature: a radical critique

Much critical scholarship on the financialization of nature builds on Marx's biological analogy of the human species' metabolic relations with nature, which stands in stark contrast to the anthropocentric and narrow framework underlying Environmental Finance. Central to Marx's position is recognition that non-human labor (e.g. weather systems, hydrological cycles, soil nutrient cycles) is inherently entwined with human labor in the process of social development. Social values have traditionally mediated socio-natural relations, including the unintentional or harmful environmental outcomes from human activities. However, the historical and class-based transition to urbanized capitalist society dramatically altered the value framework mediating this metabolic relationship (Clark & Hermele, 2013).

Early political economists, including Marx, articulated capitalism's altered metabolic relationship through the *Labor Theory of Value*, which essentialized the production of all economic value as arising solely from human labor and market exchange. With the onset of the capitalist division of labor and the enclosure of common land and resources, the production of nature has been organized purely around the never-ending production of market value. Orthodox economic theory thus subsumes all value into exchange values as market prices, and market prices are expressed in terms of money, which is itself a commodity and abstract expression of "universal" value. Money and exchange value thus become the mechanism through which capital interacts with, frames, and values nature. Non-human labor is deemed a "free gift" of nature, and the ecological impacts of capitalist production are largely externalized. Marx, however, adopted the term "abstract labor" to include all the various kinds of *social and non-human labor* involved in the production of value. The co-production of value in socio-natural assets is itself an abstraction of the socio-natural metabolic relations produced through conjoined human and non-human labor (Ekers & Prudham, 2017: 10; Huber, 2017: 3–7; Smith, 2008; Walker, 2017: 2–3).

Polanyi (2001: 75–80) argued that value creation under capitalism requires the continuous development of three fictitious commodities: land/nature, labor, and money. The endless and rapacious production of value from these commodities comes up against the limits of their own reproduction. The transition to a pure credit-money system, combined with neoliberal globalization and world population growth, enabled the reproduction of money and human labor as near endless commodities. The reproduction of non-human labor, especially the reproductive spatial-temporal rhythms of biophysical and geological processes, are less amendable to the short-term demands of capitalist value production. The capitalist division of labor and uneven development alter and disrupt socio-natural metabolic relations, and drive capitalism toward the persistent under-production (reinvestment) of nature and labor. The logic of capital accumulation thus sews its own crises through the tendency to under-produce the conditions necessary for the future production of value (i.e. to invest in the reproduction of nature and labor). Financial and economic crises, degraded human health and quality of life, a deteriorating environment and "metabolic rifts", and inadequate public infrastructure are ecologically linked with systems and processes of non-human labor that are critical to socio-natural reproduction, and therefore essential in creating the conditions necessary for value production and capital accumulation (Nugent, 2015: 2467; O'Connor, 1998).

Debtfare states and finance capital converge in "fixing" the crises and risks of over-accumulation by managing the reproduction of land/nature, labor, and money as key capitalist commodities. Neoliberal globalization has been described as an "ecological regime" predicated on exploiting these crises of accumulation through spatially expanding the plunder, monetization, and extraction of rents from human and non-human assets. Financial crises intensify the need to absorb surplus capital through growing credit/debt, capital switching, and spatial-temporal fixes tied to long-term fixed capital projects, like infrastructure and land, or into speculative investments and other rent-based financial activities. Financial "fixes" can be external and include value-capturing through the privatization and monetization of non-human labor, or internal, like the production of value through new socio-natural assets, like GMOs. The under-production of nature (e.g. soil degradation, pollution) in one moment can create a landscape of investment opportunities and "fixes" (e.g. brownfield clean-up, carbon markets) elsewhere. While finance capital assumes an overtly socio-ecological role in managing socio-natural relations via monetization and financialization, all such relations are framed and regulated exclusively through the lens of markets, value, risk management and investment strategies (Castree, 2010; Ekers & Prudham, 2018; Loftus & March, 2015; McCarthy, 2015; Moore, 2017; Ouma, 2018; Yasin, 2017).

Conclusion

Environmental Finance assumes that capital markets can solve the very socio-natural problems that capitalist markets, in general, produce. Markets are neither adequate mechanisms for efficiently allocating socio-natural assets or neutral governance mechanisms for mediating socio-natural relations. Economic incentives and financial speculation drive short-term accumulation strategies at the expense of the long-term development and sustainability transitions needed to contend with socio-environmental degradations, and the time-sensitive requirements for addressing climate change. Historically capitalist markets have been unable to efficiently manage nature, labor, and money as key commodity inputs of value production and capitalist development. The capitalist propensity toward the over-production of value infers a tendency to over-accumulate value in the commodity form of money, which comes at the expense of neglecting investments into the reproduction of nature and labor. The degradation of social and environmental conditions under capitalism therefore are not only a resulting failure of capitalist markets, they are the cumulative results of its historical tendency to under-produce the conditions necessary for future production. While the externalization of production costs and the monetization of non-human labor have long been "free gifts of nature", the expropriation of value through rentierism in non-human value streams has become an increasingly essential short-term "fix" to the crises of over-production/under-consumption (i.e. until the next financial crisis hits!).

The spectacle of Environmental Finance and its discourse of "green" and "sustainable development" obscures the roles of class struggle inherent in the co-production of value, and the uneven reproduction of nature, labor, and money under capitalism. By legitimizing the fallacy that credit/debt-based consumption is key to addressing environmental problems, Environmental Finance panders the fantasy that unfettered capital accumulation via short-term speculation and rentierism is compatible with long-term environmental sustainability and the temporal reproduction of nature and earth's natural systems. These compelling stories of "selling nature to save it" are more about delivering compelling "value stories" to a financial consumer market than about adequately regulating socio-natural metabolic relations (Bigger, 2017; McAfee, 1999). Instead, the discourse of Environmental Finance should be understood as a discursive "technology of power" deployed by finance capital and neoliberal debtfare states to rationalize human and market domination over the environment and the earth's natural systems. Financialization gives finance capital ever more control over nature, and thus control over the conditions of production and life itself. Sadly, Environmental Finance cannot solve the environmental problems associated with capitalism and uneven development any more than it can overcome the spatial and temporal rhythms of the planet's natural systems.

Notes

1 The term Environmental Finance is herein used as a generalized category for market-based financial instruments, institutions, agents, and their practices and activities involved in monetizing and managing socio-natural assets.
2 Disintermediation in financial services refers to the growth of non-bank creditors and non-traditional financial service providers, like digital banking, FinTech, and service providers like PayPal, Apple Pay, and so forth.
3 Some countries re-imposed capital controls during the Great Recession, but the global trend has been to temporally limit capital controls in favor of an eventual return to liberalization.
4 The so-called "Wall Street-Washington-Treasury" regime.
5 The value of most speculative instruments, like weather derivatives and commodity futures, are predicated on rents captured through the commodification of *information* about an asset and its risks (Pollard et al., 2008).

Bibliography

Andreucci, D., Gracia-Lamarca, M., Wedekind, J., & Swyngedouw, E. (2017). "Value Grabbing": A Political Ecology of Rent. *Capitalism, Nature, Socialism*, 28(3): 28–47.

Barua, M. (2017). Nonhuman Labour, Encounter Value, Spectacular Accumulation: The Geographies of a Lively Commodity. *Transactions, Institute of British Geographers*, 42(2): 274–288.
Bigger, P. (2017). Hybridity, Possibility: Degrees of Marketization in Tradeable Permit Systems. *Environment and Planning A*, 28(5): 896–912.
Birch, K. (2016). Rethinking Value in the Bio-Economy: Finance, Assetization, and the Management of Value. *Science, Technology & Human Values*, 42(3): 460–490.
Bond, P. (2012). Emissions Trading, New Enclosures and Eco-Social Contestation. *Antipode*, 44(3): 684–701.
Bracking, S. (2012). How Do Investors Value Environmental Harm/Care? Private Equity Funds, Development Finance Institutions and the Partial Financialization of Nature-Based Industries. *Development and Change*, 43(1): 271–293.
Bryan, D., & Rafferty, M. (2007). Financial Derivatives and the Theory of Money. *Economy and Society*, 36(1): 134–158.
Bryant, G. (2018). Nature as Accumulation Strategy? Finance, Nature, and Value in Carbon Markets. *Capitalism, Nature, Socialism*, 108(3): 605–619.
Bumpus, A. G., & Liverman, D. M. (2008). Accumulation by Decarbonization and the Governance of Carbon Offsets. *Economic Geography*, 84(1): 127–155.
Callon, M., Millo, Y., & Muniesa, F. (Eds.). (2007). *Market Devices*. New York: Wiley-Blackwell.
Castree, N. (2010). Neoliberalism and the Biophysical Environment: A Synthesis and Evaluation of the Research. *Environment and Society: Advances in Research*, 1: 5–45.
Chen, J. C. (2013). Sustainable Territories: Rural Dispossession, Land Enclosures and the Construction of Environmental Resources in China. *Human Geography*, 6(1): 102–118.
Clark, E., & Hermele, K. (2013). *Financialisation of the Environment: A Literature Review*. FESSUD Papers no. 17. Lund, Sweden: Lund University.
Corson, C., MacDonald, K. I., & Neimark, B. (2013). Grabbing "Green": Markets, Environmental Governance and the Materialization of Natural Capital. *Human Geography*, 6(1): 1–15.
D'Arista, J. (2005). The Role of the International Monetary System in Financialization. In G. A. Epstein (Ed.), *Financialization and the World Economy*. Northampton, MA: Edward Elgar.
Ekers, M., & Prudham, S. (2017). The Metabolism of Socioecological Fixes: Capital Switching, Spatial Fixes, and the Production of Nature. *Annals of the Association of American Geographers*, 107(6): 1370–1388.
Ekers, M., & Prudham, S. (2018). The Socioecological Fix: Fixed Capital, Metabolism, and Hegemony. *Annals of the Association of American Geographers*, 108(1): 17–34.
Emel, J., & Huber, M. T. (2008). A Risky Business: Mining, Rent and the Neoliberalization of "Risk". *Geoforum*, 39(3): 1393–1407.
Epstein, G. A. (Ed.). (2006). *Financialization and the World Economy*. Northampton, MA: Edward Elgar Publishing.
Ervine, K. (2013). Carbon Markets, Debt and Uneven Development. *Third World Quarterly*, 34(4): 653–670.
Fairbairn, M. (2014). 'Like Gold with Yield': Evolving Intersections between Farmland and Finance. *The Journal of Peasant Studies*, 41(5): 777–795.
Felli, R. (2014). On Climate Rent. *Historical Materialism*, 22(3–4): 251–280.
Fletcher, R., Dressler, W., Buscher, B., et al. (2016). Questioning REDD+ and the Future of Market-Based Conservation. *Conservation Biology: The Journal of the Society for Conservation Biology*, 30: 673–675.
Foster, J. B., & Holleman, H. (2010). The Financial Power Elite. *Monthly Review*, 62(1): 1–19.
Funke, J. (2017). Demystifying Money: Fictions of Capital and Credit. *Human Geography*, 10(1): 20–35.
Green, S. (2000). Negotiating with the Future: The Culture of Modern Risk in Global Financial Markets. *Environment and Planning D: Society and Space*, 18: 77–89.
Gunnoe, A. (2014). The Political Economy of Institutional Land Ownership: Neo-Rentier Society and the Financialization of Land. *Rural Sociology*, 79(4): 478–504.
Hager, S. (2016). *Public Debt, Inequality, and Power: The Making of a Modern Debt State*. California: University of California Press.
Harvey, D. (2006). *The Limits to Capital* (3rd ed.). London: Verso.
Huber, M. T. (2017). Value, Nature, and Labor: A Defense of Marx. *Capitalism, Nature, Socialism*, 28(1): 39–52.
Huff, A., & Brock, A. (2017). Accumulation by Restoration: Degradation Neutrality and the Faustian Bargain of Conservation Finance. *Antipode*. https://antipodefoundation.org/2017/11/06/accumulation-by-restoration/

International Finance Corporation. (2007). *International Finance Corporation's Guidance Notes: Performance Standards on Social & Environmental Sustainability*. Washington, DC: International Finance Corporation.
Jessop, B. (1990). *State Theory: Putting the Capitalist State in its Place*. Cambridge: Polity Press.
Kallis, G., & Swyngedouw, E. (2017). Do Bees Produce Value? A Conversation Between an Ecological Economist and a Marxist Geographer. *Capitalism, Nature, Socialism*, 29(3): 36–50.
Kay, K. (2017). Rural Rentierism and the Financial Enclosure of Maine's Open Lands Tradition. *Annals of the Association of American Geographers*, 107(6): 1407–1423.
Kenney-Lazar, M. (2018). Governing Dispossession: Relational Land Grabbing in Laos. *Annals of the Association of American Geographers*, 108(3): 679–694.
Kenney-Lazar, M., & Kay, K. (2017). Value in Capitalist Natures. *Capitalism, Nature, Socialism*, 28(1): 1–6.
Klein, N. (2008). *The Shock Doctrine: The Rise of Disaster Capitalism*. New York: Picador.
Knox-Hayes, J. (2017a). Alternative Circuits of Capital: Parallel Economies of Environmental Finance. In R. Martin & J. Pollard (Eds.), *Handbook on the Geographies of Money and Finance* (pp. 499–517). Northampton: Edward Elgar.
Knox-Hayes, J. (2017b). Carbon Markets: Resource Governance and Sustainable Valuation. In G. Clark, M. P. Feldman, M. S. Gertler, & D. Wojicik (Eds.), *The New Oxford Handbook of Economic Geography* (2nd ed.). Oxford: Oxford.
Lansing, D. M. (2012). Performing Carbon's Materiality: The Production of Carbon Offsets and the Framing of Exchange. *Environment and Planning A*, 44: 204–220.
Lapavitsas, C. (2011). Theorizing Financialization. *Work, Employment and Society*, 25(4): 611–626.
Lastra, R. M. (2015). *International Financial and Monetary Law* (2nd ed.). Oxford: Oxford University Press.
Loftus, A., & March, H. (2015). Financialising Nature? *Geoforum*, 60: 172–175.
Marandola, G., & Sinclair, T. (2017). Credit Rating Agencies are Poorly Understood and the Rules Developed for Them Will Not Work. In R. Martin & J. Pollard (Eds.), *Handbook on the Geographies of Money and Finance* (pp. 478–495). Northampton: Edward Elgar.
Martin, R. (2002). *Financialization of Daily Life*. Philadelphia: Temple University Press.
McAfee, K. (1999). Selling Nature to Save it? Biodiversity and Green Developmentalism. *Environment and Planning D: Society and Space*, 17(2): 133–154.
McCarthy, J. (2015). A Socioecological Fix to Capitalist Crisis and Climate Change? The Possibilities and Limits of Renewable Energy. *Environment and Planning A*, 47(12): 2485–2502.
McLeay, M., Radia, A., & Thomas, R. (2014). Money in the Modern Economy: An Introduction. *Bank of England Quarterly Bulletin*, Q1.
McMichael, P. (2013). Value-chain Agriculture and Debt Relations: contradictory outcomes. *Third World Quarterly*, 34(4): 671–690.
Mehrling, P. (2000). *Modern Money*: Fiat or Credit? *The Journal of Post Keynesian Economics*, 22(3): 397–406.
Minsky, H. (1993). The Financial Instability Hypothesis. In P. Arestis & M. Sawyer (Eds.), *Handbook of Radical Political Economy*. Aldershot: Edward Elgar.
Moore, J. W. (2017). The Capitalocene, Part 1: On the Nature and Origins of our Ecological Crisis. *The Journal of Peasant Studies*, 44(3): 594–630.
Nadvi, K. (2008). Global Standards, Global Governance and the Organization of Global Value Chains. *Journal of Economic Geography*, 8: 323–343.
Nugent, J. P. (2015). Ontario's Infrastructure Boom: A Socioecological Fix for Air Pollution, Congestion, Jobs, and Profits. *Environment and Planning A*, 47(12): 2465–2485.
O'Brien, P., & Pike, A. (2017). The Financialization and Governance of Infrastructure. In R. Martin & J. Pollard (Eds.), *Handbook on the Geographies of Money and Finance* (pp. 223–252). Northampton: Edward Elgar.
O'Connor, J. (1998). *Natural Causes: Essays in Ecological Marxism*. New York: Guildford Press.
O'Tuathail, G. (1997). Emerging Markets and Other Simulations: Mexico, the Chiapas Revolt, and the Geofinancial Panopticon. *Ecumene*, 4(3): 300–317.
Ouma, S. (2014). Situating Global Finance in the Land Rush Debate: A Critical Review. *Geoforum*, 57: 162–166.
Ouma, S. (2018). This Can't be an Asset Class: The World of Money Management, "Society" and the Contested Morality of Farmland Investments. *Environment and Planning A*, 0(0): 1–22.
Ouma, S., Johnson, L., & Bigger, P. (2018). Rethinking the Financialization of 'Nature'. *Environment and Planning A*, 50(3): 500–511.
Parenti, C. (2015). The 2013 ANTIPODE AAG Lecture The Environment Making State: Territory, Nature, and Value. *Antipode*, 47(4): 829–848.

Peet, R. (2007). *The Geography of Power: The Making of Global Economic Policy*. London: Zed Press.
Pistor, K. (2013). A Legal Theory of Finance. *Journal of Comparative Economics*, 41(2): 315–330.
Polanyi, K. (2001). *The Great Transformation: The Political and Economic Origins of Our Time* (2nd ed.). Boston: Beacon Press.
Pollard, J., Oldfield, J., Randalls, S., & Thornes, J. E. (2008). Firm Finances, Weather Derivatives and Geography. *Geoforum*, 39: 616–624.
Sandor, R. L. (2016). *How I Saw It: Analysis and Commentary on Environmental Finance (1999–2005)*. Singapore: World Scientific Pub Co Inc.
Sandor, R. L., Kanakasabai, M., Marques, R. L., & Clark, N. (2014). *Sustainable Investing and Environmental Markets: Opportunities in a New Asset Class*. Singapore: World Scientific Publishing Co Inc.
Sassen, S. (1999). Global Financial Centers. *Foreign Affairs*, 78(1): 75–87.
Sheppard, E., & Leitner, H. (2010). *Quo Vadis* Neoliberalism? The Remaking of Global Capitalist Governance after the Washington Consensus. *Geoforum*, 41(2): 185–194.
Smith, N. (2008). *Uneven Development: Nature, Capital, and the Production of Space* (3rd ed.). Athens, GA: University of Georgia Press.
Soederberg, S. (2014). *Debtfare States and the Poverty Industry: Money, Discipline and the Surplus Population*. London and New York: Routledge.
Streeck, W. (2017). *Buying Time: The Delayed Crisis of Democratic Capitalism* (2nd ed.). London: Verso.
Sullivan, S. (2013). Banking Nature? The Spectacular Financialisation of Environmental Conservation. *Antipode*, 45(1): 198–217.
Tickell, A. (2000). Dangerous Derivatives? Controlling and Creating Risk in International Finance. *Geoforum*, 31: 87–99.
Toni, F. (2011). Decentralization and REDD+ in Brazil. *Forests*, 2(1): 66–85.
Vasudevan, R. (2008). Finance, Imperialism, and the Hegemony of the Dollar. *Monthly Review*, *59*(11).
Wainwright, T. (2011). Elite Knowledges: Framing Risk and the Geographies of Credit. *Environment and Planning A*, 43: 650–665.
Walker, R. (2017). Value and Nature: Rethinking Capitalist Exploitation and Expansion. *Capitalism, Nature, Socialism*, 28(1): 53–61.
Waters, C. N., et al. (2016). The Anthropocene is Functionally and Stratigraphically Distinct from the Holocene. *Science*, 351(6269).
Yasin, Z. T. (2017). The Adventure of Capital with Nature: From the Metabolic Rift to the Value Theory of Nature. *The Journal of Peasant Studies*, 44(2): 377–401.

5

COLONIALISM/POST-COLONIALISM NEXUS

An oxymoron of coloniality and globality

Johannes Tsheola

Introduction

Colonialism and globality have historically been intricately intertwined through processes of "imperial conquest, immigration, and subjugation (which) have always been global phenomena" in character (Guelke & Kay, 1996: 163). This chapter grapples with the oxymoron of colonialism and post-colonialism by asking one of the most profound questions confronting the present capitalist society and state, viz.: are there present-day empires? For Zielonka (2012: 503), there are present-day empires pursuing "civilizing missions". Johnston et al. (1994) differentiate colonialism from colonization in a manner that separates rule, domination, and subordination from mere "peopling" of spaces. Accordingly, Johnston et al. (1994: 75) state that colonialism is the "establishment and maintenance of rule, for an extended period of time, by a sovereign power over a subordinate and alien people that is separate from the ruling power", whereas colonization "involves the physical settlement of people . . . from the imperial centre to the colonial periphery". Explaining colonialism and coloniality, Maldonado-Torres (2007: 243) notes that the former is "a political and economic relation in which the sovereignty of a nation or a people rests on the power of another nation, which makes such nation an empire" whereas the latter "refers to long-standing patterns of power that emerged as a result of colonialism, but that define culture, labour, inter-subjective relations, and knowledge production well beyond the strict limits of colonial administrations". To this extent, Johnston et al.'s (1994) description of colonialism needs to be understood through a broader and deeper prism of the ruled's lived experiences, which involve a myriad of underpinnings of knowledge, mind-sets, values, cultures, beings, social relations, and so on. Drawing from these explanations and understanding, Maldonado-Torres (2007: 243) concludes that "as modern subjects, we breath coloniality all the time and everyday". Perhaps, a question needs to be asked thus: have these characteristic features of colonialism ceased to exist in the modern world of perceived post-colonialism?

Colonial independence involved reflexivity of colonialism as colonial rule transformed itself to accommodate its criticisms by formalizing and legitimizing the same characteristic features of coloniality. It is for the same reasons that Flenley (1996: 595) concludes that whereas the old and modern "colonialism was justified on religious grounds", that of the 21st century justified itself on the basis of the so-called "new religion – ecology". That is, post-colonialism remains illusive for all of the world largely due to the old-standing development-environment nexus

DOI: 10.4324/9780429450310-7

contestations. Whereas it is true that the "world has continued to change, at an ever-increasing pace", there is equally an undeniable degree of continuity (Drakakis-Smith, 1996: 706). This chapter concedes that post-colonialism will remain colonial in terms of the present world's lived experiences, primarily because of the paradoxes of the state's reason for existence, which is the same institution that colonialism required in order to entrench itself globally. The chapter asserts that colonialism and post-colonialism exude with the oxymoron of continuities of globality and coloniality; and, it is presented in seven sections, inclusive of the introduction and conclusion. The next section discusses colonialism and post-colonialism through the time and space philosophical perspective. The third section analyzes colonialism in the intersections of development and the environment. Colonialism and post-colonialism, as drivers of global transformation, are discussed in the next section from a variety of dimensions, inclusive of the evolution of people, ecosphere, urbanization and state as the institution of governance, among others. The oxymoron of continuities of global coloniality in colonialism and post-colonialism is examined in the fifth section. The sixth section deals with the matrices of coloniality that are rendered illusive in the current capitalist societies through particular knowledges, minds, and beings of domination, subjugation, and extraversion.

Old and modern colonialism in time and space: a philosophical overview

Efforts to explanations and understanding of colonialism and post-colonialism in time and space as well as the characteristic features of these processes have produced vast literature (Johnston et al., 1994). However, there is universal convergence on the idea that there was old colonialism which was later replaced by modern colonialism. The age of colonialism began in the 15th century, generally recognized as old colonialism, that "erupted from the contradictions of European feudalism and lasted for three centuries", and was disrupted in the 18th century by rapid advancement of industrial capitalism in England and "an expansionary world capitalism", ushering in a second phase of colonialism (Johnston et al., 1994: 76). Modern colonialism (dis) order evolved between 1820 and the First World War (1914–1918), and it was "backed by complete European hegemony over world trade, finance and shipping, and by new forms of political and military authority sustained by technology, applied science and information systems" (Johnston et al., 1994: 76). During this phase, Britain, France, Germany, Belgium, Italy, USA, and Japan have each increased their colonial landholding in sheer square kilometers.

Overall, colonialism was geared exclusively for effective commercial operations that secured "maximum profits possible" for shareholders (Guelke & Kay, 1996: 167); and, this motive has remained the same in the present capitalist international relations that are supposedly "new". The most profound question to be framed of the 21st century, therefore, is whether or not the post-Second World War colonial independence ushered in post-colonialism. Taylor (1993: 40) concludes that "the nature of structural power" established through colonialism is such that "winning political independence is a hollow victory while economic poverty prevails". The present global status quo of alienation, disenfranchisement, estrangement, and disadvantage of the majority of the population of the world suggests that post-colonialism exists only in theory and that the state has continued to collude with capital to sustain the status quo.

Colonialism and the development-environment nexus

Colonizers and the colonized have equally understood "their environments as accessible to human use for subsistence, wealth, and exchange; yet both had a significantly more spiritual or magical approach to their landscape" (Guelke & Kay, 1996: 164). Natives too "saw Nature as sentient and often animate" (Guelke & Kay, 1996: 164). Historically, the world's quest for development coincides with that for colonialism and the "peopling" of spaces (Mkandawire, 2011) as well as environmental considerations, stewardship, and/or neglect. Goodall (1987: 78) notes that colonialism "facilitated the expansion of capitalism" and that it, therefore, epitomizes uneven economic development. Colonialism-facilitated capitalism meant that "the encouragement of free enterprise and the pursuit of private wealth" were paramount and that land and all its resources were "potentially marketable commodities" (Powell, 1996: 281). The same was true even under old colonialism wherein "Imperialism exported more or less the same stressors around an ever-shrinking globe" (Powell, 1996: 281).

In essence, the economic consequences of colonialism and colonization

> were to enlarge and rationalise the use of the earth's resources, to take advantage of new crops and animals, and to make possible the exploitation of new areas of the world for raising crops and animals not readily grown in Europe.
>
> *(Guelke & Kay, 1996: 169)*

Colonizers extended their domain by clearing lands and creating towns and cities, which involved (re)shaping the development trajectories and the environment. To be precise, colonialism forged "a distinctively European world-image which was generally anthropocentric and exploitative" (Powell, 1996: 281). Hence, "By the end of the seventeenth century, assisted by advances in science and technology, a characteristically European idea of 'progress' had emerged", exercising "a powerful leverage during the next three centuries", though with contradictions and inconsistencies (Powell, 1996: 278). Hence, Huggett & Robinson (1996c: 479) state that "people are best able to reflect and to influence, not merely the environment of life as it exists, but the environment of life as it might become". Inevitably, the "peopling" of spaces and colonialism have historically been intricately interlocked with development and the environment which, in return, came to be "powerfully affected by egocentric constructions of ethics and moralities that were related in some complex ways to a need for security" (Huggett & Robinson, 1996c: 479). Whereas different cultures had "natural shrines and sacred places", colonizers often imposed their views of "native environmental beliefs as incompatible with Christianity, and therefore (as) superstitions (that have) to be abolished" (Guelke & Kay, 1996: 164).

Historicity of the development-environment nexus involves complex matrices of slavery, colonialism, neo-colonialism, capitalism, neoliberalism, and globalization (Grosfoguel, 2007; Maldonado-Torres, 2007; Ndlovu-Gatsheni, 2007, 2013; Mkandawire, 2011). Inescapably, the development-environment nexus has always agitated for suspicions, manifesting in cognitive descriptors such as "the myth of development", "development in crisis", "the rise and fall of development", "the impasse of development", and "post-development" (Mkandawire, 2011: 5), among others. Colonialism was indeed predicated on the notion of attacking, conquering, and subduing the indigenous peoples as well as capturing their goods and territories, transferring their lands and properties to ownership of the colonizers, whilst undermining their human dignity by reducing their existence to slavery. The colonizers were driven by their sense of superiority and need for resources (Braudel, 1981 cited in Guelke & Kay, 1996), which historically focused "powerfully on the preservation, protection, or improvement" of the living environments for all institutions, inclusive of peoples (Huggett & Robinson, 1996c: 479). The idea that colonizers "had the right to settle new lands and displace and enslave peoples who were not like themselves was generally taken for granted and provided a basis for many colonial enterprises that were brutal and destructive" (Guelke & Kay, 1996: 164). From the development-environment perspective, colonizers drew more privilege from their overseas possessions than from their homelands; and, the same political economy persists in the present global capitalist society, rendering the notion of post-colonialism untenable as an analytical tool.

According to Mkandawire (2011: 5), the "homogenising thrust" of colonialism and "compulsions of the market" have "rechristened as globalisation" where "emulation is a consequence of all human exchange". For the majority of the world, post-colonialism is an illusion. To be precise, developmentalism, environmentalism and the pursuit of sustainable development have been equivalent to "groping in the dark", and they continue to be dismissed as disasters of social engineering, accession to Western superiority, denigration of local knowledges and innovation, as well as repressive meta-narratives that are designed to perpetuate privilege of those defined to hold merit and oppression of those recognized as enthroning mediocrity (Mkandawire, 2011). That is, the present globalist world of capitalism wherein everything else is political cannot be devoid of coloniality or, at the minimum, post-colonialism disorder.

Colonialism and global transformation

This chapter does not seek to present a comprehensive account of global transformation attendant to colonialism. Overall, colonialism "produced a global transformation that increasingly integrated continents and peoples" (Guelke & Kay, 1996: 163). Colonialism and "peopling" of spaces meant that in "the process of enlargement, cultures inevitably came into conflict. . . (out of which) they began slowly to homogenise" (Huggett & Robinson, 1996b: 159). Colonizers took with them both their advancements and deficiencies into the colonized lands and peoples. Collectively, colonizers' social ills and deficiencies as well as "technological and military sophistication and organisational skills" served as their "powerful engine for the domination or destruction of many native societies" (Guelke & Kay, 1996: 166). Advanced and sophisticated organizational skills of the colonizers, which facilitated the mobilization of capital and people, have simultaneously promoted colonial political and administrative purposes as well as commercial motives. According to Claval (1996: 312), "None of the traditional forms of political organization were able to withstand the imperialist expansion of Western states". At the end of the First World War when a search for "a rational world political order" was in currency, the "only form of political organization which appeared legitimate was the Western nation-state" because with colonialism, "the modern forms of state administrative control" had already been imposed "all over the world by the end of the nineteenth century" (Claval, 1996: 312). In modern society "the primary structures of value formation as well as of alienation and domination are defined by state power and capital circulation" (Harvey, 1996: 211), which are the direct legacies of colonialism (Claval, 1996: 312).

Notwithstanding the divergent explanations and understanding of its evolution, colonialism involved experiences of "both resistance and adaptation to colonial rule" (Johnston et al., 1994: 77) during which "all the relationship between people and land was transformed" (Clarke, 1996: 252). In essence, colonialism and anti-colonialism have been fundamental forces in transforming the rest of the planet and in the making of the world (Johnston et al., 1994; Guelke & Kay, 1996; Powell, 1996). The first Great Transformation of the planet arose from "the evolution of people" (Huggett & Robinson, 1996a) wherein, as Harvey (1996: 349) puts it, relationships between peoples were mediated through "market functions and state power". By the 18th century, the Industrial Revolution "transformed social economies by creating new goods and also by creating new systems for their exchange and valuation, new markets for their consumption, and new means for their transport over greater distances and at greater speed" (Huggett & Robinson, 1996b: 160). Also, demographic transformation was integral to "the industrial transformation of the earth", resulting in vastly increased demand of agricultural output, housing, employment and income (Huggett & Robinson, 1996b: 160).

As Drakakis-Smith (1996: 705) notes, colonialism involved "the imperialism of industrial capital combined with aggressive territorial colonialism", resulting in "structured exploitation" between the mid-19th and mid-20th centuries. Also, colonial "political power structures evolved within the colony" as technical changes, intense industrial colonialism and colonial urbanization were incrementally introduced "amidst existing cultural contrasts" (Drakakis-Smith, 1996: 705). In that context and as Johnston et al. (1994: 77) put it,

> the shift from informal spheres of influence to formal colonial rule in the nineteenth century is rooted in a new phase of capitalist transformation (sometimes called the second industrial revolution) in which inter-capitalist rivalry and the growth of transnational forms of industrial and finance capital promoted a search for raw materials and new investment opportunities.

However, "political structures were becoming essential for the expansion of economies" under colonialism (Claval, 1996: 312). Huggett & Robinson (1996a: 11) caution that whereas the Great Transformation was one of the most radical changes to the ecosphere, there have been other equally significant changes.

This chapter does not seek to be comprehensive in scope nor to recite any of them in detail. It should suffice to note that a wide range of legacies were produced in urbanization, industrialization, land-use structures, socio-economic relations, administrative institutions, cultural manifestations, and so on, that came to define the spatial and temporal extent of colonialism. Beyond the built environment, legacies of colonialism "relate also to the regulation of social space" (Drakakis-Smith, 1996: 705). A variety of colonizers' phenomena, such as "languages, religions, social customs, states, systems of government, methods of communication, economic activities, commercial organisations, diseases, medicine and health, styles of architecture" and so on, were disseminated around the world in what Roberts (1985 cited in Clarke, 1996: 252) describes as "The Triumph of the West".

Also, political and economic transformation of all societies began to accelerate with "the evolution of urban industrial cultures, and powered by differentials in industrial output", resulting in emphatic separation and variations among countries that " 'had' and those that 'had not'" (Huggett & Robinson, 1996b: 160). In equal measure, "Economic changes also led to extensive diffusion of European peoples and cultures which transformed economies, polities and societies world-wide" (Clarke, 1996: 252). Essentially, "everything had become political" with colonialism (Claval, 1996: 313; Drakakis-Smith, 1996: 706). As colonial investment policies were biased in favor of urban economic development and industrialization, former colonies continue to be trapped in "sustained urban growth rather than sustainable urban development" (Drakakis-Smith, 1996: 709), largely characterized by "planetary urbanisation", exclusionary planning and fragmentary governance (Tsheola & Mmotlana, 2018). The latter phenomenon has meant that city growth created nuanced "problems of government, of organisation, of transport, of planning, of security and the like, on a scale previously unknown" (Huggett & Robinson, 1996b: 161). These societal inequalities and divisions have remained stark. Colonizers established, as it were, "new societies in which there would be unequal rights for people based on religion, ethnicity, class, culture and gender" (Guelke & Kay, 1996: 164–165). It is tenable to argue that colonialism epitomizes the global development landscape because as a variant of imperialism, it shaped societies and their lands as it was in return shaped through anti-colonialism. This chapter asserts that matrices of colonialism were entrenched for endurance and virtual permanence with planetary transformation over time and space.

Colonialism/post-colonialism governance: a coloniality oxymoron

Colonialism required states, historically carved "from the flow of multiple intersecting spatial processes" (Harvey, 1996: 262), and their territorial sovereignty for raising revenue to finance war and the administration of their affairs (Langton, 1996). For this reason, even mercantilist policies "laid burdens of taxation and military conscription on the peasantry" as an "indispensable resource of the state" in addition to "trade or capture" (Anderson, 1974 and Jones, 1981 cited in Langton, 1996: 213). The success of the colonialism ventures required state power; and, centralized "absolutist states" served this purpose by sucking "money, people and produce into the capital cities from which they operated", allowing for "royal splendor" (Langton, 1996: 214–215).

In the same way as processes resultant from colonialism "varied in their spatial and temporal extent" (Drakakis-Smith, 1996: 705), states too continue to be "bounded and isolated as entities from their environments" and they continue to acquire a degree of permanence through institutions that assure their "character and internal integrity" as they did under colonialism (Harvey, 1996: 262). States can, therefore, be reasonably construed as "'victory of space over time', enabling strategies to be formulated within their confines to assure internal discipline and legitimacy while pursuing external geopolitical strategies" (Harvey, 1996: 262). Just as colonialism was ambivalent and inadequate, states too, as Harvey (1996: 262) maintains, engage in "geopolitical strategies . . . that produce space . . . while simultaneously being perpetually undermined by the very processes of space creation that enabled them to be constituted as permanences in the first place". That is, whereas no single state in the world is completely independent, it has been instrumental, as an institution of governance, in the consolidation of colonialism as it has continued in the present world to sustain coloniality. It is also true that a post-independence state was a mere inheritance of an untransformed colonial state (Guelke & Kay, 1996).

It is pertinent to understanding the oxymoron of colonialism and post-colonialism to emphasize that state "operations have become much more strongly disciplined by money capital and finance" and the role of the state as an institution of governance has been reduced to "finding ways to promote a favorable business climate, which frequently means exercising strong discipline over the labor force" (Harvey (1996: 423). It is true that a state "come[s] into existence" as "an institution whose existence is defined by its relation to other 'states'" (Wallerstein, 1984: 29). The "diversity of the character of the population of a state" is overwhelmingly bound together by "the state-idea" (Taylor, 1993: 151). According to Taylor (1993: 151), "Every state has a *raison d'etre* (a reason for existence)" and the strength of the state-idea is "closely associated with nationalism" and countering of disruptive forces. With the state as an institution of governance, nationalism has become "a mechanism both of imperium/integration and of resistance/liberation" (Wallerstein, 1984: 20). As an institution of governance, the state has been historically "both indispensable and inadequate" (Deutch, 1981: 331 cited in Taylor, 1993: 190), and it has continued to be "available to be manoeuvred", rather than serve as a neutral umpire adjudicating societal disputes and conflicts (Taylor, 1993: 183). In that context, the key concern has to be "who has decisive or critical say" in the "ongoing decisions of a state-machinery" rather than just "who has some say?" (Wallerstein, 1984: 31). Notwithstanding the myriad of institutions of governance, those who controlled the state have always become successful in "the assertion of power" (Wallerstein, 1984: 32). Indeed, the democratic politics of post-colonialism, as it were, are inconsequential in the context of the globality of the matrices of coloniality that proceed through the control of state power.

Coloniality proceeds through unending negotiations of the dominant and the dominated, which are discriminatory, ambivalent, ambiguous, and contradictory (Bhadha, 1994; Ndlovu-Gatsheni, 2007). The present-day globality too, just like the old notorious empires, evolved through combinations of coercion, negotiations, mimicry, and hybridity (Bhadha, 1994). However, contemporary "civilizing missions" purport to be serving the cause of freedoms and democracy, promotion of good governance; advocating for human rights and dignity; securing global order, stability and security; establishing a harmonious world; spreading enlightenment and reason; combating barbarism and terrorism as well as resistance of domination (Zielonka, 2012). Zielonka's (2012: 503) contention is that the imperial paradigm holds the potential of shedding better insight about "several important aspects of international relations such as governance". Unsurprisingly, former colonies are regularly content with contesting rotational seats at the United Nations Security Council. The independent state in post-colonialism has in reality remained colonial but reflexively accommodated popular nationalist aspirations for the new state and the new elite of power (Ake, 2000: 33 cited in Omotola, 2011: 59). With nominal and narrow democracies, the independent state has largely been a "totalistic" "[l]aw and order state, predicated upon the use of force (largely covert) to enforce its decisions . . . (and to exert) dominion over the society in every facet of life" (Omotola, 2011: 64). Hence, Omotola's (2011: 70) assertion that the post-colonialism democratization process has gradually facilitated "the de-institutionalisation of the people to become mere clients, onlookers and/or consumers, instead of acting as the primary stakeholders of democracy", has credence.

Indeed, as Taylor (1993: 24) asserts, "political processes do not begin and end with states" because "all social institutions have their politics". However, formal power resides with the states because they are "responsible for the laws that define the practices of all other institutions", which "facilitate and constrain the behaviour of individuals by laws, rules, customs and norms" (Taylor, 1993: 26, 27). In the contemporary world, the state does "maintain a privileged position", largely because it "controls society", notwithstanding the fact that its apparatuses are open to control by political parties and/or corporations (Harvey, 1996: 204, 236). But the state is universally accepted as "the *arena*" and "the locus of formal politics" (Taylor, 1993: 31) and the power to control society is accessed through the state. For these reasons, this chapter asserts that the state as an institution of governance has been instrumental in perpetuating colonialism in post-colonialism, which is a coloniality oxymoron of existence of the present capitalist societies.

Coloniality of knowledge, mind and being in contemporary globality

Mignolo (2003: 669) describes the spatio-political economy of science as "knowledge and wisdom", wherefrom "coloniality of power and knowledge engendered the coloniality of being". Hence, as Ndlovu-Gatsheni (2013: 11) puts it, decoloniality is "a political-cum-epistemological liberatory project" because it is sustained not only "by colonial matrices of power but also by pedagogies and epistemologies of equilibrium that continue to produce alienated" peoples and spaces. According to Mkandawire (2011:19), "knowledge matters in development"; and by extension, it matters in colonialism, post-colonialism, and globality. This chapter accepts that coloniality is readily reinvented in present globality because of their intricate historical association with colonialism. As Comaroff and Comaroff (1991: 34 cited in Ndlovu-Gatsheni, 2007: 177) note, "the struggles of the dominated were purely ideological for they necessarily involved an effort to control the cultural terms by which the world was ordered and power legitimised". Even notions of *Pax Americana*, *Pax Britannica*, and *Pax Africana* (Mazrui, 1967; Ramutsindela, 2009) are all imbued with undertones of domination and resistance associated with the lived ambivalence of colonialism in post-colonialism of modern globality. Whereas *Pax Americana* and *Pax Britannica* are perceived to be driven by neo-colonialism motive, *Pax Africana* is assumed to provide a formidable counterweight in favor of post-colonialism (Mazrui, 1967; Ramutsindela, 2009).

According to Gibson in Ramutsindela, Yamazaki, Gibson and Mamadouh (2010: 11), "the culture of colonialism is also about discourses that structure encounters, and infuse behaviors and practices much as ink might soak into chalk". That is, "the ideas and techniques that make domination possible" (Gibson cited in Ramutsindela et al., 2010: 11) are important to the explanations and understanding of the continuities of colonialism in a world of perceived post-colonialism. The present world of neo-libertarian globality consists of complex processes of "intentionality" and "purposefulness" that disguise coloniality of knowledge, mind, and being (Grosfoguel, 2007; Maldonado-Torres, 2007; Ndlovu-Gatsheni, 2007, 2013; Mkandawire, 2011). Through the hype associated with the ascendency of globality, the attendant coloniality was tacitly normalized and universalized as "a natural state of the world" (Ndlovu-Gatsheni, 2013: 11). The "coloniality of globality", as well as the "globality of coloniality", is necessary in explaining and understanding the ambivalence, ambiguity and contradictions of contemporary capitalist societies.

Coloniality draws from an ongoing articulation of new and old ideologies, therefore creating continuities and disjunctures, as well as understanding of the meaning underpinning "complex historical episodes" of relations of "material phenomena and ideological controversies" (Marks, 1986 cited in Ndlovu-Gatsheni, 2007: 176). As Maldonado-Torres (2007: 243) posits, coloniality persists through contemporary global capitalism and "it is maintained alive in books, in the criteria for academic performance, in cultural patterns, in common sense, in the self-image of peoples, in aspirations of self, and so many other aspects of our modern experience". Inevitably, the contemporary world-economy consists of subverted continuities relating to processes and legacies of slavery, imperialism, colonialism, exercise of power, and reflexivity of capitalism, wherein coloniality of knowledge, mind, and being are reinvented (Grosfoguel, 2007; Maldonado-Torres, 2007; Ndlovu-Gatsheni, 2007, 2013). In essence, "coloniality of globality" and "globality of coloniality" are equally trails of "ambiguous legacies of (re)invented traditions" (Ndlovu-Gatsheni, 2007: 175).

Conclusion

This chapter has argued that the notion of post-colonialism is problematic in that the untransformed institution of governance that sustained colonialism, the state, has continued to be indispensable, inadequate, and ambivalent for contemporary global capitalism. The chapter has asserted that in championing global transformation, colonialism has entrenched matrices of coloniality for its permanences through the state as the primary institution of governance. Thus, interventions intended for transition to post-colonialism have uniformly become vulnerable to mechanisms of reflexivity, reinventing characteristic features of colonialism. The chapter concludes, therefore, that there is no better exposition of the reality of the adage that "as things change, they remain the same" with more continuities than discontinuities, than experiences of colonialism, development-environment interface, global capitalism, the state as institution of governance, coloniality, and post-colonialism.

References

Bhadha, H. (1994). *The Location of Culture*. London: Routledge.

Clarke, J.I. (1996). Changes in global demography. In Douglas, I., Huggett, R. & Robinson, M. (eds.). *Companion Encyclopedia of Geography: The Environment and Humankind*. New York: Routledge, pp. 249–273.

Claval, P. (1996). From a "cultural" world to a "political" one. In Douglas, I., Huggett, R. & Robinson, M. (eds.). *Companion Encyclopedia of Geography: The Environment and Humankind*. New York: Routledge, pp. 310–326.

Drakakis-Smith, D. (1996). The nature of third world cities. In Douglas, I., Huggett, R. & Robinson, M. (eds.). *Companion Encyclopedia of Geography: The Environment and Humankind*. New York: Routledge, pp. 702–729.

Flenley, J.R. (1996). The tropical rain forest. In Douglas, I., Huggett, R. & Robinson, M. (eds.). *Companion Encyclopedia of Geography: The Environment and Humankind*. New York: Routledge, pp. 573–598.

Goodall, B. (1987). *Dictionary of Human Geography*. London: Penguin Books.

Grosfoguel, R. (2007). The epistemic decolonial turn. *Cultural Studies*, 21(2–3): 211–223.

Guelke, L. & Kay, J. (1996). European settlement, 1450–1750. In Douglas, I., Huggett, R. & Robinson, M. (eds.). *Companion Encyclopedia of Geography: The Environment and Humankind*. New York: Routledge, pp. 162–181.

Harvey, D. (1996). *Justice, Nature and the Geography of Difference*. Massachusetts: Blackwell Publishers.

Huggett, R. & Robinson, M. (1996a). Introduction: A differentiated world. In Douglas, I., Huggett, R. & Robinson, M. (eds.). *Companion Encyclopedia of Geography: The Environment and Humankind*. New York: Routledge, pp. 11–14.

Huggett, R. & Robinson, M. (1996b). Introduction: A world transformed by the growth of a global economy. In Douglas, I., Huggett, R. & Robinson, M. (eds.). *Companion Encyclopedia of Geography: The Environment and Humankind*. New York: Routledge, pp. 159–161.

Huggett, R. & Robinson, M. (1996c). Introduction: A world of questions. In Douglas, I., Huggett, R. & Robinson, M. (eds.). *Companion Encyclopedia of Geography: The Environment and Humankind*. New York: Routledge, pp. 479–481.

Johnston, R.J., Gregory, D. & Smith, D.M. (1994). *The Dictionary of Human Geography* (3rd edn.). Oxford: Blackwell.

Langton, J. (1996). The origins of the capitalist world economy. In Douglas, I., Huggett, R. & Robinson, M. (eds.). *Companion Encyclopedia of Geography: The Environment and Humankind*. New York: Routledge, pp. 206–227.

Maldonado-Torres, N. (2007). On the coloniality of being. *Cultural Studies*, 21(2–3): 240–270.

Mazrui, A.A. (1967). *Towards Pax Africana: A Study of Ideology and Ambition*. London: Trinity Press.

Mignolo, W. (2003). *The Darker Side of the Renaissance: Literacy, Territoriality, and Colonisation* (2nd edn.). Ann Arbor: The University of Michigan Press.

Mkandawire, T. (2011). Running while others walk: Knowledge and the challenge of Africa's development. *Africa Development*, 36(2): 1–36.

Ndlovu-Gatsheni, S.J. (2007). Re-thinking the colonial encounter in Zimbabwe in the early twentieth century. *Journal of Southern African Studies*, 33(1): 173–191.

Ndlovu-Gatsheni, S.J. (2013). Why decoloniality in the 21st century? *The Thinkers*, 48: 10–15.

Omotola, S. (2011). Explaining electoral violence in Africa's "new" democracies, pp. 51–73. DOI: 10.4314/ajcr.v1oi3.63320. www.researchgate.net/publication/272540716.

Powell, J.M. (1996). Origins of modern environmentalism. In Douglas, I., Huggett, R. & Robinson, M. (eds.). *Companion Encyclopedia of Geography: The Environment and Humankind*. New York: Routledge, pp. 274–292.

Ramutsindela, M. (2009). Gaddafi, continentalism and sovereignty in Africa. *South African Geographical Journal*, 91(1): 1–3.

Ramutsindela, M., Yamazaki, T., Gibson, C. & Mamadouh, V. (2010). Interventions on the meanings of the Obama presidency for US relations with global regions. *Political Geography*, 29: 5–16.

Taylor, P.J. (1993). *Political Geography: World-Economy, Nation-State and Locality* (3rd edn.). Essex: Longman Group.

Tsheola, J.P. & Mmotlana, L. (2018). Politics of land investments, informalities, exclusionary planning and risk exposure in South African urbanism. *African Journal of Public Affairs*, 10(2): 163–182.
Wallerstein, I. (1984). *The Politics of the World-economy: The States, the Movements and the Civilizations*. Cambridge: Cambridge University Press.
Zielonka, J. (2012). Empires and the modern international system. *Geopolitics*, 17(3): 502–525.

6
ECOSOCIALISM
Historical roots and current movements

Salvatore Engel-Di Mauro

> [I]t is only in recent decades, with the rise of ecosocialism, that we have seen serious efforts to integrate the fundamental intuitions of ecology into the body of Marxian theory.
>
> *(Lowy and Mage, 1998)*

As with the building of any movement, sustained or recurring mass organizing, revolts, and other direct action precede the articulation of a concept or label to describe them. The word socialism, for example, came into being after centuries of organized and spontaneous uprisings (Cole, 1953). The same can be argued for ecosocialism. The past few decades have witnessed the increasingly explicit linkages made between development, social justice or self-determination, and environmental preservation. Ecosocialism is associated with a great diversity of left-wing political currents involved in such linkage-making, some of which also contrast each other in crucial ways. Within these currents one may find variants of Marxism, feminism, anarchism, Indigenism, decolonizationism (national liberationism), syndicalism, and cooperativism, and sometimes aspects of certain strands of social democracy, technocratic statism, and religious philosophies, like liberation theology, Buddhism, and Taoism (Baer, 2018; Pepper, 1993, 2010; Wallis, 2018).

In other words, ecosocialism is an inheritor to multiple forms of struggle. As such, it is little wonder that ecosocialism should be as splintered as socialism generally has always been from its very beginnings. Within the large umbrella that is ecosocialism one may find not a few debates and some fundamental differences over objectives. However, in broad terms, ecosocialism represents the crystallization of long-standing efforts to explain and formulate solutions to address environmental degradation from socialist perspectives, to reconcile socialist and environmentalist concerns, especially in light of actually existing practices and outcomes within state-socialist countries, and to provide alternatives to predominant capitalism-friendly or otherwise socially acritical or reactionary forms of environmentalism. This includes much of ecological economics, technocratic theories of "natural capital", environmental economics, and populationism. At a minimum what different ecosocialist approaches share is the striving to contribute to practices and/or to establish theoretical foundations for the attainment of a society founded on the cooperative (or at least coordinated, socially planned) use of resources for the benefit of all

DOI: 10.4324/9780429450310-8

and in ways that do not compromise the ecological conditions of our existence. This is also one way to describe the meaning and practice of the multifarious and rambunctious nature of "red-green" politics.

Ecosocialist perspectives

Intellectually, ecosocialism is the outcome of sustained anti-capitalist critiques of the mainstream left, which, barring some exceptions, long ignored the problem of environmental degradation generally and as it impinges on development issues or prospects for the improvement of material well-being. This is what underpinned the writing of the *Ecosocialist Manifesto* in 2001,[1] made to coincide with the World Social Forum. The political effect has been appreciable. Since 2006, activists led by campesino revolutionary Hugo Blanco, who participated in the signing and propagation of the *Manifesto*, have founded an organization and publication called *Lucha Indígena*,[2] dedicated to documenting and analyzing ecosocialist struggles, especially in Peru. More recently still, since 2011, Kurdish-led communities in Rojava (Northern Syria) have adopted ecosocialist principles in a Democratic Confederalism, according to the thought of the imprisoned Kurdish leader Abdulah Öcalan (in part by way of Social Ecology). These communities have devised and implemented, as much as feasible, ecologically sensitized and ecofeminist ideas in the midst of war (Engel-Di Mauro, 2015).

I herein concentrate on the major ecosocialist frameworks that have been having international influence at least in the English- and Spanish-speaking worlds. It bears mention that the term ecosocialism has been around since at least the 1980s, but in different contexts and for different reasons (e.g. Wolf, 1986). For instance, in the United Kingdom, a comic-strip pamphlet appeared from the Socialist Environment and Resources Association that introduced a class-based analysis of environmental degradation, with an active role played by Raymond Williams (Writers and Readers Publishing Cooperative, 1980; Williams, 1995; see also Ryle, 1988). However, much of the linking between ecological destruction and capitalist exploitation has developed out of resistance movements, for example, in Brazil, Peru, and other parts of South America by the early 1980s, as explicitly formulated by the likes of the assassinated syndicalist rubber tapper Chico Mendes (Löwy, 2011).

It is primarily these resistance movements, described briefly later, who have informed the intellectual circles who gave rise to present ecosocialist theories.[3] The major ecosocialist currents can be divided into Eco-Marxist and Social Ecology varieties. The former involves at least three overlapping, but distinctive approaches. One is a broadly Marxist approach associated with and/or derived from the writings of James O'Connor. Building from Marx's original insights, he posited the "Second Contradiction" thesis as a pivotal addendum to classical Marxist ideas, whereby capitalist relations and forces of production undermine the conditions of production (O'Connor, 1988). As O'Connor was among the principal founders and editor of *Capitalism Nature Socialism* (*CNS*), the journal itself is usually associated with this thesis and overall approach, but it has officially shifted to an explicitly and broadly ecosocialist orientation since 2003. Viewed through a political economy lens, O'Connor's can be viewed as an underproductionist thesis showing how capitalist overproduction tendencies (via the first contradiction between the relations and forces of production) cause eventual underproduction problems, devastating simultaneously in social and ecological terms.

This understanding resembles the Eco-Marxism of John Bellamy Foster, Paul Burkett, and others who propose the "Metabolic Rift" reinterpretation of Marx's work, associated as well with *Monthly Review*. In this view, Marx and Engels already understood the inherently destructive tendencies of the capitalist mode of production. The matter was posed as a rupture in the metabolism between society and nature. Initially involved in *CNS*, Foster eventually parted ways with O'Connor over interpretive differences. Foster surmises that Marx developed deep ecological principles that were central to his work and directly contributed to the development of ecological thought. Consequently, there is no need to add a Second Contradiction thesis because Marx already had conceptualized the problem as one of what Foster has named "Metabolic Rift". O'Connor's thesis is that capitalist relations undermine the social and ecological conditions of production and Foster's thesis is that capitalist relations cause a metabolic rift between "human production and its natural conditions" (Foster, 1999, 370). Where O'Connor sees a need to correct for Marx's oversights or underestimations, Foster insists on a need to recover a Marx that has been apparently lost to most Marxists. It seems unlikely anyone in the mainstream would be able to tell the difference between these positions, especially as they more often than not converge over the identification of main causes (capitalism) and, more obviously over the past decade, political objectives (ecosocialism).

The contrast is rather more pronounced with Social Ecology. It is a mixture of social anarchist and some Marxist ideas that attempts to supplant both, which are considered as failing to address environmental issues adequately, if at all. The framework is centered about the works of Murray Bookchin (1982), who with Janet Biehl developed ideas encapsulated in the term libertarian municipalism (Biehl and Bookchin, 1997). There are also different currents of thought within this perspective. One has contributed much to and is closer to the perspectives espoused in ecosocialism and *CNS* in particular. John Clark's more recent dialectical approach to Social Ecology, for example, recuperates the work of anarchist Communard Elisée Reclus and strands within classical Buddhist and Taoist philosophical thought. This enables him to rework and go beyond Social Ecology and develop a materialist dialectical understanding of ecological relations that opens avenues for practicable alternatives to the capitalist mode of production focused on (re)building communities that transcend capitalist ideologies and practices (Clark, 2004, 2014).

Finally, there is the ecosocialism represented in the *Ecosocialist Manifesto*. This has developed out of a confluence of thinkers coming from diverse backgrounds, mainly Marxist, but also ecofeminist, anarchist, and Indigenist, as well as from within groups linked to *CNS*. The approach may be largely Marxist and materialist ecofeminist in orientation, but is open to and also adopts perspectives from other intellectual currents and socialist movements. Unlike other ecosocialist currents that adhere to one or another Marxist tendency, this politically and philosophically wider-ranging version of ecosocialism can contribute to transcending long-standing divides among anti-capitalist movements. I will therefore refer to this line of thinking simply as ecosocialism because of its greater inclusiveness.

The basic understanding of this line of thought starts from multiple inspirations, including and especially Eco-Marxist ones. Hence, as Eco-Marxists insist, the capitalist mode of production needs to be grasped as a set of social relations and effective cause of worldwide and current historical environmental destruction (Kovel, 2007a). The Social Ecology notion that this is all due to the problem of hierarchy in general is considered misplaced, as it obfuscates specificities of power relations and separates social justice from freedom, thereby leading to elitist and desultory political action (Clark, 2008; Kovel, 1997). Yet, as much as they are helpful, Eco-Marxist interpretations are also restrictive in that they fail to give central importance to struggles associated with social reproduction (hence the fundamental importance of gender relations), decolonization, and those Peasant and Indigenous Peoples' practices that are ecologically sustainable and socially egalitarian. In this light, it is understood that women involved in social reproduction and subsistence-based communities show, even if not exclusively, the way to feasible ecosocialist practices (Bennholdt-Thomsen and Mies, 1997; Salleh, 2009; Ricoveri, 2013).

For these reasons, specifically materialist ecofeminist understandings and approaches are viewed as foundational to ecosocialism and this is what further differentiates ecosocialism from at least some forms of Eco-Marxism (Salleh, 2010). The Eco-Marxist focus on democratic planning and recuperating insights from Marx and Engels is also very useful and shared by most of those advocating for ecosocialism (Löwy, 2007; Kovel, 2011a). However, with an overwhelming focus on political economy in leftist circles, including Eco-Marxism, there is also very little engagement with developing the sort of ecocentric ethics, arguably inchoate in Marx, necessary to overcome capitalist ideologies, or, rather, delusions (Kovel, 2007a; Löwy, 2005). These are the basic viewpoints espoused and promulgated by Ecosocialist Horizons (founded in 2011), picking up from the *Ecosocialist Manifesto*, the defunct Ecosocialist International Network (2007–2013), which produced the Belém Ecosocialist Declaration in 2008 (for a critique, see Tanuro, 2009). More recently, an expressly indigenist-oriented ecosocialist framework imbues the First Ecosocialist International (founded in Veroes, Venezuela, 3 November 2017) and is represented in part within the newly constituted Global Ecosocialist Network (founded in 2019–2020).

As this brief survey of ecosocialism suggests, several schools of ecosocialist thought have evolved over the past four decades and with substantive differences relative to political strategy. Little clarity is gained by confusing such diversity with stages of intellectual-political development and persisting in ascribing to Marx ecological understandings that were at best embryonic and fragmentary (Burkett, 2006; Foster and Burkett, 2016). As Löwy (2017) has remarked, Marx and Engels' insights were incomplete, but they form the basis for an ecological understanding of socialism today. This, concurring with Schwartzman (1996), cannot be achieved by insisting on starting from Marx – rather than the natural sciences and Peasants' and Indigenous Peoples' knowledge systems – in formulating ecological approaches, because social theories are inappropriate to studying and much less explaining biophysical processes (see Engel-Di Mauro, 2014). An overhaul in political philosophy is much needed. The renewal of Marxism (towards ecosocialism)

> requires that Marxists undertake a deep critical revision of their traditional conception of "productive forces," and that they break radically with the ideology of linear progress and with the technological and economic paradigm of modern industrial civilization.
>
> *(Löwy, 2005, 16)*

For this overhaul to be feasible, one must also draw from the great wealth of socialist experiences and ideas. The writings of Marx and Engels, as crucially useful as they still are, should be understood as a basis among many for ecosocialism, for there have also been other thinkers much more directly concerned with ecological questions already in the 1800s, like Elisé Reclus (Clark, 2004). This is aside from the great strides that had been made in the USSR, including beyond its early days (Gare, 1993; Foster, 2015), and especially in Cuba over the past few decades (Fernandez et al., 2018; Levins, 2005), as well as from struggles by diverse egalitarian anti-capitalist movements since then.

Historical roots of ecosocialist movements and intellectual currents

Ecosocialism, as emphasized earlier, is a product of recent, changing material conditions (environmental destruction, undermining the health and lives of millions of people) and of the many struggles for a healthy livable future in many different communities, combining egalitarianism and the pursuit of a classless society with environmentalism or social and environmental justice. Socialist ideas have infused environmentalist movements to some extent since the late 1980s. This is in spite of early sensibilities evident in the writings of Engels, Kropotkin, Marx, and Reclus, as well as in some political currents during the early Bolshevik period in the USSR (Gare, 1993). Several theorist-activists and dissidents in state-socialist countries began to reassess (and in many cases prematurely or opportunistically reject) leftist approaches and objectives from an ecological standpoint and, in some cases, to mine classical works for ideas, until then mostly ignored, on people-environment relations.

Arguably, more advanced ecological understanding in socialism emerged in areas of the world where struggles for self-determination or sheer survival (e.g. decolonization) involve the protection of ecosystems, such as forests. The intellectual roots of ecosocialism should be understood as situated in the material reality of combined social devastation and environmental degradation, which is experienced especially in places that are marginalized or used for resource extraction and export. Hence, the Brazilian Seringueiros movement led by the aforementioned Chico Mendes was among the first to combine ecological and socialist approaches in the 1980s. Chico Mendes joined the rights of rubber-tappers with the defense of the Amazonian forests against the rapacity of large landed and financial concerns. For such organizing against logging companies, he was assassinated in 1988, as other such activists have been in different South and Central American countries, including Berta Cáceres in Honduras (Méndez, 2018). The relatively late interest in ecological thought within socialist movements (and later in some parties, too) is only partly explainable by the predominance of socialist state industrialization prerogatives, in their varied garbs. The preoccupations of the majority of socialists lay firmly within the social. After all, it was the social question that lay at center stage in the very origins of socialism. Socialists of all stripes were not immune to the society-nature dualism and other ideological constructs typical of capitalist societies.

Historically, it has therefore been challenging to overcome such predominant worldviews. The 1990s, however, saw the bridging of what have been often called red and green perspectives, congealed in emerging and spreading red-green movements. In some cases, green parties, for example, have incorporated some traditional socialist issues as part of their platforms, as in the UK. Conversely, anarchist communist, Trotskyist and other communist parties and organizations have increasingly adopted environmental issues as their own, such as in the Fourth International. Efforts have also continued to be made to draw red-green movements closer to the numerous struggles of Indigenous Peoples, especially as a result of the latter's worldviews and everyday practices being traditionally more constructive or holistic relative to the rest of nature. Such major self-critical renewals are also behind the development of what have become known as ecosocialist perspectives.

Aside from movements taking up the already discussed ecosocialist ideas, including Kovel and Löwy's *Ecosocialist Manifesto* (2001) or Bookchin's municipal socialism, there have been several political formations and even state institutions where ecosocialism is increasingly being incorporated as platforms and policies. What is novel to these developments in socialist movements and institutional political formations is their attentiveness to and promotion of gender egalitarian outlooks, viewed as crucial to ecosocialist transformation. Ecosocialist ideas have been adopted as part of party programs and even legal frameworks.

In Iceland, the Left-Green Movement is an established party since 1999 that is a democratic socialist feminist and environmentalist party. It is part of the Nordic Green Left Alliance, founded in 2004 and which, since 2021, is a member of The Left in the European Parliament (https://left.eu/). Member parties of this grouping are present throughout Europe and can even boast 39 seats in the European Union Parliament, as of the time of writing. In the UK, Green

Left was established as a current within the Green Party in 2006 and ran a weekly newspaper called *The Ecosocialist* (see also Huan, 2010a). The Brazilian PSOL (Partido Socialismo e Libertade) was founded in 2004 through the joining of dissident forces from Lula's Workers Party with other left-wing groups. In 2011, an ecosocialist section was formed within the party, who succeeded in turning the party into an ecosocialist formation. In Greece, a party called Ecosocialists Greece was founded in 2007 and participates in parliamentary elections and in the many demonstrations and revolts against European Union policies of pauperization. The Fourth International declared itself ecosocialist in its 2010 congress. This signifies not only a major shift for a Trotskyist organization, but also a breakthrough for Marxism-inspired political action and a greater geographical reach for ecosocialist ideas. The French Parti du Gauche, founded in 2009, has adopted an ecosocialist agenda, and there are other such examples in Spain, Portugal, and Italy. In the US, the Green Party declared itself ecosocialist in 2016, though internal divergence and an anti-socialist "deep green" current threaten at the very least to undermine coherence of action (Fitz, 2019; Green Party Collective, 2019). Notably, the Democratic Socialists of America, who have US Congress representation, recently established an Ecosocialist Working Group, but their role remains advisory so far.

There are also governmental actions that are not unrelated to the development of ecosocialist objectives, even if not the result of ecosocialist parties or movements. On the 15th of October 2013, the Bolivian government passed a law called "Ley Marco de la Madre Tierra y Desarrollo para Vivir Bien" (Basic Law of Mother Earth and Development for Living Well) that reinforces another law passed two years earlier on the rights of Pachamama (Nature). In March 2015, after much pressure from below, President Maduro announced the creation of a Ministry of Eco-Socialism, the first of its kind anywhere. The Ministry is already charged with (re)formulating and implementing national water management and reforestation projects. Venezuela has actually become a hotbed of ecosocialist radicalism since the days of President Hugo Chávez. As a country where socialist principles have been developed and implemented with sensitivity to the Venezuelan context (i.e. the specifically Bolivarian nature of the concept of "Socialism for the 21st Century"), the Venezuelan case could set an example from which the Communist Party of China, for example, could draw in their quest to realize Ecological Civilization (Gare, 2012, 2017; Pan and Pan, 2012).

These are some salient examples, but such institutional political groupings do not seem much coordinated, except the Fourth International parties and now possibly the nascent Global Ecosocialist Network (in part a resuscitation of an earlier attempt, the Ecosocialist International Network, which dissipated by 2012). The fact that there is a dearth, if not lack of dialogue among these parties and organizations suggests that to some extent fundamental differences exist among them. Still, what this also shows is that even within state institutions there are currents that are at least sympathetic to ecosocialist causes. Organizations that declare themselves ecosocialist are still a small minority compared to the number of environmental and socialist movements and parties worldwide, but there has been a steady rise over the past decade and there is reason to be confident that ecosocialist ideas will continue spreading beyond explicitly ecosocialist organizations. Nevertheless, ecosocialist schools of thought are becoming more widely known and considered, especially within countries like China and Venezuela (Huan, 2010a, 2014; Ke, 2013; Salleh, 2008; Wang et al., 2015).

Existing struggles that are close to ecosocialist objectives

It is, after all, as Marx so often underlined, out of material circumstances imposed violently by capitalist relations that anti-capitalist movements are born. One way through which ecosocialist frameworks have been developing is by inspiration from conflicts and struggles in many parts of the world. There are movements that, while not expressly ecosocialist, nevertheless represent responses to capitalism that can help build egalitarian alternatives worldwide in agreement with ecosocialist objectives and principles. In India, there are uprisings against large mines, special economic zones, and large-scale dams. One of the most well-known movements against dams is the Narmada Bachao Andolan (Save the Narmada Movement). But revolts against the building of dams are much more widespread and deadly. In Assam, a mass movement has risen up against the building of a network of large-scale dams in the Brahmaputra valley. Assam and much of the northeast of India has been under virtual military rule since the early 1990s, if not earlier, and at least 30,000 people have been killed in conflicts there since then.

Then there is the Niger Delta, where transnational corporations and central government gain huge profits from decades of onshore oil and gas extraction. The levels of soil and water contamination are so high as to make for permanent health hazards and the destruction of local livelihoods. Local people have seen barely a trickle of the huge profits made, so the matter has been met since the 1970s by mass protests, legal action, and armed struggles. Violent government repression and assassinations are the order of the day there, along with guerrilla organizing and actions (Watts, 2007). Such struggles over environment and health and military violence affect many places in the African continent. To a relatively less intense and mass-murderous degree, such struggles have also erupted in South Africa. There, expressly ecosocialist grassroots organizing, like the Socialist Green Coalition (founded in 2009), has emerged out of environmental justice concerns to counter the current capitalist regime's continuation of apartheid-era environmental pollution and denial of access to basic amenities in majority communities. Some also surmise that there are potentials for ecosocialism to resonate with traditional communitarian and egalitarian ethics (Ubuntu) shared by many peoples within the continent and, one could add, in diaspora communities (Cock, 2018; Ngwane, 2009; Satgar, 2018; Terreblanche, 2018; Traub and Conway, 2009).

In Colombia, there continue to be struggles led by local Indigenous Peoples, Mestizos, and Afro-Colombians culminating in national strikes, involving unions as well, and despite hundreds of assassinations of community leaders and environmentalists. Among the demands are not just for governments to make good on promises of peaceful resolution to the conflict with the FARC and the dissolution of paramilitaries, but also to block large-scale mining and land grabs (Asher and Ojeda, 2009; Huezo, 2019). For example, the Quimbo Hydroelectric Project has been of great concern and involves the transnational Emgesa (an affiliate of Endesa-Enel). The project would destroy some 10,000 ha of farmland and the livelihoods of hundreds of displaced communities. In Argentina and Chile, the movement against large-scale mining is combined directly with Mapuche self-determination and livelihood struggles (Haughney, 2012; Ulloa, 2017). One could make similar connections between destruction of communities and environments in Palestine as well. Because Israeli state and settler colonist terror has involved from the beginning the ransacking and destruction of ecosystems, the Palestinian struggle is also one of protecting resources like groundwater, rangeland, and orchards in existence for hundreds of years (al-Butmeh et al., 2019; Kovel, 2007b).

In spite of the development of many ecosocialist ideas among US-based scholars and activists, the presence of ecosocialism in the US is at most indirectly felt. It is ensconced in the various Green Parties of the US and there is an explicitly ecosocialist organization active in Québec, Canada. Some municipalities have mayors with more or less ecosocialist orientation, such as in Galesburg, Illinois. There is very little ecosocialist presence in the mainstream and only some socialist organizations like Solidarity have adopted ecosocialist ideas and have begun promoting them. The North American context is particularly difficult as a result of a long history of persecution against leftists generally and the successful inculcation of anti-socialist ideas in the population at large. There is also a history of colonialism to which socialists have either contributed actively or have been largely indifferent (see Engel-Di Mauro, 2013).

Nevertheless, there are promising developments especially in Jackson, Mississippi. Cooperation Jackson have taken the lead in developing grassroots-based efforts at laying the material and educational basis for local African-American self-determination, decolonization, and the development of ecosocialism. They draw especially from decades of experiences in the Republic of New Afrika and the Black Panther Party for Self-Defense, political organizations that were violently suppressed by the racist US government by means of assassination, frame-ups, and incarceration of leaders (Akuno and Nangwaye, 2017). Not far from Jackson is also the La Terre Institute,[4] founded by anarchist dialectical ecologist John Clark. The Institute serves as a venue for many organizations, including ecosocialist ones, to develop new ecologically sustainable practices and radical political ideas in a forest preserve of about 30 hectares.

There is otherwise much anti-capitalist resistance existing in North America that often acquires characteristics similar to ecosocialist perspectives. Indigenous Peoples' struggles against settler colonialism have always involved the issue of land and water resources. For instance, the Akwasasne Mohawk or Kanien'keha community have had to deal with the take-over of reservation land and destruction of the aquatic ecosystems on which they based their livelihood when the St. Lawrence River was channelized and major industries used the river as their dumping ground, including GE and Reynolds Aluminum, in the 1950s. The current Kanien'keha struggles (within the community as well) to restore traditional livelihoods is therefore at the same time an effort to restore ecosystems (Mascarenhas, 2012).

In other contexts of North America, since the late 1980s, there has been the environmental justice movement, which still tends to reduce problems to one of civil rights. However, it is a very promising development that issues of environmental degradation are being discussed in terms of people's health, economic inequality, and racism (Pulido, 2016). This has recently led to the Idle No More movement in Canada and the forging of alliances between Native American and White rural communities (the Cowboy Indian Alliance) to resist the Keystone XL Pipeline. Interweaving decolonization and Indigenous liberation with explicitly leftist anti-capitalist ends, the Red Nation's Ten-Point program (The Red Nation, 2015) provides among the most comprehensive thinking in this realm. A recent convergence of frameworks between the Red Nation and the Democratic Socialists of America over the Green New Deal[5] could show the way forward in developing the basis for tying environmental concerns to the dismantling of settler colonialism and capitalism in North America. Perhaps the more mainstream movements for environmental justice, taking cues from anti-capitalist Native American organizing, will turn into something like an ecosocialist movement when it will become evident even to the most stalwart leftward leaning US Democratic Party supporter that the problem is systemic and that the system has a name, and that it is the capitalist system.

The role and future of ecosocialist organizing

The point of describing these and the aforementioned struggles as consonant with ecosocialism is not to pretend that they can or will be ecosocialist. These struggles present occasions for building political affinities and connecting with existing movements. Ecosocialists should and often already do have a role to play in such political projects. The challenge is to establish coordination among existing and rising ecosocialist formations so as to contribute to the forces fighting to overcome capitalism worldwide and replace it with egalitarian societies. As Daniel Tanuro (2011) states, "Every struggle of the workers, even the most immediate, must be supported and considered as an opportunity to increase consciousness and orient it toward a socialist perspective". I would translate this into every struggle by all sorts of workers, including the unemployed or unpaid, being an opportunity to raise awareness and direct it towards ecosocialist perspective and action. One way of achieving this is by establishing organizations devoted to diffusing such a message and facilitating consciousness-raising, to borrow and learn from feminist strategies.

What Tanuro observes is therefore among the central objectives for the First Ecosocialist International. This is to advance ecosocialism as a worldview and as a movement capable of offering practicable answers to the multi-dimensional social and ecological damage caused by capitalism, starting mainly from Venezuela, Bolivia, and the US, and drawing inspiration and collaborating with explicitly ecosocialist organizations and Indigenous Peoples in both of these settler-colonial countries. The focus is on developing an integrated approach that has three components: (1) providing the basis for international cohesion of ecosocialist movements and for the formation of durable alliances with affine movements; (2) educating in producing creative work, including those aiming at establishing self-sufficient or maroon communities, and to organizing events and actions, especially in solidarity with decolonization movements; (3) organizing convergences to advance diverse struggles towards ecosocialist ends everywhere, according to place-specific conditions, cultural frameworks, and histories. Although party structures are downplayed in this scheme, relationship to parties is not denied as necessary. However, the preferred strategy is the emergence from below of a loosely centralized ecosocialist political organ, which could be called a world ecosocialist party (Kovel, 2007a), which answers to grassroots initiatives and dictates, much as soviets were originally intended to be and in some ways along the lines of Zapatismo principles as practiced in Chiapas, México, and of Democratic Confederalism as actualized (but presently threatened with annihilation) in Rojava. These objectives originate from decades of work, the kind described before, by people in struggle, including organizers and intellectuals, as well as from reflections on what so far struggles for social and environmental justice have managed to achieve in different parts of the world. In this understanding, the relative success in implementing the aims outlined in this chapter and in advancing ecosocialism worldwide will not come out of writings and intellectual debates, but out of concrete actions and the dialectical (iterative, self-critical) grasp of theory and practice as ecosocialist struggles evolve.

Notes

1 See http://environment-ecology.com/political-ecology/436-an-ecosocialist-manifesto.html
2 See www.luchaindigena.com/
3 This is rather explicit already in the first issues of the journal *Capitalism Nature Socialism*, where direct reports were presented and discussed on struggles in Brazil, Mexico, and Nicaragua, among others.
4 See www.laterreinstitute.org/
5 See http://therednation.org/wp-content/uploads/2020/04/Red-Deal_Part-I_End-The-Occupation-1.pdf; www.dsausa.org/statements/dsa-red-deal-statement-of-endorsement/

References

Akuno, Kali, and Ajamu Nangwaye. 2017. *Jackson Rising: The Struggle for Economic Democracy and Black Self-Determination in Jackson, Mississippi*. Montréal: Daraja Press.

al-Butmeh, Abeer, Zaineb al-Shalalfeh, Mahmoud Zwahre, and Eurig Scandrett. 2019. "The Environment as a Site of Struggle against Settler-Colonisation in Palestine." In *Environmental Justice, Popular Struggle and Community Development*, edited by Anne Harley, and Eurig Scandrett, 153–172. Bristol: Policy Press.

Asher, Kiran, and Diana Ojeda. 2009. "Producing Nature and Making the State: Ordenamiento Territorial in the Pacific Lowlands of Colombia." *Geoforum* 40 (3): 292–302.

Baer, Hans. 2018. *Democratic Eco-Socialism as a Real Utopia. Transitioning to an Alternative World System*. New York: Berghahn.

Bennholdt-Thomsen, V., and M. Mies. 1999. *The Subsistence Perspective: Beyond the Globalised Economy*. London: Zed Books.

Biehl, Janet, and Murray Bookchin. 1997. *The Politics of Social Ecology: Libertarian Municipalism*. Montréal: Black Rose Books.

Bookchin, Murray. 1982. *The Ecology of Freedom: The Emergence and Dissolution of Hierarchy*. Palo Alto: Cheshire Books.

Burkett, Paul. 2006. "Two Stages of Ecosocialism?" *International Journal of Political Economy* 35: 23–45.

Clark, John. 2004. "The Dialectical Social Geography of Elisée Reclus." http://72.52.202.216/~fenderse/Reclus.htm#nh26

Clark, John. 2008. "Domesticating the Dialectic: A Critique of Bookchin's Neo-Aristotelian Metaphysics." *Capitalism Nature Socialism* 19 (1): 82–97.

Clark, John. 2014. "It is What It Isn't! A Defence of Dialectic." *Review 31*. http://review31.co.uk/essay/view/7/it-is-what-it-isn%27t-a-defence-of-the-dialectic

Cock, Jacklyn. 2018. "The Climate Crisis and a 'Just Transition' in South Africa: An Eco-Feminist-Socialist Perspective." In *The Climate Crisis: South African and Global Democratic Eco-Socialist Alternatives*, edited by Vishwas Satgar, 210–230. Johannesburg: Wits University Press.

Cole, George D.H. 1953. *A History of Socialist Thought: Volume I – The Forerunners 1789–1850*. London: MacMillan.

Engel-Di Mauro, Salvatore. 2013. "Bridging Indigenous and Socialist Perspectives: An Introduction to the Special Issue." *Capitalism Nature Socialism* 24 (3): 141–146.

Engel-Di Mauro, Salvatore. 2014. *Ecology, Soils, and the Left: An Ecosocial Approach*. New York: Palgrave Macmillan.

Engel-Di Mauro, Salvatore. 2015. "Rojava." *Capitalism Nature Socialism* 26 (1): 1–15.

Fitz, Don. 2019. "Green Party Debates Green New Deal." www.counterpunch.org/2019/10/03/green-party-debates-green-new-deal/

Fernandez, M., J. Williams, G. Figueroa, G. Graddy-Lovelace, M. Machado, L. Vazquez, N. Perez, L. Casimiro, G. Romero and F. Funes-Aguilar. 2018. "New Opportunities, New Challenges: Harnessing Cuba's Advances in Agroecology and Sustainable Agriculture in the Context of Changing Relations with the United States." *Elementa Science of the Anthropocene* 6: 76.

Foster, John Bellamy. 1999. "Marx's Theory of Metabolic Rift: Classical Foundations for Environmental Sociology." *American Journal of Sociology* 105 (2): 366–405.

Foster, John Bellamy. 2015. "Late Soviet Ecology and the Planetary Crisis." *Monthly Review* 67 (2): 1–20.

Foster, John Bellamy, and Paul Burkett. 2016. *Marx and the Earth: An Anti-Critique*. Boston: Brill.

Gare, Arran. 1993. "Soviet Environmentalism: The Path Not Taken." *Capitalism Nature Socialism* 4 (4): 69–88.

Gare, Arran. 2012. "China and the Struggle for Ecological Civilization." *Capitalism Nature Socialism* 23 (4): 10–26.

Gare, Arran. 2017. *Philosophical Foundations of Ecological Civilization: A Manifesto for the Future*. New York: Routledge.

Green Party Collective. 2019. "Deep Green Declaration." www.counterpunch.org/2019/12/27/deep-green-declaration/

Haughney, Diane. 2012. "Defending Territory, Demanding Participation: Mapuche Struggles in Chile." *Latin American Perspectives* 39 (4): 201–217.

Huan, Qingzhi. 2010a. "Eco-Socialism in an Era of Capitalist Globalisation: Bridging the West and the East." In *Eco-Socialism as Politics. Rebuilding the Basis of Our Modern Civilisation*, edited by Huan Qingzhi, 1–14. Dordrecht: Springer.

Huan, Qingzhi. 2014. "Development of the Red – Green Environmental Movement in China: A Preliminary Analysis." *Capitalism Nature Socialism* 25 (3): 45–60.

Huezo, Alexander. 2019. "Contested Natures: Coca, the War on Drugs, and Ecologies of Difference in Colombia's Afro-Pacific." *Journal of Political Ecology* 26 (1). https://journals.uair.arizona.edu/index.php/JPE/article/view/23060/22189

Ke, Jinhua. 2013. "Special Issue: Ecological Civilisation and Beautiful China. Introduction." *Social Sciences in China* 34 (4): 139–142.

Kovel, Joel. 1997. "Negating Bookchin." *Capitalism Nature Socialism* 8: 3–36.
Kovel, Joel. 2007a. *The Enemy of Nature. The End of Capitalism or the End of the World?* London: Zed Books.
Kovel. Joel. 2007b. *Overcoming Zionism: Creating a Single Democratic State in Israel/Palestine*. London and Ann Arbor: Pluto Press.
Kovel, Joel. 2011a. "On Marx and Ecology." *Capitalism Nature Socialism* 22 (1): 4–17.
Kovel, Joel,and Michael Löwy. 2001."AnEcosocialistManifesto." http://ecosocialisthorizons.com/2001/09/an-ecosocialist-manifesto/.
Levins, Richard. 2005. "How Cuba Is Going Ecological." *Capitalism Nature Socialism* 16 (3): 7–25.
Löwy, Michael. 2005. "What is ecosocialism?" *Capitalism Nature Socialism* 16 (2): 15–24.
Löwy, Michael. 2007. "Eco-Socialism and Democratic Planning." In *Socialist Register 2007. Coming to Terms with Nature*, edited by Leo Panitch, and Colin Leys, 294–309. London: Merlin Press.
Löwy, Michael. 2011. *Écosocialisme. L'Alternative Radicale à la Catastrophe Écologique Capitaliste*. Paris: Mille et Une Nuits.
Löwy, Michael. 2017. "Marx, Engels, and Ecology." *Capitalism Nature Socialism* 28 (2): 10–21.
Lowy, Michael, and Shane Henry Mage. 1998. "Globalization and Internationalism: How Up-to-Date is the 'Communist Manifesto'?" *Monthly Review* 50 (6): 16–27.
Mascarenhas, Michael. 2012. *Where the Waters Divide: Neoliberalism, White Privilege, and Environmental Racism in Canada*. Lanham: Lexington Books.
Méndez, María José. 2018. " 'The River Told Me': Rethinking Intersectionality from the World of Berta Cáceres." *Capitalism Nature Socialism* 29 (1): 7–24.
Ngwane, Trevor. 2009. "Socialists, the Environment and Ecosocialism: A View from South Africa." www.cadtm.org/Socialists-the-environment-and,5088
O'Connor, James. 1988. Prospectus. Capitalism Nature Socialism. A Journal of Socialist Ecology. *Capitalism Nature Socialism* 1 (1): 1–6.
Pan, Xing Lai, and Qian Ying Pan. 2012. "On the Construction of Eco-Agriculture in China." *Capitalism Nature Socialism* 23 (4): 27–34.
Pepper, David. 1993. *Eco-Socialism: From Deep Ecology to Social Justice*. London: Routledge.
Pepper, David. 2010. "On Contemporary Eco-Socialism." In *Eco-Socialism as Politics. Rebuilding the Basis of Our Modern Civilisation*, edited by Huan Qingzhi, 33–44. Dordrecht: Springer.
Pulido, Laura. 2016. "Flint, Environmental Racism, and Racial Capitalism." *Capitalism Nature Socialism* 27 (3): 1–16.
Ricoveri, Giovanna. 2013. *Nature for Sale. The Commons versus Commodities*. London: Pluto Press.
Ryle, Martin. 1988. *Ecology and Socialism*. London: Century Hutchinson.
Salleh, Ariel. 2008. "Eco-Socialism and 'Ecological Civilization' in China." *Capitalism Nature Socialism* 19 (3): 123–129.
Salleh, Ariel, ed. 2009. *Eco-Sufficiency and Global Justice Women Write Political Ecology*. London: Pluto Press.
Salleh, Ariel. 2010. "How the Ecological Footprint is Sex-Gendered." In *Eco-Socialism as Politics. Rebuilding the Basis of Our Modern Civilisation*, edited by Huan Qingzhi, 141–150. Dordrecht: Springer.
Satgar, Vishwas. 2018. "The Climate Crisis and Systemic Alternatives." In *The Climate Crisis: South African and Global Democratic Eco-Socialist Alternatives*, edited by Vishwas Satgar, 1–28. Johannesburg: Wits University Press.
Schwartzman, David. 1996. "Solar Communism." *Science & Society* 60 (3): 307–331.
Tanuro, Daniel. 2009. "Some Criticisms of the Belem Ecosocialist Declaration." https://climateandcapitalism.com/2009/03/24/some-criticisms-of-the-belem-ecosocialist-declaration/
Tanuro, Daniel. 2011. "Foundations for an Ecosocialist Strategy." www.internationalviewpoint.org/spip.php?article2269&var_recherche=tanuro
Terreblanche, Christelle. 2018. "Ubuntu and the Struggle for an African Eco-Socialist Alternative." In *The Climate Crisis: South African and Global Democratic Eco-Socialist Alternatives*, edited by Vishwas Satgar, 168–189. Johannesburg: Wits University Press.
The Red Nation. 2015. "Native Liberation Struggles in North America: The Red Nation 10-point Program." *Capitalism Nature Socialism* 26 (2): 1–7.
Traub, Norman, and Terry Conway. 2009. "Socialist Green Coalition Launched in South Africa." https://climateandcapitalism.com/2009/03/04/socialist-green-coalition-launched-in-south-africa/
Ulloa, Astrid. 2017. "Perspectives of Environmental Justice from Indigenous Peoples of Latin America: A Relational Indigenous Environmental Justice." *Environmental Justice* 10: 175–180.
Wallis, Victor. 2018. *Red-Green Revolution. The Politics and Technology of ecosocialism*. Toronto: Political Animal Press.

Wang, Zhihe, He Huili, and Fan Meijun. 2015. "The Ecological Civilization Debate in China." *Monthly Review* 66 (6): 37–59.

Watts, Michael. 2007. "Petro-Insurgency or Criminal Syndicate? Conflict & Violence in the Niger Delta." *Review of African Political Economy* 34 (114): 637–660.

Williams, Raymond. 1995 [1982]. "Socialism and Ecology." *Capitalism Nature Socialism* 6 (1): 41–57.

Wolf, Frieder Otto. 1986. "Eco-Socialist Transition on the Threshold of the Twenty-First Century." *New Left Review* 158: 32–42.

Writers and Readers Publishing Cooperative. 1980. "Eco-Socialism in a Nutshell." http://climateandcapitalism.com/wp-content/uploads/sites/2/2013/11/ecosocialism-in-a-Nutshell.pdf

PART 2

Global development, environment, and resources

7
FOOD, DIGITAL LIFE, AND NEW ENVIRONMENT-DEVELOPMENT DYNAMICS

Alistair Fraser

Introduction

All stages or components of the global food system – food production, food processing, food distribution, and ultimately consumption – are in the purview of well-established, and now many new, firms looking to use digital devices and services to generate and capture value across a system infused with digital technologies. A new, digital, and more integrated food system is in formation (Bronson and Knezevic, 2016; Fraser, 2019a). Consider Brazil. Against a wider backdrop in which new food delivery firms are emerging, such as Deliveroo, JustEats, and UberEats, the Brazilian food delivery company iFood announced in late 2018 that it had raised $US500 million in funds (Mari, 2018). iFood runs a smartphone "app", which links consumers in Brazil, Argentina, Mexico, and Colombia with restaurants and arranges deliveries. The money will be invested to improve logistics – it currently processes almost 400,000 orders per day but has room to expand – and develop the firm's underlying technologies, such as artificial intelligence (AI) and voice recognition. The funding is a boost to iFood and complements its successful social media strategy, which cuts across Facebook, Twitter, YouTube, and Instagram and connects it with 2.6 million "followers" or "subscribers" to whom it can send messages, write posts, or create media such as videos or photographs. By using social media in these ways, iFood joins numerous other food firms such as Coca-Cola and McDonald's who work on platforms such as Facebook to market "ultra-processed" food products (e.g. see Monteiro et al., 2013) via "videos, celebrities, gift offers and rewards" (Horta et al., 2018, 1516). Unlike marketing via television, the digital landscapes of social media platforms offer food firms continuous and round-the-clock access to consumers. Digital life – played out on smartphones and using a wide and growing range of services – alters how food is marketed, distributed, and indeed produced, with many of Brazil's agricultural systems incorporating digital technologies for "real-time quality monitoring in vineyards, fruit crops, and coffee as well as in the transportation of food products" (Pivoto et al., 2018, 26). There is still significant scope for further integration of digital technologies if hindrances are addressed, such as a poor telecommunications infrastructure. Digital technology is undoubtedly a growing factor in Brazilian food systems.

Although there are peculiarities about the Brazilian case, the picture emerging therein calls attention to new connections between digital technologies and food that can be found all across the world. In "majority world" contexts and in wealthier regions such as North America and

 DOI: 10.4324/9780429450310-10

Western Europe, new arrays and combinations of diverse digital technologies are taking shape to alter, if not transform, food systems. Autonomous machines, opaque algorithms, and smartphone apps are in play alongside consumers and the trillions of clicks, taps, swipes, and "likes" they make. In turn, new environment-development dynamics are called forth and demand critical attention.

In response, I begin this chapter by arguing that the integration of digital technologies in food systems should be seen against the light of an environment-development nexus engineered by mainstream development theorists, which assumes that an embrace of digital technologies in and across food systems will boost "development" and reduce the environmental impact of "modern" agricultural practices. I subsequently use materials on emerging configurations in the digital-infused food system – a system guided by data, controlled by artificial intelligence (AI), and conceivably generating profits for giant food firms; a system involving so-called "smart farming" on the land and numerous other digital practices beyond the farm – to question these assumptions. Finally, as I outline in the concluding section of the chapter, I argue that digital life interacts with food systems in some ways that depart from a strict business-as-usual interpretation. Of central importance here is the contingent nature of digital life. Indeed, digital life takes shape, in part, via "data curation" by digital subjects, which includes actions by critical subjects capable of blending technical proficiencies to create alternative, conceivably more just, digital futures. Forthcoming conditions are still, as yet, unwritten. Using data curation as a form of resistance stands some chance of success, despite the powerful array of forces trying to use digital technologies to create a predictable and closed-off future.

Development, food, and environment

Dominant mainstream development theorists – whether they are calling for interventions per Rostow's modernization theory or deregulations per Friedman's version of neoliberalism – agree about the need for technological development in food systems. One component in this consensus is the existence of "yield gaps" between capital-intensive agrarian systems in North America or Western Europe and "traditional" systems in majority world contexts (e.g. see World Bank, 2007). The core assumption is that traditional systems and their associated environments require transformation, for example via an embrace of "Green revolution" technologies such as hybrid seeds, irrigation, and synthetic fertilizer; or via use of "gene revolution" techniques to alter the inherited properties of plant DNA. A related component concerns integrating food systems within value chains, for example by linking producers of non-traditional agricultural exports with foreign supermarket chains, thereby promoting a "permanent global summertime" (Blythman, 2004). In this view, the production and re-production of a *global* food system requires the development of standards, new knowledge flows, and ultimately an ongoing overhaul of technical practices that transform socio-ecological relations.

If an expanded food supply is one key outcome of all these moves (Food and Agriculture Organization, 2018a, 46), another is growing corporate control over the food system. There are still producer-run cooperatives, nationalized agricultural marketing boards, publicly funded agriculture and food research laboratories, and extensive systems of state subsidies. But the general drift has been towards the formation of a "corporate food regime" (McMichael, 2005) in which trade, regulatory, and subsidy arrangements are designed or heavily influenced by, and inevitably disproportionately benefit, a small group of large and powerful firms (for an up-to-date analysis, see IPES-Food, 2017). In public pronouncements – often while proclaiming the value of their corporate social responsibility initiatives – these firms emphasize how the environment and human health are their first priorities. However, critics point out that the corporate food regime expands via ongoing efforts to privatize the commons (e.g. see Kloppenburg, 2004), displace traditional practices (seed saving, mixed cropping, etc.) (e.g. see Wittman et al., 2010), and simplify complex environments and ecologies (e.g. see Weis, 2007).

Digital technologies come into play here as simply the logical next step in a broader process of deepening technical sophistication and system-wide integration and coordination led by food firms. Their potential is celebrated by development theorists because digital technologies should increase the food supply, thereby reducing inflation, and their use will boost "development" as measured by increased investment, greater technical proficiency among populations, or simply growth in Gross Domestic Product. Digital technologies might also help food production keep pace with global population growth and rising incomes. Official measures of absolute poverty and (albeit, crude) figures on GDP per capita in "developing" countries demonstrate that incomes have improved since the 1970s and 1980s. With even the poorest people on the planet enjoying more disposable incomes, there has been a concomitant increase in demand for animal protein, which expands the "meatification" of human diets, a major characteristic of dietary shifts in wealthier societies over the last half a century (Weis, 2013). With many observers concluding that the planetary agricultural system will need to rapidly increase outputs (e.g. Food and Agriculture Organization, 2018b), digital technologies are proclaimed as part of the solution.

Then, with regards to critiques of the corporate food regime – specifically, that reliance on "industrial" agriculture pivots on the over-use of agri-chemicals such as pesticides or the use of antibiotics in livestock farming – proponents of digital technologies suggest there is potential to create adjusted, conceivably more sustainable practices that reduce negative impacts (e.g. see Balafoutis et al., 2017). At the same time, the incorporation of digital technologies in food systems – to pursue, as the European Union describes it, "digitalisation" (European Commission, 2019) – is presented by some observers as a way to improve efficiencies, for example by reducing food waste, or to increase the nutritional content of food by empowering firms to know their customers and thereby produce products that can satisfy a desire for balanced diets. In view, therefore, is a food system infused by sensors, "bots", and artificial intelligence; a system in which digital technologies are mobilized and exploited to maximize the potential of the "circular economy", which recycles resources across production, distribution, and consumption (Jeffries, 2018). In a celebration of the market's ability to (eventually) strike a balance between the environment and development, some scenarios even depict a food system infused with digital technologies, such as blockchain, that *regenerates* environments, while simultaneously boosting the food supply (e.g. see Nori, 2019). At issue are emerging dynamics to the environment-development nexus engineered by mainstream development theorists: the core claim is that digital technologies can reduce the environmental impact of "modern" agricultural practices.

But the emergence and growth of digital technologies raises the prospect of entirely new dynamics taking shape that call into question a lot of what we know about the relationship between environment and development. On the technological horizon are advanced waves of artificial intelligence using techniques such as "unsupervised predictive learning" (Anthes, 2017, 20) that will move far beyond the actions occurring mostly behind the scenes today in business processes. AI looks set to gradually begin occupying the main stage of social life due to its

> ability to extract value from unstructured data [and thereby] help us find answers to some of our biggest problems, such as fighting disease, providing food, energy and water security, keeping well ahead of the effects of climate change and managing ever more complex algorithm-centric economies.
>
> *(Curioni, 2018, 10)*

Amplifying the extant and emerging drive to compute and control space via algorithmic machinations intended to boost profits (e.g. see Zuboff, 2019), new forms of AI will look to reform the environment and re-position the lives of all populations (Tegmark, 2017). Per the technological optimists, certain environments could flourish if AI deems it beneficial to protect or conserve species or biodiversity (Curioni, 2018). Such a vision goes far beyond the hope that digitalization will lead to reduced use of agricultural pesticides, say. Instead, it posits that AI will see a more abstract picture of the planet's future and accordingly activate socio-technical relations in ways to create an idealized equilibrium of human and non-human life.

Per some critics, however, far too many humans could occupy a marginal position in these scenarios (e.g. see Häggström, 2018). The obvious danger with regards to food systems pertains to the position of smallholders or peasant producers producing food on small portions of land and remaining relatively independent of corporate control. In play here is an emergent "planetary cognitive ecology" (Hayles, 2017) underpinned and driven by AI, which will mobilize and mold cognitive processes occurring within machines and people with a view to constructing a more predictable world. "Development" shifts gears to become a matter of expanding AI's knowledge about the future. The environment, as a space of non-human action, which is to be drawn upon by humans operating within social structures, is a source of uncertainty and contingency that should be populated by sensors and modeled and adjusted by algorithms. To the extent the environment can interrupt action and flows within the purview of AI, its latent capacities should as much as possible be held within certain parameters.

Digital technologies are therefore foundational to the expanded subordination of the environment to development. The basis for this effort consists of technocratic dreams (e.g. on the dream of "solutionism", see Morozov, 2013); but in contrast to the visions of *possible* future economic growth that modernization theorists or practitioners might have had when they drew up plans to construct dams or build new port cities, the drive to expand AI is about eliminating possibility and creating *certainty*. The soil bulldozed to make dams or reclaim land from the sea in the 1960s or 1970s – and all of the other environments altered for the sake of development – was moved in the hope it would pay off. It was development over environment but without certainty, because the environment-development relationship *here* in *this* place was unfolding in relation to a wider geography. The "chance of space" (Massey, 2005, 111–117) intervened and many projects went awry (e.g. see Stiglitz, 2002). In 1960s Afghanistan, for example, the Helmand Valley Authority implemented an ambitious dam project funded by the United States, which led US planners to dream of creating "an America-in-Asia" (Arnold Toynbee, quoted in Cullather, 2010, 108). However, subsequent Green Revolution developments, with irrigated high-yield seeds fed by fertilizer, were confounded and ultimately overcome by the local topography, national politics, international geopolitics, and global climatic patterns (Cullather, 2010, 108–133). Development negotiated unpredictability. The difference today is that the soil to be moved now or in the future, or the environments to be transformed for the sake of development, will be underpinned by a drive not to negotiate but rather to *eliminate* the chance of space. Digital life permits firms to collect and analyze data; to understand patterns and make appropriate predictions. The dream is to increase knowledge about the future; to chase after "guaranteed outcomes" (Zuboff, 2019).

Digital life and food systems: emerging configurations

In light of the issues raised in this chapter, it is helpful to consider some of the emerging configurations. I first examine smart farming before addressing the impact of digital technologies on food marketing.

Smart farming

Incorporation of digital technologies in agricultural systems gives rise to the term "smart farming", which is intended to be the rural corollary of the "smart city" (e.g. see Shelton et al., 2015). It is a development of earlier rounds of investment under the label of "precision agriculture", which involved using technologies to maximize the effectiveness and efficiency of inputs (Wolf and Buttel, 1996). The major factor that converts precision agriculture into smart farming is the integration, rather than occasional inclusion, of digital technologies. The underlying idea is that entire systems are (amenable to be) populated by digital technologies in the form of sensors, in particular, and wider monitoring and reporting systems tied into off-farm agricultural technology providers (ATPs). In theory, smart farming systems should lead to higher yields per unit of land as well as higher yields per unit of inputs such as fertilizer or pesticide.

In terms of its geography, smart farming is a growing feature of capital-intensive farm operations globally. But a crucial dynamic here is that the underlying technologies consist of devices and services that need not be beyond the reach of even some of the world's poorest farmers. Economies of scale in manufacturing are driving down the sale price of sensors. Moreover, some of the software farmers might use is available in free or open source formats developed by non-profits or volunteers. A final element is that the business model of software firms demonstrates the value of providing services to customers for free but under the condition that firms can collect and use data generated by users with a view to improving services and developing products. Even some of the world's poorest farmers will be able to access relatively useful digital services if they own a digital device such as a smartphone. As such, the emerging configurations of smart farming point toward a planetary development.

To the extent that smart farming involves the production of socio-ecological data by diverse food producers and their subsequent capture, distribution, and analysis by ATPs (or other firms pursuing and benefiting from these new waves of technological development), a core analytical consideration is the flow of data in numerous directions across smart farming systems. Data become a significant new farm output beyond crops or animal protein. As a consequence of data flows and their centrality to smart farming systems, ATPs within the overall agricultural system are now beginning to occupy prominent positions as aggregators and processors of data. Elsewhere in debates about digital life, firms that accumulate and control the value emerging from data have been theorized as engaged in "data colonialism" (Thatcher et al., 2016) insofar as they pursue accumulation by dispossession via colonizing *social* life. With respect to smart farming, however, ATPs pursue accumulation by dispossession of social *and* environmental data. The life of the farmer matters along with the biological or chemical conditions of their (and surrounding) land. Smart farming brings data colonialism into new territory and dimensions. Indeed, much like "bioprospecting" (e.g. see Bingham, 2008), whereby wealthy research institutes collect biological material from poor countries for analysis and potential use in the development of medicines, the planetary dynamics of smart farming is tantamount to "data-prospecting" insofar as the collectors and analysts in ATPs or other technology firms create mechanisms, via the pervasive use of digital technologies, for mining or more simply grabbing environmental data. Powerful new arrangements are now on the horizon, as Mooney (2018, 392; my emphasis) points out:

> Since the 1980s, farm equipment companies have been accumulating weather and market data, along with ever-improving satellite information, to understand crop and livestock production and markets. More recently, they have added a battery of sensors – linked to drones and satellites – that can record everything put into and taken

> out of the field, measured every few centimetres. This gives the equipment companies unprecedented and unchallengeable detail down to the plant variety (including fertilizer and pesticide seed dressings and injections) and, ultimately, the yield quality. Linked to the equipment companies' historic data on weather and markets, and their ability to advise producers and processors, *this detail gives them enormous power.*

Although equipment companies are engaged in a data grab that makes use of specialist knowledge of their markets and customers, they are not plowing these furrows alone. Of note here is that leading farm equipment company John Deere's investment in research and development, which reached $US2.6 billion in 2018 (Deere and Company, 2019), is only a fraction of Amazon's total expenditure of $US28.8 billion (Amazon, 2019), which is undoubtedly directed at a wider range of contexts but nevertheless connects with food systems in numerous opaque ways. Then there are firms such as Alphabet (owner of Google) that try to deploy machine learning techniques in new artificial intelligence configurations that depend on accessing huge quantities of data, including the type of environmental data collected via data-prospecting. In short, therefore, data colonialism becomes a strategy within smart farming configurations that can yield profits for ATPs and other firms based on the accumulation and subsequent dispossession of data about people and the non-human worlds they live alongside. It follows that more and more data about the planet's diverse environments will be established and eventually stored in local and private, as well as global and conceivably public, data reserves (e.g. see Cheney-Lippold, 2017). The analytical question is what it will mean for the environment-development nexus when machine learning algorithms and artificial intelligence make use of these datasets and begin introducing new calculations about smart farming operations. A major concern is that ATPs or other firms in the technology sector will be empowered to create and structure corporate strategies that displace smallholders and peasant producers standing in the way of smart agricultural systems controlled by automated arrangements of drones and robots.

Digital life and food systems beyond the farm

Digital technologies also infuse food systems beyond the farm, with some new practices and emergent effects coming into view in diverse settings. The food system's environmental impact is an active component in these developments. As is the case with the emergence of digital technologies in smart farming operations and systems, there is potential to create new arrangements across food processing, marketing, distribution, and consumption that ameliorate the food system's environmental problems. For example, one possibility is for applications to emerge that reduce total food waste by connecting separate parts of the value chain and exploiting the possibilities of the circular economy (Jeffries, 2018). Then there are signs from projects and practices across the world that electric or low-carbon vehicles will reduce the food system's greenhouse gas emissions (e.g. on the Indian food delivery firm Ola's use of electric rickshaws, see Shu, 2018).

However, digital technologies are also amplifying some problems. Consider the matter of ultra-processed food (Monteiro et al., 2013), the "production, distribution, marketing and consumption [of which leads] to damage to culture, social life, *and the environment*" (Horta et al., 2018, 1517; my emphasis). One component here is that ultra-processed food tends to be sold in plastic packaging, which is often discarded and thereby pollutes diverse environments such as beaches (e.g. see Andrades et al., 2016). Another is that the low sale price of ultra-processed food also invites waste, much like the phenomenon of throw-away or "fast" fashion (e.g. see Bhardwaj and Fairhurst, 2010). But the absolutely crucial consideration is the link between consumption of ultra-processed food and malnutrition (e.g. see Monteiro et al., 2017). To understand how digital life comes into play, a key analytical factor concerns changes in the notion of the so-called "digital divide", which once simply referred to differences between those with or without access to the Internet but is now more a matter of the quality or speed of access (e.g. see Pierce, 2019); if it was once possible to imagine digital technologies as mainly a matter of life in wealthier societies, it is much harder to do so today because access has expanded considerably in the last decade, matched by a significant growth in the number of connected devices, including many relatively low-cost handheld devices such as smartphones. The effect is widespread, although still uneven, enrollment of consumers in digitally mediated chains of social relations worked out across devices, inside web sites, or within social media platforms.

Consider now the world's largest food processing firm, Nestlé, which made sales in 2018 of $US90.6 billion across 190 countries and operated factories in 86 countries (Nestlé, 2019). Digital life presents Nestlé (and firms like it) with a challenge and opportunity. The *challenge* entails finding ways to drive up e-commerce sales, which were worth $US6.7 billion to Nestlé in 2018 and had grown by 18% from the previous year. One factor is the emergence of new competitors, such as the online retailer Amazon, which purchased the US supermarket Whole Foods in 2017 for $US13.2 billion (Amazon, 2019). Amazon is a competitor with a reputation for aggressive practices, including the development of its own-brand product lines. For food processing firms such as Nestlé, therefore, the challenge is effectively interacting with Amazon. In this regard, it is instructive that Nestlé has tapped into a platform developed by Boomerang Commerce which

> uses machine learning to scrape data signals and uses sales, marketing, operations and competitor data to automate brands' programmatic advertising on Amazon. Brands can also use it to identify high-value keywords on consumer searches in their categories and estimate their competitors' spend and share of search on Amazon Marketing Services.
>
> *(Dua, 2019)*

Navigating digital life drives the development of new digital solutions.

Meanwhile, the *opportunity* for Nestlé is to develop "digitally-centric business models" that use data to deliver "personalized messaging, services and products to consumers at scale" (Nestlé, 2019, 8). Public health and nutrition scholars are tracing outlines of what this type of personalized strategy might mean in practice. On social media applications, for example, firms target children and adolescents using celebrity endorsements, peer endorsement, prizes, and games and use tracking software embedded within Internet browsers to target them "based on their demographic characteristics and interests" (Kent et al., 2019, 6). Indeed, in a Canadian study of 54 million food ads on children's top ten preferred websites from June 2015 to May 2016,

> a large proportion of the five most advertised products, accounting for 45% of total ads on all ten websites, featured advertising techniques that expressly appeal to children, including, among others: the use of branded or licensed characters (e.g. Frosted Flakes and McDonald's Happy Meal), the promotion of a prize giveaway (e.g. Pop Tarts) and the use of cartoons and animation (e.g. Red Bull energy drink and Lunchables).
>
> *(Kent and Pauzé, 2018, 1614)*

The vast number of ads means "children and adolescents view on average 0.14 and 0.26 instances of branded [food marketing] content per minute" (Kent et al., 2019, 5), which translates to an exposure to food marketing of 30 and 189 times per week respectively. Crucially, "more than nine out of ten products advertised were ultra-processed with the most frequently advertised food categories being restaurants, cakes, cookies and ice cream, and cold cereal" (1614).

Online spaces must be understood, therefore, as arenas in which food firms market, target, and track consumers all over the world. Interaction with and use of digital technologies empowers firms to pursue strategies to boost sales of ultra-processed food. In Brazil, then, when firms producing ultra-processed food or drink products such as Coca-Cola or McDonald's curate Facebook pages, the consumers who visit and/or like content therein (e.g. see Horta et al., 2018) generate data that Facebook can share with or sell to those firms, thereby providing knowledge that can inform future product or marketing development. As such, the core conclusion to make is that the spread and prevalence of digital technologies – in the form of devices such as smartphones and services such as social media platforms – alters the dynamics of food systems beyond the farm by amplifying the speed at which "the toxic food and activity environment [is] spreading around the world" (Brownell, 2010, 718).

Conclusion: contested digital foodscapes

Corporate-driven uses of digital technologies can amplify extant societal problems. There are good reasons to be alarmed, given the likely outcomes are reduced sovereignty, new forms of opaque governance, and heightened concentrations of economic power. Ultimately, as noted by Mooney (2018, 394), "Whether the final winner in the battle to control the food chain is Bayer, John Deere, Walmart, or Amazon, there is no reason to trust them – or their theoretical technologies – to take care of our food security". However, a curious and significant feature about digital life is that arrangements can also emerge that empower resistance. Relative to the specific issues raised in this chapter, therefore, I argue there are possibilities to use digital technologies in "food sovereignty construction" (Schiavoni, 2017).

Food sovereignty is often used as a contrasting concept to the concept of "food security" which governments promise but regularly fail to actually deliver. Whereas the pursuit of food security has tended to prioritize and justify large-scale, quasi-industrial agriculture using inputs purchased from transnational corporations (TNCs) such as synthetic fertilizer and agri-chemicals, food sovereignty emphasizes small-scale, traditional approaches to food production which eschews inputs sold by TNCs and demands that food producers have sovereignty over the food they produce, distribute, and sell (Wittman et al., 2010). Food sovereignty construction builds an alternative food system around notions of justice, democracy, equality, women's rights, and indigenous rights. Might the use of digital technologies bolster food sovereignty construction?

One pertinent concern is "data curation", which involves presenting, positioning, and translating data in a similar manner to the way an art curator presents, positions, and translates an artifact in a museum or art gallery (Fraser, 2019b; citing Villi, 2012). Data curation for food sovereignty construction can involve using social media to campaign for policy change, for example when movements such as La Via Campesina use hashtags or tagging to build momentum around a particular action (Schiavoni, 2017). Another example is when civil society organizations use e-newsletters to raise awareness about unjust practices in the food system (Schneider et al., 2017). Then there are instances where environmental conditions are tracked and recorded by grassroots organizations using open-source software and thereby informing broader publics of changing circumstances and potentially criminal activities affecting marginalized peasant producers (Gutierrez and Milan, 2017). Finally, data curation for food sovereignty construction can entail the use of commercial platforms such as YouTube to share hacks or forms of knowledge about agroecology among peers, in much the same way as the Cuban *campesino-a-campesino* approach to expanding agroecological knowledge (Rosset et al., 2011), only in the context of digital life peer-to-peer networks for food sovereignty construction exploit the possibilities of data curation. The fact is that:

> almost all the world's farmers either have cell phones or have access to cell phones, and are constantly exchanging production and research information – including news and descriptions of advancing pests and diseases and new techniques. Even with language barriers, the information flow has grown massively over the last decade, and there is a constant scientific exchange among farmers' organizations and through direct seed exchanges from country to country and continent to continent.
>
> *(Mooney, 2018, 395)*

As such, it might be the data curated by small-scale farmers or peasants constructing food sovereignty that will ultimately come to matter in the coming years. Certainly, despite all the talk of smart farming and its potential, it remains unclear if all the investment and actual practice of the pertinent technologies will truly pay off. The expectation might be that smart farming will generate new rounds of consolidation that will increase the power of ATPs or other firms and enable them to create the automated farms of the future; and there are plenty of reasons to expect food processing firms such as Nestlé to expand their operations and thereby drive up sales of ultra-processed food. But an alternative scenario – perhaps bolstered by efforts such as the International Panel of Experts on Sustainable Food Systems and its vision of a Common Food Policy in Europe (IPES-Food, 2019), which might encourage positive developments elsewhere – could involve a more democratic, even a post-capitalist, food system in which the "smart" in smart farming describes the ability of food producers to use digital technologies while eschewing data colonialism and the ATPs or other technology firms driving its expansion. By effectively curating data to oppose dominant, and then to build alternative, arrangements of digital technologies, critical subjects in the food system might take advantage of the affordances of digital life to create new relationships between environment and development defined by notions of social or environmental justice.

References

Amazon. 2019. *Amazon Annual Report 2018*. Accessed May 16, 2019. https://ir.aboutamazon.com/static-files/0f9e36b1-7e1e-4b52-be17-145dc9d8b5ec

Andrades, R., A. S. Martins, L. M. Fardin, J. S. Ferreira, and R. G. Santos. 2016. Origin of marine debris is related to disposable packs of ultra-processed food. *Marine Pollution Bulletin* 109 (1): 192–195.

Anthes, G. 2017. Artificial intelligence poised to ride a new wave. *Society* 60 (7): 19–21.

Balafoutis, A., B. Beck, S. Fountas, J. Vangeyte, T. van der Wal, I. Soto, M. Gomez-Barbero, et al. 2017. Precision agriculture technologies positively contributing to GHG emissions mitigation, farm productivity and economics. *Sustainability* 9 (8): 1339. https://doi.org/10.3390/su9081339

Bhardwaj, V. and A. Fairhurst. 2010. Fast fashion: Response to changes in the fashion industry. *The International Review of Retail, Distribution and Consumer Research* 20 (1): 165–173.

Bingham, N. 2008. Bioprospecting and the global entanglement of people, plants and pills. In *Material Geographies: A World in the Making*, eds. N. Clark, D. Massey, and P. Sarre, 105–160. London: Sage.

Blythman, J. 2004. *Shopped: The Shocking Power of British Supermarkets*. London, UK: Harper Collins.

Bronson, K. and I. Knezevic. 2016. Big data in food and agriculture. *Big Data & Society* 1–5. https://doi.org/10.1177/2053951716648174

Brownell, K. D. 2010. The humbling experience of treating obesity: Should we persist or desist? *Behaviour Research and Therapy* 48 (8): 717–719.

Cheney-Lippold, J. 2017. *We are Data: Algorithms and the Making of Our Digital Selves*. New York: New York University Press.

Cullather, N. 2010. *The Hungry World.* Cambridge, MA: Harvard University Press.

Curioni, A. 2018. Artificial intelligence: Why we must get it right. *Informatik Spektrum* 41 (1): 7–14.

Deere & Company. 2019. *Deere & Company Annual Report 2018.* Accessed May 15, 2019. http://q4live.s22.clientfiles.s3-website-us-east-1.amazonaws.com/253594569/files/doc_financials/annual_proxy/2018/2018_John-Deere-Annual-Report.pdf

Dua, T. 2019. Big brands like Kimberly-Clark, Kellogg's, and Nestlé are tapping into this machine-learning startup to help them crack Amazon. *Business Insider*, February 14. Accessed May 17, 2019. www.businessinsider.com/brands-are-tapping-into-this-startup-to-help-them-crack-amazon-ads-2019-2?IR=T&_ga=2.19232632.1178243265.1558370586-744537387.1558370586

European Commission. 2019. *Digitalisation.* Accessed May 20, 2019. https://ec.europa.eu/growth/tools-databases/dem/monitor/tags/digitalisation

Food and Agriculture Organization. 2018a. *World Food and Agriculture – Statistical Pocketbook 2018.* Rome: FAO.

Food and Agriculture Organization. 2018b. *The Future of Food and Agriculture – Alternative Pathways to 2050.* Rome: FAO.

Fraser, A. 2019a. Land grab/data grab: Precision agriculture and its new horizons. *The Journal of Peasant Studies* 46 (5): 893–912.

Fraser, A. 2019b. The digital revolution, data curation, and the new dynamics of food sovereignty construction. *Journal of Peasant Studies.* https://doi.org/10.1080/03066150.2019.1602522

Gutierrez, M. and S. Milan. 2017. "Technopolitics in the Age of Big Data." In *Networks, Movements & Technopolitics* in *Latin America: Critical Analysis and Current Challenges*, eds. F. S. Caballero and T. Gravante, 95–109. Basingstoke: Palgrave Macmillan. SSRN: https://ssrn.com/abstract=2935141.

Häggström, O. 2018. Remarks of artificial intelligence and rational optimism. In *Should We Fear Artificial Intelligence?: In-Depth Analysis*, eds. P. J. Bentley, M. Brundage, O. Häggström, and T. Metzinger, 19–26. Brussels: European Parliament. Accessed May 21, 2019. www.europarl.europa.eu/RegData/etudes/IDAN/2018/614547/EPRS_IDA(2018)614547_EN.pdf

Hayles, N. K. 2017. *Unthought: The Power of the Cognitive Nonconscious.* Chicago: University of Chicago Press.

Horta, P., F. T. Rodrigues, and L. C. dos Santos. 2018. Ultra-processed food product brands on Facebook pages: Highly accessed by Brazilians through their marketing techniques. *Public Health Nutrition* 21 (8): 1515–1519.

IPES-Food. 2017. *Too Big to Feed: Exploring the Impacts of Mega-Mergers, Consolidation, and Concentration of Power in the Agri-Food Sector.* Accessed May 22, 2019. www.ipes-food.org

IPES-Food. 2019. *Towards a Common Food Policy for the European Union.* Accessed May 21, 2019. www.ipes-food.org/_img/upload/files/CFP_FullReport.pdf

Jeffries, N. 2018. *A Circular Economy for Food: 5 Case Studies.* Accessed May 15, 2019. https://medium.com/circulatenews/a-circular-economy-for-food-5-case-studies-5722728c9f1e

Kent, M. P. and E. Pauzé, 2018. The effectiveness of self-regulation in limiting the advertising of unhealthy foods and beverages on children's preferred websites in Canada. *Public Health Nutrition* 21 (9): 1608–1617.

Kent, M. P., E. Pauzé, E. Roy, N. de Billy, and C. Czoli. 2019. Children and adolescents' exposure to food and beverage marketing in social media apps. *Pediatric Obesity* 14 (6): e12508, https://doi.org/10.1111/ijpo.12508

Kloppenburg, J. R. 2004. *First the Seed: The Political Economy of Plant Biotechnology* (2nd ed.). Madison, WI: University of Wisconsin Press.

Mari, A. 2018. Brazilian startup iFood raises $500m in Latin America's largest funding round. *Forbes*, November 13. Accessed May 15, 2019. www.forbes.com/sites/angelicamarideoliveira/2018/11/13/brazilian-startup-ifood-raises-500m-in-latin-americas-largest-funding-round/#54c170f74b22

Massey, D. 2005. *For Space.* London: Sage.

McMichael, P. 2005. Global development and the corporate food regime. *New Directions in the Sociology of Global Development: Research in Rural Sociology and Development* 11: 269–303.

Monteiro, C. A., G. Cannon, J. C. Moubarac, R. B. Levy, M. L. C. Louzada, and P. C. Jaime. 2017. The UN decade of nutrition, the NOVA food classification and the trouble with ultra-processing. *Public Health Nutrition* 21 (1): 5–17.

Monteiro, C. A., J. C. Moubarac, G. Cannon, S. W. Ng, and B. Popkin. 2013. Ultra-processed products are becoming dominant in the global food system. *Obesity Reviews* 14 (Suppl. 2): 21–28.

Mooney, P. 2018. What's cooking for climate change? Techno-fixing dinner for 10 billion. *Bulletin of the Atomic Scientists* 74 (6): 390–396.
Morozov, E. 2013. *To Save Everything, Click Here*. London: Allen Lane.
Nestlé. 2019. *Nestlé Annual Review 2018*. Accessed May 20, 2019. www.nestle.com/asset-library/documents/library/documents/annual_reports/2018-annual-review-en.pdf
Nori. 2019. *Nori: A Blockchain-Based Marketplace for Removing Carbon Dioxide from the Atmosphere*. Accessed May 21, 2019. https://nori.com/white-paper
Pierce, J. 2019. Digital divide. In *The International Encyclopedia of Media Literacy*, eds. R. Hobbs and P. Mihailidis, 1–8. DOI: 10.1002/9781118978238.ieml0052
Pivoto, D., P. Dabdad Waquil, E. Talamini, C. P. Spanhol Finocchio, V. F. Dalla Corte, and G. de Vargas Mores. 2018. Scientific development of smart farming technologies and their application in Brazil. *Information Processing in Agriculture* 5 (1): 21–32.
Rosset, P. M., B. Machín Sosa, A. M. Roque Jaime, and D. R. Ávila Lozano. 2011. The Campesino-to-Campesino agroecology movement of ANAP in Cuba: Social process methodology in the construction of sustainable peasant agriculture and food sovereignty. *Journal of Peasant Studies* 38 (1): 161–191.
Schiavoni, C. M. 2017. The contested terrain of food sovereignty construction: Toward a historical, relational and interactive approach. *Journal of Peasant Studies* 44 (1): 1–32.
Schneider, T., K. Eli, A. McLennan, C. Dolan, J. Lezaun, and S. Ulijaszek. 2017. Governance by campaign: The co-constitution of food issues, publics and expertise through new information and communication technologies. *Information, Communication & Society* 22 (2): 172–192.
Shelton, T., M. Zook, and A. Wiig. 2015. The 'actually existing smart city'. *Cambridge Journal of Regions, Economy and Society* 8 (1): 13–25.
Shu, C. 2018. Ola will add 10,000 electric rickshaws to its Indian fleet over the next year. *Techcrunch*. Accessed May 22, 2019. https://techcrunch.com/2018/04/15/ola-will-add-10000-electric-rickshaws-to-its-india-fleet-over-the-next-year/
Stiglitz, J. E. 2002. *Globalization and Its Discontents*. Norton: New York.
Tegmark, M. 2017. *Life 3.0: Being Human in the Age of Artificial Intelligence*. London: Allen Lane.
Thatcher, J., D. O'Sullivan, and D. Mahmoudi. 2016. Data colonialism through accumulation by dispossession: New metaphors for daily data. *Environment and Planning D: Society and Space* 34 (6): 990–1006.
Villi, M. 2012. Social curation in audience communities: UDC (user-distributed content) in the networked media ecosystem. *Participations: The International Journal of Audience and Reception Studies* 9 (2): 614–632.
Weis, T. 2007. *The Global Food Economy: The Battle for the Future of Farming*. London, UK: Zed Books.
Weis, T. 2013. *The Ecological Hoofprint: The Global Burden of Industrial Livestock*. London, UK: Zed Books.
Wittman, H., A. A. Desmarais, and N. Wiebe. 2010. *Food Sovereignty: Reconnecting Food, Nature & Community*. Oxford, UK: Pambazuka.
Wolf, S. A. and F. H. Buttel. 1996. The political economy of precision farming. *American Journal of Agricultural Economics* 78 (5): 1269–1274.
World Bank. 2007. *World Development Report 2008: Agriculture for Development*. Washington, DC: World Bank.
Zuboff, S. 2019. *The Age of Surveillance Capitalism*. London: Profile Books.

8

HISTORIC-DIALECTICAL ASPECT OF ENVIRONMENT AND DEVELOPMENT

Analysis

Anand Prasad Mishra

The process of development brings in countless environmental changes with its dynamism. In fact, the concept of development came to the front on the global platform due to the natural communion of human labor and nature in the history of human civilization. The process of natural communion of humans and nature has been affected by this gradually developing knowledge and quality of skills. The persistent development of knowledge and skill of humans brings both qualitative and quantitative change in labor. The changing form of human labor transforms the relation of human kind and nature every second. In fact, the quality of labor and skill and knowledge of humans is observed by the profundity and intensity of connection with one's surroundings. The process of the development of human civilization is basically associated with the process of Land-Knowledge Integration (LKI). LKI refines the creative skill and human labor to transform the mode of production system. The process of refinement and change in the mode of production by human civilization also affects the natural form of environment and this is the departure point where the internal process of development receives its complex and dialectical form. Whereas Charles Darwin first asserted the historical rules of the organic growth of humans scientifically, on the other hand Marx logically systematized the history of human civilization. The contemporary challenges can be understood with the help of the scientific research system of Marx. Humans and their surroundings have always been in dynamic process due to the changing form of humans and their mode of production.

Marx explains, in his historical study, capitalist development in industrial conversion and modernization with new facts. He very interestingly explains the emerging relationship between modern industry and agriculture. This explanation can be helpful in understanding the contemporary environmental problems. Marx says in the first volume of his *Capital*,

> Capitalist production completely tears asunder the old union which held together agriculture and manufacture in their infancy. But at the same time it creates the material conditions for a higher synthesis in the future, viz., the union of agriculture and industry on the basis of the more perfected forms they have each acquired during their temporary separation. Capitalist production, by collecting the population in great centers, and causing an ever-increasing preponderance of town population, on

DOI: 10.4324/9780429450310-11

> the one hand concentrates the historical motive power of society; on the other hand, it disturbs the circulation of matter between man and the soil, i.e., prevents the return to the soil of its elements consumed by man in the form of food and clothing; it therefore violates the conditions necessary to lasting fertility of the soil. By this action it destroys at the same time the health of the town laborer and the intellectual life of the rural laborer. But while upsetting the naturally grown conditions for the maintenance of that circulation of matter, it imperiously calls for its restoration as a system, as a regulating law of social production, and under a form appropriate to the full development of the human race.

Human-nature (land)-environmental interconnection

Human-Soil Relationship, which was simple and natural, was the initial departure point of our rudimentary development journey. In the past, it furthered the complicating development journey through the human's knowledge of the process of land-soil integration. This context, due to historic-material-dialectical process, started making the environmental conditions complex and disintegrative. Capitalist development brought serious and complex transformation in the human-environment relationship and its negative effects on nature went from bad to worse. This complex question of nature could be understood with the help of three main facts, which are as follows:

1. First the dynamism of society needs to be analyzed in the context of historical development.
2. Mode of production has a special importance in deciding social mobility; moreover, it is a result of interconnection of production master and mode of production.
3. Capitalist production influences society and its environment by three types of processes, which are as follows:

    - **a** Separating working population from means of production and converting them into free daily wage labors.
    - **b** Bringing class division to the front by transferring means of production in monopoly private property.
    - **c** Allowing persistent development of the process of maximum profit from market and property accumulation. It achieves this by promoting ideology of profit through the means of private production.

The previously mentioned contexts, on the basis of which interrelationship of environment and development could be understood, are important elements of the Marxist study system. It is important to mention here that the social-economic structural system based on profit is expansionist, which is basically attached to its geographical expansion and multifaceted activities. The global form of capitalism needs to be highlighted in these contexts. The environmental challenges of contemporary conditions of development can be understood in the context of whole world. The triumph of the 21st-century capitalist development chariot has presented serious challenges to the environment. Today, in order to understand environmental problems, class division, economic inequality, cultural and lingual displacement, crisis of biodiversity, land knowledge and skill structure, public participation, divisive and monopolistic power of democracy, and other crises of capitalist development need to be disclosed. The continuously emerging relationship of discontent in the whole world is also associated with environmental crisis. To analyze the emerging character of development-displacement and environment development, the journey of civilization can be contextualized only by the historical context of scientific methodology.

Development-displacement and environment

The destruction of nature and environmental imbalance mainly was generated by the association of machines with industrialization. The steam engine, which was the main factor of industrial development, pitched serious challenges to the natural ecology due to its unconventional and man-made. It utilized the physical, chemical, and mechanical capacity of nature, but, as the development of the steam engine evolved, the human environment was set up against nature. The last two centuries presented countless and unparallel examples of development but human civilization had to pay a serious environmental price forit. This contradiction between development and environment is a complex and serious subject of Marxist methodology in study. Friedrich Engels, highlighting the role of labor in evolution from apes to humans, observed that in the process of origins of the animal world, the contribution of the contradictory interaction between the animal world and social environment is important. In the contemporary era, the contradictory relationship between humans and nature is proving to be complex, challenging, and destructive. The dangers looming over the social structure made by human beings have become very comprehensive and are proving to be threatening the existence of the whole world. The coalition between monopolistic capital and machines has set up a barrier against civilization, culture, individual autonomy and freedom, democracy, and local knowledge-skill.

The coalition of monopolistic power is deviating the simple form of democracy and nature in a wrong direction by its interference. In this context, the explanation of the environmental challenges related to displacement is an important element in the developing studies of the present. In fact, the concept of the struggle of existence inspires cooperation among different animal species. This form of struggle and cooperation presents the continuously emerging contradictory relationship between human beings and nature as a historical reality. Cooperation is mainly an outcome of historic and cooperative development of consciousness-generated knowledge (Noosphere) earned by human civilization. The concept of consciousness-generated knowledge (Noosphere) was brought forward by Vladimir Vernadsky. The capitalist structure of development has seriously challenged the existence of various species of flora and fauna for the sake of profit. Anil Rajimwale mentioned that Gayars (1988 and 1990) in his study found that 50,000 species of important vegetation are found in 18 places, 20% of which are concentrated only in 0.5% of the soil on earth's surface (2006, p. 26).The profiteering nature of the market economy has sown various seeds of problems for the environment. The "YarTsaGumba", a Tibetan species of vegetation which is used in the Chinese Medical System, is one such example. This is sold at a rate of Rs.60,000 to Rs.80,000 per kilogram in Pithoragarh. Thousands of people every year make the effort to collect this market-driven herb in a circumference of 4000 meters, and due to the market powers this finally reaches to Nepal and China where its price becomes 1 lakh rupees per kilogram (Down to Earth, 15 October, 2003, pp. 25, 49). As a result of the market the regional environment has to pay a very heavy price. This situation of the environment highlights the limitations of human-generated knowledge. The monopolistic tendency of development has badly destroyed forests, which has affected our green belt. In the early stages of human civilization the earth was covered with 6 billion hectares of forest. The rain-based "Tropical Forests" are 6% of the world's forest covers; nevertheless, they are home to 50% of the world's species of organisms. It is a matter of great concern that the deforestation rate of "Temperate Forests" is 6,000,000 hectares per year. According to an assessment, 20 hectares of forest is being cut every minute. At the same time, the "Humid Forest" is being cut at an alarming rate of 44 hectares per minute. In recent times, deforestation in Africa has wildly increased.

The previously mentioned facts expose the cruel face of global capitalism's tendency toward geographical expansionism and profit-minded development. This has distorted the natural relationship of human labor and nature under the pressure of financial capital. Moreover, this process of distortion has given birth to serious negative challenges in front of the development-environment relationships. The contemporary development has caused serious problems for the world of organisms. The Washington-based "Center for Applied Biodiversity Science" (CABS) has prepared a map of protected geographical areas related to the habitation of 11,000 birds, mammal, and amphibian organisms. According to this, 260 out of 4734 species of mammals do not have any kind of protection and there is a crisis of existence for 140. In the same way, 223 out of 1183 species of birds are at the verge of destruction. The most affected areas of destruction include the Andes mountain range and Indonesia (Down to Earth, 15 October, p. 25). The development of African countries, under the pressure of financial capital, is proving to be dangerous. The deforestation of 17 rain-affected African countries is suggesting emerging environmental dangers of the near future. Business interests along the important Trans-Gabon Railway of Africa have started giving birth to different crises. Ten percent of 75,000 square miles of natural rain forests of Ghana is now prone to crisis. Today the process of capitalist and expansionist development has focused on the extinction of the forests of countries like Zambia, Senegal, Togo, and Benin.

The deforestation of forests in Chile has wildly increased due to "corporate" loot and according to an assessment, the forest cover of Chile will be completely lost by 2025. Chile is significant in the world for its biodiversity. However, the fast process of environmental destruction is a serious threat for the whole biodiversity of Chile. Brazil cherishes one-fifth of the total world's forests. But in recent times, the process of capitalist development and loot has started destroying it. These forests of Brazil are known to work as lungs for this whole planet and its natural system. These forests contribute 50% of the total oxygen for whole world (New Perspectives, Helsinki, 1/1989 Sydney, 1985, p. 72). The forests of Brazil are also present as protected places of pure water on earth. These also safeguard nutrients and humidity and ensure the process of the climate cycle through proper evaporation (Times, 18, September 1989). Due to the process of market-driven capitalist development, a serious negative pressure has started to build up on the natural environmental system of Amazon and its valley region. The construction of the BR-174 highway in Brazil was undertaken with the markets of Venezuela and Caribbean countries in mind. In addition to this, in 1997 the Brazil government decided to construct human settlements on 60 lakh hectares of land on the sides of this road. With the help of a remote sensing satellite imagery system, it was known that in 1997 fire broke out at 44,734 different places in the forests of Amazon Valley. All these incidents of fire are human-made. The increasing pressure of El Niño will be the consequence of aforementioned incidents. The increasing pressure of world financial capital has changed human-land relationships, which has given birth to the serious environmental conundrums. The environmental question of today is a probable threat for the existence of the whole human civilization.

The agricultural system of Latin American countries was dangerously bent under the pressure of global food trading houses. Brazil started using environmentally sensitive regions for soya bean cultivation, which has given rise to serious environmental challenges for biodiversity.

In recent times, Neoliberal Development Policies have provided new opportunities for the expansion of market profit to multinational companies. The MNCs of the timber industry present in Malaysia, Indonesia, and other countries have now shifted attention towards Brazil. These companies invested 500 million dollars in Brazil and have occupied 450 lakh hectares of land in Amazon Valley. In the same manner, these Asian companies have also invested in

Guyana, Suriname, and Bolivia. The 80% possession of Brazilian forests is illegal. This process of capitalist demolition is seriously damaging the environment of the Amazon region. The increasing tendency of globalization, liberalization, and privatization after 1990 has given new heights to financial capital. The contemporary form of development has given birth to displacement, inequality, and cultural deviation in Asia, Africa, and Latin American countries. The naturally built environmental habitat is forced to struggle with destruction. This process of destruction has made problems of the environment more serious and the list of these problems is quite long. China, which has achieved high standards of growth, has neglected environmental concerns in its development journey. The whole of the human race will have to pay the cost for this development. There are1150 lakh hectares of forests in China, or 30% of the world's total forest cover. The per capita forest share is only one-seventh of the total world's. In 1949, there were forests on 13% of the total land cover but only 8.9% now. In recent times the expansion of global financial capital has rapidly increased in Asian countries. In India these policies of globalization, liberalization, and privatization have brought about many environmental problems and have increased development-generated problems for the future. Crony capitalist development has forcefully given birth to displacement, which is destroying the social, cultural, and economic fabric. The ancient body of tribal society in Singrauli (M.P. India) is being destroyed due to displacement. Due to the displacement, women have experienced various social problems. The displaced groups face countless difficulties at new places in their daily life. The women in displaced groups led a happy life in their old social and cultural environment, but are now the victims of theft, loot, and sexual exploitation. In the same way, the influence of alcoholism and domestic violence has increased among displaced groups of Orissa. This tendency has also been observed in people displaced due to Sardar Sarovar Dam. The displaced are becoming powerless, and children and women are scapegoats. This displacement adds various negative aspects to the social and cultural environment. In addition to this, this process of development-generated displacement is seriously harming our Land-Knowledge-Integration system. And due to this, the knowledge of Land-Skill accumulated by human civilization so far is vanishing. This is a grave challenge to our civilization and culture.

Organized capitalist development has seriously affected agricultural production and its related environmental system. After the Green Revolution, the influential social-economic section of rural life very much benefited due to the use of the high-yielding variety technique and modernization of agriculture. This benefited section always emphasized that agriculture can be improved by modernization techniques, which also suggests possible solutions to the contemporary agriculture problem. This is basically an outcome of the ideology of capitalist development. This model of agricultural development increases the separation of nature-land and human labor, which sharpens the class divide in rural society. This whole process finally gives birth to socio-cultural and land-related environmental problems. The farmer-labor violent conflicts between 1970 and 1980 in rural life have been overcome. In the past, the problems of land erosion, water corrosion, and change in the natural pH of soil in rural areas were observed. This negative change is a result of blind overuse of pesticides for the sake of monopolistic financial capital and cruel capitalist development. In the era of the Green Revolution, a substantial decrease was recorded in the fertility of soil. The deficiency of phosphorus in soil was 3% in 1980 which increased to 73% in 1995. Nitrogen increased from 89% to 91%. Potassium decreased from 91% to 81% in the same time. This happened due to the wrong crop cycle of wheat and paddy. The capitalist production pattern of agriculture expanded class division and the separation with nature rapidly. Because of this the complex problems of the environment are reflected in agricultural and rural areas.

Dialectical aspect of environment

Capitalist development has broken the natural engagement of remote rural regions and industrial urban rural land structures and has created a deep trench. The scientists of the 18th and 19th centuries highlighted and associated this capitalist process of detachment in land structure in their research. They believed that the rising influence of nitrate in world trade and the intense production of fertilizer in the industrial sector was dangerous for the environment. This globally increasing tendency of development also proved to be a reason for colonial domination. The industrial development of "fertilizer" expanded the production of chemicals useful for war and this provided extra strength to the global impact of capitalism. As a matter of fact, the process of separation of labor and soil, which is the base for capitalist development, has continuously intensified class division. This tendency toward development shattered the natural form of environment through class interference, due to which the dialectical form of nature brought various environmental problems into existence. Marx in his critical analysis of political economics brought forward these already-mentioned facts of development on the world platform. Marx described,

> It is not the unity living and active humanity with the natural, inorganic conditions of their metabolic exchange with nature, and hence their appropriation of nature, which needs explanation or is the result of historic process, but rather the separation between these inorganic conditions of human existence and this active existence, a separation which is completely posited only in the relation of wage labor and capital.

Marx, in his *Capital*, has given proper attention towards the environmental problems of development in different time periods. Marx, giving due importance to the study of famous botanist Carl Fraas, explained how class-based development of civilization is responsible for the increased process of destruction and desertification of soil. In this historical analysis Marx showed that the modern systems of labor and production are giving birth to division, duality, and environmental destruction. He also explained that this process of dialectical-destruction of historical development is unconsciously procreating socialist tendency and forcing contemporary production networks to march towards a revolutionary change. Marx's ideas offered a scientific basis for the ideology of the 21st century. The process of contemporary monopolistic capitalist development has continuously built a strong wall between nature and mediums of production and hence gave birth to separation and destruction. Moreover, this dialectical process of development has intensified the class division and has seriously dilapidated the natural fabric of the environment. In the deepening crisis of global capitalism, environmental crisis has appeared in its dangerous form before the whole of humanity. This crisis is also raising problems for capitalist development. In fact, the capitalism of the 21st century is not invincible and is also not the last alternative. Surrounded by various questions, the capitalist model of development has become the central subject matter of the study of social sciences. And, to that extent, the question of the environment is now associated with ideology rather than being technical. At the opposite end of this ideology, the Leftist revolutionaries, social scientists, and geographers are now standing with their grounded knowledge. Leftist geographers Neil Smith and Erik Swyngedouw tried to explain the emerging dualism regarding the environment through Marxist methodology. "Given Marx's own treatment of nature", Smith opines, "it may not be unreasonable to see in his vision also a certain version of the conceptual dualism of nature" (Foster and Clark, 2016, p. 8). Swyngedouw, expanding on the dualistic aspect of environment, writes, "the social and the natural may have been brought together and made historical and geographical by Marx, but he did so in ways that keep both as a priori separate domains". Bellamy and Clark further clarify the statement of Swyngedouw and state, "To overcome what he sees as Marx's dualism of society and nature, Swyngedouw proposes an all-encompassing hybridism in the form of a singular 'socionature'" (op. cit., 2016, p. 8). The critics of Marxist studies do not consider Marx more than a political-economic critic of capitalism and relate this to the irrelevance of Marxism. However this is a partial truth. In fact the dualistic aspect of principles asserted by Marx helps in developing a better understanding of the corrosion process of labor and nature and complicating form of capitalist development. And due to this the environmental-socialist movement is emerging as a material truth in world. Nowadays the Populist Movement has started trying to encompass environmental problems into itself. Therefore, it is the need of the hour to mold the Marxist analysis according to the contemporary challenges.

Capitalist power accuses nature for its domination. This allegation of dominance, to an extent, is successful in transforming the natural form of the environment. This transformation of the environment is a serious challenge for natural science. The tendency of neglecting the dialectical aspect of nature deprives logical explanation of interrelations of development and the environment. In Marxist study, some social scientists set aside the dialectical aspect of nature. Luckas (1968) disagreed with "Dialectics of Nature" elucidated by Friedrich Engels and various other scholars supported him. This resulted in the shrinking dimension of logical analysis of nature and natural sciences. However, using Marxism properly, it could be believed that Marxism presents a base for a scientific philosophy. This sentimental society contains the subject-object of scientific study dialectically and contributes in the development of humanity, the human world, and human history and science. The inhibition of Marxism with historical materialism and dialectical study patterns was only a result of increasing influence of postmodernism. Nonetheless, in recent times, in the context of the anthropocentric epoch, the "radical" thinkers provide the proper internalization of nature by society. The form of nature is seriously affected by social interaction, which can become an important aspect of environmental analysis. Marxist geographer Neil Smith believes that nature is non-existent without society. Famous environmentalist Jason W. Moore and Marxist geographer like Neil Smith added many new things on the issues of environment and development. On the basis of the various emerging views, we may conclude that capitalism appropriates nature on its own and then include in its gambit that ultimately bend nature for its own purpose. This historical activity of human society linked its dynamic contradiction and established physical characteristic of nature. So without society, it is difficult to understand nature and environment.

Society, nature, and environment

For understanding the dynamic interconnection of nature and environment, the related social mobility also needs to be integrated as a subject matter. However, many dominant thinkers in the field of environmental development give lopsided importance to the multifaceted ideological bridge. In fact society, in its sequence of historical development, becomes complex and cyclical leaving a simple and flat form. This gradual process of development, by developing class division, provides more strength to the dialectical character of society. The strong dialectical bundle, which is an outcome of Accumulated Development (SanchaiVikas), further complicates the dialectical relationship of society and nature. And in this context, the environment problem is a complex product of socially formed processes.

Through social dynamism, development, and its complexity of internal processes of development, and its complexity with its surrounding environmental ecology and concern, humans expand their accumulated knowledge. This development makes the cultural aspect multifaceted.

The dialectical relationship of humans and nature provides material resources for the durability of human life. Basically culture is neither determined by nature nor it is required to include nature for becoming durable (op. cit. p. 13). To understand the relationship between nature, humans, and culture, assessment of scientific rules and its principles is required. For the assessment of the present context, understanding the separation of humans and nature through dialectical-critical realistic analysis, new contexts can be brought forward. The dialectical relationship of the environment and nature has been exposed as a class problem due to the intensification of capitalist development.

The capitalist class, in capitalist development and the production system, neglects the price system and production only for the sake of capital accumulation. This is possible only due to the exploitation of the labor force in capitalist production. This process of exploitation of the labor force brings change in subject matter on nature. This process of change expands accumulated capital by developing class division. This tendency was highlighted by the classical, political economists in the beginning of the 19th century. They said that private capital accumulation in capitalist development is completed only through the destruction of public property. In the contemporary era, expansion of globalization, liberalization, and privatization has started affecting every aspect of development. The friendly relationship of financial capital and the capitalist class is making the class divide worse. This form of "Crony Capitalism" is further complicating the environmental issue. Moreover, the problem of environment is no more ex parte in the present situation but is associated with development, production, distribution and its dialectical background. In the past decade of 1950–1960 the environmental problem was not associated with class division but was considered to be a separate unit in a sentimental way. The ideology of postmodernism did not see social problems in the context of class and presented the problems as a partition wall with unscientific logic. The ruling class used the previously discussed problems as a weapon against class movement. Most of the environmental movements were brought about for capitalist development and its benefit. Today there is a need to combine and understand problems born due to class division.

New contexts of environmental discussions

After vivid investigation of the problem of development and the environment, it becomes clear that it is result of a dialectical relationship. The Advait philosophic wing explains development and environment in a sentimental way, which is incomplete. "Capitalism internalizes the contradiction of nature as a whole, while the web of life internalizes capitalism's contradictions" (Moore 2015).

This statement sees the global crisis of capitalism and the global challenge of environment as supplements to each other. The situation of nature is an outcome and product of integrated action of socio-cultural and political interference which is possible due to development expansion of capitalism and contradiction. According to this perspective, the environment is a challenge which is complex and full of contradictions. Its solution is not technical but associated with institutional changes. The capitalist exploitation and blind race of development is bent on making this planet a vast dump yard. Nature has started taking revenge on us due to thoughtless development. Humans and nature are closely interconnected with each other and emergence of any kind of hostile relationship between them will destroy both. There is a need to understand this emerging relationship of human and nature. According to the needs of the time, this hostile relationship should be transformed into a friendly progressive option. We will have to leave the destructive model of development and instead will have to embrace such a development path in which the environment, society, culture, and the process of natural knowledge production can be protected. The mechanical thought of development is the root cause of contemporary problems. In fact the machine is a non-cultural and non-conventional tool. It uses the physical, chemical, and mechanical powers of nature, but sometimes due to separation stands against nature. This capitalist, social kinetic of victory over nature gives birth to serious crises when it crosses its limit. A new context of the environment is now associated with the industrial context of development. The profit-minded development policy does not take this environmental context seriously. Due to this tendency the problems of the environment became uncontrollable, which is a challenge today. For the development of the future it is required to move forward by accommodating social development and environment protection in the present. And this process will be based on appropriation rather than separation. The Marxist Development study method can help in developing new formulae in this context.

References

Castree, N., 2002, False Antitheses? Marxism, Nature and Actor-Networks, *Antipode*, Vol. 34, No. 1, pp. 111–146; DOI: 10.1111/1467-8330.00228

Foster, J. B. and Clark, B., 2016, Marxism and the Dialectics of Ecology, *Monthly Review*, 68, pp. 1–17.

Laurance, W. F., 1998, A Crisis in the Making; Responses of Amazonian Forest to Land Use and Climate Change, *Trends in Ecology and Evolution (TREE)*, Vol. 13, No. 10, pp. 411–415, DOI: https://doi.org/10.1016/S0169-5347(98)01433-5

Luckas, G.,1968, *History and Class Consciousness*. Merlin, London, p. 24.

Moore, J. W., 2015, *Capitalism in the Web of Life: Ecology and the Accumulation of Capital*. Verso, London.

Rajimwale, A., 2006, *Environment versus Development*, Vol. 1. Sunrise Publication, New Delhi, p. 17.

Venkatesh, A., 2019, Das Kapital, An Immortal Work of Everlasting Relevance – 2, *People's Democracy*, Vol. XLIII, No. 15.

Translated by: Adarsh, Research Scholar, Department of English, Banaras Hindu University

9

NILA NUNANICO, THE THREAT TO OUR LANDS

Ana Sofia Solano Acuña

At the end of the 19th century, the French explorer Alphonse Louis Pinart expressed surprise at how inhospitable western Panama was for the government authorities and for the people of Panama City, but he was particularly surprised at the indigenous populations that inhabited the region. Through his writings, the Frenchman allows us to discover a natural environment, a social organization, and inter-ethnic relations that, even when they have changed, enjoy a certain permanence in some remote villages (Pinart, 1882, 1892). This image of an inhospitable, untamed, and wild frontier is found in practically all writings by travelers and scientists of the late 19th and early 20th centuries; and it is in turn an indispensable part of the discourses coming first from Bogotá[1] and later from Panama to justify the urgent need to pacify this region.

All of these writings were influenced to a great extent by the Latin American discourses of the end of the 19th century and the first decades of the 20th century, towards the representation of ethnic groups and national subjects. In addition, during this period, the practice began of criminalizing indigenous struggle and demands. Foreign capital arrived as a third front of the indigenous communities' struggle that would come to see their territory threatened and their people proletarianized. As stated by Florencia Roulet and Pedro Navarro Floria (2005), at this time the persistence of the so-called "internal borders" was then contradictory to the idea of integrity of the State; Ronald Soto and David Díaz (2007) add that it is possible to reconstruct the genesis of how the Indian was included, excluded, made invisible, resized, and operationalized.

The current indigenous struggles of western Panama are the long-term response to early projects to dismantle their collectives, which sought to erase them from the political scene as cohesive subjects, to plunder the territory as an articulating axis of the symbolic and the ordinary, and finally to tear the regional social fabric that historically has been one of the great walls against the colonizing advance. The discourse of the need to civilize the Indian in order to find progress and form a single nation contributed to the deprivation of individual and collective rights as well as to the precariousness of the indigenous population in the social structure, deeply rooted in the Old Regime.

In 1903, Panama was born into its republican life with a thought built around the mestizo as a prototype and around the inferiority of the indigenous, which in turn resulted in a racialized geographic map of the national territory. Imaginaries that would become the foundations for a national strategy based on the de-Indianization of the population, translated into policies on

 DOI: 10.4324/9780429450310-12

their congregation and Castilianization. This thought was intended to liberate (at least discursively) the indigenous people of the colonial past, only to reinsert them into a model of domination similar to the one they had experienced but now with obligations as citizens, second-class, but citizens after all.

As part of the process of cultural and political vindication that had covered all of Latin America, approximately in the mid-90s of the 20th century, the indigenous movement of the peoples of western Panama claimed their right to call themselves the Ngäbe-Buglé People, and gradually abandoned the ethnonym of "Guaymí" with which they were called from the first years of the Spanish conquest. The historical sources prior to this transition period refer to "Guaymí" and on very few or almost no occasions is it possible to make any differentiation in language.

This chapter aims to contribute to understanding the historical interpretation of the indigenous movements of struggle and resistance in the face of various dangers, such as mining and hydroelectric exploitation, and the racialization of poverty in productive sectors such as coffee and bananas, among others.

Cultural territories are understood as the spaces of history and identity that each people or society recognize as their own, since in them they not only find habitation, sustenance, and reproduction, but also the opportunity to transmit socio-cultural practices over time. Local and global territories have borders and geographic-symbolic landmarks where centers are marked; sites that imply sacred meanings for people since the irruption of psychic entities have occurred in them (Barabas, 2008).

Control and territorial identification have direct implications on processes of ethnic construction since they tie an indigenous present with a past, with ancestors, with the very genesis of a people. A geographical imagination through the location of settlements, their interconnection in cultural life, the narration of migratory routes, the historicizing of community formative processes, as well as the "imagination of the ancestral territory" reveals *space* as fundamental in the identity processes of the Ngäbe-Buglé.

One of the characteristics of a meaningful space is its condition as a "cosmic center", vital for the group that imagines it and who has endowed it with senses and therapeutic and magical powers, as well as of affectivity and trust (Barabas, 2008).

The importance of articulating the idea of territory for indigenous societies through their stories and memories does not have as its maximum value the fidelity of said story, but the ability to become an integral part of the historical consciousness and identity. From this logic, the memory mechanism takes on special relevance when the territory that is spoken of has been plundered, in which case the story becomes resistance (Santos-Granero, 2005).

The social interaction with the territorial entities in each daily or ritual encounter, although it is singular on each occasion, creates a sedimentation background of representations allowing the community to elaborate a more or less shared social imaginary about the characteristics and physiognomy of these sacred entities (Barabas, 2008).

Miguel Bartolomé (2010) proposes that ethnic territory is not delimited but neither is it indefinite; it includes diverse spaces but is not restricted to these, as they can be found and reproduced in other areas, as well as being physically and symbolically appropriated. This characteristic is of vital importance in the case of the Ngäbe-Buglé since they have been displaced from their historical territory, which never ceases to be important in their identity process, but they do appropriate and re-signify new spaces.

Methodologically, this chapter contribution is part of a broader research agenda called "Images of memory and power: The Guaymí of western Panama in the formation of the National State (1880–1925)", which is mainly a process of reviewing documentary sources in Panamanian, Colombian, and Costa Rican archives; as well as a space for reflection and co-theorization with some indigenous actors on what has been their process of irruption in the National State. I understand "co-theorization" in the same terms that Rappaport (2007) does, as the collective production of conceptual vehicles that take up both a body of anthropological theories and the concepts developed by our interlocutors.

Regarding the reliability of written and oral texts, I have carried out the process of critiquing the source to the extent possible and under the conditions required for each type of source. In both cases, a large part of what was sought were imaginations, beliefs and interpretations about the territory, about indigenous societies, about coexistence; thus the interpretation of a historical fact does not allow false oral testimonies but rather offers rich material for contextualized interpretation. In this sense, my role as a researcher has been to carry out the task of disassociating "objective facts" that are achieved by triangulating various sources and "interpretation of facts". This is because to understand the imaginary and symbolic of a narrative, the context in which it occurs must be understood.

It is thus my hope that this study provides new perspectives to understand the current indigenous societies of western Panama, their logics of struggle and their relationship with the State. Tracing through the documentation told me a different story from the one attributed to the Guaymí people in Panama and Costa Rica. The written source and the oral source helped me to identify and rescue from oblivion a dynamic society, self-reflective of its situation, interconnected despite geography, internally diverse, adaptable to the political situation, and in general terms "vigilant of being able to exist".

Panamanian legal framework and indigenous peoples

Currently Panama is one of the few countries in Latin America that has not ratified Convention 169 of the International Labor Organization, where the problems and demands of indigenous sectors are protected in national legislation and Convention 107 of this same organization. Among the arguments that the sequence of government administrations have had for not ratifying Convention 169 is mainly the issue of lands and natural resources found therein. Likewise, it has been considered since 1993 that said international instrument could harm the unity of the Panamanian people, since constitutionally the existence of the Panamanian people can only be recognized as a unit within the territory of the Republic; since the Constitution prohibits "*fueros y privilegios*" (translator's note: it abides by principles of the rule of law, or that no one is above the law) (Jacquelin-Andersen, 2018).

Regarding the land issue, it has been considered that the ratification of Convention 169 is not convenient because the provisions on indigenous lands, their control, and self-management would create conflicts with peasant sectors "who have lived among them". According to the indigenous sectors, part of the fear of the governments (and fear of economically powerful sectors), are the solid criteria for the obligatory nature of prior consultation established by those projects that affect them. Unlike the Declaration of the United Nations, the Convention is legally binding, which provides a firm legal basis for indigenous rights.

For the traditional authorities of the indigenous peoples, one of the main conflicts they face every day is the need for titling land, as in the case of the Emberá Wounaam, Maje Emberá Drua, the Tagarkunyal ancestral land, and the Bri Bri collective lands, as well as the legally secure recognition of Ngäbe territory for People in northern Santa Fe province, of central Panama. Many of the lands where indigenous families live have also remained within protected areas and others were usurped or taken over by peasants and landowners.

Currently, of the 75,517 square kilometers of the Republic of Panama, indigenous regions constitute 15,103.4 square kilometers, in addition to non-legalized indigenous territories that were left outside the official boundaries. This is equivalent to approximately 23% of the Panamanian territory being made up of indigenous territories.

The Panamanian Constitution of 1904 expressly recognized the creation of the indigenous *Comarcas* (districts[2]), a situation that changed abruptly with the Constitution of 1972 which put forth political divisions subject to special regimes or laws, a situation that has weakened the possibility of indigenous movements to enforce demands on land.

The Comarca Ngäbe-Buglé in particular was created by Law No. 10 (*Official Gazette* No. 23,242) of March 7, 1997 and indicates that the lands are collectively owned by these two indigenous peoples from the territory of Bocas del Toro, Chiriquí, and Veraguas. Unfortunately for the indigenous sectors that inhabit these lands, the process of deterritorialization to which they were subjected in the past was not enough, as the natural wealth found in their regional lands has been a curse no less.

For the year 2003, through article 10 of Law No. 18 of January 24 of that same year, the State repealed articles 63,[3] 96,[4] 98,[5] 101,[6] and 102[7] of the General Environmental Law in order to open all series of gates so that big capital, in collusion with the governments of the time, could plunder the natural resources in the Comarcas. Specifically, these events were carried out within the framework of the promotion of the Tabasará II project and all subsequent projects.

The Ngäbes and Buglés, some important characteristics

Until the 20th century known as Guaymís, today they are self-recognized as Ngäbes and Buglés. They are an indigenous people that is currently binational, located on the Costa Rica-Panama border as a result of the east-west displacement process that they have suffered mainly since the mid-19th century (see Figure 9.1). There are a total of 9,543 people living in Costa Rica (INEC-Costa Rica, 2013) and 285,033 in Panama (INEC-Panamá, 2010).

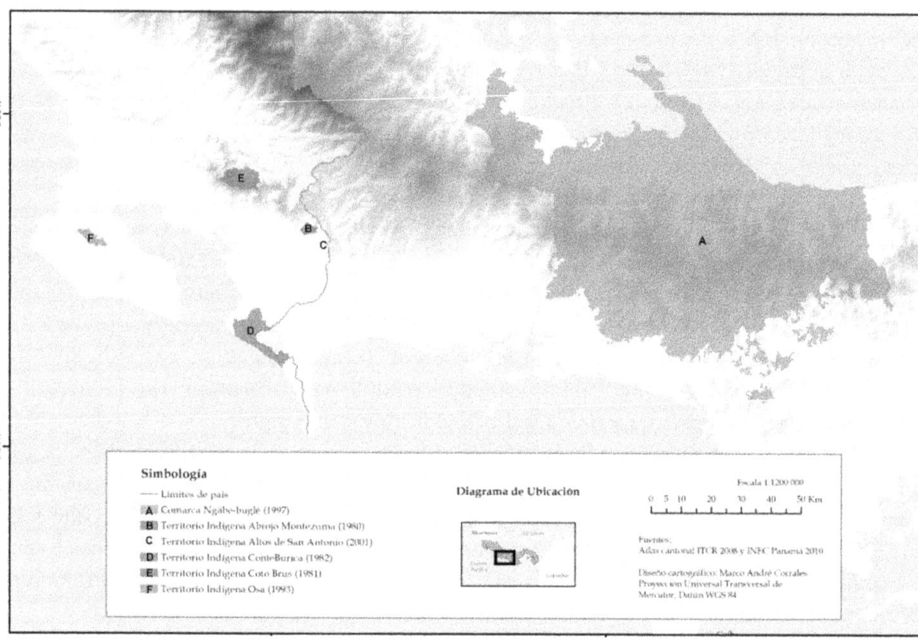

Figure 9.1 Current territory of the Ngäbes and Buglés in Costa Rica and Panama

According to modern linguistic studies, both Bokotá or Buglé as well as Movere or Ngäbe belong to the family of Chibcha languages located in the Intermediate Area. The Intermediate Area geographically extends through eastern Honduras, the Atlantic coast, and central Nicaragua; Costa Rica except the Nicoya Peninsula; Panama; the western half of Colombia; the western portion of Venezuela that surrounds Lake Maracaibo; and the highlands and coast of Ecuador (Constenla Umaña, 1991). In the ethnographic present, marital unions can be located between people who speak different languages and bilingual environments or family polyglotisms develop within the indigenous world.

In Panama they are distributed throughout Chiriquí, Bocas del Toro, Veraguas, and a significant number of the population are found in Panama City as labor migrants. In 1997, after a long struggle, through Law No. 10 (*Official Gazette* No. 23,242) the Ngäbe-Buglé Comarca was created with an area of 6,968 km^2. This Comarca was established from the territory of Bocas del Toro, Chiriquí, and Veraguas, and its headquarters is Llano Tugrí.[8]

In Costa Rica, they are located in five indigenous territories, Abrojo Montezuma (recognized in 1980), Altos de San Antonio (recognized in 2001), Conte Burica (recognized in 1982) in the canton of Corredores, and Coto Brus (recognized in 1981) in the canton of the same name extending in part over that of Buenos Aires. And finally, Alto Laguna de Osa (recognized in 1993) is located in the canton of Golfito and borders the Corcovado National Park. As in Panama, there is also a dispersed indigenous population in other parts of the country, mainly in areas where they are needed as temporary workers (coffee harvest, fruit picking, banana fields, among others), thus doubling their social vulnerability (Solano Acuña, 2019).

The founding of the territories in Costa Rica are the result of a process of displacement of family groups, mainly from Chiriquí, in the '40s of the 20th century, who were in search of new lands to settle and produce their livelihoods. Until 1991, the State of Costa Rica, through Law No. 7225, the Indigenous Registration and Identification Law, declared the State's obligation to grant Costa Rican identity cards to the Guaymí[9] (Guevara Berger, 2000). This experience was fundamental for the indigenous movement in Costa Rica, since for the first time the Constitutional Chamber decided on the basis of Convention 169 of the International Labor Organization, which together with the Declaration of the Rights of Indigenous Peoples constitute the most important bastions in indigenous legal matters in force in the country.

Some authors such as Morales and Lobo (2014) have hypothesized that the Guaymí people are "nomadic" or "mobile" as part of their worldview. However, this statement lacks anthropological support; rather, ethnohistorical study shows that such mobility has been carried out as a last resort for survival. In any case, it is worth noting the high capacity that they have as indigenous peoples to recover social fabric despite difficult social, ecological, and political conditions in which the displaced find themselves.

This ability to cope with cultural change in order to survive as an indigenous society is what Darcy Ribeiro (1971) calls ethnic transfiguration. This phenomenon consists of the possibility of maintaining one's own identity despite changes experienced (due to internal or external factors). It is possible that in the middle of the process, cultural expressions as important as language may be lost, that levels of biological interracial mixing may be high, that cultural traits of others may be assumed, but the particular identity and cohesion ties are preserved.

Ramiro Barrantes (1982) for his part offers an explanation of how the displacement and formation of new habitat spaces have come about, based on population genetics studies that have been carried out in the populations of Costa Rican territories. The author states that, in the first place, fissions of related groups occurred in Panama, which gradually migrated to the Costa

Rican localities. Here different groups merged to establish populations of around 300 inhabitants about 200 km from the place of origin. This phenomenon is known as the "fission-fusion" model and the "linear effect" model, which indicates that migration occurs between related groups. Once these groups reach the optimum level of their population, new displacement processes begin, where the main triggers are demographic structure and the physical space available.

To understand the territorial and social situation of the Ngäbes and Buglés today, I will dedicate a few lines to outlining the complex assimilation and violence to which they were subject before and after 1903, the year of Colombia's independence. For this period, the idea of "empty space" coexists and competes with that of "wild space", common adjectives to be found in the press, official documentation, and writings of intellectuals of the time in order to assess western Panama. As Arturo Taracena Arriola (1999) proposes, through the history of a region it is possible to understand not only a local historical phenomenon, but also provide elements for an analysis of national projects in crisis. Western Panama has historically been a zone of resistance since the Spanish period, throughout the period of Panama's belonging to Colombia. The attempts of penetration and organization to control the territory, the population, and natural resources have been constant but highly difficult.

Gerhard Sandner (1981) argues, in the case of Panama that after the "goldrush" and the settlement of California after 1848, the isthmus became an "intra-American" transit space and a gap-connection through Mexico/Central America and South America. For the 19th century there is a policy of penetration of the Central American isthmus by distant powers, particularly the United States. As we will see throughout the study period, such penetration could be thought of in a broad spectrum with respect to the fate of indigenous societies, or on the contrary, it is a defining element of economic, political, and even symbolic identity. Areas considered in the past as peripheral to Spanish power and republican control, in the early years of the 20th century, will become new spaces for the extraction of wealth, thus leading to new forms of exploitation and relations with the populations. In this context, the territorial plundering of rural and indigenous sectors has accelerated, not only in Panama but throughout the Central American isthmus.

Within this scenario, the importance of the Panamanian Isthmus justified organizing everything as inside and outside in order to "protect foreign interests". In Panama, the encounter between the colonial structure inherited in the economy and the global economy oriented towards capital at the end of the 19th century, transformed the organization of work and the subsistence economy, and disrupted ethnic-territorial organizations in the midst of identity and ethnic political recognition within national legislation, just to mention a couple of impacts.

Geographic spaces that were previously considered to be of little value began to be evaluated for their incorporation as part of the Nation. The ultimate purpose of this incorporation for the particular case of western Panama was to control the connection by land with the rest of Central America, the exploration and exploitation of lands suitable for large-scale crops such as banana production, the development of large farms in the most fertile lands, mineral exploration and exploitation that since the Spanish period had been of great interest, and border definition with Costa Rica, among other objectives.

It suggests that there was a process of ladinization *in situ* of the Guaymí in some regions, a process which at first glance makes us consider it to have led to a gradual disappearance of the indigenous, but actually emerges as a local identity since there is not necessarily an open renunciation of their origins but rather adaptations to the middle ground. It also suggests that the "mestizo" or "cholo" appearance is part of a resistance-survival strategy developed over the years by the Ngäbe-Buglé People (see Figure 9.2).

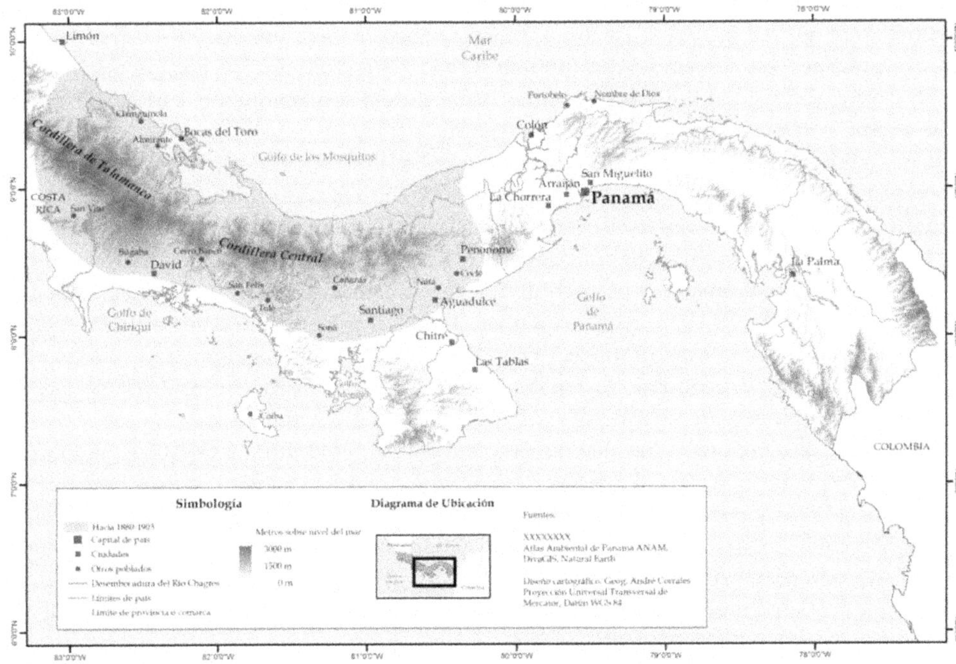

Figure 9.2 Estimated territory of the Guaymí in Costa Rica and Panama in the late 19th century

Old and new threats to the land of the *"Lords of the West"*

Western Panama's geographical location (being the bridge with Central America) and the region's physical characteristics (mines, water, healthier climate, fertile lands) have been the fate dealt to its people, mainly the poorest and most forgotten in indigenous communities. These communities have had no other option than to join racialized economic activities (cattle ranching, coffee, and bananas) that defined exponential exploitation between ancestral territory and workforce.

The impact on the population, as well as the plans of foreign capital on this portion of the isthmus were accelerated in 1967 when the Inter-American Highway was finished, a project dating from 1923. Particularly for the Ngäbe-Buglé people, the conclusion of this project resulted in the multiplication of "white peoples" (and confrontations), of trade, and in general of a greater presence of the State. Today the route has been redefined by the indigenous movement and transformed into a space of resistance, since the blockades and closures at the height of its main towns are constant tools to pressure different government regimes to meet their demands. It should be noted that the consequences of the closure of this road is of immediate effect since it paralyzes land transit between Panama and the rest of Central America.

During the 17th and 18th centuries, the entire western area of the isthmus was modified by the development of livestock activity, which impacted the use of the land and human occupation. In the 19th century, the same activity added to demographic growth and expanded the border, increasingly causing indigenous communities to withdraw towards the mountainous area and to eastern Costa Rica.

According to Omar Jaén Suárez (1998), in the case of Chiriquí in 1873 the highest livestock densities were located in David, Alanje, and San Lorenzo, where the density was between 10 and 39 cattle per 100 hectares. The foothills of the Barú Volcano, Bugaba, Boquerón, and Gualaca were occupied in the first half of the 19th century, and Penonomé, La Pintada, and El Valle de Antón at the beginning of the 20th century.

In the 19th century, the model of tropical banana plantations and the development of the coffee plantation emerged, driven to some extent by the railroad which concluded in 1855. Tropical banana plantations focused mainly in Bocas del Toro from approximately 1880, defining a region of impact of this economic activity together with the Costa Rican Caribbean.

As for mining luck, since the colonial period the Panamanian West was coveted for metals found in its geography. In 1932, geologist Robert Terry discovered the Cerro Colorado deposit, which was located at the head of the San Félix River and thus began a long history of exploitation efforts. Multiple economic feasibility studies determined that Cerro Colorado has the capacity to be exploited for a period of approximately 70 uninterrupted years.

For Panama the important metallic minerals are copper, gold, iron, manganese, aluminum, and silver. There are outcrops of zinc, lead, titanium, and other metals but they have not yet been evaluated for their viability. The mining activity is considered of public order and of social interest according to article 256 of the Political Constitution. Protected by this, the Code of Mineral Resources in article 123 proposes the expropriation of land as a possibility with the prior request of the concessionaire to the State and theoretically prior agreement with the owner.

In the case of Comarca lands, it is established that in order to take advantage of the natural resources existing in those areas, there must be coordination between the government authorities and the indigenous ones, in such a way that any activity related to the use of said assets must be done respecting cultural, religious, and social aspects, so that they are not negatively affected. On the other hand, the law establishes that the communities affected by a specific project will have the right to a fair share of the economic benefits generated by this activity, all of which is contained in Law No. 41 of July 1, 1998.

This law establishes that civil society must participate in the consultations, a situation that has been constantly denounced by the Ngäbe-Buglé General Congress and other indigenous and environmental organizations for the particular case of Cerro Colorado (excerpted from Mining, environment and wealth, its cost? *La Estrella de Panamá*. January 22, 2010. Retrieved from http://laestrella.com.pa). Another of the arguments of the indigenous opposition to the project is the environmental damage caused by this economic activity, the same for which there is already a history of other similar projects developed in the region.

> We remember the events that occurred in Cañazas de Veraguas, where the Santa Rosa gold mine was exploited, which compared to Cerro Colorado was 60 times smaller. However, there the environmental damage was tremendous. You can still see the devastation caused and the people of Cañace have cruel memories of the diseases caused by pollution. And it is not only the cyanide that is used to remove the mineral from the rocks that are extracted, but also the damage to health caused by lead, mercury, molybdenum, etc., which are naturally found in the bottom of the earth along with gold and copper. When afloat, they pollute the environment and cause diseases in humans and animals. Mines are also the cause of acid rains that carry pollution hundreds of kilometers, especially affecting agricultural plantations and sterilizing soils. We then wondered if the president of CAMIPA had asked the students of the Primer Ciclo School of Cañazas and their parents if the itching, respiratory problems, nausea and

blood cancer were minor "annoyances". Little does it matter to the miners whether or not the toxic gases permanently emanating from the leach heaps are a major "nuisance" or the long-term effect on humans of exposure to such a toxic environment. They also don't care about the permanent damage a mine causes to a major protein source (fish and shrimp). In Cerro Colorado three important hydrographic basins are born: those of the San Félix and Cuvíbora rivers that drain towards the Pacific, and that of the Cricamola River that empties towards the Atlantic.

(Excerpted from Ngäbes, Buglés y Cerro Colorado. La Estrella de Panamá. May 20, 2010. Retrieved from http://laestrella.com.pa)

The government of the President of the Republic of Panama Ricardo Martinelli (2009–2014) was characterized by coining a discourse of "doing what is necessary" so that foreign investment could enter the country even when that required changing the national legal framework in different areas. In 2010, Rogelio Moreno Cacique General declared,

> that the purpose of the government is not to improve the conditions of its people, but to include an article that allows them to exploit the resources that exist within the region, such as the mines, in this case Cerro Colorado, which has one of the largest great copper deposits of the world.

The highest authority of the Comarca affirmed that the resources that exist in the region

> belong to the Ngöbe people and we are going to defend them because we cannot easily let them take our wealth, where we only have damage and consequences to mother nature.
>
> *(Excerpted from Valley of the poor, mountain of riches. La Estrella de Panamá. June 11, 2010. Retrieved from http://laestrella.com.pa)*

In Panama for the year 2016 there were 389 applications from companies to explore and exploit metallic and non-metallic minerals in addition to the 168 mining concessions granted in past decades (Solano Acuña, 2019).

In hydroelectric matters, the projects under discussion and outright friction between the State and the Ngäbe-Buglé indigenous people are Barro Blanco, Chan I, and Chan II. The first is again an example of a project promoted by the State without consulting directly affected indigenous communities, resulting in clashes and deaths, national and international complaints, criminalization of the indigenous struggle, and in general a division of Panamanian society.

The Barro Blanco project covers 234 hectares, of which only 6.8 hectares include the neighboring areas of the region, and was granted concession in 2006. Ninety-eight percent of the lands that the project will use are private property and 2% of annexed areas of Comarcas. Part of the arguments that have been used in this case revolve around the fact that the affected lands are not within those delimited as part of the Comarca and that "nobody lives there" in areas that are modified by the riverbed. In this regard, the Deputy Chief Clementina Pérez indicated the following:

> The Government has said that there are five people. There are not five people, there are 23 thousand people here . . . we, the original peoples of these communities, will remain in the fight as required by constitutional law, Law 10 of March 7, 1997, the law of the Universal Declaration of Indigenous Peoples. We will then keep all populations on alert and also pronounce against the damage and prejudice and devastation

of natural resources. Damage to the water, floods. We are not going to allow our population that is on the banks of the Tabasará river to be evicted. Therefore the Ngäbe-Buglé population remains; we are Ngäbe-Buglé; we are a nation constituted in the Panamanian nation.

(Interview with Clementina Pérez, Deputy Chief, 28 April 2014. Kodrini Region)

According to Manolo Jiménez, leader of the April 10 Movement, there are four river basins that will be affected by the damming of the Tabasará (excerpted from El Tabasará and the repeal of regulations against indigenous people. *La Estrella de Panamá*. February 27, 2012. Retrieved from http://laestrella.com.pa). Reports from the National Public Services Authority specify that until 2015 there were 111 hydroelectric concessions pending in the country, of which 65% are in basins and rivers in the province of Chiriquí, followed by Veraguas with 22 concessions and Coclé with eight water generation projects (excerpted from Communities claim to suspend mining licenses in Panama. OCMAL. June 9, 2015. Retrieved from http://noalamina.org).

Among the great achievements of the indigenous movement in the face of a highly violent political and social scenario, it is possible to register in the first place, having achieved the paralysis of millions of dollars for the execution of extractive and hydroelectric activities. Another of the achievements of the indigenous movement is the development of complex discourses and narratives that link the past with the present, the sacred and the everyday, the defense of water and cultural identity anchored in it,[10] the right to territory, the right to belonging to a state that recognizes and respects their right to consultation and active citizenship – all of which translates into a renewed, strengthened, and reconciled cultural identity in the space of resistance against the State and foreign capital.

According to Roquebert León (2016: 188–189), particularly the experience of managing

the mining issue is a clear example of how the relationship between the State and the Ngäbe group has changed; during the first period these groups had little chance of having their opinion considered. In the current period, aware of their rights, they have exerted measures of pressure, and have managed to bring the central government to the dialogue table for a new negotiation or have even achieved in the last four years the partial paralysis of projects in their region.

This reflection "inward" to face what comes "from the outside" has delineated a process of deep reflection on the present and the future, taking as a reference that

each town defines history according to its own references and experiences passed down for hundreds, if not thousands, of years. The way that history has been configured, as the narration of events or happenings, as metaphors or allegories, is legitimate. The story is told so that other individuals know and reproduce it; it has a pragmatic value and an educational value. However, cultural changes, especially those that have derived from cultural imposition and ethnocidal policies, configure a scenario in which certain myths have lost their functionality or validity.

(Guevara Berger, 2010: 12)

The Ngäbe-Buglé indigenous movement has also managed to join the international community, establish alliances with other sectors of civil society, and reconnect with the Ngäbe-Buglé movement of the displaced peoples and migrants in Costa Rica, in the multiple fronts of struggle that are currently open, continuing to fight for almost 50 years[11] even with the wear and tear that this implies. As a country, the wave of violence, dispossession, and Ngäbe-Buglé resistance, has been the necessary stage for the discussion and generation of jurisprudence on vital issues for indigenous communities.

Some conclusions

On the eve of the commemoration of the Bicentennial of Panama's Independence from Spain, it is possible to observe how indigenous communities continue to be part of the sectors that suffer the most from marginalization and oblivion. In the particular case of the Ngäbes and Buglés, who have resisted all previous cycles of violence, the current panorama becomes absolutely discouraging, because it is not only limited to forced displacement (which they have already experienced in the past) but also to the systematic destruction of important places of memory.

In this context, the territory has become the axis of the resistance movement, creating a discourse amalgamated with the past, which has among other objectives the identification of a deeply territorially dispersed indigenous population. The idea of a new rebirth in the territory of the ancestors is a recurring image in these narratives that accompany the political and environmental struggle, where myth and historical events of the past seek to explain actions of the present. The figures of the Indian Governors, the mythical Caciques of the Spanish period, and the martyrs of the most recent confrontations reveal themselves as a messianic lineage necessary to lead the resistances in this "long night".

In general, these stories do not follow a chronological order and also draw information from other sources, external religious traditions such as Catholicism, stories from other indigenous peoples, books, press, and mainly the oral tradition (the stories) from one generation to another. These stories that are built around places of memory are a tool for the transmission of values to new generations, an opportunity to create bonds of solidarity, and a weapon to build a "we" despite nationalities, displacement, and acculturation processes.

Translated by: Patricia Solís

Notes

1 After the bloody War of the Thousand Days, where Panama had a leading role that is scarcely addressed by Colombian historiography, at the initiative of the Municipal Council of the District of Panama on November 4, 1903, the Open Town Hall proclaimed separation from Colombia. Then a few months later, the Canal Zone was handed over to the United States on May 19, 1904.

2 The Comarca is a special political division of the Panamanian territory, governed according to the law that creates it and the norms, customs, or worldview of the indigenous people that inhabit it or of the indigenous peoples that inhabit them, based on the traditional indigenous institutions created according to their spiritual, social, economic, cultural, legal, and political values, whose highest authority is the General Indigenous Congress or the Indigenous Council or the political entity created by them, whose representative is the Cacique (Chief) or a traditional authority chosen by them according to their tradition.

3 Article 63. Indigenous regions and municipalities where natural resources exist and are exploited or extracted shall have the duty to contribute to their protection and conservation, in accordance with the parameters established by the National Environmental Authority together with the indigenous authorities of the regions, in accordance with current legislation

(Gaceta oficial, Panamá-Panamá, General de ambiente de la República de Panamá, 3 de julio de 1998, 28).

4 Article 96. The National Environmental Authority shall coordinate with the traditional authorities of the indigenous peoples and communities, everything related to the environment and the natural resources existing in their areas

(Gaceta oficial, Panamá-Panamá, General de ambiente de la República de Panamá, 3 de julio de 1998, 35)

5 Article 98. The right of the Comarcas and indigenous peoples is recognized in relation to the use, management and sustainable traditional exploitation of renewable natural resources, located within the indigenous regions and reserves created by law. These resources must be used in accordance with the purposes of protection and conservation of the environment, established in the Political Constitution, this Law and other national laws

(Gaceta oficial, Panamá-Panamá, General de ambiente de la República de Panamá, 3 de julio de 1998, 36)

6 Article 101. The use for industrial or commercial purposes of the resources located on the lands of indigenous communities or peoples, by their members, requires authorization issued by the competent authority

(Gaceta oficial, Panamá-Panamá, General de ambiente de la República de Panamá, 3 de julio de 1998, 101)

7 Article 102. The lands included in the indigenous regions and reserves are unattachable, imprescriptible and inalienable. This limitation does not affect the traditional system of land transmission in indigenous communities. Indigenous communities or peoples, in general, may only be transferred from their regions and reserves, or from the lands they own, with their prior consent. In the event of the transfer, they will be entitled to prior compensation, as well as to relocation to lands comparable to those they occupied

(Gaceta oficial, Panamá-Panamá, General de ambiente de la República de Panamá, 3 de julio de 1998, 36–37).

8 In the Law of the Comarca Ngöbe-Buglés, Law No. 10 of March 7, 1997, in which the basic principles of indigenous participation in the projects to be developed in their territories were included for the first time in a county law and the Environmental and Cultural Impact Study. In accordance with the indicated legal norms, the Ngöbe-Buglé region elects 3 deputies, 7 mayors and 58 village representatives, all of them in national elections, which include that of the President of the Republic. In addition, it also has the figure of the governor, and the 57 corregidores and regidores [city and subcity level councillors] who are chosen by the central government. Like the other regional laws and following the constitutional principles established in Article 127 of the Political Constitution, Article 9 of Law 10 of 1997 establishes that the lands delimited by this Law constitute collective property of the Ngöbe-Buglé Comarca, in order to achieve the cultural, economic and social well-being of its population; therefore, the private appropriation and alienation of said lands under any title is prohibited. The modes of transmission, acquisition and modalities of use and enjoyment of the property shall be carried out in accordance with the collective norms and practices of the Ngöbe-Buglé people.

9 The entire struggle of 1991 is known as the "Cedulación de los Guaymí" (identification carding), whereas after this struggle a movement began in both countries to claim the ethnonyms "Ngäbe" and "Buglé". However, it should be noted that for some sectors of the indigenous population the "Guaymí" is a valid form of self-identification and they continue to use it.

10 Both the Cerro Colorado mining project and the Barro Blanco hydroelectric project seriously injure the Ngäbe-Buglé people because they threaten the health of the rivers of western Panama, fundamental in the mythology of this indigenous people.

11 The problem of mining in particular began during the government of Omar Torrijos in the 1970s.

References

Barabas, Alicia, "Cosmovisiones y etnoterritorialidad en las culturas indígenas de Oaxaca, Antípoda," *Revista de antropología y arqluelogía*, Bogotá- Colombia, no. 7, 2008.

Barrantes, Ramiro, "La estructura poblacional de dos grupos indígenas guaymí de Costa Rica," *Revista médica del Hospital Nacional de Niños* (1 y 2) 1982, 297–307.

Bartolomé, Miguel Alberto, "Interculturalidad y territorialidades confrontadas en América Latina," *RUNA*, Barcelona-España, no. 1, 2010.
Constenla Umaña, Adolfo, *Las lenguas del Área Intermedia: introducción a su estudio areal*. San José-Costa Rica: Editorial de la Universidad de Costa Rica, 1991.
Guevara Berger, Marcos, *Perfil de los pueblos indígenas de Costa Rica*. Informe técnico. Costa Rica: RUTA/ Banco Mundial, 2000.
Guevara Berger, Marcos, "Sobre el valor pragmático del mito apuntes desde la mitología comparada de talamancas y kunas," *Anuario De Estudios Centroamericanos*, Vol. 35, 11–35, 2010.
Instituto Nacional de Estadística y Censos (Costa Rica), *X Censo Nacional de Población y VI de Vivienda: Territorios Indígenas*. San José-Costa Rica: Editorial INEC, 2013.
Instituto Nacional de Estadística y Censos (Panamá), *Diagnóstico de la población indígena en Panamá*. Panamá-Panamá: Editorial UNFPA-INEC, 2010.
Jacquelin-Andersen, Pamela, ed, *El mundo indígena 2018*. Copenhagen-Dinamarca: Editorial IWGIA, 2018.
Jaén Suárez, Omar, *La población del Istmo de Panamá. Estudio de Geohistoria*. Madrid-España: Editorial Cultura Hispánica, 1998.
Morales, Abelardo y Lobo, Diego, *Mejorando la situación socio laboral de la población móvil Ngäbe Buglé en Costa Rica y Panamá*. Costa Rica: FLACSO, 2014.
Pinart, Alphonse, *Noticias de los indios guaymíes y de sus costumbres*. San Francisco-Estados Unidos: Editorial Imprenta de A. L. Bancroft, 1882.
Pinart, Alphonse, *Vocabulario castellano-guaymí. Dialectos move-valiente, norteño y guaymí penonomeño*. París-Francia: Editorial Ernest Leroux, 1892.
Rappaport, Joanne, "Más allá de la escritura: la epistemología de la etnografía en colaboración," *Revista Colombiana de Antropología*, Bogotá-Colombia, Vol. 43, 2007.
Ribeiro, Darcy, *Fronteras indígenas de la civilización*. México: Siglo XXI, 1971.
Roquebert León, Jorge Luis, "El recurso cuprífero de Cerro Colorado, Comarca NgäbeBuglé, Panamá (1946–2014). Factor de negociación o de represión entre el Estado panameño y el grupo Ngäbe", *Revista Secuencia. Revista de historia y ciencias sociales*, Distrito Federal-México, no. 96, 2016.
Roulet, Florencia y Navarro Floria, Pedro, "De soberanos externos a rebeldes internos: la domesticación discursiva y legal de la cuestión indígena en el tránsito del siglo XVIII al XX", *Revista TEFROS*, Río Cuarto-Argentina, no. 1, 2005.
Sandner, Gerhard, "Estructuración espacio-político-geográfica y la geopolítica en la Región Caribe", *Revista geográfica de América Central*, Heredia-Costa Rica, no. 13–14, 1981.
Santos-Granero, Fernando, "Writing history into the landscape: Yanesha notions of space and territoriality," Surrallés, Alexandre y García Hierro, Pedro, eds, *The land within: Indigenous territory and the perception of environment*. Copenhagen-Dinamarca: Editorial IWGIA, 2005.
Solano Acuña, Ana Sofía. *Imágenes de la memoria y el poder. Los guaymí del occidente de Panamá en la conformación del Estado Nacional (1880–1925)*. Tesis de doctorado, Universidad Pablo de Olavide, 2019.
Soto Quirós, Ronald y Díaz Arias, David, *Mestizaje, indígenas e identidad nacional en Centroamérica: De la Colonia a las Repúblicas Liberales. Cuaderno de Ciencias Sociales 143*. San José-Costa Rica: Editorial FLACSO, 2007.
Taracena Arriola, Arturo, "Región e historia," *Desacatos*, México-México, no.1, 1999, 1–8.

10
IS THIS LAND MADE FOR YOU AND ME?

Richard Furtick and Waquar Ahmed

Introduction

We are natural creatures that lead alienated lives – we are alienated from nature itself. Everything that is presented to us on a day-to-day basis runs through complex production cycles that leave it unrecognizable from the natural resources it was forged from. We eat food that is raised on mega-farms and killed in factories. We live in civically planned cities that flatten topographies. We sleep under constellations of artificial lights and wake up every day to pursue lives totally divorced from the natural world around us. Humanity's alienation from our own environment has become a naturalized part of life, and its impacts manifest all around us in small ways every day. Poets like Thoreau escaped to the woods, Edmund Hillary climbed mountains, and companies like North Face made a profit by helping us attempt to reconcile our natural self. As capitalism alienates people from their environment, it seeks to harness those anxieties, repackage them and exploit them. In the midst of alienation, capitalism continues to reproduce. It reproduces whatever it can in order to continue and reinforce itself as a natural process. This chapter seeks to examine how this has unfolded using the example of the American folk standard "This Land Is Your Land" and how its rhetoric has been continually reproduced by capital in order to both subvert its radical sentiments and utilize its poetic language to reinforce humanity's dominion over the natural world.

Few songs have had as convoluted a history as "This Land Is Your Land". Take, for example, when legendary folk singer and social activist Pete Seeger walked out onto the steps of the Lincoln Memorial on January 18th, 2009; he faced an estimated crowd of 400,000 people who had come to commemorate Barack Obama's forthcoming inauguration as President of the United States (Hendrix and Mummalo, 2009). Armed with his banjo, and flanked by his grandson Tao Rodriguez-Seeger and music titan Bruce Springsteen, Seeger launched into a rendition of the American anthem "This Land Is Your Land"[1] that sought to burrow deeper into the crowd then any of the songs on the previous slate of the day's entertainment and spark an emotive sweep that only a song with as byzantine and convoluted a pedigree as "This Land" could. Seeger intentionally sought to evoke that unifying emotion by playing the political verses of the song, ones that speak out against private property and poverty, that are traditionally cut in the diluted, colonized version most people learn (Marchese, 2014). Filled with poetic imaginations of America's ecological diversity and ephemeral voices singing praise of the land's unique

DOI: 10.4324/9780429450310-13

togetherness, it is hard to imagine "This Land" as anything less than the wholehearted patriotic ode. Ever since it became integrated into elementary school music curriculums, the song has become a staple in the teaching of the vastness of America to its children. "This Land" was even once sung by a nationwide chorus of children broadcast on live television (Jackson, 2002). Noted musicians, such as Gary Clark Jr. and Ryan Bingham, claim it is one of the first songs they learned as children in school and one that resonates with a powerful message (O'Rear, 2019; Zaleski, 2019). A powerful message that has become obscured with time, and through its foggy glass modernity has sought to find what it wants to hear in the song, not its intended message. In reality, the same forces the song polemicized have colonized it, twisting and distorting the song into an ironic facsimile of itself (Jackson, 2002; Santelli, 2012).

Corporations are plundering our society's music in order to benefit themselves financially, in an overt way, and, in a more subtle way, in order to radically alter the social symbology embedded in these pieces in order to rephrase the messages of these musical works into whatever they so desire. This chapter will attempt to interrogate this complex process through appreciating the dialectical relationships that protest music from the 1940s and 1960s American folk revivals had to larger social movements, with particular emphasis paid to Guthrie's "This Land", and observing how these social relationships have changed across a temporal scale, specifically how these songs have been commercialized despite their strong countercultural sentiments. Qualitative data will be gleaned from analysis done on advertisements of several corporations including Jeep and the North Face, as well as readily available interviews and the contents of the songs themselves (both authentic and colonized). In order to do this, previous literature on the subject will be amalgamated into a framework from which to unearth and analyze the various discourses embedded in these songs. This framework will attempt to merge political economic theory with a phenomenological analysis that can grapple with the ephemeral "feeling" of music so that they may be tied into the social currents of the era. The song will then be placed in its era by a section that contextualizes to its history and the intended discourses it was putting forth. The following section will utilize that novel framework to show the subtle ways these songs have been colonized through their use in various modern advertisements. Finally, this chapter will show that the colonization process is imperfect, and at the edges there are still spaces of resistance that are embedded in these songs as well.

Music as cultural inheritance and discourse

The colonization of "This Land" has notably been used to reproduce its collectivist message into one that reintegrates people with nature in all its mystical, magical, powerful, and wild representations. In order to fully conceive a framework that can outline how "This Land" has been colonized, an appreciation of why people are alienated from their natural environments should be had. In fact, the dominant Euro-Christian philosophical and theological traditions allude to the doctrines of domination, mastery, control, and husbandry of nature rather than being part of nature. Descartes (2006 [1637]) in *Discourse on the Method* had argued that general good of all humankind could best be pursued not by resort to speculative philosophy but by the attainment of knowledge that is useful in life so as to render ourselves the masters and possessors of nature. To Marx, labor is a natural process, and it is natural to use resources and alter the environment in the production process (Marx, 1975). In doing this work, a person achieves self-actualization and a realization of their purpose and abilities (Sayers, 2003). Classical political economics, as advocated by the likes of Adam Smith (2003 [1776]) and David Ricardo (1821) alluded to self-actualization and self-realization in individualized capacity in the realm of the market through choices. For Marx, however, this self-actualization project is a project of self-realization through relations with others in a collective society. As the industrial process divorces the worker from this actualization, they become increasingly divorced from their natural surroundings and the agency they have to produce their environment (Sayers, 2003). Again, to alter and produce the world is the natural way according to Marx, but in contrast capital forces seek to dominate the environment (Smith, 2009). Humanity, under capitalism, domineers the environment and naturalizes that as the true hierarchical order. This happens in spite of natural disasters, ecological devastation, and large-scale changes to our climate through global warming, all of which serve as a proof that humans are not truly in control of nature (Smith, 2009). Intensive work has been done on what nature actually means to our modern society and how, from there, alienation from our environment unfolds (see for example Castree, 1995; Castree, 2003; Smith, 2009). Many result in findings that show how capitalism commodifies the same tension that it creates in its contest with nature. Duffy (2015) outlines several examples of this, using eco-tourism and the markets for certain animals as a basis for stating that neoliberalists enter ecological conversations in order to cover contradictions within capitalism. Harvey (1993) points out the myriad of ways that environmental and ecological issues take political and commercial bends. Capitalism has repositioned nature as it continues to reproduce itself; humanity now dominates the environment and approaches it only for the furtherment of political or capital ends.

How these concepts integrate into music colonization requires an understanding of the ways music naturally springs from its environment. To best understand these impacts, it is important to reflect on the subject as not solely an artistic product. Music is also a product of the geographical and historical context that it sprung out of and becomes an encapsulation of that time to be presented to future generations as a sort of cultural inheritance. This is evident in smaller anecdotes from Springsteen about Seeger, where he remarked that Seeger was a "walking, singing reminder of all of American history" and a "living archive of America's music and conscience" (Marchese, 2014). This idea has inspired several academic works seeking to crystallize this cultural evolution process (see for example Bohlman, 1988; Belz, 1967; Reuss and Reuss, 2000; Roy, 2010) within folk music as it has developed from an oral tradition to a popular medium. Folk music is shown to have stemmed out of oral tradition, even cited by Bohlman (1988) as perhaps the most imperative feature of folk music, which serves as a symbolically rich tableau, a unique cultural transmission device that naturally reproduces itself and carries along social values and connectors nested within those reproductions. It is an internal process though, as all oral traditions are passed in more than just tangible understandings of stories and characters. Folklore, as bedrock of culture, is learned through immersion in it instead of through academic deconstruction (Dundes, 1980). In music this becomes apparent in how the structure of folk music is learned, with transmission being done through being taught songs, tethering certain meanings to melodies and note intervals, and observing the scene unfolding around you as an artist (Bohlman, 1988).

If music cannot come out of a vacuum, then to fully appreciate the dynamics of how music is created requires attention to be paid to the entirety of the variables that make up its composite. Dimaggio and Useem (1978) show class boundaries are innate to cultural consumption and exposure to the arts (which is again necessary for the transmission of cultural messages that pass on signs and symbologies) is dependent on familial income; thusly these cultural messages fissure along class lines and create different scapes of symbology. What becomes deemed "high art" is chosen on a political and economic axis, and its production becomes an area of study of students at prestigious and expensive universities, scions to economically powerful families. As music becomes seen as a constituency of social relations, how it is influenced by Gramscian hegemonies – that is, the bourgeois controlling of social constructs (knowledge, values, public perceptions) in order to dominate the working class (Gramsci, 1996) – becomes imperative to understand how bourgeois commercialized music is used to reinforce these hegemonies. Furthermore, it can then be seen how the structure of folk music can be used to project counter-hegemonies. Patterson (1975) shows that capital structures attempt to undercut the authenticity of music by commodifying it. Once the music becomes assimilated, its potent mobilizing ability becomes defanged and the ability for it to awaken a class consciousness is lost. This article even uses the example of the folk revival to show how organically created music is a constructor of counter-hegemonies. Further work done by Gottdiener (1985) merges this idea of semiotics and shows that use values, stemming from the music's symbolic values, are reliant on the audience's originating cultural scape or "scene" and prove to be inelastic when faced with market demands for music. As Clifford Geertz (2000, 99) puts it, culturally speaking, art is essentially "a sensibility, that such a sensibility is essentially a collective formation, and that the foundations of such a formation are as wide as social existence and as deep". Art in general, including music, becomes a reflexive production requiring an audience that both knows the semiotic structures of their cultural scape and can interact with the piece as it becomes ingrained into the wider culture (Geertz, 2000). This construes music as a cultural inheritance, with it authentically being passed down through being steeped within it, but gives room for explanation on how this cultural inheritance is not passed down evenly across a physical topography but instead through social filters that create more of a cultural topography.

An understanding of the melded line between artistic creation and its spatial and cultural contexts has led to a slew of geographic work on the production of music (see for example Bennett, 2017; Connell and Gibson, 2003; Hudson, 2006; Leyshon et al., 1998; Krims, 2012). This work shows that music is a product of specific geographic spaces. A good example of this would be work done in Cohen (1995) that shows how the physical and cultural topography of Liverpool actively influences its music scene. Furthermore, the chapter shows how the musical history, easily translatable to ideas of "tradition" in folk music, of the city drastically affected the culture of its music scene and the artistic networks that arise from it (Cohen, 1995). Geography ends up being integral to how people enmeshed in these different cultural scapes interact and co-relate as the spaces end up housing numerous artistic networks that span large swaths of cultural understandings. In particular, the spatial boundaries of musical performance become integral to music as it must be played within a set space, and the emotive qualities (the feeling or Geertz's idea of art as a sentiment) come out in the performance of music in these spaces. Music is an emotive art that relies on the audience believing in the authenticity of the performance, a subjective measurement that is based on numerous social attributes such as the cultural identity of the audience and the emotional content of the music as was shown in Matheson (2008), but also in the audience relating to each other during a transcendental experience. Music, folk music especially, is meant to be a shared experience that fosters a feeling of collective identity within the groups listening to the performance (Danaher, 2010). It accomplishes this because music is supposed to be a reflection of the experiences of that community. As Frith (2007, 32) puts it, "there was no distinction between folk artist and folk audience"; they were one in the same. This also allows musicians to inject messages, such as the folk revivalists' counter-hegemonies, into these songs that are absorbed by the audience with a singularity of mind as is seen in Danaher (2010). This makes music performance intently political and ideological in nature, and the class dimensions to it are not lost on leftist theory. Popular music is simply seen as a soother of the people, a distraction from their oppression. Truly reverberant art must come from the people: telling their stories, using their pastoral symbology, and awakening their collective consciousness (Denisoff, 1969).

These discourses become embedded in the music, and developing the art of tracking discourses as they evolve has led to a bevy of literature. Gill (2000) shows the depth at which discourse analysis needs to interrogate the rhetorical organization of text. Furthermore, it merges the concepts of description and evaluation so that proper analyses will produce a description of the text that is rich in critical evaluation. Rose (2016) connects work in discourse with content analysis, encouraging analysts to observe the rhetorical construction of observable materials through its visual, auditory, and textual composites. She also notes, in a very Foucauldian manner, that doing discourse analysis is a discursive act in itself. This is imperative because discourse itself is power, as van Dijk (2006) articulates when he explains that discourse can control and manipulate social cognition. With singer-songwriters like the folk-revivalists this becomes an imperative consideration, due to the fact that their work was meant to evoke social change. As Seeger once stated in an interview, "my life's been changed by music. I've received letters from people who have said their life's been changed by music" (Stanton-Rich, 1996). Music has had a profound impact on their lives and they want to inspire a similar change around themselves, but how their music is used can pollute their messages and, thusly, their intended social change. Their work can be skewed towards anything feasibly interpretable, so haphazard discourse analysis threatens polluting those messages towards the analysts' desired outcomes. Finally, Nattiez (1990) integrates discourse into musical analysis, arguing that no aspect of music can be left out of the analysis; it all conglomerates into a single rhetorical organization that is a "total social fact" and requires interpretation of all nuances.

Out of this literature emerges a novel framework for appreciating music as having numerous dimensions; it is discursive, symbolic, and organically rises out of cultural scapes that are distributed unevenly due to class restrictions. This music holds meaning from the cultural scape it comes from, and becomes integrated with the ideologies and the politics of social movements emerging from those scapes. These aspects interlace and merge into the same cultural fabric that bore them, influencing the production of further music and passing discourse and symbols onwards. Thusly, music becomes appreciable as a cultural inheritance; something that has a definite, albeit ephemeral, use value to listeners that capital hegemonies attempt to seize and reconstrue.

"This Land Is Your Land" as counter-hegemonic rhetoric

The extent to which "This Land", and other songs of a similar ilk, have been co-opted and colonized by capital forces is hard to quantify. "This Land" has a total of seven verses, but the song reached the vast majority of its listening audience through sheet music that only carried four of those verses (Jackson, 2002). To essentialize the song to those four verses would do a disservice to how mercurial the songwriting process was for Guthrie, which might be best encapsulated in the several copyrights he filed for the song as it evolved and adapted over time (Guthrie, 1956). As a folk song, it springs from the bedrock of a past tradition but reaches "the people" by nesting its message within the current geopolitical and cultural context (Roy, 2002). Guthrie composed the song to easily disseminate a political message; it utilized a common key based on an open chord (G), only required a vocal range of one octave to sing, and had a melody based on the popular Carter Family song "When the World's on Fire" (Jackson, 2002). All of these attributes can give a song the feeling that it stems from a collective history (Dunaway and Beer, 2010). This notion, "collective history", may be ephemeral, but it is powerful. It elevates these songs into a cultural inheritance that foster a sense of social unity and can be passed along from person to person (van Dijck, 2006). This feeling is an essential part of the use value that the emotive structure of music presents to a listener, and that value was utilized by protest singers like Guthrie to convey progressive ideologies in digestible ways and convince listeners to, for lack of a better phrase, "come on along" (Denisoff, 1968). Somewhere along the way, "This

Land", and the ephemeral feeling it carries with it, became subsumed to the processes of capitalist co-optation and colonization that sought to seize the song's cultural capital and integrate it into American corporate structure (Chapple and Garofalo, 1977).

Seeger disrupted this colonization by reaching back into the rich history of the song and, backed by an ethnically diverse chorus and utilizing a call-and-response style of song leading, performed it in a way that reflected how it was originally sung to the American audience during the 1940s folk revival (Evans, 1979). Seeger sang it as a communal effort, encouraging the 400,000 people in a shared space to sing in harmony as a method of producing group solidarity and to stoke emotive movement towards an ideological core (Pedelty, 2009). More importantly, he sung "This Land" to the electric crowd with the intended message that its songwriter Woody Guthrie was trying to transmit. Guthrie was not proclaiming that the United States was a topographical space rich in vague quintessences of human freedom. In fact, the song only utilized a songwriting pattern common to most American anthems as a rhetorical tool to satirize them (Jackson, 2002), but instead questions the United States as an embodiment of those values. Prior to its colonization, the song actually originated as a response to Irving Berlin's "God Bless America", and before Guthrie settled on the hook "this land was made for you and me" he finished every verse with "god blessed America for me" to directly tie the two songs together. Guthrie's echoed lines reframe the same scapes Berlin lionized in his song, essentially refuting him and his vision of America (Jackson, 2002). In its unneutered form, the true thesis of the song is revealed in the following stanza:

> In the shadow of the steeple I saw my people / By the relief office I seen my people / As they stood there hungry, I stood there asking / Is this land made for you and me?
>
> *(Guthrie, 1956)*

The political aspects of this verse are striking: juxtaposing America's endless skyways and golden valleys with its rampant poverty, and articulating the tension between the metaphor of America the beautiful emancipator and the reality that Guthrie had witnessed in his travels across the land. Guthrie, originally displaced from Oklahoma, would frequently ride railways across vast swaths of America, seeking opportunities as a troubadour (Jackson, 2002). He would come face to face with the third of the rural population that, in the 1950s, lived under the poverty line (Tickamyer and Duncan, 1990). Those experiences profoundly influenced Guthrie, and pushed him towards social activism which manifested as politically charged songwriting (Shaw, 2013). Guthrie himself wrote the phrase "all you can write is what you see" on the bottom of the original lyric sheet to "This Land" (Blair, 2019). What he saw filled his songs with dusty highways, wealth disparity, broken men and women at the foot of capital's power, and the ultimate culmination of those aspects into both the aforementioned radical verse and a second verse about disregarding private property:

> As I went walking I saw a sign there / And on the sign it said "No Trespassing". / But on the other side it didn't say nothing, / That side was made for you and me.
>
> *(Guthrie, 1956)*

These two verses rally against capital structures that subsume and suppress people underneath the auspice of the free market. Guthrie's allusions to "the other side" of the No Trespassing sign being the side "for you and me" describes the unnaturalness of divvying up land and resources (essentially owning the means of production) through legal means and land titles, and his allusions to seeing his people as they "stood there hungry" followed his thoughts on how a just and egalitarian society would not leave anyone wanting and in need of basic necessities. But those messages would fall by the wayside when Guthrie's song would be broadcast to the general listening audience sans the political verses; reconstrued as a praise of all it meant to rally against (Santelli, 2012). Guthrie was well aware of what was happening to his song, and when his ability to play began to falter he taught his young son Arlo how to play "This Land", including the incendiary stanzas, for fear that "no one would remember them" (Jackson, 2002).

What Guthrie was doing was relying on a folk tradition, passing down the rhetoric of his song to his son so that its message may be preserved in the most organic way. Once Arlo learned the song it would continue to be projected to audiences the way folk music always had, through live performance with acoustic instruments, a way that preserves the music's authenticity (Eyerman and Barretta, 1996). In that way it would remain intact, the same tune that came from "the hungry folks and they get the credit for all I pause to scribble down" would get projected the way Guthrie always played it, with the philosophy that "music is some kind of electricity that makes a radio out of a man and the dial is in his head and he just sings according to how he's feeling" (Eyerman and Barretta, 1996). Guthrie's music, and the folk revivalist movement at large, used this authenticity to ingratiate itself to the audience, but used a variety of other musical attributes to project counter-hegemonies and political rhetoric into a crowd in order to unify them and mobilize them towards action against current domineering structures. For one, the songs are led by the singer. Often during that era, in a closer knit setting like a union meeting (Denisoff, 1969), they sing the line to the crowd, who would then return the line in harmony. This rhetorical tool is evidenced in Seeger's rendition of "This Land" at the Lincoln Memorial, where he led the 400,000 attendees through the song in the same manner (Marchese, 2014).

Most of the songs of the Folk Revivalist movement also involve lyrical repetition, a rhetorical tool that internalizes the message within the receiver (Danaher, 2010). In "This Land", this is seen in the repetition of the line, "this land was made for you and me" at the end of every stanza. When paired with the vivid descriptions of landscapes that span across the country, it internalizes a social relation between the audience, the singer, and their surrounding geography. It contrasts the audience and singer with no other identifiable group. Instead, it connects the two of them, and reinforces the idea that they collectively have an absolute claim to the beautiful and ecologically diverse space around them. Twice Guthrie breaks from the repetition: first, when he slightly alters the line to be "that side was made for you and me", and second, when he rephrases it as the more interrogative "is this land made for you and me?". Each time a break occurs, it creates tension within the internalized message. In these lines, the audience remains connected to Guthrie and operate with him as a collective unit within the poetically described landscapes. Their agency over this space has been disturbed, however, and other actors (private property owners and the state) have been interjected by Guthrie as counterposed powers creating that tension. The unabridged version of "This Land" leaves the audience feeling that, as a community, they have been denied their right to the space they exist within. That they live within an ecology that they truly are divorced from, and that they have no agency over their surroundings. The unnaturalness of capital hegemonies is called to be scrutinized, and the alienation from our natural world is called to be simply ignored and bypassed.

As "This Land" grew in popularity across the nation, the last lines of the stanzas took on a symbolic value. This symbolic value rests in a composite of the whole song, with every line taking on its own sign value and allowing for variation on how it is interpreted. The act of listening and receiving these messages is filled with dynamic social relations (Fornás, 2003); factors such as where you listened to it, who played it, and how they played it all share a role in transmitting these values. With leftist folk singers, the main hook of "This Land" carried the weight of its discursive message; reutilization of the line at the end of their songs would imbue the song with that same anti-establishment vigor. This created a pastiche of resistance, a stamp on a litany of protest verses that spring from a single symbolic node. Seeger, who sang as an environmental advocate nearly as much as he sang to inspire the working class (Ingram, 2008), has notably used the line to blend different lines of protests into the overall folk revivalist movement:

> We've logged the forests, we've mined the mountains / We've dammed the rivers, but we've built fountains / We got tin and plastic and crowded freeways / This land was made for you and me.
>
> *(Jackson, 2002)*

Utilization of the Guthries' musical rhetoric continued into the United States' second folk revival as the music of songwriters like Cisco Houston, Seeger, Joe Hill, and, of course, Guthrie became mythologized by a new wave of folk protest singers (Rodnitzky, 1973). Songwriters like Janis Joplin, Bob Dylan, Country Joe McDonald, and Phil Ochs were influenced by the same tradition that built the 1940s wave, but were molded and adapted to fit the issues of the current day (Kizer, 1983). Namely these issues ended up being the hot button issues of the counterculture movement; issues such as sexual liberation, civil rights, and Vietnam found their way into new folk-inspired compositions. These singers rallied against these injustices in larger and more public concerts, ranging from playing cafes to political marches (Eyerman and Barretta, 1996), and blended these new issues into older traditions through emphatic performances reminiscent of Guthrie and Seeger, sometimes referencing them directly (perfectly exemplified in Dylan's "Song to Woody") and oftentimes playing songs of old tradition alongside compositions of their own (Jackson, 2002). Guthrie actively encouraged the adapting of his songs to fit the issues of the time, saying that "this bringing them up to date is what keeps a folk song a folk song, it says whatever needs to be said, or as much as the law allows, at the time when it needs to be said" (Jackson, 2002). So when an artist sang an anti-Vietnam war alteration of "This Land" that included:

> This land is your land but it isn't my land / From the Mekong Delta to the Pleiku Highland / When we get shot at the ARVN flee / This land was meant for the V.C.!
>
> *(Jackson, 2002)*

to Guthrie, that fit perfectly into the folk tradition and the greater social movement that folk revivalism was integrated with. The symbolic power of the hook to "This Land" is emphasized by Guthrie himself when he remarked that "the main idea about this song is, you think about these eight words all the rest of your life and they'll come a bubbling up into Eighty Jillion all union" (Jackson, 2002). The point of the song was nested in these last lines, the power of this song rests in their repetition and breaks. When a force reproduces the song without those rhetorical breaks, the song begins to internalize a far different message in the audience. It becomes a song strictly about repetition, stable and balanced, that keeps internalizing a message that America is a land of beauty, prestige, and a totalizing unity across all scales.

"This Land Is Your Land" as colonized rhetoric

In our modern era, many decades after Guthrie performed "This Land" to union halls and poor farmers, the abridged version reigns as the predominantly known song. It is sung at political campaigns and elementary schools, and it holds credence as an unofficial American anthem.

Nora Guthrie, Woody's daughter, has even received messages from "thousands of people" calling for "This Land" to be the new national anthem based on its "authentic" patriotism (Blair, 2019). The same authenticity, expressed through acoustic instruments and pastoral phrasing, that was used to mobilize rural populations was being heard as a heartfelt ode to old-fashioned American values. In 2018 the Air National Guard, a branch of the United States military, aired a television advertisement that featured a choral rendition of "This Land" playing over rapidly cut scenes of the Air National Guard fighting fires, rescuing flood victims, and aiding in disaster relief, all from the vantage point of a military helicopter (Air National Guard, 2019). The Air National Guard are shown combating numerous disasters, and the high cinematic vantage point always portrays them as a dominant force over the situation; completely in control and preserving order amidst chaos. None of this is too strange for an advertisement trying to garner enlistment to a military organization, but the inclusion of "This Land" exemplifies how the state has co-opted the song as a patriotic anthem. Aurally, the song is sparse with most of the instruments playing only on the more emphatic downbeats, so the voices of the feminine choir carry most of the melody. Unlike the single octave range needed for Guthrie's version, this version has a vocal melody that requires two octaves, much like "God Bless America" (Jackson, 2002). This crisply and professionally composed approach causes the song to feel more intrinsically anthemic. The saccharine quality of the choral leader's voice grants a sweeter overall tone to the production, one that reinforces thoughts of hope in the face of danger. The sweetness in her voice does not betray that things are awry in these disasters, yet it does reinforce the National Guard as a protector of America's land and its people. The quick cuts of National Guard members facing down natural disasters shows a domination of American land, even when the militarized control of the land is fully portrayed as benevolent. The safety of the people, and the natural world they live within, is shown to be a byproduct of America's domination of the land.

The National Guard ad serves as a general example of how Guthrie's message gets maimed by interests it diametrically opposed, but several examples also exist that warp the message in increasingly subtle ways. Ways that neuter Guthrie's points on the unnaturalness of capital, and how that unnaturalness divorces our social connection to our natural environment. Our alienation is then packaged and sold to us as a commodity through products designed to alleviate said alienation through presumed reintegration into our natural world; one that still exemplifies capitalism as a natural process. Abrasions that occur from the coexistence of the unnatural capitalist system and the natural world become perceived as solvable through market solutions.

This played out on an even bigger stage, Super Bowl XLIX in 2015 (watched by 114.4 million viewers; Palotta, 2015), where Jeep premiered an ambitious campaign around their new Jeep Renegade headed by an advertisement showing scenes of a Jeep Renegade driving around the globe set to a version of "This Land" sung by commercial songwriter Marc Scibilia (Buss, 2015). The advertisement took great care to line up the aural, visual, and textual aspects of the composite whole; featuring shots of sequoias when Scibilia sings "great redwood forests", helicopter shots of Rio de Janeiro's "Christ the Redeemer" statue when he sings "that golden valley", and so on matching lyrics to vistas across the globe (Jeep, 2015). Using this, Jeep expands the natural landscapes innate to "This Land", disassociating it from America and clarifying the entire globe as the "land made for you and me". Panoramas of the scenic vistas in the Serengeti, China, and the United Kingdom were used whenever the all-important hook was being sung. Jeep drives this home by alternating different shots of people from vastly different ethnic groups as Scibilia sings "you and me" and ends on the proclamation that "the world is a gift" and that the audience should "play responsibly" (Jeep, 2015). The world, as portrayed from a strictly American perspective, is merely a playground to romp around in while driving American cars; everything is viewed through the American lens (particularly reinforced by the car's "American" label in the beginning; Jeep, 2015) projecting it outward throughout the globe. Jeep taps into certain elements of "exoticism" to attract the American audiences' occidental gaze. Renegades are shown driving on roads that are populated by elephants, kangaroos, and other flora and fauna that give the audience the feeling they are on a safari. Strangely, Jeep still includes the line "from California to the New York Islands" which bounds the symbolic landscape of the song to strictly the United States.

Aurally, the song is sparse with a heavy focus on the soulful voice of Scibilia, leaving the audience feeling alone in the commercial; a far cry from the communal call-and-response nature of its original performances. The visual aspect of the commercial corroborates this solitary feeling by featuring several shots of empty land and often not including more than a single person in a scene at a time. Scibilia's version was picked out of 12 options composed at a songwriting camp specifically geared to teach musicians how to sell work to commercials (Morrissey, 2018). In a statement released after the commercial aired, Jeep called Scibilia's version a "unique and surprising twist" on "This Land" which they claim was "written as a crowning achievement to America's homeland" broadcasting a false history of the song (Buss, 2015), erasing the progressive ideology nested within the song by cementing the anthemic, abridged version as the true song and then further reproducing it to align with their ideology as a multinational corporation by imbuing it with a neoliberal ideology steeped in globalization. A focus on a globalized society might be a deflection for the company though. Jeep markets the car as "America's smallest, lightest SUV", which is a far cry from a domestic product (Buss, 2015). Only 22% of the car is manufactured in the United States after all, with the lion's share of 62% being made in Italy (DeBord, 2017). However, Jeep's restructuring of "This Land" worked, and managed to sell 61,000 units in 2015 by marketing a mostly Italian car as an embodiment of American globalization (DeBord, 2017).

This ad campaign was the brainchild of Fiat-Chrysler CMO Olivier François, noted for his "statement" ad campaigns that revel in spectacle in the name of "doing something different" (Buss, 2015). He's known for using celebrities like Eminem, Clint Eastwood, and even folk revival legend Bob Dylan (which could be construed as a large colonization of folk revivalism, if Dylan had not made a concerted effort throughout his career to obfuscate his relation to the movement, waffling from saying he thought all "message songs" were vulgar to later calling all his music protest material in 1984; Cott, 2017) in commercials to sell cars, but François focuses largely on the relation of music to his ads (Newman, 2018). According to his colleague Brian Monaco (CMO of Sony/ATV Music Publishing), as a former music producer François "thinks deeply about the lyric, the melody and the production and how they synchronize with the message he is trying to convey" (Newman, 2018). In interviews he philosophizes about merging brands and music in order to give each brand a tone and a distinct voice, utilizing different genres to emphasize different emotions and identities and burrow corporate messages into them (Schultz, 2018). These identities and emotions stem from music being an output of different cultural scapes; they arise out of the symbolic rich tradition of these musical spaces and are imbued with ideologies and values which make them good holders of cultural identity (Gottdiener, 1985). François seeks to use music to tie corporate identities into these cultural identities through his advertising strategy, even calling music a "trojan horse" to sneak brands in front of audiences (Schultz, 2018), colonizing the political gravity and cultural inheritance songs like "This Land" are supposed to hold.

François has found success using this strategy despite self-proclamations that he is culturally ignorant, even at one point saying that he thinks country music is just "about cowboys, beer, breakups, trucks" as he makes a concerted effort to lace Ram's identity with country music (Newman, 2018). Seeking to reproduce the music, François aired a Ram ad featuring John

Denver's "Thank God I'm a Country Boy" performed by a primarily black band dressed in country western outfits (Ram, 2018). In doing this, François is attempting to broaden the cultural scape his message operates on, interspacing scenes in classic New York sets (i.e. the Brooklyn Bridge/the subway) that feature black musicians singing the Denver song with scenes of rural America filled with white farmers loading hay bales into a Ram truck (Ram, 2018). The juxtaposed scenes begin to merge as the black musicians end up in the pastoral vignettes as the song picks up energy to hammer home François' message that American divisions all become unified under the banner of automotive consumption. This is especially evident when the final motto shows "This is Ram Nation. Population: America" emphasizing a final message of a total American harmony achieved through buying trucks. Social tensions that exist with urban/rural and racial dynamics are blurred with urban scapes being utilized in the commercial to represent black populations, and their merging with the primarily white rural area, as it is portrayed in the commercial, is seen as the harmonizer. The hegemonic structures that cause these tensions stem from capital, yet capital is seen to be the answer to these problems; not the originator of them. Modernity is not approached as unnatural, what is natural is instead shown to be consumption. In this case the purchasing of trucks is used to link groups that are normally disparate, despite their separation being intrinsically linked to racialized hegemonies supported by capital. Two days after it went live, the ad had six million views on YouTube and Ram's Facebook video of the ad was still amassing 130,000 views per hour (Courtney, 2018). François uses these kind of metrics to boast about his impact on the music industry; he uses his numerous music industry connections to strategically place cars in high profile music videos, once claiming Fiat Chrysler is the most viewed car company on YouTube with over 18 billion views (Schultz, 2018).

So far every commercial that has been analyzed has steered away from using the controversial lyrics in "This Land" but corporate interests have utilized them before. Both Johnny Walker and the North Face have aired ads using "This Land" and both used the verse about private property, but neither used the verse with the same intended message that Guthrie ingrained in the song. Both Diageo and VF (the parent companies) use that same verse to transmit different messages imbued with their own corporate rhetoric. Strictly concerning how natural alienation is commodified, the North Face ad serves as perhaps the most succinct example of how this process unfolds. The North Face ad was made during an expansion of the company's marketing plan during the 2015 year, and it featured the popular indie band My Morning Jacket performing an evocative version of "This Land" (Coffee, 2015). The aural aspects of the version are more complex, with effects added to the soundscape to create an ambient soundscape that detaches the viewer and strews them across a myriad of natural landscapes. The soundscape is placed across various shots of people, clad in North Face gear, climbing mountains, hiking, sailing, and partaking in all sorts of outdoor activities (The North Face, 2015). As the commercial begins, the North Face logo appears and quickly fades over a forested valley emphasizing how the abstract concepts of North Face, a corporation, and the natural world are tightly linked (The North Face, 2015). Jim James, lead singer of My Morning Jacket, sings softly in the first third of the advertisement but enters a more animated style of singing when he suddenly transitions to the private property verse which visually is mimicked by visuals of people crossing rusted gates and entering into strange territory (The North Face, 2015). The advertisement continues to build, and James gets a choral effect as it reaches "this land was made for you and me" and the viewing audience is treated to scenes of increasingly physical outdoor pursuits encouraging them to get out there and adventure; the world was just out the door and any place was truly explorable (The North Face, 2015). This rephrases Guthrie's verse to mean that there are no boundaries to exploration and that any outdoor pursuit is possible in North Face ware. Instead of being about overcoming oppressive institutions like private property, the verse becomes about overcoming one's own inhibitions and appreciating the ecological diversity Guthrie had described in previous verses. It becomes a call to adventure, a call to climb a mountain or ford a river all while wearing North Face jackets. The natural world only plays the role of a setting for these adventures, the people in the advertisement do not coexist with the environment; instead they actively seek to domineer the environment. The North Face ad shows people conquer the natural world through rhetorical images like topping a mountain or sailing an arctic ocean (The North Face, 2015); people in this ad, aided by devices construed from natural resources, venture forth into hostile landscapes just to show that people can make it there. The refrain of "This Land" further gets reproduced under this context, as the landscapes described in the song become used to symbolize the domination of humanity over their environment.

The Johnnie Walker Black Label ad featured another version from another popular indie band, the LA-based Latino band Chicano Batman, as it was part of a broader advertising campaign titled "Keep Walking America" that celebrated diversity in America and emphasized the various narratives at play within ethnic groups in the United States (Walker, 2017). In this respect, the ad approaches how it reconstrues Guthrie's song in a significantly different manner than the previous ads. It attempts to reproduce the refrain towards a starkly different political message, but a message still tethered to neoliberal, corporate ideology. The Chicano Batman version is up-tempo, maintaining a funky but upbeat spirit to the vocal melody, as the visual spectrum of the composite is filled with different people of different ethnicities going through their day-to-day lives in a nondescript American setting (Walker, 2017). Chicano Batman's version includes the controversial lyrics about private property, but the rhetoric of the high wall takes on a wholly new symbolic meaning. Instead of standing in for capitalist hegemony, it reproduces the lyrics to be about immigration policy and recasts the message of the verse to another political message. This is further emphasized when Chicano Batman begins singing the rest of the song, sans relief office verse, in Spanish while Johnnie Walker finishes the ad by telling America to "keep walking" (Walker, 2017). Despite its generally positive sentiment, Johnnie Walker does not seek to posit a world free from borders in the ad, nor does it seek to truly make any radical statements at all. The ideology of the ad still returns to the neoliberal, globalized core that simply seeks to sell their whiskey to as many diverse audiences as they can. Reproducing Guthrie's line about "no trespassing" to call for immigration reform neuters its intended message as a polemic against private property. It makes the song a rallying call to fix an issue that conceptually is only symptomatic of capital structures such as private property. In doing so, the ad also naturalizes an unnatural system. For their efforts, Diageo won a U.S.H. Idea Award for best marketing idea in the Hispanic market, further reconstruing the intended recipients of its hopeful message into a market to be tapped (Wentz, 2017). In the campaign, Johnnie Walker calls on America to keep walking towards a greater future; one more united, racially harmonious, and one more drunk on Johnnie Walker scotch. The American people responded well to that last call, with Johnnie Walker Black Label increasing sales by 5% (Diageo, 2018). The North Face garnered similar attention, leading to a 50% growth in their marketing budget and the beginning of a partnership with the US Department of Interior and the Conservation Corps through an injection of $100,000 of capital from these entities (Coffee, 2015).

Spaces of resistance

It would seem that any hope of reinvigorating "This Land" as a standard bearer for the folk revivalist movement has drastically depreciated with time. However, in spite of the level of influence marketing executives have on how the song is disseminated, corporatism cannot fully colonize the rhetoric of "This Land" as it still organically springs from a vibrant cultural scape that appreciates the song with its intended message. Strangely, My Morning Jacket's version of the song serves as a good example. Even though their version is tethered to a corporatist ideology that commissioned it, the band composed a full version of the song that falls closer in line to Guthrie's authentic message than covers such as Scibilia's version. The aural qualities, highly structured to project the North Face's message in the ad, are looser in the full song. The introduction, which covers the first four verses, is sparser and contains only James' voice in an ambient soundscape, but a rest occurs after the anthemic verses end and the full band jumps into celebrate and emphasize the controversial lyrics that James proceeds to sing (My Morning Jacket, 2015). The style of composition follows traditional folk revivalists by adding in harmony and singing as a collective unity when it gets to the controversial lyrics. On top of that, My Morning Jacket did not make any money off the recording, instead donating all proceeding to the Conservation Corps (Newman, 2014). As research was being undertaken on this chapter, more and more spaces of resistance kept appearing that used "This Land" as a progressive rhetorical symbol. It occurs to the scale that would warrant further investigation of how music retains its rhetoric over time, and interrogating the subject will most certainly lead to enlightening outcomes. Just a few for example, Gary Clark Jr., who earlier mentioned it was one of the first songs he learned in school, based his defiantly anti-racist song "This Land" on the hook of the song (O'Rear, 2019). Bruce Springsteen, who cites folk revivalists like Seeger as huge influences that encouraged him to peer into the folk tradition and write symbolically rich songs like "Ghost of Tom Joad" (Marchese, 2014), was stated to be the main celebrity Olivier François could not successfully court. In François' words, "He's not for sale . . . he's not for rent. And there's nothing you have that he wants" (Schultz, 2018). David Rawlings, noted for his pairing with singer Gillian Welch, often performs "This Land" in medley with another protest song, "I Hear Them All", to close out many of their shows, shows that have been described as follows:

> The band including the oft-ignored verses about The People being relegated to the relief office and the defiance that wells up at a No Trespassing sign – and the air went goosebump-electric, the house knowing those verses, loving their preservation, and screaming, whooping, raising fists. The audience as well as the musicians were people who know their roots, who are moved in spirit by the music and the message, and that is a deeply gratifying community to be part of.

Organic, authentic resistance is still vibrantly alive in music, feeding off the traditions that bore it and keeping the folk cultural scape one that is rich and symbolically dynamic. That final quote echoes a reunification of the artist and audience as one mind, just as it was in the farmlands and union halls that served as the music's Fertile Crescent.

Concluded remarks

As is evident from the last section, the subtle ways that capital colonizes songs like "This Land" is imperfect and allows for fissures along the edges in which resistance to co-optation blooms. However, the commodification process still trundles onward, and as it moves it absorbs whatever it can reproduce. Not care about any originating contexts, it will take a satirical, lively, anti-commercial song like Janis Joplin's "Mercedes Benz" and place it in a sterile advertisement selling the very car the song lampoons (Wollenberg, 1995). The commodification process that seeks to sell an alienated populace their natural wholeness back did this to Woody Guthrie's "This Land". It takes a radical song that rails against poverty and injustice and uses it to naturalize the very system that spawns those issues. The hook to the song gets mangled in different advertisements to stand for a myriad of disparate meanings, all resonating from the same neoliberal core. The reproductions of the song do not fully drown out the message of Guthrie's original tune. The same traditions that created the song have preserved it, and helped transmit to audiences a salient idea; that this land really is made for you and me.

Note

1 Which henceforth will be referred to as just "This Land" for convenience sake.

Bibliography

Air National Guard. (2019). *This Land* [Television commercial]. Retrieved from www.ispot.tv/ad/IdmK/air-national-guard-this-land

Belz, C. I. (1967). Popular music and the folk tradition. *The Journal of American Folklore, 80*(316), 130–142.

Bennett, A. (2017). *Music, Space and Place: Popular Music and Cultural Identity*. New York, NY: Routledge.

Blair, E. (2019, March 14). *How 'This Land is Your Land' Roamed and Rambled Its Way Into American Life*. Retrieved from www.npr.org/2019/03/14/702792467/woody-guthrie-this-land-is-your-land-american-anthem

Bohlman, P. V. (1988). *The Study of Folk Music in the Modern World*. Bloomington, IN: Indiana University Press.

Buss, D. (2015, February 1). *Jeep's 'This Land is Your Land' Ad and It's Multifaceted Globalism*. Retrieved from www.forbes.com/sites/dalebuss/2015/02/01/jeeps-this-land-is-your-land-super-bowl-ad-why-well-be-talking-about-it/#3bfeb39d3a3d

Castree, N. (1995). The nature of produced nature: materiality and knowledge construction in Marxism. *Antipode, 27*(1), 12–48.

Castree, N. (2003). Commodifying what nature? *Progress in Human Geography, 27*(3), 273–297.

Chapple, S., & Garofalo, R. (1977). *Rock'n'Roll is Here to Pay: The History and Politics of the Music Industry* (p. 262). Chicago, IL: Nelson-Hall.

Coffee, P. (2015). *Mekanism Fills in All the Blanks in First Global Campaign as North Face Lead*. Retrieved from www.adweek.com/agencyspy/mekanism-fills-in-all-the-blanks-in-first-global-campaign-as-north-face-aor/94452/

Cohen, S. (1995). Sounding out the city: music and the sensuous production of place. *Transactions of the Institute of British Geographers*, 434–446.

Connell, J., & Gibson, C. (2003). *Sound Tracks: Popular Music Identity and Place*. New York, NY: Routledge.

Cott, J. (Ed.). (2017). *Bob Dylan: The Essential Interviews*. New York, NY: Simon and Schuster.

Courtney, W. S. (2018, January 17). *2019 Ram 1500 Breaks the Internet With Patriotic, Cross-Cultural "Thank God I'm a Country Boy" Ad*. Retrieved from www.thedrive.com/watch-this/17736/2019-ram-1500-breaks-the-internet-with-patriotic-cross-cultural-thank-god-im-a-country-boy-ad

Danaher, W. F. (2010). Music and social movements. *Sociology Compass, 4*(9), 811–823.

DeBord, M. (2017). *Check Out the All-American, Made-in-Italy Jeep Renegade SUV*. Retrieved from www.businessinsider.com/jeep-renegade-review-2017-2

Denisoff, R. S. (1968). Protest movements: Class consciousness and the piopaganda song. *The Sociological Quarterly, 9*(2), 228–247.

Denisoff, R. S. (1969). Folk music and the American left: a generational-ideological comparison. *The British Journal of Sociology, 20*(4), 427–442.

Descartes, R. 2006 (1637). *A Discourse on the Method*. Oxford: Oxford World's Classics.

Diageo. (2018). *Diageo Annual Report 2017*. Retrieved from www.diageo.com/PR1346/aws/media/6212/b0000391_diageo_ar-2018_interactive.pdf

DiMaggio, P., & Useem, M. (1978). Social class and arts consumption. *Theory and Society, 5*(2), 141–161.

Duffy, R. (2015). Nature-based tourism and neoliberalism: concealing contradictions. *Tourism Geographies, 17*(4), 529–543.

Dunaway, D. K., & Beer, M. (2010). *Singing Out: An Oral History of America's Folk Music Revivals*. New York, NY: Oxford University Press.

Dundes, A. (1980). *Interpreting Folklore*. Bloomington, IN: Indiana University Press.

Evans, D. (1979). Folk revival music. *The Journal of American Folklore, 92*(363), 108–115.

Eyerman, R., & Barretta, S. (1996). From the 30s to the 60s: The folk music revival in the United States. *Theory and Society, 25*(4), 501–543.
Fornäs, J. (2003). The words of music. *Popular Music & Society*, *26*(1), 37–51.
Frith, S. 2007. *Taking Popular Music Seriously*. London: Routledge.
Geertz, C. 2000. *Local Knowledge: Further Essays in Interpretive Anthropology*. United States of America: New York: Basic Books.
Gill, R. (2000). Discourse analysis. *Qualitative Researching with Text, Image and Sound*, 172–190.
Gottdiener, M. (1985). Hegemony and mass culture: a semiotic approach. *American Journal of Sociology*, *90*(5), 979–1001.
Gramsci, A. (1996 [1975]). *Prison Notebooks*. New York, NY: Columbia University Press.
Guthrie, W. (1956). "This land is your land." *Bound for Glory*, Smithsonian Folkways.
Harvey, D. (1993). The nature of environment: dialectics of social and environmental change. *Socialist Register*, *29*(1).
Hendrix, S., & Mummalo, J. (2009). Jammin' on the Mall for Obama. *The Washington Post* [Washington, DC]. 19 January 2009 Published. Print.
Hudson, R. (2006). Regions and place: music, identity and place. *Progress in Human Geography*, *30*(5), 626–634.
Ingram, D. (2008). "My dirty stream": Pete Seeger, American folk music, and environmental protest. *Popular Music and Society*, *31*(1), 21–36.
Jackson, M. A. (2002). Is this song your song anymore?: revisioning Woody Guthrie's "this land is your land". *American Music*, pp. 249–276.
Jeep. (2015). *Beautiful Lands* [Television Commercial]. Retrieved from www.ispot.tv/ad/7x75/jeep-renegade-super-bowl-2015-beautiful-lands
Kizer, E. J. (1983). Protest song lyrics as rhetoric. *Popular Music & Society*, *9*(1), 3–11.
Krims, A. (2012). *Music and Urban Geography*. New York, NY: Routledge.
Leyshon, A., Matless, D., & Revill, G. (Eds.). (1998). *The Place of Music*. New York, NY: Guilford Press.
Marchese, D. (2014, January 28). *Bruce Springsteen's Moving Tribute to Pete Seeger*. Retrieved from www.rollingstone.com/music/music-news/watch-bruce-springsteens-moving-birthday-tribute-to-pete-seeger-231982/
Marx, K. (1975). *Early Writings*. Harmondsworth, UK: Penguin.
Matheson, C. M. (2008). Music, emotion and authenticity: A study of Celtic music festival consumers. *Journal of Tourism and Cultural Change, 6*(1), 57–74.
Morrissey, J. (2018, October 14). *Need a Song for Your Commercial? Try These Campers*. Retrieved from www.nytimes.com/2018/10/14/business/media/need-a-song-for-your-commercial-try-these-campers.html
My Morning Jacket. (2015). "This land is your land" *This Land is Your Land*. ATO Records.
Nattiez, J. J. (1990). *Music and Discourse: Toward a Semiology of Music*. Princeton, NJ: Princeton University Press.
Newman, J. (2014, October 14). *Jim James on the Magic of This Land is Your Land*. Retrieved from www.rollingstone.com/music/music-features/my-morning-jackets-jim-james-on-magic-of-this-land-is-your-land-81804/
Newman, M. (2018, December 13). *Fiat Chrysler's Olivier François on Why Music is 'The Best Investment' and the One Artist Who Said No (After the Ad Was Shot)*. Retrieved from www.billboard.com/articles/business/8490183/fiat-chrysler-olivier-francois-music-is-the-best-investment-interview
O'Rear, C. (2019, February 19). *Gary Clark Jr.: All or Nothing*. Retrieved from https://americansongwriter.com/2019/02/gary-clark-jr-all-or-nothing/
Palotta, F. (2015, February 2). *Super Bowl XLIX Posts the Largest Audience in TV History*. Retrieved from https://money.cnn.com/2015/02/02/media/super-bowl-ratings/
Patterson, T. (1975). Notes on the historical application of Marxist cultural theory. *Science & Society*, 257–291.
Pedelty, M. (2009). This land: Seeger performs Guthrie's "lost verses" at the inaugural. *Popular Music and Society*, *32*(3), 425–431.
Ram. (2018). *Thank God I'm a Country Boy* [Television Commercial]. Retrieved from www.youtube.com/watch?v=TMS_m_CltX8
Reuss, R. A., & Reuss, J. C. (2000). *American Folk Music and Left-Wing Politics, 1927–1957* (No. 4). Lanham, MD: Scarecrow Press.
Ricardo, D. (1821). *On The Principles of Political Economy and Taxation*. London: The Electric Book Company Ltd.

Rodnitzky, J. L. (1973). The Mythology of Woody Guthrie. *Popular Music & Society, 2*(3), 227–243.

Rose, G. (2016). *Visual Methodologies: An Introduction to Researching with Visual Materials.* Thousand Oaks, CA: Sage Publications.

Roy, W. G. (2002). Aesthetic identity, race, and American folk music. *Qualitative Sociology, 25*(3), 459–469.

Roy, W. G. (2010). *Reds, Whites, and Blues: Social Movements, Folk Music, and Race in the United States* (Vol. 45). Princeton, NJ: Princeton University Press.

Santelli, R. (2012). *This Land is Your Land: Woody Guthrie and the Journey of an American Folk Song.* New York City, NY: Running Press Adult.

Sayers, S. (2003). Creative activity and alienation in Hegel and Marx. *Historical Materialism, 11*(1), 107–128.

Schultz, E. J. (2018, July 22). *What Fiat Chrysler CEO Change Means CMO Olivier Francois.* Retrieved from https://adage.com/article/cmo-strategy/fiat-chrysler-ceo-change-means-cmo-olivier-francois/314335

Shaw, J. (2013). *This Land That I Love: Irving Berlin, Woody Guthrie, and the Story of Two American Anthems.* New York, NY: Public Affairs.

Smith, A. (2003 [1776]). *The Wealth of Nations.* New York: Bantam Classics.

Smith, N. (2009). Nature as accumulation strategy. *Socialist Register, 43*(1).

Stanton-Rich, D. (1996). Being inducted into the Rock and Roll Hall of fame is no thrill for Pete Seeger. *Sassafras,* Spring 1996, 14-15.

The North Face. (2015). *Your Land.* Retrieved from www.ispot.tv/ad/7smZ/the-north-face-your-land-song-by-my-morning-jacket

Tickamyer, A. R., & Duncan, C. M. (1990). Poverty and opportunity structure in rural America. *Annual Review of Sociology, 16*(1), 67–86.

Van Dijck, J. (2006). Record and hold: popular music between personal and collective memory. *Critical Studies in Media Communication, 23*(5), 357–374.

Van Dijk, T. A. (2006). Discourse and manipulation. *Discourse & Society, 17*(3), 359–383.

Walker, J. (2017). *This Land is Your Land* [Television Commercial]. Retrieved from www.ispot.tv/ad/A0k2/johnnie-walker-this-land-is-your-land-featuring-chicano-batman

Wentz, L. (2017, October 13). *The Winners are Here: The 5 Best Ads From U.S. Hispanic Awards.* Retrieved from https://adage.com/article/creativity/5-creative-ideas-u-s-hispanic-agencies/310885

Wollenberg, S. (1995, March 8). *Mercedes-Benz Dusts off Janis Joplin Song for Commercial.* Retrieved from www.apnews.com/ad4de31302e74c1389b3b09c3911fe4e

Zaleski, A. (2019, February 15). *Interview: Ryan Bingham Returns Strong With American Love Song.* Retrieved from https://theboot.com/ryan-bingham-american-love-song-interview-2019/

11

CONTESTING INVISIBILITY OF IMMIGRANT DETENTION LANDSCAPES IN TEXAS

Alicia Danze, Rebecca Maria Torres, and Caroline Faria

Only a few months after President Trump's election, it was announced that a new immigrant detention facility, the first under the new administration, would soon be constructed in Conroe, Texas – a small city just north of Houston (Hoffman April 13, 2017). The choice to place the facility in Conroe was not surprising, given that Montgomery County has decades-long ties with the privately contracted prison facility operator the GEO Group, the world's second largest for-profit prison corporation (Shahani March 25, 2011). It is no coincidence that the same company also runs the Montgomery County jail, Joe Corley Detention Center, and Montgomery County Mental Health Facility, all in Conroe. However, what was surprising was the mayor's response to this announcement, in which he rejects the plans for a second detention facility in Conroe, stating, "I don't think that we need any more detention centers here in Conroe" (Jordan, April 15, 2017). His statement points to the existence of tensions within the community that have developed since the establishment of detention facility in Conroe nearly two decades earlier. What was first celebrated as a viable economic development strategy providing jobs and revenue for the county had since become the focus of debate regarding prison privatization and expansion without community consultation. In response to growing local concerns over the optics of housing such facilities within the burgeoning community, officials attempted to keep these prisons "hidden" from view, out of sight of the local residents. This invisibility in turn serves to block administrative legibility and accessibility for families of the detained with sometimes dire consequences for migrants and asylum seekers.

This research employs a feminist geopolitical approach to investigate the local and intimate discourses surrounding detention practices in Conroe since the insertion of private prison contracting in the community in 2004 (Thackeray, December 18, 2004). We supplement conclusions drawn from this discourse analysis with interviews with local inhabitants, business owners, and government officials, reflecting on the relative (in)visibility of these facilities and their local reception. Operating within a feminist geopolitical framework, this research explicitly seeks to expose uneven relations of power at diverse scales, as well as the everyday impacts of state control on differently situated actors.

The discourses explored operate in two spheres – one emerging out of debates amongst political representatives and local residents, especially apparent in local online news (*The Courier of Montgomery County*) and supplemented with fieldwork interviews; the other emerges out of conversations amongst family members of the detained, especially apparent in social media

DOI: 10.4324/9780429450310-14

postings ("Joe Corley Detention Center" Facebook page)[1] and interviews with immigration advocates. While opposition to detention operations might be shared in these distinct discursive spheres, the sentiments and motivations grounding these conversations vary. However, common to both spheres of discourse are the problematics of the (in)visibility of these facilities. These interlocutors voice concerns over: (1) the opacity of the financing, development, and management of these facilities, (2) their physical inaccessibility and lack of administrative transparency and clarity for families of the detained, and (3) the optics of their placement, as well as the diverging desires and sense of place and belonging they inspire amongst differently positioned viewers.

This concern with visibility corroborates other feminist geopolitical scholars' theorizations that detention practices are intentionally obscured by the state (Mountz, 2012), erased from public view. This invisibility works to isolate detainees from community and human rights networks, and to hide unsavory lived realities of immigration enforcement from the voting public. As the state with the highest percentage of the US-detained immigrant population as well as the greatest number of facilities in the country (Ryo & Peacock, 2018; Royall, July 8, 2018), Texas is ground zero for studying current trends in immigrant detention operation. Given recent record highs in the number of immigrants placed in detention facilities (Cullen, 2018), as well as the Trump administration's recent push to continue expanding the practice of immigrant detention, especially in Texas (McCullough & Essig, Aug. 2, 2018), it is urgent that immigration scholars investigate how immigrant detention centers insert themselves into local communities, with and without residents' consent, and sustain their clandestine operation without disruption.

Geographers and other social scientists conducting spatial analyses have made significant contributions to recent studies of immigration control across the world. While some scholars have focused on the historical development of the policies generating the infrastructure of immigration enforcement (Coleman, 2007, 2009, 2012a, 2012b; Coleman & Kocher, 2011; Coleman & Stuesse, 2016; Varsanyi, 2007) as well as the state of exception that permits such immigration policing regimes (Dikec, 2009; Klein & Williams, 2012; Mainwaring, 2012), other scholars have focused on politicians' and political parties' deployment of anti-immigrant rhetorics, ultimately shaping cultural values and conceptions of national identity (Carter & Merrill, 2007; Nagel, 2016; Nespor, 2014). Quantitative methods have been used to analyze variability in immigrant apprehension and detention rates across space (Moinester, 2018), as well as to estimate determinants in rates of Assisted Voluntary Return (Leekes et al., 2017).

While there have been a limited number of studies examining prisons as a rural economic development strategy (Bonds, 2006; Glasmeier & Farrigan, 2007) and community response (Che, 2005), there is little on the proliferation of for-profit immigrant detention centers in rural and suburban communities. Immigrant detention potentially generates different community responses and impacts than prison operation, given both the specific political discourses circulating on the topic of immigration (i.e. women and children crossing the border, family detention, etc.), and detainees' and their families' impeded access to legal resources due to their precarious migratory status.

Feminist geographic approach to studying immigrant detention

This article takes a feminist geographic approach to studying the local and intimate discourses surrounding immigration detention in Conroe, Texas. Such an approach entails explicitly political commitments, with attention to the ways power operates between differently situated bodies, across different scales (Nelson & Seager, 2005). It works to make visible the often naturalized processes that beget social inequality, often manifested in "interlocking violences" (Faria & Mollett, 2018). Additionally, feminist geographers have noted the ways that uneven power relations are *spatially* as well as bodily produced; that is to say, *place* matters in the positioning of bodies in social hierarchies (Nelson & Seager, 2005, 8).

Feminist geopolitical geographers have worked to trouble binary oppositions often deployed in conventional geopolitical research, instead arguing that such conceptual spheres are inseparable from one another (Hyndman, 2001; Mohammad, 2001; Pratt & Rosner, 2006). In complicating the dualistic conception of "global" and "local", feminist geopolitical geographers can trace the way power moves across scales, shifting analytical gaze away from "big", abstract political actors (i.e. the nation-state) where power is expected to reside, to concrete and particular places, moments, and bodies in the realm of everyday life (Dowler & Sharp, 2001; Massaro & Williams, 2013; Mountz, 2010; Mountz & Hyndman, 2006). Such an approach acknowledges the material, emotion-laden manifestations of differential power relations between various people, communities, and places, tracing the contours of uneven power distribution across spaces and scales (Tolia-Kelly, 2004, 2008).

Feminist geopolitical geographers have made significant inroads in tracing the impacts and material mechanics of immigration enforcement policies and geopolitical processes on the ground, as well as the intersecting discourses that facilitate their enactment. For instance, borders have been reconceived by feminist geopolitical geographers as outsourced locations, beyond the national territory, where refugees, detainees, and other prisoners might be confined indefinitely, out of sight of the national public and out of reach of rights and citizenship (Loyd et al., 2016; Dickson, 2015; Mountz, 2015). Conversely, in her article on the US-Mexico border, Williams (2014) illuminates the ways that the state strategically reconciles calls for greater humanitarian care for migrants with its own agenda for increased securitization of the region, ultimately remaking the border as an increasingly militarized space through the intersection of these discourses.

Borders have been theorized not only as spaces that mark or buffer the line between nation-states, but also as temporal moments, actively produced, in state bureaucratic processes and institutions (Carte & Torres, 2014; Conlon, 2011; Mountz, 2010; Torres, 2018). In these labyrinthian journeys, migrants' bodies, emotions, and narratives are often the price of passage. This focus on bodies, embodiment, and mobility is also apparent in feminist geopolitical scholars' work on identity and surveillance, in which borders shift as particular raced and gendered bodies move across spatial thresholds, perpetually positioned outside of full citizenship and rights (Hyndman & de Alwis, 2005, 2007; Fluri et al., 2015).

Feminist geopolitical geographers have also contributed significantly to scholarship on immigrant detention, which lies at the intersection of literature on state-bordering projects as well as prison geographies (Bonds, 2009; Bosworth, 2010; Gilmore, 2007; Martin & Mitchelson, 2009). Within this literature, immigrant detention is often discussed as a state strategy to isolate and obfuscate the presence of detainees, an instrument of neoliberal policy that has spread around the world (Belcher & Martin, 2013; Dickson, 2015; Flynn, 2014; Gill, 2009; Hyndman & Mountz, 2008; Mountz, 2012; Mountz et al., 2013). Some scholars have attempted to illuminate the inner workings of detention centers, paying attention to the intimacies, micro-economies, and politics that emerge through and within these facilities (Bosworth, 2014; Bosworth & Slade, 2014; Bruzzone, 2016; Conlon & Hiemstra, 2014), as well as the emotions and imaginings of detainees in relation to their sense of home, family, and belonging (Hiemstra, 2012; Lietart et al., 2014).

In this article, we highlight the ways that detention centers are produced not only in the abstract, through state policies of isolation and invisibility (Mountz, 2012, 2015), but also in everyday debates and discourses, circulating on local and intimate levels amongst community residents as well as family members of the detained. This study demonstrates how the immigrant detention centers in Conroe, while not located in a rural or as remote setting like many immigration control facilities and prisons studied (Bonds, 2009; Gilmore, 2007; Hyndman & Mountz, 2008; Lawson et al., 2008, 2010; Mountz, 2012), do remain out of sight (with regard to Conroe residents) and out of reach (with regard to detainees' family members, attorneys, and advocacy groups), in part due to their increasing privatization over time, as well as the desires of local residents to minimize Conroe's association with carceral industries. This research expands on the work of feminist geographers who work to locate and ground state practices in specific places, with regard to their impacts of specific communities (Coleman & Stuesse, 2016; Hyndman & Mountz, 2008). Through the juxtaposition of disparate discourses surrounding detention construction and operation in Conroe, this research offers a glimpse as to the ways that tactics of concealment and privatization facilitate the acceleration of immigrant detention in Texas.

Placing Conroe, Texas on the immigrant detention landscape

Although people driving through Conroe along the I-45 highway might see the community as merely an extension of Houston's monstrous suburban sprawl, residents describe Conroe as a city in its own right, with an estimated population of over 80,000 residents[2] and a thriving local economy (US Census Bureau, Conroe, TX). The city was founded in the 1800s, originally the site of lumber industries, a sawmill and railroad stop on the International-Great Northern Railroad's Houston-Crockett line (Jackson, 2010). In the 1930s, discovery of oil in the area boosted the then economically depressed economy, and following the construction of highway I-45, more oil-related manufacturing and petrochemical businesses came to the area (Jackson, 2010). Today, the city markets itself as an ideal retirement or second home destination, citing its proximity to wildlife (Sam Houston National Forest and Lake Conroe), its historical significance, and its budding entertainment and service industries as incentives for purchasing property there.[3]

Unlike many detention hosting sites in Texas, Montgomery County's poverty rate is lower than the national average, and in 2016 Conroe was one of the fastest growing cities with over 50,000 residents in the US (US Census Bureau, 2017; Rhor & Harden, May 25, 2017). In interviews, several residents described the county as economically robust and diverse, naming the manufacturing, oil and gas, and service industries as the largest opportunities of employment for local residents. From an economic standpoint, Montgomery County does not match assumptions about the siting of carceral facilities in extremely impoverished, neglected rural hinterlands.

Demographically, Conroe is majority white/non-Hispanic (estimated at 55%, according to US Census Bureau statistics), with about 33% of residents identifying as ethnically Hispanic or Latino, 9% identifying as black, and the remainder identifying as Native American, Asian, or two or more races (US Census Bureau, Conroe, TX). Nearly 29% of residents surveyed reported that they spoke a language other than English at home, and about 18% self-reported as being foreign-born. From a political standpoint, Montgomery County is considered very conservative: in the 2016 presidential election, 74% of county residents voted for Trump, and in the 2018 midterm election, 78% of straight ticket votes went to the Republican party (*Politico* , 2016 Presidential Election Results; Montgomery County Cumulative Report, 2018).

In June 2008, the Joe Corley Detention Facility opened in Conroe. The facility was initially owned and operated by the county, but following its sale in 2013, it is now owned and managed by the GEO Group, a for-profit incarceration company that already had strong ties in the county (Roden, May 6, 2013). GEO, which has prison management contracts across the globe, has also diversified its business to include ownership and management of mental health treatment centers, supervised release programs, and rehabilitation programs in addition to immigrant detention (Shahani, March 25, 2011). The GEO Group has earned a reputation for rampant mismanagement and neglect of incarcerated populations, which has incited numerous lawsuits against the company in recent years (ibid.; Holpuch, Nov. 28, 2018). In a 2016 report, the US Department of Justice found that privately run prisons, including the GEO Group, are far more likely than government run facilities to have high incidences of guard use-of-force, inmates' self-harm, and inmates' suicide (US DOJ, 2016).

At different times, the Joe Corley facility has housed both Bureau of Prisons (BOP) inmates and Immigration and Customs Enforcement (ICE) detainees, as it shares some infrastructure with the BOP-contracted Montgomery County jail next door (also operated by the GEO Group). The Joe Corley facility is located in what is referred to as the "law enforcement complex" of Conroe – an isolated yet central area of the city that houses the aforementioned county jail as well as the Montgomery County Mental Health Treatment Center (MCMHT) (also owned and managed by the GEO Group) (Flake January 22, 2011). The so-called "law enforcement complex" is situated along Hilbig Road, a dead-end street just south of Loop 336; the short road lies perpendicular to the city's old railroad tracks as well as the Stewarts Creek watershed. The complex is shrouded with tree coverage, invisible from main roads despite its proximity to them.

Although initially welcomed to the community over a decade ago as a means to create local jobs and generate county revenue (Roden Feb. 10, 2007), recently there has been significant local debate regarding the privatization of these carceral facilities, the opacity of their operation, as well as the unsavory optics of their placement in Conroe. These debates intensified following the announcement that a *second* detention center – the Montgomery Processing Center – would be constructed in the same "law enforcement complex" in the city, this time without any county funds (Jordan April 15, 2017). This 1,000 bed facility would be the first new detention center under Trump's administration, a monument to the Executive Office's expanded immigration enforcement plans.

Discursive analysis of community response to immigrant detention facilities

We use critical discourse analysis as our principal method for research. This approach "primarily studies the way social power, abuse, dominance and inequality are enacted, reproduced, and resisted by text and talk in the social and political context" (van Dijk, 2015, 466). We follow Rose's assertion that discourse exposes "a particular knowledge about the world which shapes how the world is understood and how things are done in it" (2001, 136). According to Rose, "[d]iscourses are articulated through all sorts of visual and verbal images and texts . . . and also through the practices that those languages permit" (2001, 136). Intertextual meaning can be drawn through a multitude of forms or texts, in circulation together. Both the structure of "statements" as well as the social positioning of the person enacting them are analyzed in discourse analysis:

> [i]n relation to written or spoken discourses, discourse analysis is interested in, for example, how a particular discourse describes things (although the power of discourse means that it produces those things it purports to be describing), in how it constructs

> blame and responsibility, in how it constructs stake and accountability, in how it categorizes and particularizes.
>
> (*Rose, 2001, 150)*

Discourse analysis of Facebook posts, specifically, has been used by scholars in research of public political discourse in the US and abroad (Chiluwa & Ifukor, 2015; Felicia, 2018; Robertson et al., 2010). Scholars interpret key or repeated phrases or themes in context, conducting a qualitative analysis of posts and comments on specific Facebook pages.

Drawing on the example of these discourse scholars, we identify narrative patterns and rhetorics deployed in two discursive spheres: online news articles from *The Courier of Montgomery County*, the local daily county newspaper currently owned by *The Houston Chronicle* parent company (Lewis, July 29, 2016), which prints both English and Spanish language articles, and first-person accounts regarding access to the detention facilities in Conroe, posted to the "Joe Corley Detention Center" Facebook page, which includes comments and posts in both Spanish and English. The latter website is *not* an "official" Facebook page managed by the GEO Group or its employees; rather, it is an "unofficial" page where anyone can post, comment, or ask questions of other Facebook users. Many of the users on the "Joe Corley Detention Center" page are Spanish-speaking family members of the detained, immigration attorneys, and occasionally former detainees and employees of the facilities.

In our analysis of local news discourse, we reviewed 162 articles from 2004 to 2017, using search terms "The GEO Group" and "immigrant detention" to locate relevant publications within the news platform. Our analysis of the social media discourse was less structured, as the platform encourages sporadic, fragmented conversations, often evolving simultaneously on multiple threads on the page, making narratives difficult to trace from "beginning" to "end". The earliest posts begin in August 2011, and continue to the present. Statements reviewed include posts to the "Joe Corley Detention Center" page, comments from other users on those posts, as well as "reviews" of the facility. Analyzing posts with high levels of engagement from other users, we were able to identify trends in posts from family members of detainees, many of whom were concerned with how they might contact or visit their loved ones in the Joe Corley facility. In this chapter, original names have been changed to pseudonyms to protect the identities of the interlocutors.

Finally, we supplement these discourse analyses with interviews with local residents, business owners, and government officials from fieldwork in Conroe, Texas in the summer of 2018. These interviews corroborate findings in the discourse analysis, in some cases adding further explanation for the diverging motivations and sentiments between the discursive spheres.

Contesting invisibility of immigrant detention landscapes in Conroe, Texas

Based on our discourse analysis described earlier we identified three main discursive themes related to the (in)visibility of the detention facilities: (1) concern over the opacity of the facilities' privatization, management, and development over time (apparent in the online news discourse); (2) apprehension over lack of transparency in physically accessing and administratively navigating the detention facility (apparent in the Facebook page postings); and (3) unease over the optics of housing carceral facilities within the community (apparent in the online news discourse). While there is overlap among different community stakeholders regarding the lack of transparency and accessibility in the first two themes, the third marks a major difference between the positionings, values, and perspectives of the interlocutors in these two discursive spheres. Interviews with local residents, business owners, and government officials offer further insight into the divides between interlocutors in the online newspaper discourse, which tends to focus on the concerns of county residents, and the "Joe Corley Detention Center" discourse, which tends to focus on the concerns of family of the detained.

In the following section, we trace the debates circulating in online news and interviews related to the opacity of detention privatization and management in Conroe. We then explore the sentiments, complaints, and concerns with facility inaccessibility, as expressed in the intimate discourses between family members and friends of detainees, posted to the "Joe Corley Detention Center" Facebook page, and supplemented with interviews from immigration advocates. Finally, we juxtapose the conversations circulating in these two discursive spheres, examining their overlaps and divergences, identifying the tensions that characterize differences in interlocutors' sentiments and motivations.

Opacity of detention facility development, management, and privatization in Conroe

In 2005, plans were first developed and announced to allow the GEO Group to take over management of the Montgomery County jail, as well as the soon-to-be-constructed Joe Corley Detention Facility. At the time, local news articles in *The Courier of Montgomery County* framed the decision as a means to "help the county save money" (April 27, 2005), and as a possible new "revenue-producing avenue" (Thackeray April 25, 2005). Earlier statements also indicate that "a major lure for the county is the fact that the private companies are fully insured, freeing the county of financial responsibility" (Thackeray Dec. 18, 2004). By 2008, when the new Joe Corley facility was completed, it was understood that GEO's management of the facility would be a "very good deal for local taxpayers [as stated by County Commissioner Breazeale]" (Meyer June 26, 2008).

However, in early 2009, *The Courier* began publishing articles from other prison-hosting towns in Texas where riots over prisoner mistreatment had broken out, ultimately causing the contracts with BOP and ICE agencies to be revoked. That year the paper published an article stating:

> A bill making its way to the House floor would provide much-needed transparency at privately run jails and correctional facilities in Texas. . . . While there have been no problems at the new Conroe facility so far, we believe the bill opens the door for accountability when GEO's contract comes up for renewal in two years. . . . Montgomery County citizens have a right to know what is going on behind those fences, who is watching the criminals and in whose pockets the money going back and forth ultimately lands.
>
> *(May 3, 2009)*

At the same time that concern over the operation of these privately run facilities developed, parallel concerns over their financing began to surface. In 2012, discussions circulated regarding the need to refinance – and ultimately to sell – the Joe Corley Detention Facility after its tax-exempt status was revoked. In response to this possibility, another editorial in *The Courier* was published, stating,

> We don't believe the county should sell out so soon. We want a county that is responsible for – and responsibly handles – its facilities and the financial matters that go with

> them. . . . It is up to our commissioners to make the necessary corrections and improvements in order to ensure that these facilities are run effectively and above-board.
>
> *(September 22, 2012)*

Shortly thereafter, it was announced that GEO purchased the Joe Corley Facility from Montgomery County, paying $10 million over the original asking price (Roden May 4, 2013). It was suspected that other "unnamed properties" were included in GEO's purchase proposal, although that "information [was] exempted from public release" (April 27, 2013). The news discourse indicates that the public had not been informed of all the details of GEO's proposal for purchase of these facilities, and that such information was desired by county residents before the sale was finalized (April 27, 2013).

Despite this concern, it was only after the sale was approved by the county commissioners' court that the public learned that the GEO Group had "propos[ed] the development and operation of a [second] federal detention facility – with as many as 1,500 beds – located on a 23.75-acre site adjacent to the Joe Corley facility", purchased from a private estate in July 2010 (Roden May 18, 2013). According to the news discourse, the county's pledged support for this second detention center came as a surprise to county residents: "Montgomery County commissioners voted their support for a second federal prison in Conroe, but the facility didn't receive a ringing endorsement from at least one county resident" (Roden May 20, 2013). One county resident and vice president of the Texas Patriots PAC is quoted as stating,

> I don't think the citizens of this county would have ever agreed to this had we honestly come forward and told them we would build a jail for a third party with no control over the type of prisoners in the jail [Joe Bauman].
>
> *(Roden May 20, 2013)*

He goes on to say that the sale of the Joe Corley facility and agreement to support GEO's construction of a second detention center is "really outrageous and not very transparent" (Roden May 20, 2013).

As political controversy over the misrepresentation of county financing of the construction of the Joe Corley Detention Facility ensued, in a guest-authored column, one writer posed the question,

> Who won here? Not the county tax payers, as they are facing a bill that they shouldn't have had to pay [the debt from construction of the Joe Corley facility]. Not the county commissioners, unless GEO executives or GEO acting through its PAC contributed to the commissioners. . . . The straight answer is the big winner is GEO. Getting approval to build privately owned correction facilities in the middle of a suburban community nearby to a city as large as Houston is difficult if not impossible. GEO will now own two. This will inure to their economic benefit.
>
> *(O'Sullivan Nov. 20, 2014)*

This suspicion of the decision to allow GEO further entrance into Conroe is expressed in a statement by the Conroe city mayor ("I don't think that we need any more detention centers in Conroe"; Jordan April 15, 2017), and as well as in Letters to the Editor, published after the official announcement that a $110 million contract from ICE had been awarded to GEO. One county resident writes: "I am disturbed that the federal government is imposing a $110 million detention center here in our county without first informing our senators. . . . Conroe mayor or sheriff" (April 24, 2017). Another asks, "How many residents *were aware* that Montgomery County would become entangled with a for-profit prison system with a history of abuse and mismanagement[?]" (April 26, 2017).

These statements point to troubling trends in detention center development: the intentional obscuring of information regarding their sale, construction, and management, leaving local populations out of the loop, without means to contest such projects until they are already underway. This corroborates Mountz's assertion that, on an abstract level, immigration enforcement can be hypervisible in the public imaginary, even while its mechanics and material impacts (on local communities, and especially on the people detained) are consistently hidden from view (Mountz, 2015, 2017). Although Mountz's conclusion is based on her research of detention centers physically removed from mainland populations through their construction on islands, the case of immigrant detention in Conroe points to the process of obfuscation through *privatization*. The GEO Group slowly obtained more control in Conroe, at first managing the county's jail and the Joe Corley Detention Facility, later purchasing "Joe Corley" from the county, and ultimately choosing to build a *new* immigrant detention facility on private land despite noticeable pushback from county residents.

In an interview, a mayoral assistant compared the development of the new detention center to that of a shopping center, or other private development: because it would be constructed without county funding, the GEO Group did not need local permission to construct it. Similarly, when asked about local tensions surrounding the construction of the new immigration detention facility, the Chief of Staff of Judge Doyal, the Chief Executive Officer of county government and head of the County Commissioners' Court, stated that in Montgomery County a company or developer can choose to construct whatever type of facility he or she wants on private lands, as long as the development complies with safety building codes. He also emphasized that there is no zoning in Montgomery County, and thus a developer might build whatever type of business he or she wants on private property. These statements reveal the ways that public discourses surrounding detention in Conroe are redirected through rhetorics supportive of private enterprise.

Originally touted as a means of promoting local community development, the immigrant detention industry ultimately becomes a point of contention in Conroe, shielded from public critique because of its private development status. The movement towards privatization of carceral management and ownership demonstrates a type of "neoliberal restructuring" that ultimately fails to promote community development (Bonds, 2006). Furthermore, the focus on the economic potential (or failure) of carceral industries in Conroe serves to direct discourse away from the lives of migrants held within the facilities, ignoring an ugly reality that undergirds some locals' rejection of the facilities' placement in Conroe. As in the case of prison siting (Bonds, 2012), immigrant detention center placement depends upon the erasure of the bodies housed in such facilities, framing discussion in terms of local economic gains and private enterprise rather than in terms of human rights.

Opacity of access and information for navigating detention facility administration

Although grounded in distinct perspectives and positionings, the discourses posted on the "Joe Corley Detention Center" Facebook page reveal parallel frustrations with the opacity of detention facility management and operation, albeit on a more intimate level. In the "Reviews" section of the Facebook page, one woman remarked that she was having difficulties visiting her brother due to the facilities' requirements to be placed on a "list" by detainees, prior to visitors' arrival (Maria Karli, n.d.) She asked for advice on where to find the address of the detention center in order to send a letter to her detained relative. In the ensuing comments posted by others, she wrote that she had been unable to get useful or accurate information by calling the facilities directly or by looking at the detention center website (which, she had been informed, does not list correct information regarding communication or visiting procedures).

Others responding to her post posed similar questions and complaints. Another woman stated that she also wanted to write a letter to someone detained, but lacked the corresponding "A-number",[4] and thus her letter was returned to her (Idelsa Santiago, n.d.). Someone else inquired how she might write to her son, also detained at Joe Corley; she stated that she had no way of getting to the facility, and has not been able to find the appropriate telephone number to call to make a visit (Estela Vasquez, 2015). Another woman in the same thread posted that she has received phone calls from a detained friend, but has been unable to speak with her because a "code" is required to facilitate the call. She was uncertain what the code might refer to, or how or where to retrieve the appropriate information (Marisol Beatriz, n.d.).

On the whole, posted questions and exchanges of advice, phone numbers, and addresses on the Facebook page point to a lack of clarity and difficulty in navigating the Joe Corley Detention Facility administration. Information regarding procedures for visiting, making, and accepting phone calls with detainees, as well as the movement of detainees from one site to another are not learned easily. Additionally, information that *is* made publicly available – such as the information posted on the facility's website – was reported among these interlocutors to contradict information gathered elsewhere. The obscuring of information from family and friends of detainees works to isolate the persons detained from community and advocacy networks (Mountz, 2012). As Mountz writes, "migrants are hidden from view in multiple ways, not only locked behind bars, but omitted from or hidden within records and statistics" (2012, 95). By making access to detainees confusing and difficult, detention practices continue with minimal monitoring, and detainees are left without legal, financial, and emotional support of their loved ones. Carte (2014) terms these subtle, insidious, and sometimes extra-official tactics of withholding information, misinformation, and arbitrary bureaucratic barriers, as "everyday restriction", which can have devastating consequences for detainees and their families.

Converging contestation and diverging emotional encounters with immigrant detention landscapes: optics vs. attachments

The two discursive spheres we examined – that of local residents and that of family members of the detained – converge in their contestation of the invisible development and administration of detention landscapes in Conroe. They intersect with the first group's concern regarding the opacity of detention development and management, and the latter's apprehensions over detention inaccessibility to support detainee family members. Nevertheless, the interlocutors diverge significantly in their perspectives, positionings, and emotional attachments to the detention landscape discussed and encountered. As previously stated, the discourse among local residents primarily focuses on economic development, and private management of the facilities; the lived realities of migrants are seldom referred to, except in terms of the disrepute of the GEO Group in managing carceral facilities. On the other hand, discourse among detainees' families seeks to breach the emotional isolation imposed by detention walls; they work to form connections within their dispersed community, offering emotional support, encouragement, and a space to voice frustration and anguish with the violence of separation through detention and deportation.

Among county residents, outrage with lack of transparency regarding the growth of GEO's influence in the community is frequently voiced. However, it is primarily coupled with a concern for town reputation rather than for the populations who are detained. For instance, in her Letter to the Editor, Bates writes that

> the GEO Group already owns and operates the Joe Corley Center, which on April 13 was the subject of a front page article in the Houston Chronicle for the abhorrent treatment of detainees. Such a facility and the reputation it brings with it does not enhance the reputation of The Woodlands, Conroe, or Montgomery County.
>
> *(Bates April 24, 2017)*

In his letter, another city resident asks,

> Why should the people of Montgomery County go along with this attempt to turn the city into a prison town? Why should the people of Montgomery County accept the depressed property values and corresponding loss to the tax base that will surely result?
>
> *(Miller May 3, 2017)*

These opinions echo earlier statements made in a *Courier* editorial regarding the "failure to release information" regarding the GEO Group's offer to build a second facility: "It's not the type of growth Conroe residents should want, or the city of Conroe needs. . . . What a selling point to those interested in buying one of the half-million-dollar homes being built in Conroe. . . . What will Conroe become? Con-vict-roe?" (June 3, 2013).

Mirroring county residents' concern with the optics of becoming a "prison town", which might depress property values and mar the communities' profile as a "playground" destination (Lawson et al., 2010) are detention center advocates' claims that the facilities are tastefully kept out of sight. In response to the many objections to the new facilities constructed, one pro-detention center Conroe resident writes, "How many people in Conroe and Montgomery County are even aware that this facility exists and if told to go there how many would know how to get there? This facility is in a very remote area and is a threat to no one" (Martelon April 29, 2017).

Not only are the facilities described as "out of the way" and low-profile, low-impact, but the detainees are also framed as non-threatening, in part due to their transitory status. In one article, Montgomery County Judge Doyal is quoted as stating, "These are immigrant detainees. It's not like a prison where you have hardened criminals there" (Jordan April 15, 2017). In his Letter to the Editor, Martelon stated something similar, writing that "this is a federal detention center, not a jail or a prison. . . . Most of the detainees spend a few months at Joe Corley awaiting hearing and/or trial and then are released or transferred" (Martelon April 29, 2017).

These counter-arguments for the continued use of detention facilities pander to concerns over the optics of developing a prison industry in Conroe. While city residents do occasionally voice concern for the mistreatment of detainees at the hands of private management, they couple these complaints with concern for depreciating property values associated with carceral landscapes. Unlike the intimate discourses between family and friends of detainees, for whom the emerging detention landscape inspires feelings of isolation and desired connection to those detained, the discourses of county residents focus on Conroe as territory conquered by federal or private interests, against the will of local residents, resulting in the "pollution" of the city's prosperity through the presence of "criminal" inmates or "alien" detainees (Sundberg, 2008). Discourses like this rhetorically work to distance Conroe residents from detention landscapes as "dumping grounds" – spaces "trashed" by the industries they contain (Lawson et al., 2010; Sundberg, 2008).

Encountering the same detention landscape, marked by opacity and inaccessibility for both groups of interlocutors, the two discourses diverge in the emotional significance and connections that motivate and direct opposition to the practice of immigrant detention. These diverging emotional encounters might be attributed to the social positioning of the speakers, and their perceived relationship to the detained. Implicit in county residents' statements are notions that the people detained in these facilities – disproportionately black and brown populations – do not belong in Conroe. One woman interviewed in Conroe stated that she believed that most Conroe residents would not be sympathetic to the plight of people detained in these facilities, citing the conservative political leanings of the county. A mayoral assistant stated that he believed residents may have resisted the presence of detention facilities because it would bring more people "from outside" into their community. Both a fear of conflation with the carceral facilities, and fear of the facilities' spillover into the community, motivate residents' rejection of their operation.

Conclusion

By examining detention centers from the local and intimate perspectives of those just outside their walls, we can understand how state tactics of invisibility and dislocation (Mountz, 2012, 2015) are enacted through detention centers' privatized ownership and management, opaque administration, as well as the diverging discourses they generate among different stakeholder groups. These tactics work in tandem to keep local residents in host communities, as well as friends and families of detainees, at a distance, without opportunities to interfere or disrupt dominant narratives of immigrant detention.

Although it is possible to trace overlaps between these two groups of interlocutors – county residents frustrated with the privatization and growth of the detention industry in Conroe, on the one hand, and family and friends of detainees, navigating an opaque detention system, on the other – it is important to remember the differences in their emotional registers regarding the practice of immigrant detention. While the discourses of family members of the detained center on accessing the detention facilities and connecting with those who are incarcerated, county residents' discourses focus on the imposition of outsider's infrastructure in their community, as well as the optics of placing such unsavory business in their prosperous suburban community (Sundberg, 2008; Lawson et al., 2010). Implicit in their rejection of the new detention facility is a fear of "other" bodies – predominantly black and brown bodies – who, in the view of some residents, run the risk of turning Conroe into "Con-vict-roe" (June 3, 2013).

Is it possible that an advocacy opportunity exists at the intersection of these discourses, or are positionality, interests, and motivations between the two groups of speakers too diverging for any shared outcome? Further research might be conducted to assess the potential for recasting anti-detention arguments, centering on the people detained rather than the infrastructure imposed.

Acknowledgements

This research was made possible through the generous support of a Veselka Fieldwork Fellowship from the Department of Geography and the Environment at the University of Texas at

Austin. We are grateful for the thoughtful comments on earlier drafts provided by Dr. Sarah Lopez and her Migratory Urbanism seminar, as well as the Feminist Geographies class taught by Dr. Caroline Faria.

Notes

1 See "Joe Corley Detention Center" Facebook page at www.facebook.com/pages/Joe-Corley-Detention-Center/141224869247031.
2 Based on US Census Bureau estimates for year 2017.
3 See City of Conroe's tourism website for more: www.visitconroe.com/.
4 An A-Number, also known as an Alien Registration Number or Alien Number, is a unique eight- or nine-digit number assigned to a noncitizen at the time their A-File is created. An A-Number is the letter "A" followed by eight or nine numbers (such as A012345678). (see USCIS website for more)

References

(2005, April 27). Our view: competition might just work – even without jail privatization. *The Courier of Montgomery County*. Retrieved from www.yourconroenews.com/neighborhood/article/Our-View-Competition-might-just-work-even-9809617.php

(2009, May 3). Bill to provide accountability at privately run jails. *The Courier of Montgomery County*. Retrieved from www.yourconroenews.com/opinion/article/Bill-to-provide-accountability-at-privately-run-9248849.php

(2012, September 22). Commissioners must fix woes at treatment, detention centers, without selling out. *The Courier of Montgomery County*. Retrieved from www.yourconroenews.com/opinion/article/Commissioners-must-fix-woes-at-treatment-9268512.php

(2013, April 27). Questions remain over GEO's proposals for county facility. *The Courier of Montgomery County*. Retrieved from www.yourconroenews.com/opinion/article/Questions-remain-over-GEO-s-proposals-for-9259913.php

(2013, June 3). Conroe doesn't need, shouldn't want another federal prison [Editorial]. *The Courier of Montgomery County*. Retrieved from www.yourconroenews.com/opinion/article/Conroe-doesn-t-need-shouldn-t-want-another-9280913.php

(2014, August 12). Letters to the editor. *The Courier of Montgomery County*. Retrieved from www.yourconroenews.com/opinion/article/Praying-for-the-best-along-the-border-9492142.php

(2017, April 24). Letters to the editor. *The Courier of Montgomery County*. Retrieved from www.yourconroenews.com/neighborhood/woodlands/news/article/Letters-to-the-editor-Township-should-approve-11096072.php

(2017, April 26). Letters to the editor. *The Courier of Montgomery County*. Retrieved from www.yourconroenews.com/neighborhood/moco/opinion/article/Letter-Show-some-courage-Repeal-Obamacare-11102053.php

Beatriz, M. (n.d.) *[Facebook Thread Response Posted on "Joe Corley Detention Center" Page]*. Retrieved from www.facebook.com/pages/Joe-Corley-Detention-Center/141224869247031.

Belcher, O., & L. Martin. (2013). Ethnographies of closed doors: conceptualizing openness and closure in US immigration and military institutions. *Area*, 45(4), 403–410.

Bonds, A. (2006). Profit from punishment? The politics of prisons, poverty and neoliberal restructuring in the rural American Northwest. *Antipode*, 38(1), 174–177.

Bonds, A. (2009). Discipline and devolution: constructions of poverty, race, and criminality in the politics of rural prison development. *Antipode*, 41, 416–438.

Bonds, A. (2012). Building prisons, building poverty: Prison sitings, dispossession, and mass incarceration. In J. Loyd, M. Mitchelson, & A. Burridge (Eds.), *Beyond Walls and Cages: Prisons, Borders, and Global Crisis* (pp. 129–142). Athens: University of Georgia Press.

Bosworth, M. (2010). The new detention: securing the border. In *Explaining U.S. Imprisonment*. Los Angeles: Sage.

Bosworth, M. (2014). *Inside Immigration Detention*. New York: Oxford University Press.

Bosworth, M., & G. Slade. (2014). In search of recognition: gender and staff-detainee relations in a British immigration removal centre. *Punishment & Society*, 16(2), 169–186.

Bruzzone, M. (2016). On exterior and interior detention regimes: governing, bordering, and economy in transit migration across Mexico. In N. Hiemstra & D. Conlon's (Eds.), *Intimate Economies of Immigrant Detention: Critical Perspectives* (pp. 105–120). London, UK: Routledge.
Carte, L. (2014). Everyday restriction: Central American women and the state in the Mexico-Guatemala border city of Tapachula. *International Migration Review*, 48(1), 113–143.
Carte, L., & R. Torres (2014). Role playing: a feminist-geopolitical analysis of the everyday workings of the Mexican state. *Gender, Place & Culture*, 21(10), 1267–1284.
Carter, D., & H. Merrill. (2007). Bordering humanism: life and death on the margins of Europe. *Geopolitics*, 12(2), 248–264.
Che, D. (2005). Constructing a prison in the forest: conflicts over nature, paradise, and identity. *Annals of the Association of American Geographers*, 95, 809–831.
Chiluwa, I., & P. Ifukor. (2015). 'War against our children': stance and evaluation in *#BringBackOurGirls* campaign discourse on Twitter and Facebook. *Discourse and Society*, 26(3).
Coleman, M. (2007). Immigration geopolitics beyond the Mexico – US border. *Antipode*, 39(1), 54–76.
Coleman, M. (2009). What counts as the politics and practice of security, and where? devolution and immigrant insecurity after 9/11. *Annals of the Association of American Geographers*, 99(5), 904–913.
Coleman, M. (2012a). The "local" migration state: the site-specific devolution of immigration enforcement in the U.S. south. *Law & Policy*, 34(2), 159.
Coleman, M. (2012b). Immigrant il-legality: geopolitical and legal borders in the US, 1882-present. *Geopolitics*, 17(2), 402–422.
Coleman, M., & A. Kocher. (2011). Detention, deportation, devolution and immigrant incapacitation in the US, post 9/11. *The Geographical Journal*, 177(3), 228–237.
Coleman, M., & A. Stuesse, (2016). The disappearing state and the Quasi-event of immigration control. *Antipode*, 48(3), 524–543.
Conlon, D. (2011). Waiting: Feminist perspectives on the spacings/timings of migrant (im)mobility. *Gender, Place and Culture*, 18, 353–360.
Conlon, D., & N. Hiemstra. (2014). Examining the everyday micro-economies of migrant detention in the United States. *Geographica Helvetica*, 69, 335–344.
Cullen, T. (2018, March 13). ICE released its most comprehensive immigration detention data yet. It's alarming [Press Release]. *National Immigrant Justice Center*. Retrieved from: https://immigrantjustice.org/staff/blog/ice-released-its-most-comprehensive-immigration-detention-data-yet.
Dickson, A. (2015). Distancing Asylum seekers from the state: Australia's evolving political geography of immigration and border control. *Australian Geographer*, 46(4), 437–454.
Dikec, M. (2009). The 'where' of Asylum. *Environment and Planning D: Society and Space*, 27, 183–189.
Dowler, L., & J. Sharp. (2001). A feminist geopolitics? *Space and Polity*, 5(3), 165–176.
Faria, C., & S. Mollett. (2018). The spatialities of intersectional thinking: fashioning feminist geographic futures. *Gender, Place & Culture*, 25(4), 565–577.
Felicia, O. (2018). Discursive strategies in Nigeria's 2015 Facebook campaign discourse. *Discourse & Society*, 29(5).
Flake, N. (2011, January 22). Commissioners set to OK top mental health facility officials. *The Courier of Montgomery County*. Retrieved from www.yourconroenews.com/news/article/Commissioners-set-to-OK-top-mental-health-9244952.php
Fluri, J., P. Jackson, & D. Paudel. (2015). A new development technology? South Asian biometrics and the promise of state security and economic opportunity. *Geography Compass*, 9, 539–549.
Flynn, M. (2014). There and back again: on the diffusion of immigration detention. *Journal on Migration and Human Security*, 2(3), 165–197.
Gill, N. (2009). Governmental mobility: the power of the movement of detained asylum seekers around Britain's detention estate. *Political Geography*, 28(3), 186–196.
Gilmore, R. W. (2007). *Golden Gulag: Prisons, Surplus, Crisis and Opposition in Globalizing California*. Berkeley, CA: University of California Press.
Glasmeier, A. K., & T. Farrigan. (2007). The economic impacts of the prison development boom on persistently poor rural places. *International Regional Science Review*, 30(3), 274–299.
Hiemstra, N. (2012). Geopolitical reverberations of US migrant detention and deportation: the view from Ecuador. *Geopolitics*, 17, 293–311.
Hoffman, M. (2017, April 13). Immigrant center to be built in Texas for $110M. *U.S. News*. Retrieved from www.usnews.com/news/best-states/texas/articles/2017-04-13/texas-getting-first-immigrant-lockup-built-under-trump

Holpuch, A. (2018, November 28). Private prison companies served with lawsuits over using detainee labor. *The Guardian*. Retrieved from www.theguardian.com/us-news/2018/nov/25/private-prison-companies-served-with-lawsuits-over-usng-detainee-labor.

Hyndman, J. (2001). The field as here and now, not there and then. *Geographical Review*, 91(1–2), 262–272.

Hyndman, J., & M. de Alwis. (2005). Performing the pass: conflict, mobility and displacement in Sri Lanka. In B. P. Lorente, N. Piper, S. Hsiu-Hua, & B. S. A. Yeoh (Eds.), *Asian Migrations: Sojourning, Displacement, Homecoming & Other Travels* (pp. 25–49). Singapore: Asia Research Institute.

Hyndman, J., & M. de Alwis. (2007). Bodies, shrines, and roads: violence, (im)mobility and displacement in Sri Lanka. *Gender, Place & Culture*, 11(4), 537–557.

Hyndman, J., & A. Mountz. (2008). Another brick in the wall? Neo-refoulement & the externalization of Asylum by Australia and Europe. *Government & Opposition*, 43(2), 249–269.

Jackson, C. C. (2010). Conroe, TX. *Handbook of Texas Online*. Retrieved from www.tshaonline.org/handbook/online/articles/hec03

Jordan, J. (2017, April 15). Mayor: no need for new immigration detention facility in Conroe. *The Courier of Montgomery County*. Retrieved from www.yourconroenews.com/news/article/Mayor-No-need-for-new-immigration-detention-11076005.php#photo-12727778

Klein, A., & L. Williams. (2012). Experiences of migrants released from detention centres in the UK. *Population, Space and Place*, 18(6), 741–753.

Lawson, V., L. Jarosz, & A. Bonds. (2008). Building economies from the bottom up: (mis)representations of poverty in the rural American Northwest. *Social & Cultural Geography*, 9(7), 737.

Lawson, V., L. Jarosz, & A. Bonds. (2010). Articulations of place, poverty and race: dumping grounds and unseen grounds in the rural American Northwest. *Annals of the Association of American Geographers*, 100(3), 655–677.

Leekes, A., R. van Os, & E. Boersema. (2017). What drives 'soft deportation'? Understanding the rise in assisted voluntary return among rejected asylum seekers in the Netherlands. *Population, Space and Place*, 23(8).

Lewis, A. (2016, July 29). Hearst purchases community newspapers across Houston's suburbs. *Houston Chronicle*. Retrieved from: www.chron.com/neighborhood/katy/news/article/Hearst-purchases-community-newspapers-across-9882614.php

Lietart, I., E. Broekaert, & I. Derluyn. (2014). The lived experiences of migrants in detention. *Population, Space and Place*, 21(6), 568–579.

Loyd, J. M., E. Mitchell-Eaton, & A. Mountz. (2016). The militarization of islands and migration: tracing human mobility through US bases in the Caribbean and the Pacific. *Political Geography*, 53, 65–75.

Mainwaring, C. (2012). Constructing a crisis: the role of immigration detention in Malta. *Population, Space and Place*, 18(6), 687–700.

Maria Karli, (n.d.). *[Facebook Review Posted on "Joe Corley Detention Center" Page]*. Retrieved from www.facebook.com/pages/Joe-Corley-Detention-Center/141224869247031.

Martelon, J. (2017, April 29). Chaplain: get the facts about Corley detention facility [Letter to the editor]. *The Courier of Montgomery County*. Retrieved from www.yourconroenews.com/opinion/article/Chaplain-Get-the-facts-about-Corley-Detention-11109238.php

Martin, L., & M. L. Mitchelson. (2009). Geographies of detention and imprisonment: interrogating spatial practices of confinement, discipline, law, and state power. *Geography Compass*, 3(1), 459–477.

Massaro, V., & J. Williams. (2013). Feminist geopolitics. *Geography Compass*, 7(8), 567–577.

McCullough, J., & C. Essig. (2018, August 2). The Trump administration is making plans to detain more immigrants in Texas. Here's where they would be held. *The Texas Tribune*. Retrieved from www.texastribune.org/2018/08/02/trump-administration-texas-migrant-detention-facilities-map/

Meyer, B. (2008, June 26). Putting out the welcome mat: county officials look to lock up federal contracts for new jail annex. *The Houston Chronicle*. Retrieved from www.chron.com/neighborhood/article/Putting-out-the-welcome-mat-9379431.php

Miller, C. (2017, May 3). Letters to the Editor. *The Courier of Montgomery County*. Retrieved from www.yourconroenews.com/opinion/article/Letters-Is-using-druglord-s-money-really-best-11119366.php

Mohammad, R. (2001). 'Insiders' and/or 'outsiders': positionality, theory and praxis. In M. Limb & C. Dwyer's (Eds.), *Qualitative Methodologies for Geographers: Issues and Debates* (pp. 101–117), London, UK: Arnold.

Moinester, M. (2018). Beyond the border and into the heartland: spatial patterning of U.S. immigration detention. *Demography*, *55*(3), 1147–1193.

Montgomery County Joint Election Cumulative Report. (2018, November 7). *Montgomery County*. Retrieved from http://legacy.mctx.org/election/pdf/GEN110618/2018_November_Joint_Election_Cumulative_Report.pdf?ts=86895
Mountz, A. (2010). *Seeking Asylum: Human Smuggling and Bureaucracy at the Border*. Minneapolis, MN: University of Minnesota Press.
Mountz, A. (2012). Mapping remote detention dis/location through isolation. In J. Loyd, M. Mitchelson, & A. Burridge (Eds.), *Beyond Walls and Cages: Prisons, Borders, and Global Crisis*. Athens: University of Georgia Press.
Mountz, A. (2015). In/visibility and the securitization of migration: shaping publics through border enforcement on islands. *Cultural Politics*, 11(2), 184–200.
Mountz, A. (2017). Island detention: Affective eruption as trauma's disruption. *Emotion, Space and Society*, 24, 74–82.
Mountz, A., K. Coddington, R. T. Catania, & J. M. Loyd. (2013). Conceptualizing detention: mobility, containment, bordering, and exclusion. *Progress in Human Geography*, 37(4), 522–541.
Mountz, A., & J. Hyndman. (2006). Feminist approaches to the global intimate. *Women's Studies Quarterly*, 34(1), 446–463.
Nagel, C. (2016). Southern hospitality?: Islamophobia and the politicization of refugees in south Carolina during the 2016 election season. *Southeastern Geographer*, 56(3), 283–290.
Nelson, L., & J. Seager. (2005). Introduction. In *A Companion to Feminist Geography*. Oxford, UK: Blackwell Publishing Ltd.
Nespor, J. (2014). Discursive geographies: politics and public narratives of immigration in a Midwestern US city. *Journal of Language & Politics*, 13(3), 490–511.
O'Sullivan, B. (2014, November 20). Where are we now with the jail debacle? *The Courier of Montgomery County*. Retrieved from www.yourconroenews.com/neighborhood/woodlands/opinion/article/Where-are-we-now-with-the-jail-debacle-9852966.php
Politico. (2016). Texas presidential election results. *Politico*. December 13, 2016. Retrieved from www.politico.com/2016-election/results/map/president/texas/
Pratt, G., & V. Rosner. (2006). Introduction: the global & the intimate. *Women's Studies Quarterly*, 34(1), 13–24.
Rhor, M., & J. D. Harden. (2017, May 25). Conroe booming as America's fastest growing city. *Houston Chronicle*. Retrieved from www.houstonchronicle.com/news/houston-texas/houston/article/Conroe-booming-as-America-s-fastest-growing-city-11174748.php
Robertson, S. P., K. Ravi, & R. Medina. (2010). Off the wall political discourse: Facebook use in the 2008 U.S. presidential election. *Information Polity*, 15, 11–31.
Roden, H. (2007, February 10). County hopes to make profit from new jail. *The Courier of Montgomery County*. Retrieved from www.yourconroenews.com/neighborhood/article/County-hopes-to-make-profit-from-new-jail-9573092.php
Roden, H. (2013, May 4). Joe Corley sale expected Monday. *The Courier of Montgomery County*. Retrieved from www.yourconroenews.com/neighborhood/bayarea/news/article/Joe-Corley-sale-expected-on-Monday-9416050.php
Roden, H. (2013, May 6). Joe Corley jail sold for $65M. *The Courier of Montgomery County*. Retrieved from www.yourconroenews.com/news/article/Joe-Corley-jail-sold-for-65M-9260409.php.
Roden, H. (2013, May 18). GEO Group to build second federal jail. *The Courier of Montgomery County*. Retrieved from www.yourconroenews.com/news/article/GEO-Group-to-build-second-federal-jail-9261495.php
Roden, H. (2013, May 20). Commissioners creating 'jail industry' in Montgomery County. *The Courier of Montgomery County*. Retrieved from www.yourconroenews.com/news/article/Commissioners-creating-jail-industry-in-9279561.php
Rose, G. (2001). *Visual Methodologies: An Introduction to the Interpretation of Visual Materials*. London: Sage.
Royall, E. (2018, July 8). Tierrabyte: what we know about immigrant detention centers in Texas. *The Rivard Report: Nonprofit Journalism for a Better San Antonio*. Retrieved from https://therivardreport.com/tierrabyte-what-we-know-about-immigrant-detention-centers-in-texas/.
Ryo, E., & I. Peacock. (2018). *The Landscape of Immigration Detention in the United States*. American Immigration Council. Retrieved from http://americanimmigrationcouncil.org/sites/default/files/research/the_landscape_of_immigration_detention_in_the_united_states.pdf.
Santiago, I. (n.d.). *[Facebook Thread Response Posted to "Joe Corley Detention Center" Page]*. Retrieved from www.facebook.com/pages/Joe-Corley-Detention-Center/141224869247031

Shahani, A. (2011, March 25). What is GEO group? *National Public Radio (NPR)*. Retrieved from www.npr.org/2011/03/25/134852256/what-is-geo-group
Sundberg, J. (2008). 'Trash-talk' and the production of quotidian geopolitical boundaries in the USA-Mexico borderlands. *Social & Cultural Geography*, 9(8), 871–890.
Thackeray, S. (2004, December 18). The privatized jail option. *The Courier of Montgomery County*. Retrieved from www.yourconroenews.com/neighborhood/article/The-privatized-jail-option-9737037.php
Thackeray, S. (2005, April 25). Sheriff speaks against hiring private company to operate jail. *The Courier of Montgomery County*. Retrieved from www.yourconroenews.com/neighborhood/article/Sheriff-speaks-against-hiring-private-company-to-9742967.php
Tolia-Kelly, D. (2004). Landscape, race and memory: biographical mapping of the routes of British Asian landscape values. *Landscape Research*, 29(3), 277–292.
Tolia-Kelly, D. (2008). Motion/emotion: picturing translocal landscapes in the *Nurturing Ecologies* research project. *Mobilities*, 3(1), 117–140.
Torres, R. (2018). A crisis of rights and responsibility: feminist geopolitical perspectives on Latin American refugees and migrants. *Gender, Place & Culture*, 25(1), 13–36.
US Census Bureau. (2017) *Southern Cities Growing Quickly*. Retrieved from https://www.census.gov/library/visualizations/2017/comm/cb17-81-cities-growing.html
US Census Bureau. *Quick Facts: Conroe city, Texas*. Retrieved from www.census.gov/quickfacts/fact/table/conroecitytexas/PST045217.
US Census Bureau. *Quick Facts: Montgomery County, Texas*. Retrieved from www.census.gov/quickfacts/fact/table/montgomerycountytexas/PST045217.
US Department of Justice (US DOJ). (2016). *Review of the Federal Bureau of Prisons' Monitoring of Contract Prisons*. Retrieved from https://oig.justice.gov/reports/2016/e1606.pdf.
van Dijk, T. (2015). Critical discourse analysis. In D. Tannen, H. E. Hamilton, & D. Schiffrin's (Eds.), *The Handbook of Discourse Analysis* (2nd edition). John Wiley & Sons, Inc.
Varsanyi, M. (2007). Rescaling the "Alien," rescaling personhood: neoliberalism, immigration, and the state. *Annals of the Association of American Geographers*, 98(4), 877.
Vasquez, E. (2015). *[Facebook Review Posted to "Joe Corley Detention Center" page]*. Retrieved from www.facebook.com/pages/Joe-Corley-Detention-Center/141224869247031
Williams, J. (2014). The safety/security nexus and the humanitariansiation of border enforcement. *Geographical Journal*, 182(1), 27–37.

12

SMALLHOLDER FARMERS' LIVED EXPERIENCES OF WEATHER PERTURBATION IN MALAWI

Park Muhonda

Introduction

While some people (particularly in the developed world) are still skeptical about climate change, experiences of rural smallholder farmers in southern Africa, especially Malawi, attest to the saying that the one who feels it knows it. Increased frequency of occurrence of weather perturbations is negatively affecting livelihoods in southern Africa. As a rain-fed, agro-based economy, Malawi is especially more vulnerable to adverse conditions. Single season (November to April) rain-fed agriculture constitutes the source of livelihoods for about 85% of the population. The rain-fed agriculture in Malawi is finely tuned to climate as it relies on the timely onset of rainfall and its regular distribution through the rainy season (Arthur, 2003; Sivakumar et al., 2005). In this chapter, I will explore the lived experiences and perceptions about weather perturbations among local farmers in Malawi.

In the recent past, Malawi has been experiencing frequent occurrence of adverse weather conditions which have negatively impacted livelihoods and often rendered the nation food insecure (UN-OCHA, 2016). In the past 15 years, the government has declared a national disaster four times (i.e. February 2002, October 2005, January 2015, April 2016) as a result of adverse weather conditions. Enhancing and broadening our understanding of the temporality of rainfall in Malawi is thus central to improving/supporting people's livelihoods.

Although climate and rainfall variability is a significant topic of interest in Malawi and has received widespread attention from donors, NGOs, and the government, literature on Malawi's rainfall in relation to livelihood is still lacking. While Ngongondo et al. (2011) investigated the spatial and temporal rainfall patterns and Nicholson et al. (2014) analyzed the interannual rainfall variability, few attempts have been made to understand the actual lived experiences and perceptions about climate variability among local farmers in Malawi.

As the literature on farmers' perceptions on climate change has matured, the validity/accuracy of the perceptions has been debated. Panda (2016) observed that there is a dearth of concrete findings that link farmers' perception of changing climate and meteorological data. Literature on farmers' perceptions has also highlighted the importance of perceptions about climate change on farmers' adaptation actions. Deressa et al. (2011) found that in Ethiopia, farmers who perceived decreasing rainfall undertook adaptation actions. According to Ogalleh

 DOI: 10.4324/9780429450310-15

et al. (2012), farmers use their perceptions to make decisions on coping and adapting. Farmers' perceptions do not only reflect local concerns but also inform adaptation strategies. They shape the conceptual framework in which farmers are or are not willing to accept adaptation strategies (e.g. Byg and Salick, 2009; Yaro, 2013; Campos et al., 2014).

Understanding farmers' perceptions of the changing climatic conditions is essential to policy makers and to adaptation research (Le Dang et al., 2014). Studies on farmers' perception enable policy makers to design relevant policies as they understand what farmers adapt to. As asserted by Teye et al. (2014), effective adaptation to climate change is contingent on the perception of farmers. Boissiere et al. (2013) also argued that knowing how farmers perceive climate change is helpful in developing strategies to support adaptation to climate change. According to Le Dang et al. (2014), any attempt to understand adaptive behavior patterns should come after understanding how climate variability is perceived by farmers. Thus, perceiving climate variability is the first and critical step in the process of adaptation.

This chapter draws on household surveys, focus group discussions, and meteorological observations from Balaka in southern Malawi and Nkhata Bay in northern Malawi to understand the changing rainfall patterns at the local level. The main findings are (1) local farmers in both Balaka and Nkhata Bay are sensitive and knowledgeable of climatic conditions in their areas and have observed that rainfall patterns have changed; (2) local farmers' perceptions of climate change are consistent with meteorological observations. These two areas make an ideal case for studying changing rainfall patterns and farmers' perceptions because they are suitable for comparison as they represent geographies with contrasting natural environment conditions and social-cultural and livelihood systems. The chapter addresses the following specific questions: is the local climate of Balaka and Nkhata Bay changing? Do the local farmers perceive the changes? To answer these questions, the discussion is organized in two main sections. The first section provides the background of the study, data collection, and analysis. This is followed by the second section which discusses how smallholder farmers perceive climatic changes that have occurred for the past 46 years against meteorological observations.

Background to the study, data collection, and analysis

The study area is located between latitudes 9–17°S and longitudes 32–36°E) in Malawi. With a total area of 118,484 km^2 (29,604 km^2 of which is water, mainly Lake Malawi) and an estimated population of 18 million (2016), Malawi is the most densely populated country in southern Africa. It experiences semi-arid tropical wet and dry climate, also called savanna. The rainy season is from November to April and the dry season from May to October. Mean annual rainfall is in the order of 800 mm to over 1600 mm (Nicholson et al., 2014). The average daily minimum and maximum temperatures in November, the hottest month, are 17°C and 29°C respectively; those in July, the coolest month, are 7°C and 23°C. Due to variations in altitudes there are wide differences in climate across the country. According to Ngongondo et al. (2011), the climate of Malawi is mostly influenced by the north-south migration of the intertropical convergence zone (ITCZ) and its topography. Rainfall is erratic; floods and droughts occur more often.

The economy is undiversified and highly dependent on rain-fed agriculture. Agriculture accounts for more than 30% of Gross Domestic Product (GDP), 85% of the labor force, and 83% of foreign exchange earnings. Agricultural production is predominantly smallholder and subsistent, characterized by low levels of input and output. Smallholder farmers produce about 80% of Malawi's food. The main agricultural products grown by smallholder farmers are maize, tobacco, cassava, groundnuts, pulses, sorghum and millet, rice, bananas, sweet potatoes, and cotton. Livestock accounts for less than 7% of the agricultural GDP in Malawi. Only 4% of the households own cattle. A tiny fraction, about 2.3% of arable land is irrigated. And only around 3.3% of all rural households are beneficiaries of the irrigation schemes. Given the lack of irrigative infrastructure, agriculture is highly dependent on suitable weather conditions and is thus particularly highly vulnerable to weather shocks such that success of crop production is almost entirely linked with weather conditions.

As in most of southern Africa, rural livelihoods in Malawi depend heavily but not solely on rain-fed agriculture. At the intersection of interests to promote the private-sector-driven system, to achieve food security in the short term, and to push hybrid seed as an appropriate technology for smallholder farmers, the once diversified smallholder farming has gradually become specialized, with significant emphasis on maize. Estimates show that maize farming covers more than 70% of the arable land (Stevens and Madani, 2016). And according to Chinsinga et al. (2012), an estimated 97% of smallholder farmers across the county grow maize, and more than half of households grow no other crops. The maize monoculture (agricultural) system has made rural livelihoods fragile and more vulnerable to weather perturbations. Cassava production is relatively high in Karonga and Nkhata Bay districts in the northern region. With comparatively low requirement for water, labor, and inputs, cassava helps smallholder farmers ensure local food security in the event of maize crop failure and maize price increases. Most smallholder farmers draw their income from sale of crops, mainly tobacco, cotton, and also food crops. The market for the two major cash crops, tobacco and cotton, is unpredictable and is affected by international factors beyond the country's control. The market for crops is mainly in the country's urban centers, so is predictable and is relatively stronger in the south and center than in the north. Supply, however, tends to exceed demand everywhere and farmers' returns are therefore generally very low. Consequently, rural household incomes are very low. Rural households also draw income from off-farm sources of casual labor (*ganyu*), sale of firewood, charcoal, and from migrant remittances from family members working elsewhere. In the lakeshore areas especially, southern region rural households generate income from fishing, including fishing *ganyu* and fish trading. In the Shire Highlands, especially the Thyolo-Mulanje Tea Estate area, most households are estate workers.

Two case study sites were selected, Balaka in the south and Nkhata Bay in the north (Figure 12.1), to explore and gain insight into actual lived experiences and local perception about the weather perturbations. These two study areas were purposively selected because they represent geographies with contrasting natural environment (biophysical conditions), livelihood systems, and cultural and socioeconomic characteristics. They are ideal case studies for exploring climate variability as they epitomize a range of socio-ecological environments applicable to most parts of Malawi.

Nkhata Bay is located in the northern region of Malawi. It is characterized by high rainfall and rocky soils and comprises an escarpment leading down from the Viphya Mountains to the lakeshore. The average annual rainfall is 1598 mm. The average temperature is between 23.5°C. Besides having a fairly moist climate; rainfall is also spread out over a longer period than in other parts of the country. With high rainfall but poor soils, cassava is the dominant crop in the area. Other crops grown in the area include maize, sweet potatoes, rice, and bananas. Animal traction and livestock production are also limited. Local livelihoods vary according to ecological factors like topography and ecosystems in which the communities are located. Those along the lake are engaged primarily in fishing. Fishing is undertaken throughout the year, but peaks from mid-November to March following the rains. Generally, the area is described as "food-rich" but "cash-poor". Like most parts of the northern region, Nkhata Bay area is largely rural (not urbanized) and commercially isolated with poor road network and there are limited markets for the agricultural produce. There are few sources of income available besides the sale of crops.

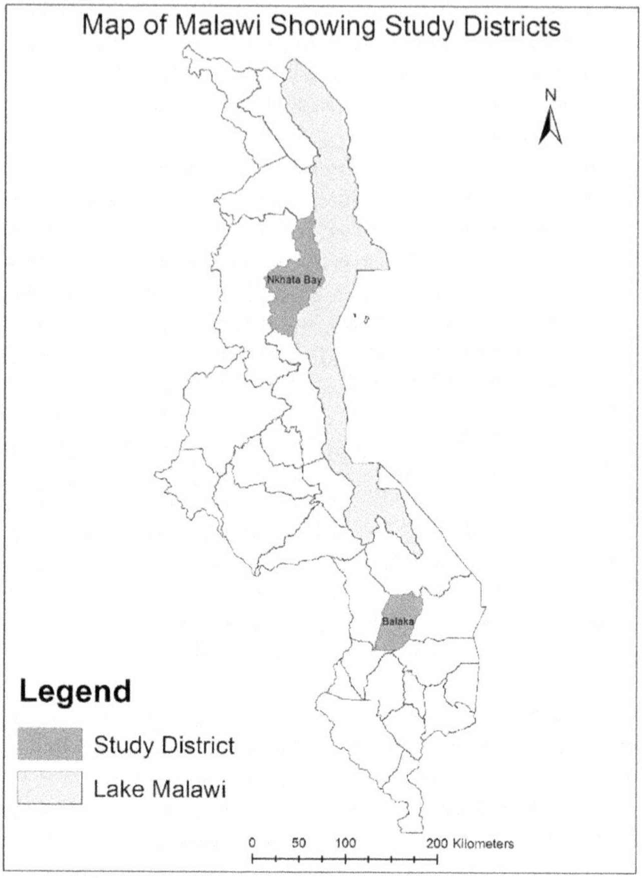

Figure 12.1 Map of Malawi showing study districts

On the other hand, Balaka is located in the southern region of Malawi. It has a relatively dry climate and is ecologically fragile and susceptible to frequent droughts. The average annual rainfall is 971 mm. The average temperature in Balaka is 23.1°C. The area is characterized by near-subsistence farming, with fishing on a small scale amongst those living close to the Shire River. Crop production is relatively low due to agricultural droughts. The main crops include cotton, maize, and sweet potatoes. Unlike Nkhata Bay, Balaka is highly populated and connected to markets in the country's region with biggest urban population as well as its largest commercial sector. The "poor" households earn income through *ganyu*, petty trade, firewood and charcoal burning and other collection activities, feeding the market in the southern region's urban centers. Market prices of staple food vary seasonally. Prices are lowest during the harvesting period and highest during the "hunger" season, which is between December and February. Most local markets are managed by private traders, following liberalization.

Data collection

Based on our knowledge of the study districts through in-depth interviews in the localities we identified three communities in each study area. The communities were purposively selected to ensure adequate representation of the study districts. We then randomly selected households in each community. A mixed methods approach was used to collect data. To draw on their complementarities, we conducted quantitative household surveys and also focus group discussions to explore and explain survey results. Six communities, three in Balaka and three in Nkhata Bay, were purposively selected to gather data on local farmers' perceptions of weather perturbations. In order to ensure adequate representation and minimize idiosyncrasies, study communities were selected to include all different socio-economic as well as biophysical characteristics. A systematic sampling technique was used within the communities to select the households. The survey was designed to include both male- and female-headed households. The heads of household were interviewed. If the household had the father and mother around, both were interviewed. A total of 515 households participated in the surveys. In the Nkhata Bay 298 households were interviewed: 103 females and 193 males. In Balaka 217 households were interviewed: 88 females and 129 males. The interviews were conducted by trained field assistants in the local languages of Tumbuka in the north and Chichewa in the south.

A total of 20 focus group discussions were conducted to gain deeper insights into farmers' perception regarding climate conditions. The focus group discussions included people of different social status: traditional leaders and ordinary men and women, young and elderly, natives and immigrants to capture the different perceptions. The number of participants in each focus group discussion (FGD) ranged from 8 to 14. A total of 244 participated in the focus group discussions:106 males and 138 males. A discussion guide was used to facilitate and moderate the discussions which generally took about two–three hours.

The climatological data used in the study was obtained from the Malawi Department of Climate Change and Meteorological Service. It included the continuous 46 years (i.e. for the period between 1970 and 2016) monthly and daily rainfall observations collected from Balaka and Nkhata Bay meteorological stations. Quality control of the data was largely carried out by the Malawi Department of Climate change and Meteorological Service. However, we noticed about five randomly missing entries in the daily rainfall data for Nkhata Bay which were reported as blanks. We applied the more flexible methods of group mean and regression to replace the missing entries. The two methods gave almost similar values for the missing entries.

Data analysis

Data analysis was conducted to compare and validate farmers' perceptions with meteorological observations, looking for agreement or disagreement. Exploratory and summary statistics were used to explore and analyze quantitative (household survey) data on perceptions of farmers on climate variability and change. Qualitative data was analyzed thematically. The transcribed discussions were rigorously coded, ordered, and structured into specific themes. Key themes included among others: irregular timing and distribution of rainfall – change of the rainfall calendar, in which the rains start late and end early; frequent occurrence of dry spell; rains not coming at the right stage of crop production; reduced amount of rainfall; increased occurrence of lightning and thunderstorms; and increasing temperature.

Qualitative data was analyzed thematically. Through an iterative process, transcribed focus group discussions were explored, queried, visualized, and coded in NVivo 12. Word frequency query was conducted as preliminary analysis to get an overview of the discussions' highlighted rains; changes; timing – start, end, early, late; planting; heavy, severe intensity – as most occurring themes, as displayed in Figure 12.2. These were then coded into specific themes which

Figure 12.2 Word frequency from qualitative interviews

included: timing of rainfall; distribution of rainfall; change of planting/farming calendar; changes in the amount of rainfall; occurrence of extreme event; impact on farming activities among others.

Climatological data analysis involved the use of normalized precipitation anomaly series and the Mann-Kendall test to understand temporal rainfall characteristics. The normalized rainfall anomaly series were used to assess inter-seasonal and annual rainfall variability. The Mann-Kendall (MK) test was used to assess temporal trends in rainfall data (meteorological observations) at annual, seasonal, monthly, and daily scales. The MK is a non-parametric test, which is robust, insensitive to missing data and outliers and is recommended by the World Meteorological Organization (WMO) for trend analysis in meteorological data (e.g. Ngongondo et al., 2011; WMO, 1988). Several studies have employed MK to assess trends in rainfall and temperature data (e.g. Taylor and Loftis, 1989; Yu et al., 1993; Basistha et al., 2009). MK was complemented with Sen's slope estimation to quantify the slope of the trends.

The acf () and pacf () functions in R, which compute the autocorrelation and partial autocorrelation corresponding to the time series, were used to visually investigate the presence of serial correlation among the annual, seasonal, monthly, and daily precipitation levels. Inspection of the serial correlation was conducted prior to applying the MK test to ensure that the right measures are taken, specifically to use the MK test in conjunction with bootstrapping in the case of the presence of significant serial correlation to correct for the P-value of the test for serial correlation.

Results

Climate change and variability and farmers' perception

Although farmers from both locations (i.e. Balaka and Nkhata Bay) had difficulties defining the term "climate change", the household surveys and the focused group discussions showed that they are more aware and sensitive to the changes in rainfall patterns that have occurred in their areas affecting their livelihoods when the term "rainfall changes" is used in questions. The household surveys show that almost all the surveyed smallholder farmers from both locations have observed changes in climatic conditions, particularly rainfall. In Balaka, 208 (96%) households indicated that rainfall has changed. In Nkhata Bay, a large share of households 290 (97%) also reported that rainfall has changed.

Despite the geographic, biophysical, and social-cultural differences between the regions, the farmers' perceptions of the changing climatic conditions were comparable in the two case studies. In Balaka many smallholder farmers, about 41%, observed that compared to the past rainfall has become more erratic, meaning there is no predictable and reliable rainfall calendar any more. "In the past", farmers observed, "rains used to begin and fall consistently, now there are frequent and extended dry spells". This was followed by perceptions that the rainy season has shortened or reduced (i.e. 18%), and that there is a late onset of rainfall (13%), and that rainfall events were oftentimes more intense (12%). Farmers reported that rainfall used to start in October and fall through to May. Unlike the past when they could grow two crops in a season as the season was long, now the rain ends well before crops mature. A small number (8%) reported that rainfall has decreased. In Nkhata Bay a large number of smallholder farmers (31%) perceived that rainfall is often intense and associated with strong winds, lightning and thunder. This was followed by perceptions that the rainfall is erratic (25%), and that the rainy season has shortened (17%), and that rainfall starts late (11%). A considerable number (7%) also reported that rainfall has decreased (Figures 12.3 and 12.4). To a large extent men and women stressed similar concerns about the changing rainfall patterns.

Through focused group discussions, smallholder farmers were able to expound on their actual lived experiences with regard to their perceptions about climate change. In the focused group discussions, farmers explained that they used to receive regular and reliable seasonal rainfall – the right amount of rainfall and on time. They mentioned that in the past rainfall would start in October or early November and they would plant by November, but now rains start in December. In Kavuzi, Nkhata Bay, for example, one farmer highlighted:

> It used to be that the very early rains called "*chicocola nyoni*" would fall in June and July, and then by mid-October we would receive the first rains locally called "*chizimya lupya*". The main rains called "*kuoloka*" would fall from November to January; in March and April we would have heavy rains called "*zandi*" and the rivers would also flood [old man].

Farmers further emphasized that the rainy season was shorter now and the frequency of dry spells has increased. Farmers explained that they used to have rains starting after mid-October through to late April or early May which would amply support more than one crop in a season. Nowadays, the rainy season is short, rain starts in December, often punctuated with severe dry

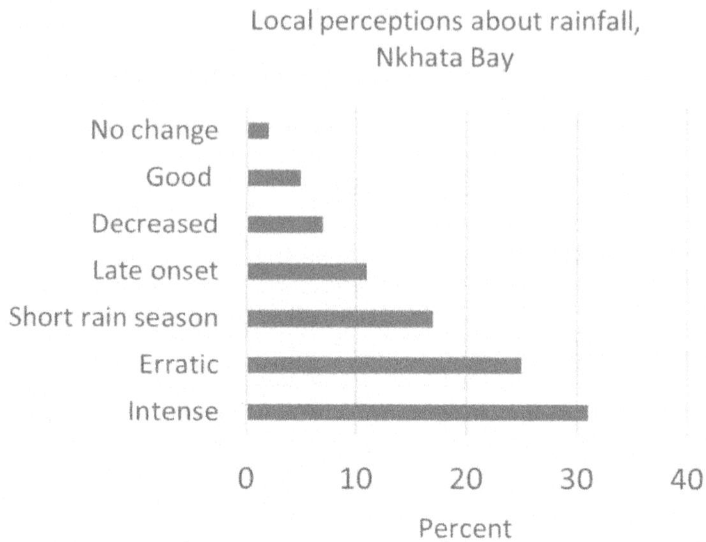

Figure 12.3 Weather changes observed, Nkhata Bay

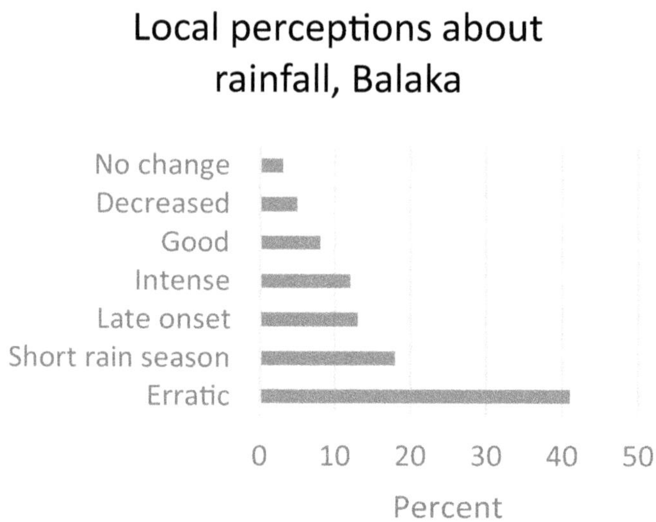

Figure 12.4 Weather changes observed, Balaka

spells, and ends early March. One older man in Mpamba, Nkhata Bay cited the dry spell that occurred in the 2014/2015 season as an example:

> This rainy season we had first rains on the 16th of October, so we planted because we are used to planting with early rains and we are also advised by agricultural extension

> officer to plant early. But then there were no rains up to the 21st December and the temperature was very high in between [very old man].

Farmers mentioned that the reduced amount but also the irregular timing and distribution of rainfall is negatively affecting their farming activities. One woman lamented: "even when the rain comes, it does not help us as it does not come at right stages of crop production, particularly maize crop". Some farmers complained about incessant rainfall at harvest time, when they expect no rain, which destroyed the little crop that managed to mature.

Farmers further commented that unlike in the past when they used to have a proper amount of rain, almost every month in the rainy season now the rains are unpredictable. Often when they are expecting rain, they instead experience lightning and thunder and very strong winds that blow the clouds away, which stops the rains. In Sanga, Nkhata Bay (along Lake Malawi) the locals mentioned that frequent occurrence of strong winds over the lake in the rainy season negatively affects their fishing livelihoods.

Apart from changing rainfall patterns farmers also mentioned the increasing temperature. Farmers in Mpamba, Nkhata Bay, for example, recounted:

> We were blessed with good climate here in Mpamba, it used to be cool in the winter months, but it is no longer the same, temperatures get very high all year round such that we don't even sleep with a bedsheet.

Climate change and variability: meteorological observations

Results of the analysis of the monthly mean rainfall (for the period 1970–2016) reveal that there are differences in the distribution of rainfall between the two study sites. In Balaka, rainfall is concentrated in the months of November to March. The peak rainfall occurs in January. In Nkhata Bay, on the other hand, rainfall is spread in the months of November to May, with maximum rainfall occurring in March. Nicholson et al. (2014) in their study of mean rainfall climatology of Malawi also observed similar patterns that along the northern lakeshore region maximum rainfall occurs in March, and it is one of the wettest parts of the country with the longest rainy season. For the southern regions, they also observed that peak rainfall occurs in January. Mean rainfall analysis also revealed high year-to-year rainfall variability for both locations, Balaka and Nkhata Bay (Figures 12.5, 12.6, 12.7, and 12.8).

The results of the MK test on rainfall series in Balaka show that rainfall has decreased over the period 1970–2015 (see Table 12.1). Annual and seasonal (November through April) rainfall series demonstrate significant decrease in annual and seasonal rainfall trends at average rates of -5.8mm/year and -5.9 mm/season, respectively. Monthly analysis of seasonal rainfall series (not shown here) shows significant downward trends in November, December, February, March, and April at average rates of -1.1, -2.0, -1.7, -1.7, and -0.7, respectively. Only the month of January demonstrated a positive trend, but it was not statistically significant. Daily analysis of the rainfall series also demonstrates statistically significant decreasing trends in the number of rain days annually and seasonally as well as for the months of November, March, December (marginal), and April (marginal) (not shown here).

Statistical analysis of meteorological data from Nkhata Bay also reveal a decreasing trend in precipitation between 1970 and 2015. Annual and seasonal (November through April) rainfall series demonstrates decreasing annual and seasonal rainfall trends at average rates of -1.8mm/year

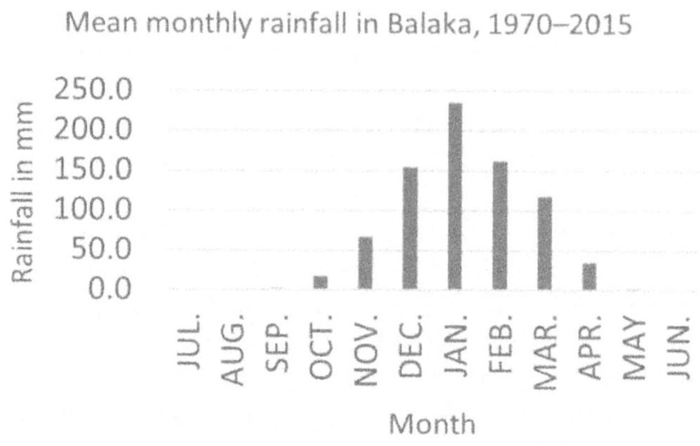

Figure 12.5 Monthly rainfall, Balaka

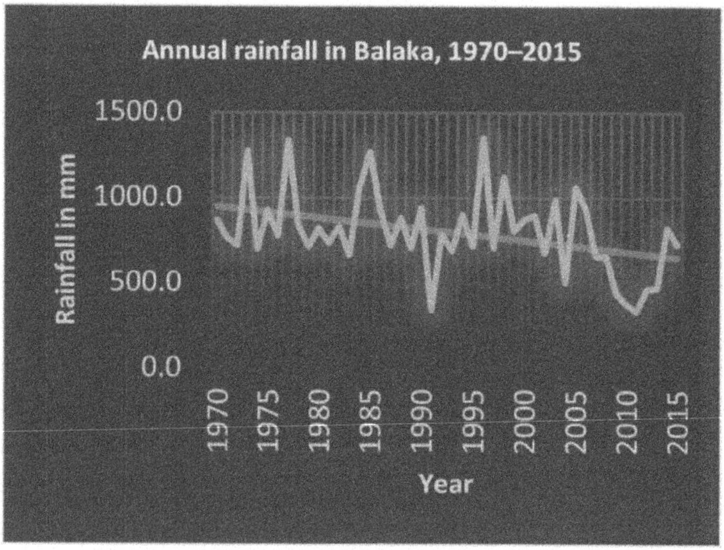

Figure 12.6 Annual rainfall, Balaka

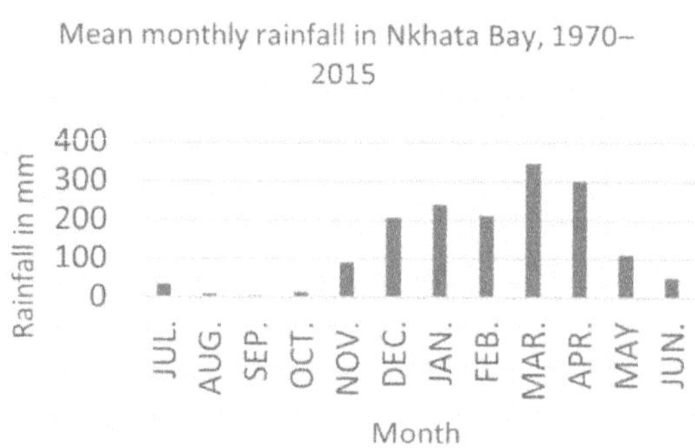

Figure 12.7 Monthly rainfall, Nkhata Bay

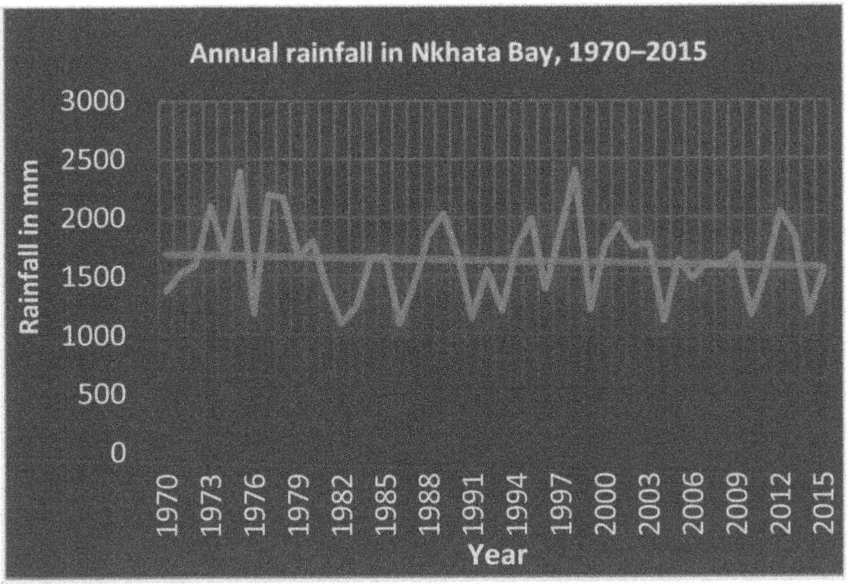

Figure 12.8 Annual rainfall, Nkhata Bay

Table 12.1 Trends in the annual and seasonal rainfall

Study site	*Annual/Seasonal*	*MK*	*P value*	*Sen's slope*
Balaka	Annual	-0.233	0.023*	-5.756
1970–2015	Seasonal (November–April)	-0.233	0.023*	-5.896
Nkhata Bay	Annual	-0.044	0.677	-1.781
1970–2015	Seasonal (November–April)	-0.037	0.712	-1.281

* Significant at 5% level

and -1.3 mm/season, respectively. While the annual and seasonal rainfall series show decreasing trends, this downward trend is not statistically significant at 5% level. A statistically significant decreasing trend, however, is demonstrated for the monthly analysis of seasonal rainfall series for December, March, and May (marginal) at average rates of -2.4, -2.98, and -1.3, respectively. The month of March also revealed a significant decreasing trend in the magnitude of rain events. Statistically significant decreasing trends in the number of rain days were demonstrated for the month of April.

In order to get an overview of the start and end of rainy season from the meteorological observations, the definitions of onset and end of a rainy season by Tadross et al. (2009) were followed. According to Tadross et al. (2009), the onset of a rainy season is once 25 mm of rainfall has accumulated within 10d, without 10 consecutive dry days (< 2 mm) occurring afterward. The end of a rainy season is defined as 3 consecutive dekads (after February 1) of < 20 mm each. The statistical analysis of the meteorological daily rainfall observation between 1970 and

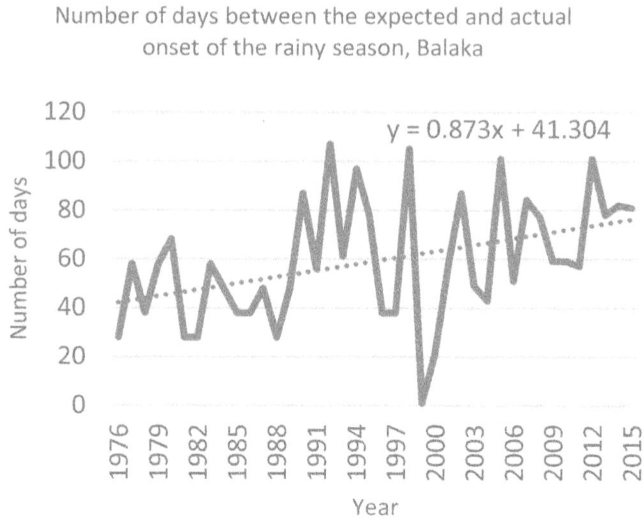

Figure 12.9 Onset of rainy season, Balaka

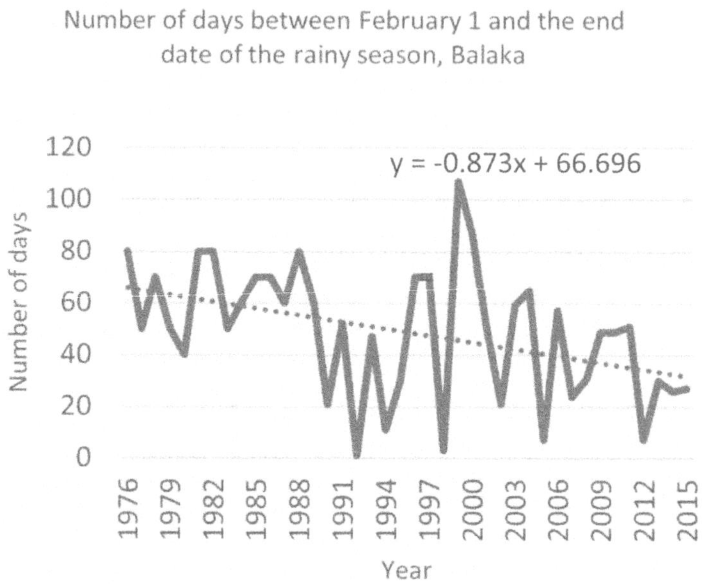

Figure 12.10 End of rainy season, Balaka

2015 demonstrates that onset and end of the rainy season in Balaka and Nkhata Bay have shifted. Figures 12.9, 12.10, 12.11, and 12.12 show an increasing trend in the number of days between actual day of onset and the expected date of onset (i.e. after mid-October) and decreasing trend in the number of days between February 1 and the end date of the rainy season for both locations, Balaka and Nkhata Bay. The changes however are statistically significant only in Balaka and not in Nkhata Bay.

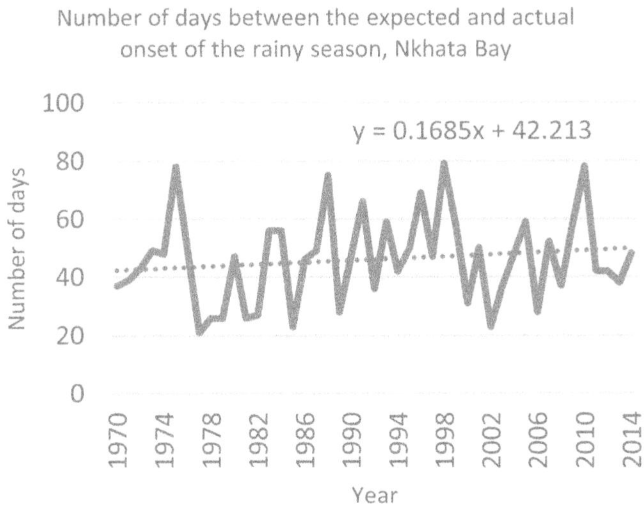

Figure 12.11 Onset of rainy season, Nkhata Bay

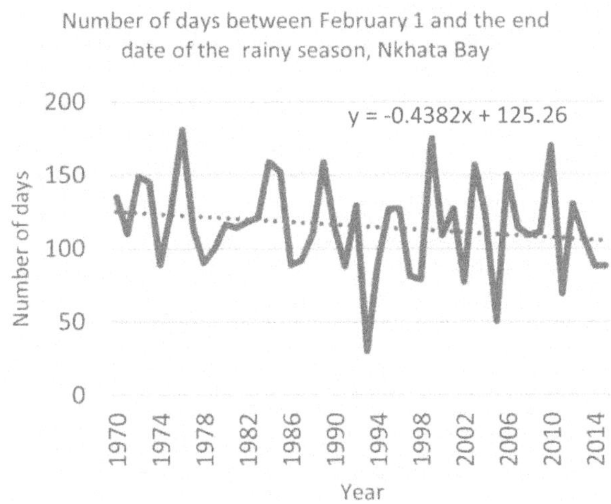

Figure 12.12 End of rainy season, Nkhata Bay

Climate change and variability: perception and meteorological observations

Farmers' perceptions about climate variability and change has been considered less reliable (e.g. Held et al., 2005; Weber, 2010). This assertion is confirmed by studies showing inconsistencies between local perceptions and meteorological data and questioning the validity of local perceptions (Ovuka and Lindqvist, 2000; Hansen et al., 2004; Sivakumar et al., 2005; Maddison, 2006; Van Aalst et al., 2008; Sanchez-Cortes and Chavero, 2011). However, this is in sharp contrast to the results reported in this chapter that show that local farmers are sensitive and aware of the changes in the climatic conditions occurring in their areas.

The study reported in this chapter supports observations that local farmers are knowledgeable of the changes in the climatic conditions occurring in their areas. The results of our study show that local farmers' perceptions about rainfall variability and change in Balaka and Nkhata Bay are comparable with the meteorological observations. Farmers' perceptions that rainfall has decreased closely corroborate with actual rainfall data from the Department of Climate Change and Meteorological Services, which show a decreasing trend in precipitation between 1970–2016. Our findings of downward trend in precipitation resonates with the trends established by Ngongondo et al. (2011), who observed decreasing trend in precipitation. Consistent with meteorological observations local farmers also correctly perceived that the rainfall pattern has changed such that the rainy season has shortened with rains starting late and ending early so the number of days in the rainy season have decreased. Furthermore, in line with farmers' perceptions, statistical analysis (SPI) of actual rainfall data show that frequency of dry spells in the rainy season have also increased over the period 1970–2016, with the increase mostly after 1990.

While statistical analysis demonstrates that all the observed changes in precipitation are statistically significant at 95% level for Balaka, not all observed changes in precipitation are statistically significant in Nkhata Bay at 95% level. Farmers' perceptions of decreased precipitation might have been influenced dominantly by the statistically significant downward rainfall trend for the month of December. For the most part local farmers in the north plant their crops in December and bearing in mind that planting is tied to adequate rainfall, it is not surprising that local farmers interpreted the rainfall changes that have occurred in December as decreasing seasonal or annual rainfall. A significant decline of rainfall in March might have also influenced farmers' perceptions. As observed earlier, and also as established by Nicholson et al. (2014), March is the month when seasonal rainfall is at its peak in Nkhata Bay. Failure to see river flooding or at least flowing to their full capacity made farmers perceive that rainfall has decreased.

The consistency of local perceptions with meteorological observation sturdily demonstrates that local perceptions about climatic conditions are reliable and valid. For centuries, from generation to generation, smallholder farmers in Malawi have mainly depended on rainfall for their livelihoods. Farmers have, over the years, developed in-depth knowledge of their local climatic conditions – no wonder they are able to correctly notice changes in rainfall taking place in their areas. The results of our study thus agree with Maddison (2006), who also argued that people who depend on rainfall for their livelihoods are more likely to notice changes in climatic conditions. With low coverage (low network of rain gauges) meteorological stations, local perceptions serve, in many areas, as the only detailed source of information about climate variability change. Furthermore, together meteorological observation and local perceptions can provide more valuable insights.

Even though climate change is bringing climatic conditions beyond what local farmers have experienced in the past (i.e. past climate perturbations), local farmers usually draw from their knowledge and perceptions to respond to climate shocks. As pointed out by Boissiere et al. (2013) farmers use local knowledge to interpret and respond to environmental perturbations. External intervention for facilitating local adaptation to climate change in Balaka and Nkhata Bay thus are more likely to be successful if they build on this existing knowledge of rainfall by the locals. Local knowledge can be useful for designing policy for dealing with impact of climate change. The strategies and initiatives to support local adaptation should be tailored according to local knowledge. With better and deeper understanding of local situations the local players (traditional leaders and politicians), government departments, as well as international organizations and civil society could be more effective and better leveraged to mediate local interventions, including rural livelihood coping and adaptation strategies, safety net programming, and food security and policy formulation. In essence, local knowledge is central to shifting from the usual and dominant unrealistic and palliative "top-down" perspective and measures driven by global climatic model scenarios to working from the "bottom-up".

Summary and conclusion

This chapter examined farmers' perceptions of changes in precipitation and compared them with meteorological rainfall observations. The analysis aimed to assess if rainfall in the study areas is changing and if farmers are able to perceive the changes correctly. The study revealed that almost all the farmers in both locations, Balaka and Nkhata Bay, are sensitive and knowledgeable of the climatic conditions in their areas and observed that rainfall has changed. Although local farmers did not have written records but rather based their perceptions on experience, when perceptions were compared with meteorological observations, the analysis indicated that what local farmers observed in both locations was consistent with meteorological observations. In Nkhata Bay it seems farmers' perception of declining rainfall was largely shaped and influenced by changes observed in rainfall amount in December when they expect to plant their crops and March when they expect rivers to flood. The correctness at which local farmers perceived the changes in rainfall sturdily support the assertion that local perception of climate change is reliable and consistent with meteorological observations.

Local farmers provided more detailed information about changes in climatic conditions, especially rainfall, in their areas. We posit policy makers can and should harness such detailed, extensive, and correct knowledge of local weather condition that local farmers possess to facilitate effective and successful adaptation. To ricochet the main argument of this chapter as local farmers react to change that they observe, adaptation strategies should therefore be specific to locations' needs. Thus, while climate change is global, adaptation policies and initiatives or strategies that are not tailored to local need are likely to fail. Building adaptation programs on local perceptions ensures local agreement and participation. Since local perceptions are an essential component of a local response, it is our recommendation that policy makers tap into local perceptions across the country such as unearthed in this study in order to facilitate and achieve sustainable and successful adaptation to climate change.

References

Arthur, A.M. (2003) Targeting Climate Forecasts for Agricultural Applications in Sub-Saharan Africa: Situating Farmers in User-Space. *Climatic Change* 58: 73–92.

Basistha, A. et al. (2009) Analysis of Historical Changes in Rainfall in the Indian Himalayas. *International Journal of Climatology* 29: 555–572.

Boissiere, M. et al. (2013) Local Perceptions of Climate Variability and Change in Tropical Forests of Papua, Indonesia. *Ecology and Society* 18(4): 13.

Byg, A., and Salick, J. (2009) Local Perspectives on Global Phenomenon-Climate Change in Eastern Villages. *Global Environmental Change* 19: 156–166.

Campos, M. et al. (2014) Land-Users' Perceptions and Adaptations to Climate Change in Mexico and Spain: Commonalities Across Cultural and Geographical Contexts. *Regional Environmental Change* 14: 811–823.

Chinsinga, B. et al. (2012) *Climate Change and Agricultural Policy Processes in Malawi*. Working Paper 046. Brighton: Future Agricultures.

Deressa, T.T. et al. (2011) Climate Change and Agriculture Paper Perception of and Adaptation to Climate Change by Farmers in the Nile Basin of Ethiopia. *Journal of Agricultural Science* 149: 23–31.

Hansen, J. et al. (2004) Earth's Energy Imbalance: Confirmation and Implications. www.sciencexpress.org/28 April 2004/Page 1/10.1126/science.1110252

Held, T.L. et al. (2005, December 13) Simulation of Sahel Drought in the 20th and 21st Centuries. *PNAS* 102(50): 17891–17896. https://doi.org/10.1073/pnas.0509057102

Le Dang, H. et al. (2014) Farmers' Perceptions of Climate Variability and Barriers to Adaptation: Lessons Learned from an Exploratory Study in Vietnam. *Mitigation and Adaptation Strategies for Global Change* 19: 531–548. DOI: 10.1007/s11027-012-9447-6

Maddison, D. (2006) *The Perception of and Adaptation to Climate Change in Africa. CEEPA*. Discussion paper No. 10, Centre for Environmental Economics and Policy in Africa, University of Pretoria, Pretoria.

Ngongondo, C. et al. (2011) Evaluation of Spatial and Temporal Characteristics of Rainfall in Malawi: A Case of Data Scarce Region. *Theoretical and Applied Climatology* 106: 79–93.

Nicholson, S. E., Klotter, D., and Chavula, G. (2014) A Detailed Rainfall Climatology for Malawi, Southern Africa. *International Journal of Climatology* 34: 315–325.

Ogalleh, S.A. et al. (2012) Local Perceptions and Responses to Climate Change and Variability: The Case of Laikipia District, Kenya. *Sustainability* 4: 3302–3325. DOI: 10.3390/su4123302

Ovuka, M., and Lindqvist, S. (2000) Rainfall Variability in Murang'a District, Kenya: Meteorological Data and Farmers' Perception. *Geografiska Annalwer* 82(1).

Panda, A. (2016) Exploring Climate Change Perceptions, Rainfall Trends and Perceived Barriers to Adaptation in a Drought Affected Region in India. *Natural Hazards* 84: 777–796.

Sanchez-Cortes, M.S., and Chavero, E.L. (2011) Indigenous Perception of Changes in Climate Variability and Its Relationship with Agriculture in a Zoque Community of Chiapas, Mexico. *Climatic Change* 107: 363. https://doi.org/10.1007/s10584-010-9972-9

Sivakumar, M.V.K. et al., (2005) Impacts of Present and Future Climate Variability and Change on Agriculture and Forestry in the Arid and Semi-Arid Tropics. *Climatic Change* 70: 31–72.

Stevens, T., and Madani, K. (2016) Future Climate Impacts on Maize Farming and Food Security in Malawi. *Scientific Reports* 6: 36241. https://doi.org/10.1038/srep36241

Tadross, M. et al. (2009) Growing-Season Rainfall and Scenarios of Future Change in Southeast Africa: Implications for Cultivating Maize. *Climate Research* 40: 147–161.

Taylor, C.H., and Loftis, J.C. (1989) Testing for Trend in Lake and Ground Water Quality Time Series. *Water Resources Bulletin* 25(4) American Water Resources Association.

Teye, J.K. et al. (2014/2015) Local Farmers' Experiences and Perceptions of Climate Change in the Northern Savanna Zone of Ghana. *International Journal of Climate Change Strategies and Management* 7(3): 327–347.

UN-OCHA (United Nations Office for the Coordination of Humanitarian Affairs). (2016) *Annual Report 2016*, p. 48. https://www.unocha.org/sites/unocha/files/2016ochaannualreport.pdf

Van Aalst, M.K. et al. (2008). Community Level Adaptation to Climate Change: The Potential Role of Participatory Community Risk Assessment. *Global Environmental Change* 18: 165–179. DOI: 10.1016/j.gloenvcha.2007.06.002

Weber, E.U. (2010) What Shapes Perceptions of Climate Change? *Wires Climate Change* 7(1): 125–134.

WMO (World Meteorological Organization). (1988) *Analyzing Long Time Series of Hydrological Data with Respect to Climate Variability*. WCAP-3, WMO/TD No. 224, p. 12.

Yaro, J.A. (2013) The Perception of and Adaptation to Climate Variability/Change in Ghana by Small-Scale and Commercial Farmers. *Regional Environmental Change* 13(6). DOI: 10.1007/s10113-013-0443-5

Yu, Y.-S., Zou, S., and Whittemore, D. (1993) Non-Parametric Trend Analysis of Water Quality Data of Rivers in Kansas. *Journal of Hydrology* 150(1): 61–80. ISSN 0022-1694. https://doi.org/10.1016/0022-1694(93)90156-4.

13

TERRA SACER

Water infrastructure and core-periphery reconfiguration in Dallas/Fort Worth

James-Eric Simon and Waquar Ahmed

Introduction

In 1945, the US Congress approved the damming of Denton Creek, a tributary of the Trinity River, to facilitate flood control and navigation, with the ancillary benefit of urban water supply. The resulting artificial lake – Grapevine Reservoir – had initially been proposed as a municipal water source for the City of Dallas, Texas. The city's rapid growth was already straining the capacity of local streamflow, and additional water was urgently needed to avoid economic stagnation. Dallas civil and business leaders therefore developed a plan to impound Denton Creek water near the tiny rural Town of Grapevine, approximately 20 miles from the growing city. With the subsequent assistance of the US federal government, Denton Creek was urbanized.

Urbanization of water presents two apparently competing possibilities for exurban residents: a promise of progress, and a threat of domination. The Denton Creek watershed was discursively produced as a useful space for urban interests in Dallas and Grapevine. The instrumental rationalities and network spatial imaginary of modernity combined to collapse the meaning of this territory from a rural expanse to a dimensionless node: an economic point-source that could be internalized into the broader network of urban accumulation. Unlike many paradigmatic case studies of urban water appropriation (Reisner, 1993; Celio et al., 2010), the Grapevine case seems to be a win-win situation that benefited both the metropolis and the village. The promise of progress was fulfilled – residents of tiny Grapevine were integrated into the space of urban flows, rather than being peripheralized by the omnivorous population of a remote urban core. This mutually beneficial network proliferation seems to suggest a benign, non-exploitative form of colonization. However, this chapter develops the argument that modernity's networking project did not disrupt the core-periphery relationship between urban and rural space. Through the process of spatial collapse and selective internalization, urban interests merely reconfigured the city geographically. The effect of the intervention was not to peripheralize the Town of Grapevine, but rather to "core-ify" it.

This chapter seeks to uncover and understand the covert colonialism that pervaded urban discourse surrounding the construction of Grapevine Reservoir. We argue that discursive practices of DFW's water networking project were substantially the same as those of overtly extractive colonialization, and characterize DFW-area water resource development as a process of settler colonialism. Finally, we examine representations of hydrosocial space with an eye to that

DOI: 10.4324/9780429450310-16

which was left behind by the process of collapse: territory that was articulated with the city, but not fully internalized. Drawing from Agamben's (1998) concept of *homo sacer* and spatializing it, we frame the urbanization of Denton Creek water as the creation of a space of ambiguity: both inside and outside the urban ordering of the world. Whereas modernity codes space in binary – urban *or* natural – we argue that the harnessed watershed became both.

Background

The core-periphery model was developed to account for patterns of uneven development and economic power (Jennings, 2006; Krugman, 2002, 2011; Friedmann, 1963, 1966). Resonating with von Thunen's (Essletzbichler, 2011) classic model of urban spatial organization, this model describes a zone of concentrated power that exploits a distant, undeveloped expanse: the periphery. Due to its lack of development, the periphery cannot effectively produce value-added manufactured goods, and is forced to export raw resources to the core (Cardoso and Faletto, 1979). The fundamentally competitive nature of capitalism requires expansion into peripheral areas in search of new resources and new markets, in order to maintain capital flows and prevent stagnation (Marx and Engels, 2002 (1888); Harvey, 1996). Colonial themes of water appropriation are threaded through the political ecology literature.

In their analysis of water grabbing by the state of Israel at the expense of Palestinians, Gasteyer et al. (2012) take the view that this new form of hydro-colonialism is better understood through the lens of "old" colonialism, with water appropriation construed as a continuation of historical colonial extraction. Drawing from Lefebvre (1991) and Harvey (1996), Gasteyer et al. identify a salient characteristic of the colonial project: state appropriation of space and the redefinition thereof in terms of its value as a means of production. A pervasive justificatory doctrine is *terra nullius*, that is, unoccupied "marginal" or "barren" land that is potentially exploitable by the colonizers (Whatmore, 2002; Ahmed, 2008; Lane, 2013). In dry environments, the essential component for productivity is water, particularly for irrigation, and often for the global agricultural market. This makes water a vital strategic resource to encourage colonization by settlement, and justifies the use of state appropriation and military control within the colonial context. Historically, colonial settlements followed this pattern of a narrative of development for "barren" landscapes, settlement for agricultural production, displacement of the local population, and expansion onto the land designated for that population over time.

Pritchard (2012: 595) picks out a similar pattern in French North Africa. According to the French colonial narrative, the goal was "to transform the uncultivated lands of this country [Morocco] into immense fields of grain". This required a solution for that which French administrators and hydrologists referred to as "the problem of water". The solution was undertaken by a military-oriented hydrological mapping program, and the development of networks of wells to enable troop mobility. On Pritchard's view, the overarching project was to encourage colonial settlement in North Africa, which would then supply France with agricultural products. This pattern of linked land/water grabbing for agricultural development of *terra nullius* pervades settler colonialism.

Water for agriculture, however, is not the only vector of hydraulic reach to be seen in colonial (state-driven) water capture. Celio et al. (2010) analyze the appropriation of water *from* agriculture by Hyderabad, a rapidly developing city in Andhra Pradesh, India. Hyderabad successfully appropriated agricultural water by manipulating government institutions such as reservoir regulations, and by shaping the discourse such that the government of Andhra Pradesh concluded that drinking water (for the city) had to be prioritized over the needs of agriculture. In this case, the relative wealth of Hyderabad resulted in its gain of water resources at the expense of the upstream region: the economically disadvantaged and water-stressed Rayalseema. Swyngedouw's (2004b) examination of Guayaquil, Ecuador's expanding water footprint gives another example of direct state intervention for the sake of urban water supply. At the beginning of the 20th century, the development of the city's water supply in support of growth was declared to be in the national interest, and therefore the state acted directly to reallocate water from agriculture to municipal uses. Not only was water appropriated, but agricultural producers, first of cocoa and later of bananas, were taxed directly in order to fund urban water infrastructure.

These are examples of state-driven water extraction from an undeveloped periphery to benefit a colonial core. Where water appropriation is contested, it is enforced by decree backed by the repressive apparatuses of the state (Althusser, 2001), including use of "legitimate" violence that the state has monopoly over. However, this model of hydroimperialism – Gasteyer's "old colonialism" – is less and less common globally. In its place has arisen a new form of colonialism: neoliberalism (Harvey, 2003). Instead of extracting water though sovereign authority (at least directly), neoliberalism determines reach through the mechanism of the market. But while the mode of power has shifted, the modes of extraction remain distinctly similar.

With respect to water, neoliberalism makes allocation and distribution a function of market value (Bakker, 2010; Perreault, 2005). On its face, it represents the retreat of government from decision-making with respect to resource allocation. Under the market framework, water is no longer conceptualized as a common resource, but rather as a rare commodity with economic value (Perreault, 2006). Hierarchical state decision-making is replaced by private contracts, which is said to increase social equity and provide water more sustainably by reducing the waste caused by government subsidy and inefficiency (Bakker, 2010). In potential markets (such as post-socialist and developing nations), the market paradigm gains purchase by shaping and controlling discourse in a Foucauldian sense (Peet, 2007).

For Foucault, a discourse is a language structure that makes it nearly impossible to speak – or even to think – beyond (McEvoy and Wilder, 2012). Discourse, therefore, is a domain of power (Hook, 2001). By refusing to attribute meaningfulness to descriptions of water beyond the narrow rubrics of the physical sciences, colonizers pre-empt objections to the implicit human/nature dichotomy that justified spatial domination. A second example is market discourse. Peet (2007) draws from Gramsci (1971) the concept of ideological hegemony – a specific cultural-institutional worldview that becomes so thoroughly dissolved into the cultural imagination that it becomes a nearly invisible determinant of everyday action. Thus metaphysical and ethical beliefs force entry into economics as commonsense ways of viewing the world. Even to think outside of this cultural imaginary becomes difficult (similar to the constraint imposed by discourse). For Peet, this phenomenon is regionally specific, since it represents both an interpenetration of society and environment, but also since these dissolved beliefs must be signified and transmitted through interaction (which also makes them dynamically responsive). An economy is not merely its material manifestation of these beliefs, but above all a way of producing human subjects and social orders. In the colonial view of land discussed earlier, market discourse construes water as an economic good. Within the market discourse of neoliberalism, water represents capital – a means of production. Its allocation, therefore, is subject to evaluation based on economic efficiency, that is, the distribution of economic costs and benefits deriving from its use (Ward, 2007). Marketization reduces water to a proxy currency, valuable only insofar as it represents capital accumulation. Hence Ahmed's (2010) argument that neoliberalism represents the clearing away of impediments to capital flows (cf. also Harvey, 2003, 1996). As Perrault (2006) points out, this comes into conflict with resource discourses based upon locality and community: it represents entirely different logics of resources and rights.

Under a regime driven by market discourse, access to water is determined by economic power above all other means or needs. Once water is marketed, capital can bring an array of tactics to bear. Contrary to neoliberal discourse, the literature is broadly in agreement that market-driven policies increase the disparity of political-economic power between the neocolonist and the neocolonized (Simmie and Martin, 2010; Bakker, 2010; Perreault, 2006; Swyngedouw, 2004b; Roberts and Emel, 1992). Since water allocation becomes purely a matter of political-economic power relations, the disempowered can be crowded out. Without political structures in place to mediate this inequality – if the only water ethic is economics – the urban core is unconstrained, and the periphery is left with an ever-decreasing resource pool (Gasteyer et al., 2012; Scott and Pablos, 2011; Reisner, 1993).

Broadly speaking, the water colonialism literature loosely organizes itself around a concept that Harvey (1999) identifies as the *spatiotemporal fix*. Because capitalism can never fully resolve its internal contradictions, it is forced both to expand and to reconfigure the world geographically. Paradigmatic water grabbing cases like the Owens Valley affair follow the colonial trajectory indicted by Fanon: vampiric projects intended "to extract and ship our wealth to the metropolis" (1963: 58). Whether through hard power or soft, these cases represent domination of the periphery by the core. This model, however, does not seem to account for the dynamics of the Grapevine Reservoir case. Harvey outlines a second "fix" that may provide a more appropriate framework for the Grapevine Reservoir case: settler colonialism. Rather than treating the periphery as an external "sacrifice zone" for rapacious extraction (Klein, 2015), the core seeks to establish permanent outposts there. However imperfectly or incompletely, the process of settlement internalizes selected parts of the periphery to the core. Instead of dominating the periphery from afar, settler colonialism aims at domesticating it – making a home there. Given the extensive cooperation between Dallas and Grapevine populations during the proposal and construction of the reservoir, the apparent symbiosis offered by the project, and the subsequent urbanization of Grapevine, settler colonialism seems to offer a more promising lens through which to view the case. Whereas the critical resource study literature has examined water appropriation *within* settler colonialism (cf. Pritchard, 2012 and Gasteyer et al., 2012), the process has not commonly been used as a conceptual model with which to analyze water resource development. We now attempt to connect the two.

In his examination of the concept of settler colonialism as an object of inquiry, Veracini (2013: 313) offers a simple gloss that is compatible with the majority of postcolonial scholarship: "colonizers 'come to stay' and to establish new orders for themselves, rather than to exploit native labour. . . . Settler colonialism is about turning a place and a specific human material into something else". Rather than a historical event, this form of colonization is a permanent structure and an ongoing process (Mikdashi, 2013; Wolfe, 1999) with which settlers attempt to reproduce the society from which they emerged (Veracini, 2013). As opposed to extractive colonialism – which only appropriates space in order to accumulate surplus value (Goldstein, 2008) – settler colonialism aims at space itself; settlers attempt to carve out room for their new (or extended) society. Postcolonial scholars have generally critiqued the settlement process along three axes: the justificatory validity of *terra nullius*, the production of identities, and cultural homogenization.

In his study of Israeli settlement in Eretz, Aaronsohn (1996) distinguishes between the spatial practice of colonization and the social practice of colonialism, and suggests that the appropriation of space need not entail dispossession: that "colonization without colonialism" is possible. This argument resonates with the colonial legacy of *terra nullius*: the doctrine that unoccupied land is legitimately available to be claimed. However, in colonial history, very little land has actually been empty at the time of its seizure (Veracini, 2013; Whatmore, 2002). As Whatmore points out in her examination of aboriginal Australia, settlers simply declined to recognize indigenous claims, since those claims were not legitimated according to European forms of governance. Without a state-sanctioned legal apparatus to "make possession proper" (Whatmore, 2002: xx), the indigenous Mabo were not recognized as owning property at all (cf. (Mikdashi, 2013). As far as British law was concerned, the land was empty and available for settlement. The influence of this doctrine is not limited to the distant past, though the context has changed somewhat. Instead of land empty of people, the theme of *terra nullius* has been enfolded into economic discourse in order to justify the appropriation of lands empty of economic development in India (Padel and Das, 2010) and Nevada (Lane, 2013).

A second line of critique derives from Fanon's (1963) "Manichean world" of the colony, and Said's (1979) concept of orientalism. Within settler colonialism, the most fundamental identities are those of the colonist and the colonized: indigenous and non-indigenous (Mamdani, 2001). Settler discourse produces the ethnic "native" as exotic, disorderly, or primitive in an effort to justify the exercise of power and dispossession. Because indigenous peoples often do not conform to Western values, knowledge systems, and modes of social organization, settlers take the view that natives *should be* ruled (Maddison, 2013) or ejected (Stanger-Ross, 2008; Edmonds, 2010). While the postcolonial literature typically examines the imposition of power along racial fault lines, recent scholarship has revealed a reflexive discourse of indigeneity emerging from the Occupy movement (Barker, 2012) and backwater Appalachian communities (Pearson, 2013). These studies reveal a phenomenon of "self-indigenization", where disempowered whites deploy Amerindian symbols in order to identify as a colonized class.

Contrasting with this identity of indigeneity, the settler represents himself as a kind of hyper-masculine pioneer, valiantly pushing back the frontier by progressively ordering the world (Veracini, 2013). Turner's (Turner and Simonsen 1963) frontier thesis established a pioneer mythos – creating order from chaos; carving society from nature – as foundational to the American national identity. In order to accomplish this spatial reconfiguration, "traditional" (i.e. non-modern) ways of knowing are erased by the invader, and supplanted with discourses that covertly self-validate (Mikdashi, 2013; Barker, 2012; Hoxie, 2008). Pearson (2013: 169) echoes Fanon, pointing out that settlers control history as a knowledge-structure: "the unqualified 'beginning' signifies the point at which history begins – that is, with the arrival of Europeans". Traditional ways of understanding and inhabiting the physical world are replaced by instrumental modern rationalities (Barker, 2012), and policed by Foucault's "community of experts": the arbiters of knowledge (Burchell, 1991; cf. Worster, 1985). This links with the broader postcolonial literature, especially Escobar's (2001) analysis of developmental discourse as a form of systemic violence. By asserting control over the structures of knowledge and language – the ways in which meaning can be legitimately conveyed – colonizers can systematically peripheralize native values, interests, and points of view because they cannot be validly encoded into core-approved discourse. In sum, settler colonialism features a fundamental settler/native differentiation in the same way that modernity bifurcates the world into society and nature, and for the same reason: capital accumulation.

This foundational differentiation of identities opens into a third line of critique: homogenization. Hoxie (2008: 1159) describes a progression in New World colonialism from differentiation to assimilation: "efforts to incorporate the indigenous survivors of the dispossession into their new nations". Mikdashi (2013: 24) glosses the intent of the settler as "making white men out of natives" by destroying existing modes of organization and reproduction of native society. Perversely, the indigenous subject is encouraged to dissolve her identity into that of the society that had invaded her own space – to renounce her native "primitivism" and to allow herself to be internalized by the "enlightened" structure of settler society (Barker, 2012; Goldstein, 2008).

Modernity offers its promise of progress if she does so, but unless she abandons her indigenous identity completely, the colonized native will not be fully internalized. Instead, she will occupy a position both inside and outside of the colonial order. She will be defined and regulated by the colonist's web of bureaucratic disciplinary structures (Mikdashi, 2013), without the benefits associated with full membership in a society that lacks conceptual structures to account for her, but nonetheless seeks to make use of her (Maddison, 2013). In short, settlers will treat her in the same way that they treat nature.

The idea in play here resonates with the Frankfurt School's contention that modernity's attempt to control nature is a covert attempt to control persons (Nelson, 2011). Within settler colonialism, the spatial doctrine of *terra nullius* and the social differentiation/homogenization of indigenous identities are not separate, but rather are different expressions of the same foundational dichotomies of modernity. *Terra nullius* is the othering of space, and differentiating the native from the settler is the externalization of nature from society. For the modernist Turnerian pioneer who seeks to penetrate the frontier and create order from chaos, the native *is* nature.

If settler colonialism – modernity *par excellence* – treats natives and nature interchangeably, then the spatial and social processes of modernity must be analogous (at least to some degree). From this perspective, the spatial logic of networks and spatial practice of networking seem to take on a new and more ominous character. The construction of Grapevine Reservoir entailed the enclosure of the Denton Creek watershed, followed by the collapse of that territorial space into a dimensionless node in the urban network topology. Whereas the reservoir project harnessed 712 square miles of territory, capital interests only internalized space that could be densely interconnected with the network of urban accumulation (waterfront real estate, commercial farms, new sites for urban development, etc.). The remainder of that territory – harnessed but not internalized – is analogous to the native in settler colonialism. It is both inside and outside the urban scheme of the world, disciplined by urban structures without accruing the benefits of urban accumulation.

The conceptual move that we are attempting to make is based on Agamben's (1998) concept of *homo sacer*: the subject who occupies an ambivalent position of simultaneous interiority and exteriority with respect to the law. "Sacred man" was a judicial status resulting from an offense against the Imperial Roman social order. *Homo sacer* could be killed by anyone without legal penalty, but was deemed unfit for sacrificial purposes – he was excluded from any formal role in society, even to die for it. Thus he was constrained by Rome, yet not internalized to the empire. Agamben deploys this concept to distinguish between "bare life", that is, merely living bodies, and life that possesses political agency. On Agamben's view, the most basic form of political power is the production of these two categories: the included and the excluded. On its face, this concept appears to be specifically social, but Agamben extends the discussion to the spatial vis-à-vis "the camp": "a zone of radical indistinction in which human communities find themselves cut adrift from the institutional and legal frameworks underpinning modernity" (Gandy, 2006: 502). Taken together with the framework of settler colonialism, this concept may help shed light on the reconfiguration of core and periphery implicated in the construction of Grapevine Reservoir.

Whereas Gandy follows Agamben closely in examining how these zones of indistinction produce urban subjects, the following sections investigate the discursive production of such a space, by examining contemporaneous regional newspaper accounts of the project in the core (*The Dallas Morning News or* DMN) and periphery (*The Grapevine Sun or* GS), as well as at the national scale in 1941 US Congressional hearings on Rivers and Harbors (R&H). We argue that modernity's networking project does not result in a "flat" geography, and that a binary spatial imaginary of networks is inadequate to account for these spaces. We analyze the construction of Grapevine Reservoir and its associated infrastructure as a networked form of settler colonialism. Urban interests conceptually collapsed the expanse of the Denton Creek watershed into a dimensionless node in the space of capital flows. This process selectively internalized space that was useful to the network of urban accumulation, but left behind other territory. We characterize this marginal land – harnessed by the city, but not internalized to the city – as *terra sacer*: bare space. Though the proliferation of the DFW-area hydro-social network may appear benign – colonization without colonialism – we seek to show that the underlying discourses merely reproduced the logics of empire.

Landlocked empire

> You have got a sort of landlocked empire in there; all you want is the water to turn it loose?
>
> *(1941 Congressional hearing on Trinity River Development)*

The intimate relationship between hydraulic networks and colonialism saturates discourse intersecting with the Grapevine Reservoir project. In his framework of empire building in the New World, Hoxie (2008) outlines a trajectory of riverine colonization. Eager invaders penetrate upriver into the interior, seeking to establish outposts for extraction by displacing natives. Over time, emphasis shifts from extraction to settlement, as the invaders begin to form their own localized societies based on the fruits of primitive accumulation. In 20th century North Texas, this colonial regime of resource extraction and local development was often portrayed in cheerfully explicit imperial language. The harnessing of Denton Creek was represented as "Texas' great chance to open a vast empire" (DMN 1943/01/22).

This nascent empire was, however, landlocked. The Grapevine Reservoir proposal evolved from a single-use water supply for Dallas and Park Cities in the 1920s to a multi-use reservoir and part of a basin-wide federal river improvement scheme in the 1940s. The primary purposes of the federal project were flood control and soil/water conservation. While these things were certainly desirable for urban interests, the overarching goal of the Trinity Improvement Association (TIA) was of much larger scope. According to the TIA, North Texas had nearly (even in 1941) reached the limits of its economic potential because it was not sufficiently connected. In order to continue to expand, the area needed a navigable waterway for efficient access to national and global markets. The grand vision of the TIA's Trinity Master Plan was to transform the river into a canal by constructing 26 locks and dams, allowing waterborne navigation for cargo vessels from the Gulf of Mexico upriver to the Dallas area. Grapevine Reservoir would be used to impound floodwater, which could later be released in order to ensure sufficient depth in the canal during periods of reduced precipitation and streamflow. Simply stated, the TIA's goal was "opening a new industrial empire in the Southwest" (DMN 1943/01/22).

At the heart of the TIA's argument before the 77th Congress House Committee on Rivers and Harbors was that railroads possessed a monopoly on freight transport. Without the competition of water transport, Dallas merchants were unduly burdened by cargo rates. Proponents of the scheme claimed that "natural resources are right there on the banks of the river, waiting for the development" (R&H). Particular attention was given to potash and gypsum – vital for many manufacturing processes – and it was repeatedly pointed out that Texas was the national leader in petroleum production. Getting those resources into the global market efficiently and economically was a prime motive for canalization. While they were valuable commodities for trade, raw resources were not the only consideration:

> Centering the Southwest, in this territory, Dallas ranks first in retail sales, first in wholesale sales, first in the number of manufacturing plants, first in the value of industrial production. . . . I quote these facts in order to indicate the size of the market and the great volume of trade that would be favorably affected by the improvement of the Trinity River.
>
> *(R&H)*

The TIA's logic was simple: growth of the Dallas economic network was imperative, but was impeded due to poor connectivity through the railroad network. The solution was to connect directly to broader markets via the Trinity River network to the Intracoastal Waterway, and from there to the oceans and ports of the world.

Although urban discourse had changed emphasis, the underlying rationality was identical to the original 1920s proposal: Denton Creek space was merely a container for urban water. Locally, the reservoir would provide commodity water for municipal metabolism and growth in Dallas, as well as protecting capital investment from flooding. At supra-local scales, the project would regularize the unreliable pulses of the river, thus allowing more efficient acceleration of raw resources into the national and global space of urban flows. This nested-scale regime of extraction was claimed to be able to do for DFW and Texas that which previous hydrological networking had done for eastern cities.

> Dallas now stands in the position of Chicago before the Erie Canal was built. That canal's construction provided a practical, low-cost trade route from the harbors of the Hudson River into the undeveloped hinterland. As a result, New York became the greatest port in the New World. Chicago became its greatest inland city.
>
> *(DMN 1948/02/15)*

In short, the Grapevine Reservoir project was a means of regularizing or disciplining space in order to reproduce the metropole.

This disciplinary regime was justified by a tacit version of *terra nullius*. Unlike classical colonial doctrine that authorized the appropriation of peripheral land because it was unoccupied, water resource development in the dry American West often justified the seizure of land that was unproductive. DFW reservoir discourse made heavy use of the productivity concept, but in a somewhat different way than the imperative to "make deserts bloom" that often drove irrigation projects in arid regions (Lane, 2013; Gasteyer et al., 2012). Instead of framing the problem as an absence of productivity in rural North Texas, urban-based reservoir advocates focused on water as a limiting factor for production in the city, and treated Denton Creek space as a remedy – as latent productivity. Agricultural production was seldom mentioned in *Dallas Morning News* accounts, and never seriously engaged in the data.

The *Grapevine Sun* data reveal contradictory tendencies. The importance of agriculture was heavily emphasized in a weekly section on soil conservation and numerous technical articles, along with moral pieces on the value of farming and the values of farm life. However, other than occasional mentions of the reservoir's potential role in soil conservation (treated as an unqualified good), the Grapevine newspaper account is nearly silent with respect to the proposal's influence on agricultural productivity until long after construction had begun. The reservoir project first appeared on the *Sun*'s front page September 6, 1946, more than a year after the proposal had been approved by Congress, and over 25 years after Dallas municipal authorities began planning for the appropriation of Grapevine space. Another year elapsed before the first full article explicitly connected the lake with agriculture, describing the auctioning of herds and "evacuation" of local farms to make way for construction (GS 1947/10/10).

Paradoxically, the Grapevine newspaper's first connection of the reservoir to local agriculture was not of local origin; the story was reprinted from a previous week's issue of the *Dallas Morning News* (1947/10/02). The article mentions the stoppage of 4300 pounds per day of milk deliveries from a displaced local dairy to Dallas milk processing company Cabell Incorporated (the corporate parent of Minit Market, claimed by its founders to be the archetype for contemporary convenience stores). However, this passing reference to spatial repurposing was treated uncritically, with no discussion of rural livelihoods, autonomy, or ways of life vis-à-vis urbanization. As with the few mentions of spatial power processes like eminent domain (DMN 1949/10/09, 1951/03/09, GS 1947/11/21), the displacement of local agriculture was not treated as an urban encroachment, but rather as an inevitable consequence of progress for all.

On one level, such discursive treatment could be construed as a paradigmatically modern conceptual shift from an agriculture-based society to an industrial/commercial one, rather than a camouflaged attempt to justify colonialist expansion. Indeed, the paucity of explicitly spatial language in the newspaper data appears on its face to preclude *terra nullius* as a relevant concept. Unlike the colonial and neocolonial case studies reviewed earlier, DFW reservoir advocates did not make overt use of concepts like barrenness, backwardness, or lack of productivity with respect to Denton Creek space. However, Schneider (2013: 151) calls attention to a related concept embedded in the settler mindset: "the assumption arising from European views of property, or land, as generic: in other words, the tendency to treat one acre of forest as equivalent to any other acre of forest when organizing space". She argues that whereas the material processes of settler colonialism are intimately bound up in a site-based geography of place, settlers (and even some critiques) tend to conceptually distribute the commons (i.e. the expanse beyond the settlement) in abstract and isotropic space. This juxtaposed place/space duality resonates strongly with the concept of *terra nullius*. From this perspective, material nodes and links of the urban network are situated or created in preferred places, beyond which extends latent territorial space that is only meaningful to modernity insofar as it can be connected and exploited. The Grapevine Reservoir project was the production of a network place in generic territorial space, and the impacts of the project were externalized to the latter: an abstract, isotropic periphery.

In lake discourse, this null space was conceived as merely a substrate for modernity, awaiting the inscription of urban circuitry. Municipality was the central organizing principle around which claims, impacts, and justification were centralized. According to consulting engineer W.J. Powell, the Denton Creek site was preferable because "there are no real towns or communities in that immediate section" (DMN 1921/12/14). Though the Town of Grapevine was incorporated, the population at that time was less than 1000 persons (many of whom were former Dallas residents). Incorporation – the formalization of a municipality as a legal entity – was seen as the only means for small communities to achieve legal standing against involuntary annexation by the City of Dallas. Though an incorporated town could only be annexed by voter consent, even officials of already-incorporated towns expressed concern that the water issue could be used as a political tool by city interests "to get us into Dallas" (DMN 1946/04/13). Unincorporated space had no protection from urban annexation. According to A.O. Hickman, mayor of the tiny hamlet of Hutchins, "Dallas growth was overflowing onto our community. We had many problems and there was no vested authority through which to solve them. Incorporation was the only way out" (DMN 1949/11/20). This resonates powerfully with patterns of settler colonialism, within which peripheral populations are denied property and agency because they are not self-codified according to legal frameworks sanctioned by the core. In 20th-century North Texas, the path to spatial agency was urbanization.

Perhaps the clearest manifestation of this urban-place/rural-space dualism can be found in the optimistic promises of lake boosters. In all of the data – *Dallas Morning News*, *Grapevine Sun*, and the 1941 Congressional hearings – proponents of the lake gave exhaustive accounts of advantages that would be conferred by reservoir construction. The overwhelming majority of these claimed benefits specifically related to urban accumulation. The DMN repeatedly cited industrial and commercial benefits of lake construction, ranging from general economic growth to the attraction of specific industries such as chemical plants, aircraft manufacture, and even an asbestos factory. "Dallas at the threshold of a new year has become a knee-pants Chicago. It's a booming, gangling industrial-distribution center growing like a boy in short pants" (DMN 1951/12/23). The *Grapevine Sun* routinely invoked the growth of local industry, along with the attraction of commerce, real estate, recreational revenue, and the possibility of attracting major projects like the proposed US Air Force Academy. Pitching the reservoir proposal before Congress in 1941, TIA General manager Fouts cited a highly detailed list of industrial, commercial, and mineral extraction opportunities that would be opened up by connecting Denton Creek to the urban space of flows. Against these exhaustive accounts of specifically urban advantages, very little was said or written about exurban benefits beyond vague references to flood prevention and occasional implications of trickle-down economics. In short, the promises of the project attached to urbanization: the process of network placemaking in the generic, bare space of the periphery.

On its face, this idea does not seem intrinsically problematic. The production of a social place necessarily involves the formation of a nexus in an amenable or hospitable space, and a city can be construed as a place of very dense interconnection. But the process becomes troubling when connected to the historical legacy of *terra nullius* in colonial practice. However amenable, such spaces are seldom empty when the placemakers arrive – something must be pushed out of the way. In the Grapevine Reservoir case, urban infrastructure and development displaced rural land use. Within the framework of modernity, productivity of this type is perceived as attaching directly to specific places – nodes of capital flow – where the exchange of money can be tabulated. As a node, Grapevine Reservoir derived its productivity and value from the space (the Denton Creek watershed) within which it was situated. However, that space was construed as generic farm or pasture land. Since development was the only thing that mattered in urban discourse, one acre of undeveloped land was treated as more or less interchangeable with any other acre of undeveloped land. Thus rural land use could be pushed aside to permit placemaking by urban settlers as they reproduced the core, under the assumption that rural activities could simply relocate within the remaining isotropic space. Rural space in North Texas was directly appropriated by procedural means (condemnation, annexation) and through political-economic power:

> When Dallas homebuilders paid Garland Mayor D. Cecil Williams $1000 an acre for 432 acres of worn farmland, it became clear that King Cotton is fighting a losing battle in rural Dallas County. You can't profitably grow crops on land that costs $1000, $500, or even $300 an acre.
>
> *(DMN 1951/12/23)*

In sum, whereas neither the doctrine nor language of *terra nullius* were explicitly invoked by urban interests, the proliferating Dallas network imprinted itself on Denton Creek space by making use of a constellation of similar spatial concepts, with similar spatial outcomes. Grapevine Reservoir was constructed as a site of value extraction, and the Town of Grapevine became an urban settlement in the midst of a generic expanse of colonized space.

Discourse, development, and domination

The functioning of the settlement – along with the rest of the colonial network – depended on efficient metabolism of Trinity River water in order to propel continued urban accumulation. This required a massive engineering effort to regularize the pulse of the river: to domesticate nature. As the scope of the reservoir proposal progressively expanded from a municipal water source to basin-wide flood prevention to canal-borne commerce with state and national significance, so did requirements for capital, technology, and expertise to execute the plan. Worster's (1985) examination of irrigation in the American West notes the same trajectory. Adapting and updating Wittfogel's *hydraulic society* thesis, Worster argues that within the modern capitalist state, water is increasingly controlled by two mutually reinforcing centers of power: an economically powerful private sector (agriculture, in Worster's research), and a public sector of bureaucrats and elected officials. Because the linkages between political-economic power, regulatory power, and policymaking authority are diffused, "power becomes faceless and impersonal, so much so in fact that many are unaware it exists" (1985: 52). Nonetheless, through the control of water, these [institutions] combine to produce a regime of covert (or at least obscured) social control.

Perhaps the most commonly deployed concept justifying the Grapevine Reservoir project was development. During the early stages of the proposal in the 1920s, when the reservoir was conceived as a purely a municipal supply concern, proponents usually advocated in terms of "growth". At the time, this term was treated as an unqualified good; growth for Dallas was construed as inevitably advantageous to the city's economy, and sufficient water would enable beneficial expansion. During the war years, and particularly after Bretton Woods, the more nuanced term "development" began to replace "growth" as a valorized concept. Rather than merely pursuing expansion, Dallas and Grapevine aspired to climb "the ladder of development" (GS 1946/05/10). This was a more comprehensive concept, deeply sourced in the modern framework of the world. Unlike bald growth, "development" implied a kind of emergent maturation: the progressive realization of human potential through increasingly refined economic and socio-natural relations. In Dallas, development meant a symbiotic agglomeration of industry, commerce, and population. In Grapevine, development entailed flourishing relationships among the recreation, residential real estate, and business sectors. Grapevine leaders explicitly rejected the idea of growth that entailed "honky-tonk operators, shanty dwellers, or big-city gamblers", in favor of "the right kind of development" amenable to "business and professional men" (DMN 1947/08/08). The discursive shift from growth to development can be seen as a move from pursuing bald quantitative increase to prioritizing a proliferating web of spatial relationships that extend between city and countryside. This comprehensive spatialization of urban development would result in more efficient accumulation over the long term, but it required a comprehensive re-working of nature. In short, the river was to be disciplined: reduced to a machine.

The spatiality of development is most clearly evident in the public discussion surrounding canalization of the Trinity. In the early 1920s – during the period of "growth" discourse – the reservoir proposal merely offered a commodity source for Dallas, along with increased recreational opportunities and enhanced property values for landowners in Grapevine (many of whom were current or former Dallas businessmen). As such, it represented a primitive level of economic connectivity between two proximate municipalities. The jump to federal scale involved a broader re-working of nature in order to provide flood control and soil/water conservation. These aims were certainly important to the US government – they were the reason the reservoir proposal was ultimately approved in the Rivers and Harbors act of 1945. However, development as a distinct spatio-economic concept only emerged in relation to the canalization aspect of the project. Navigation and inexpensive cargo transport offered more than just local growth – they promised the kind of relational, self-sustaining emergence that would allow Dallas into the ranks of the most prosperous cities in the nation.

These developmental benefits, though, did not admit of partial or incremental measures. In order to provide them, the Trinity would have to be completely reconfigured – to lose its character and identity as a river, and to be trussed in miles of concrete channel, elevation locks, and dams. The magnitude of such an intervention would be well beyond the scope of local or regional capacities. Not only would construction require enormous technical expertise and engineering competence, it would also demand the kind of vast initial capital outlay that could only be mobilized at the federal level. Thus the project was bound up both in capitalism and in politics; in Worster's (1985) public and private centers of power.

Urban interest groups exercised power and discursively constructed "the problem of water" and produced Denton Creek space in such a way that the Grapevine Reservoir proposal appeared to be the only viable solution. The logics and network spatial imaginary of modernity offered the promise of a "flat" world without spatial stratification (cf. Friedman, 2005) to those who were willing to articulate with the network. A postcolonial lens, however, orients on the territorial space *between* the mesh of the urban network, and reveals a familiar pattern of domination. In order to promote the reservoir project, the interdependent, urban-oriented public and private sectors combined to impose the same relations upon rural space that colonizers impose upon indigenous peoples. The undeveloped, "natural" Trinity was differentiated or "othered" (cf. Said, 1979) as a wild and non-modern space, in order to permit the imposition of homogenizing or regularizing power over it, such that the space could be assimilated into the circuits of capital accumulation. In short, just as colonialism produces "the native", North Texas urban development produced the periphery.

This chain of reasoning invokes Escobar's (2011: 41) discourse of development, in which:

> The system of relations establishes a discursive practice that sets the rules of the game: who can speak, from what points of view, with what authority, and according to what criteria of expertise; it sets the rules that must be followed for this or that problem, theory, or object to emerge and be named, analyzed, and eventually transformed into a policy or a plan.

Within public discourse surrounding the Grapevine Reservoir project, the question of "who can speak" reveals a very clear pattern of power relations. First, rural residents were almost never given a voice. Farmers were often written *about*, but seldom got the opportunity to speak for themselves. For example, a 1947 article that originally appeared in the *Morning News* (and was subsequently reprinted in the *Sun*) detailed the abandonment of the valley floor by farmers whose land was to be inundated. Despite a thorough enumeration of people, livestock, and equipment to be displaced, along with acreage to be abandoned, no print space is devoted to comments by the farmers (DMN 1947/10/02, GS 1947/10/10). This absence of input is perhaps unsurprising for the Dallas paper, but it is somewhat startling that Grapevine-area farmers would not be seriously engaged by their own local newspaper. Paradoxically, the relationship between farmers and the reservoir received greater coverage in the *Morning News* than in the *Sun*. Even on the potentially incendiary topic of land condemnation through eminent domain, farmers' voices are noticeably absent (DMN 1949/10/09, 1951/03/09, GS 1947/11/21). It was not until the City of Grapevine began to request water from the reservoir (all of which was allocated elsewhere) that a spark of resistance appeared. Grapevine banker D.E. Box stated "Grapevine gave up 12,800 acres of its best valley farmland for this lake, so we think we certainly should be given a share of the water *for municipal use*" (DMN 1950/06/30, emphasis added). This echoes a comment by Grapevine Mayor B.R. Wall – a displaced farmer himself – who mused "I just hope that this expected lake shore development will take up all the slack" (DMN 1947/08/08). Both of these comments imply a sense of uneasiness about the transformation of Grapevine from a primarily agricultural space to a more urbanized one. On the whole though, farmers were not among those "who can speak" within the discourse of North Texas urban development.

Those who *were* given voices in the public discussion fell almost exclusively into three categories. The first was a constellation of businessmen, who were taken to have action-based expertise on commercial and industrial matters. The second was a variety of municipal and state bureaucrats and elected politicians, who commented extensively on broad-based growth and development. The third (to whom the most epistemic authority was assigned and the most space was given) were scientists and engineers. In the words of TIA general manager John Fouts before Congress, "They are technical men" (R&H). Making use of the instrumental economic and techno-scientific rationalities (Peet, 2007), these men[1] constructed the problem of water in such a way that a radical reconfiguration of Denton Creek space seemed not only preferable, but inevitable. Further, the scope of the project meant that the reworking of nature and the resulting hydrosocial space *must* be controlled by a community of experts. Their control of knowledge-structures was passively enforced by an implied scale of epistemic deference that manifests in the newspaper data.

In general, those "who can speak" divided into economic needs/impacts-based and physical/engineering-based discourses. The entry level (so to speak) of discursive merit was assigned to businessmen who discussed growth as it applied to their own particular sector, with abstract references to broader issues. These included professionals like realtors, merchants, and manufacturers, who looked at municipal growth as providing an increased labor pool, expanded market, or both. Newspaper stories offered the statements of men like these as something like respected testimonials about business concerns. A substantially higher level of deference was paid to those with broader economic visions: public officials and private entrepreneurs who invoked ideas of development. The center of advocacy for Grapevine Reservoir was Fouts, a former municipal water commissioner for Dallas who later joined the private Trinity Improvement Association booster group to advance the cause for canalization. A charismatic spokesman with experience in both the public and the private spheres, Fouts could draw on his background to argue the need for and advantages of the reservoir. From his entry into the data in 1941, he emphasized regional development rather than merely municipal growth. As a former city official and a continuing high officer in the TIA, Fouts' statements were treated by journalists as expert pronouncements to be absorbed, rather than merely testimonials to be considered. Whereas elected officials were generally trusted to comment on economic needs, and bureaucrats were generally trusted on water resource issues, the organs of public discourse conferred upon Fouts the implied authority to comment definitively on both.

But by far the greatest level of epistemic deference was paid to engineers: the men who commanded the requisite knowledge to execute the physical reconfiguration of the river. The rank of "engineer" is consistently valorized throughout the data, and the pronouncements made by these "technical men" were never questioned. Deference manifested in several ways. The first was sourced in a fetishization of scientific/technical ability – a phenomenon that Worster

(1985) points out. Because modernity organizes around science, and modern science around metabolizing nature (to allow for personal emancipation), the water shortage in Dallas was presented as an "engineering problem" (DMN 1921/12/15, cf. Swyngedouw, 2004a). Because they were seen as having the power to determine future human-nature relationships and economic growth, the rank of engineer took on a heroic mien. Engineers "disclosed" knowledge (1921/12/15) in response to "pleas" from the population (1946/04/13). Their plans were "thorough" and "comprehensive" (1946/04/16). As the arbiters of the most cherished rationality within modernity, and as the means by which the aims of modernity could be actualized, engineers as a class were treated as experts *nonpareils*.

In addition, a second mode of deference applied to reservoir engineers, due to the institution of which many of them were a part: the highly respected US Army Corps of Engineers (USACE). Whereas the reservoir was not a new idea in Dallas, the first exploratory work and proposals were formalized in the aftermath of the First World War. More intense involvement came with the federal program of river and harbor improvement during the early 1940s, when patriotism and pro-military sentiment were swelling. By March 1945 – when the Rivers and Harbors Act was approved by Congress – the US military was steadily advancing in both the European and Pacific theaters of operations, and preparatory work for construction began several months after the final victory of the Allies.

A sense of martial triumph seems to attach to the USACE during the reservoir project. The *Grapevine Sun* includes a long article celebrating the 175th anniversary of the unit, stating "The World War II achievements of the US Army Engineers have been summed up as 'the American Miracle'" (GS 1950/06/01). The paper takes a conciliatory tone even when discussing land condemnation, emphasizing the Corps' efforts to come to beneficial settlements with landowners subject to eminent domain activity (1947/11/21). The DMN was equally respectful, especially in a long story about the USACE's reverent handling of a cemetery relocation to make way for the reservoir (1949/09/09). Admiration for the USACE ran so deeply that Fouts publicly supported the Corps when its dam-building rival – the Bureau of Reclamation – sought to take over the public works component of the Corps' mission.

> Such a change, if made, would remove the finest and best trained engineers and executives from direction of the Trinity program. It would put this program deep in politics. Why disturb such a great organization as the efficient and hard-working Army Corps of Engineers?
>
> *(DMN 1950/05/19)*

In sum, the engineers working on Grapevine Reservoir were seen as paragons of modernity: virtuosos of science, highly skilled technicians, and self-sacrificing public servants. This combination of virtues made them an idealized choice to lead the community of experts that would internalize Denton Creek to the urban network.

Discussion: spatial discipline, periphery, *terra sacer*

> I have viewed the peoples of the world and have come to the conclusion that the English speaking race is the highest ideals of Justice, Liberty and Peace. Therefore, I shall devote the rest of my life to God's purpose and make the world English.
>
> *(John Cecil Rhodes British prime minister of Cape Colony, South Africa, 1890–1895)*

The picture that emerges from Grapevine Reservoir discourse resonates powerfully with images of settler colonialism: "turning a place and a specific human material into something else" (Veracini, 2013: 313). Postcolonial critiques have tended to prioritize the transformation of human material as an object of analysis. As a very broad gloss, scholars like Said and Veracini examine the colonizer's application of power upon people in an effort to control space. Through a process of differentiation, homogenization, and assimilation, the indigenous population is disciplined in order to allow the colonization of their territory. While we make use of the same concepts, our analysis inverts the frame. Closer to Wittfogel and Worster, we examine the application of these disciplinary processes to *nature* in order to colonize the *natives*.

From this perspective, the colonial legacy of Grapevine Reservoir becomes quite clear. Modernity's networking project is fundamentally a process of settler colonialism: a reproduction of the urban core in undeveloped *terra nullius*. The Upper Trinity River basin was discursively produced as *other*: a disorderly, primitive space that could be domesticated with the proper application of modern rationalities and technical puissance. Nonurban "natives" could be pushed aside or involuntarily annexed due to absence of municipal legal standing or deficiency of political-economic power. This made way for the land to be disciplined and transformed – reduced to a reliable machine that would make wealth for the (expanded) core. Grapevine natives could enjoy the fruits of progress that diffused outward from the core, but only if they allowed themselves to be homogenized and assimilated into the modern (i.e. urban) space of flows. Rather than "making white men out of natives" (Mikdashi, 2013: 24), the Grapevine Reservoir project served to transform rural space and population into urban.

Viewed in this light, the effort to colonize Denton Creek space did not peripheralize the tiny town of Grapevine. Instead, the town was transformed into an outpost of the core, and its subsequent growth and prosperity can be attributed to the benefits of development that diffused along the network. However, the network spatial imaginary intrinsic to modernity – and embraced by various contemporary geographers – seems inadequate to account for the full range of spatial dynamics involved in the case. In short, the urban network *is* the core, so a network view seems to be necessarily incomplete (just as nighttime satellite photographs show the urban network clearly, but the surrounding space darkly). Of concern are the expanses between the illuminated nodes and flows of urbanism, and this is where Agamben's concept of *home sacer* is useful.

Grapevine Reservoir was produced as a node in the urban network. As such, it derived its value (to modernity) from the catchment surrounding it. However, city interests internalized only that space which could be closely articulated with the urban economy. The remaining territory was worthy of consideration only insofar as it supplied the reservoir. In short, the core-ification of Grapevine produced the surrounding space as a periphery. This periphery can be seen as the spatial equivalent of *homo sacer*, in that it occupies a position of ambivalence with respect to urbanization. As a catchment for urban water, it is inside the urban system of law (and governed by the community of experts). Yet because it was seen as nothing more than a commodity collection area, it also occupies a marginal position beyond the narrow network view of colonial urbanism. As in settler colonialism, it was perceived as generic, bare space – the undifferentiated darkness between city lights – that didn't possess true *being* in the modern urban scheme of the world. Because it is not covered with asphalt and concrete, it is not city, but the city way of life depended upon harnessing it. In other words, though this periphery is not urban, it is profoundly urbanized.

The implications of this peripheral space – to which we refer as *terra sacer* – manifest in multiple ways. The first resonates with Agamben's most basic exploration of the concept: the externalized convict who could be killed without penalty, but not internalized through sacrifice.

Within the liberal capitalist framework of the US, anyone with sufficient funds can purchase land within the Denton Creek watershed and build a structure (figuratively "killing" nature). However, to *sacrifice* it (internalize it through development) would damage its value to the core/network as an ecosystem service provider. Furthermore, under the framework that we have been developing, core and periphery are not fixed in space. Instead, settlement and development reproduce the core, and the core produces the periphery. *Terra sacer*, therefore, cannot be fully internalized – it can only be settled and transformed into the core/network, which in turn produces a new periphery.

Second and more importantly, we return to the interplay between spatial discipline and social control. The scope of the Grapevine Reservoir project – and the scale of its impacts – demanded extensive involvement and intensive regulation by a community of experts who acted as the gatekeepers for meaningful knowledge-structures. Attitudes of the contemporaneous North Texas population were shaped by a discursive hegemony driven by businessmen (who perceived water resources through a filter of profit), political officials (who perceived the problem through the filter of economics), and engineers (who perceived the problem as a physical obstacle to be overcome). The project created a hydro-social configuration that would require continuing social discipline of *terra sacer* from these same sectors, in the form of conservation groups, zoning, water rights allocation, environmental regulation, and so on. The bio-political idea in play is that bare space produces bare life. "The city emerges, then, as the primary locus for these new strategies of disciplinary control and the development of new interactions between different bodies of professional knowledge and expertise" (Gandy, 2006: 500). The engineering and economic discourses that drove the construction of Grapevine Reservoir produced a set of power relations that covertly colonized Denton Creek space, produced it as *terra sacer*, and inevitably tightened urban control over all space.

Note

1 Women's voices were not included.

References

Aaronsohn, R. 1996. Settlement in Eretz Israel – A Colonialist Enterprise? Critical Scholarship and Historical Geography. *Israel Studies*, 1, 214–229.

Agamben, G. 1998. *Homo Sacer: Sovereign Power and Bare Life*. Stanford, CA: Stanford University Press.

Ahmed, W. 2008. Indigenous Communities and their Marginalized Space in India. In *Studies in Indian Economy*, ed. K. R. Gupta, 188–218. New Delhi: Atlantic Publishers and Distributors (P) Ltd.

Ahmed, W. 2010. Neoliberalism, Corporations, and Power: Enron in India. *Annals of the Association of American Geographers*, 100, 621–639.

Althusser, L. 2001. Ideology and Ideological State Apparatuses. In *Lenin and Philosophy and Other Essays*, 85–126. New York: Monthly Review Press.

Bakker, K. 2010. *Privatizing Water: Governance Failure and the World's Urban Water Crisis*. Ithaca: Cornell University Press.

Barker, A. J. 2012. Already Occupied: Indigenous Peoples, Settler Colonialism and the Occupy Movements in North America. *Social Movement Studies*, 11, 327–334.

Burchell, G. 1991. Peculiar Interests: Civil Society and Governing 'the System of Natural Liberty'. In *The Foucault Effect: Studies in Governmentality*, eds. G. Burchell, C. Gordon & P. Miller, 119–150. Chicago: The University of Chicago Press.

Cardoso, F. H. & E. Faletto. 1979. *Dependency and Development in Latin America*. Berkeley, CA: University of California Press.

Celio, M., C. A. Scott & M. Giordano. 2010. Urban – Agricultural Water Appropriation: the Hyderabad, India Case. *The Geographical Journal*, 176, 39–57.

Edmonds, P. 2010. Unpacking Settler Colonialism's Urban Strategies: Indigenous Peoples in Victoria, British Columbia, and the Transition to a Settler-Colonial City. *Urban History Review/Revue d'histoire urbaine*, 4–20.
Escobar, A. 2001. Culture Sits in Places: Reflections on Globalism and Subaltern Strategies of Localization. *Political Geography*, 20, 139–174.
Escobar, A. 2011. *Encountering Development: The Making and Unmaking of the Third World*. Princeton: Princeton University Press.
Essletzbichler, J. 2011. Locating Location Models. In *The Sage Handbook of Economic Geography*, eds. A. Leyshon, R. Lee, L. McDowell & P. Sunley, 23–38. London: SAGE Publications Ltd.
Fanon, F. 1963. *The Wretched of the Earth*. New York, NY: Grove Press.
Friedman, T. L. 2005. *The World Is Flat: A Brief History of the Twenty-First Century*. New York: Macmillan.
Friedmann, J. 1963. Regional Planning as a Field of Study. *Journal of the American Institute of Planners*, 29, 168–175.
Friedmann, J. 1966. Poor Regions and Poor Nations: Perspectives on the Problem of Appalachia. *Southern Economic Journal*, 32, 465–473.
Gandy, M. 2006. Zones of Indistinction: Bio-Political Contestations in the Urban Arena. *Cultural Geographies*, 13, 497–516.
Gasteyer, S., J. Isaac, J. Hillal & K. Hodali. 2012. Water Grabbing in Colonial Perspective: Land and Water in Israel/Palestine. *Water Alternatives*, 5, 450–468.
Goldstein, A. 2008. Where the Nation Takes Place: Proprietary Regimes, Antistatism, and US Settler Colonialism. *South Atlantic Quarterly*, 107, 833–861.
Gramsci, A. 1971. *Selections from the Prison Notebook*. London: Lawrence & Wishart.
Harvey, D. 1996. *Justice, Nature and the Geography of Difference*. Oxford: Blackwell.
Harvey, D. 1999. *The Limits to Capital*. London: Verso.
Harvey, D. 2003. *The New Imperialism*. Oxford and New York: Oxford University Press.
Hook, D. 2001. Discourse, Knowledge, Materiality, History: Foucault and Discourse Analysis. *Theory and Psychology*, 11, 521–547.
Hoxie, F. E. 2008. Retrieving the Red Continent: settler Colonialism and the History of American Indians in the US. *Ethnic and Racial Studies*, 31, 1153–1167.
Jennings, J. 2006. Core, Peripheries, and Regional Realities in Middle Horizon Peru. *Journal of Anthropological Archaeology*, 25, 346–370.
Klein, N. 2015. *This Changes Everything: Capitalism vs. The Climate*. New York: Simon & Schuster Paperback.
Krugman, P. 2002. *Development, Geography, and Economic Theory*. Cambridge, MA: MIT Press.
Krugman, P. 2011. The New Economic Geography, Now Middle-Aged. *Regional Studies*, 45, 1–7.
Lane, K. M. D. 2013. Reading Boulder Dam: Landscape Alteration as National Transformation in 1930s America. *Aether: The Journal of Media Geography*, XI, 102–126.
Lefebvre, H. 1991. *The Production of Space*. Cambridge, MA: Blackwell Publishers.
Maddison, S. 2013. Indigenous Identity, 'Authenticity' and the Structural Violence of Settler Colonialism. *Identities*, 20, 288–303.
Mamdani, M. 2001. Beyond Settler and Native as Political Identities: Overcoming the Political Legacy of Colonialism. *Comparative Studies in Society and History*, 43, 651–664.
Marx, K. & F. Engels. 2002/1888. *The Communist Manifesto*. London: Penguin Classic.
McEvoy, J. & M. Wilder. 2012. Discourse and Desalination: Potential Impacts of Proposed Climate Change Adaptation Interventions in the Arizona – Sonora Border Region. *Global Environmental Change*, 22, 353–363.
Mikdashi, M. 2013. What is Settler Colonialism? *American Indian Culture and Research Journal*, 37, 23–34.
Nelson, E. S. 2011. Revisiting the Dialectic of Environment: Nature as Ideology and Ethics in Adorno and the Frankfurt School. *Telos-New York*, 105.
Padel, F. & S. Das. 2010. *Out of This Earth: East India Adivasis and the Aluminium Cartel*. New Delhi: Orient BlackSwan.
Pearson, S. 2013. "The Last Bastion of Colonialism": Appalachian Settler Colonialism and Self-Indigenization. *American Indian Culture and Research Journal*, 37, 165–184.
Peet, R. 2007. *Geography of Power: Making of Global Economic Policy*. London: Zed Books.
Perreault, T. 2005. State Restructuring and the Scale Politics of Rural Water Governance in Bolivia. *Environment and Planning A*, 37, 263–284.

Perreault, T. 2006. From the Guerra Del Agua to the Guerra Del Gas: Resource Governance, Neoliberalism and Popular Protest in Bolivia. *Antipode*, 38, 150–172.
Pritchard, S. B. 2012. From Hydroimperialism to Hydrocapitalism: 'French' Hydraulics in France, North Africa, and Beyond. *Social Studies of Science*, 42, 591–615.
Reisner, M. 1993. *Cadillac Desert: The American West and Its Disappearing Water.* New York, NY: Penguin.
Roberts, R. S. & J. Emel. 1992. Uneven Development and the Tragedy of the Commons: Competing Images for Nature-Society Analysis. *Economic Geography*, 68, 249–271.
Said, E. W. 1979. *Orientalism.* New York, NY: Vintage Books.
Schneider, E. 2013. Reservoirs make comeback in parched Texas. Available from https://texasclimatenews.org/2013/03/04/reservoirs-make-comeback-in-parched-texas/ Vol. 2014 The Texas Tribune.
Scott, C. A. & N. P. Pablos. 2011. Innovating Resource Regimes: Water, Wastewater, and the Institutional Dynamics of Urban Hydraulic Reach in Northwest Mexico. *Geoforum*, 42, 439–450.
Simmie, J. & R. Martin. 2010. The Economic Resilience of Regions: Towards an Evolutionary Approach. *Cambridge Journal of Regions, Economy and Society*, 3, 27–43.
Stanger-Ross, J. 2008. Municipal Colonialism in Vancouver: City Planning and the Conflict Over Indian Reserves, 1928–1950s. *Canadian Historical Review*, 89, 541–580.
Swyngedouw, E. 2004a. Modernity and Hybridity: Nature, *Regeneracionismo*, and the Production of the Spanish Waterscape, 1890–1930. In *Reading Economic Geography*, eds. T. J. Barnes, J. Peck, E. Sheppard & A. Tickell, 189–204. Malden, MA: Blackwell Publishing.
Swyngedouw, E. 2004b. Scaled Geographies: Nature, Place, and the Politics of Scale. In *Scale & Geographic Inquiry*, ed. E. Sheppard & R. B. McMaster, 129–153. Malden: Blackwell Publishing Ltd.
Turner, F. J. & H. P. Simonsen. 1963. *The Frontier in American History.* New York: Frederick Ungar Publishing Co.
Veracini, L. 2013. 'Settler Colonialism': Career of a Concept. *The Journal of Imperial and Commonwealth History*, 41, 313–333.
Ward, F. A. 2007. Decision Support for Water Policy: A Review of Economic Concepts and Tools. *Water Policy*, 9, 1–31.
Whatmore, S. 2002. *Hybrid Geographies: Natures Cultures Spaces.* London: Sage.
Wolfe, P. 1999. *Settler Colonialism and The Transformation of Anthropology: The Politics and Poetics of an Ethnographic Event.* London: Cassell.
Worster, D. 1985. *Rivers of Empire: Water, Aridity, and the Growth of the American West.* New York, NY: Pantheon Books.

14
SUSTAINABLE DEVELOPMENT
Quo Vadis Africa

Sagie Narsiah

Introduction

If development was modernity's finest hour then it may be argued that sustainable development is the apogee of development as evidenced by the Sustainable Development Goals (SDGs) which have been almost universally adopted. The Sustainable Development Goals (SDGs) succeeded the Millennium Development Goals (MDGs), which are largely viewed as being unmet in general and an outright failure in particular instances (Baumgartner and Korhonen, 2010; Holden et al., 2017). Yet, even as the Enlightenment aspired to the very best of human endeavor it transformed into its opposite – domination and control – a suppression of human freedoms. Instead what has transpired is a channeling of development ideals, the outcome of the Enlightenment, to ever decreasing minority interests intent on plunder and pillage. To reiterate: there has been a movement from emancipation and freedom to domination and control. While this is apposite in general, it is relevant to the periphery, Africa in particular. And, within Africa, it is sub-Saharan Africa that exhibits entrenched poverty and underdevelopment. This chapter will confine itself, in the main, to the African context. While the focus of the chapter is on Africa in general, the relationships between the global and intra-continental scales will be interrogated. Furthermore, there is due cognizance that there is an unevenness which informs the geography of sustainable development in general (Liverman, 2018). Specific examples will be used to illustrate the general argument.

The move to sustainable development has brought with it an increasing focus on issues of governance and public participation. Sustainable development strives to be inclusive, including those who have been systematically marginalized and excluded. Its focus is thus global. Governance is key to the success of sustainable development initiatives in general and the Sustainable Development Goals (SDGs) in particular. While these two concepts are concepts on their own they are not mutually exclusive. In many ways over the past three decades or so there has been an increasing coalescence of governance and public participation. The relationship between the two is quite peculiar during the neoliberal era. These two concepts and the nature of their relationship will be examined in the chapter. Furthermore, an example of the failure of governance and accountability, in a word, corruption, will be examined.

One of the key objectives of sustainable development and the SDGs in particular is to eradicate poverty and reduce inequality (Goals 1 and 10, respectively). These goals are linked.

 DOI: 10.4324/9780429450310-17

A reduction in rates of poverty will address inequality as well. The neoliberal system which is hegemonic at present has to be dismantled in order for these goals to be met substantively. The role of neoliberalism in actively increasing poverty and inequality will be discussed in the chapter.

Given that there a number of theoretical positions which inform sustainable development, the theoretical perspective adopted in this chapter is briefly outlined in the following section. Thereafter a very brief summary of the historical roots of sustainable development will be presented. An examination of governance and public participation and their relation to sustainable development will follow. As a means of interrogating the failure of governance and public participation, a case-study of South Africa will be presented. The chapter concludes with an examination of the relationship between neoliberalism and sustainable development.

Theoretical position

Sustainable development has been criticized from a number of theoretical perspectives (see for example Lélé, 1991; Redclift, 2002; Fernando, 2003; Sneddon et al., 2006; Escobar, 2011. Perhaps the most trenchant critiques have derived from the postdevelopmental and Marxist positions. The postdevelopmental position argues that sustainable development is structurally flawed and is merely another form of development which is intent on domination and control. Furthermore, the modernist epistemological and ontological bases of sustainable development militate against the realization of human emancipation and its teleology is inherently destructive. Thus, for the postdevelopmentalists sustainable development has caused more harm than good and is essentially a failed project. Moreover, for the postdevelopmentalists sustainable development ought to be done away with altogether. Those critics of sustainable development writing from a Marxist perspective don't question its raison d'etre, but ask the question "whose interests does sustainable development serve?" This is a rhetorical question because for the Marxists sustainable development serves the interest of capital. Sustainable development is therefore merely another tool that is harnessed in the service of capital to ensure its continued survival. Furthermore, the linkage between environment and development in the sustainable development discourse legitimizes exploitation of the environment in the service of capital. In other words the environment is subjected to "capitalization" (Watts and Peet, 2004; Smith, 2010). Given that there is a fundamental relationship between nature and human beings (nature/labor) the legitimation through sustainable development as a discursive device merely accentuates the contradiction. The model geared to heightened production and consumption, in other words the limits referred to by Harvey and the Club of Rome albeit from opposing positions (one Marxist and the other Malthusian) remains stubbornly in place. In this chapter the laws as applied to capital by Marx forms the basis for our interpretation of the dynamic which informs sustainable development:

> In his magisterial work, Capital, Karl Marx outlined the operation of the capitalist mode of production. In explaining the process of primitive accumulation and theory of modern colonisation, Marx suggested that geographical expansion was a logical outcome of the development of the forces of production. New lands untouched by the capitalist mode of production were drawn into the web of capitalist relations through colonisation. But, in order for this process to evolve there was a need for a separation between the owners of the means of production and the means of production. There was a need for the private appropriation of the means of production. Thereafter, the processes characteristic of a capitalist mode of production would evolve. Marx

> suggested that the process evolves in the following way. Pre-capitalist modes of production are subsumed under the capitalist mode of production through a process of formal subsumption. By formal subsumption he meant capital's acquisition of direct control of the labour process – a penetration and appropriation of the means of production. Out of the process of formal subsumption, the process of real subsumption arises whereby there is a transformation of the "nature of the labour process and its actual conditions" (Marx, 1977, p. 1034). While these processes are taking place other processes are also in operation. A process of concentration occurs as new branches of capitalist production open up. Out of this, a process of centralisation occurs as capital becomes concentrated in the hands of fewer and fewer capitalists.
>
> *(Narsiah, 2002: 4).*

The theoretical position adopted in this chapter, therefore, argues that sustainable development has suffered the same fate as development in general and has turned into its opposite. Sustainable development acts in the interest of capital. It follows the general dynamic of capital in terms of drawing in hitherto untouched swathes of space into the process of capital accumulation. Indeed, this process has led to the transformation of nature into environment. Moreover, nature is produced, packaged, privatized, and sold to generate capital.

Notes on the historical roots of sustainable development

The conventional view is that sustainable development entered the popular lexicon during the late 1980s with the publication of the World Commission on Environment and Development (WECD) monograph, Our Common Future. The report is also popularly known as the Brundtland Report, so named after former Norwegian Prime Minister Gro Harlem Brundtland. There is a huge literature which traces the entomology of the term, and there is much contestation around its meaning (Koester, 1990). Suffice to say there is hardly any consensus on the meaning of the term (Pezzoli, 1997; Jabareen, 2008). This chapter will not attempt to enter this debate. It bears mentioning though that sustainable development means different things to different constituencies and the evolution of the concept over the past three decades or more has rendered it into something that resembles, in many ways, a "contentless abstraction" (cf. Lélé, 1991; Pearce and Atkinson, 1998; Jabareen, 2008). That said though, this has not prevented sustainable development from occupying a very influential if not powerful position on the global development agenda (Sneddon et al., 2006). This chapter will attempt to present a conceptualization of sustainable development within a broader developmental context and in so doing assess its utility in addressing the major challenges facing the African continent in particular.

The key, oft-quoted definition of sustainable development is one that appears in the WCED report (1987:43): "Sustainable development is development that meets the needs of the present without compromising the ability of future generations to meet their own needs". The prefix sustainable brought together for the first time, conceptually, development as an economic concept and the environment (Lélé, 1991; Redclift, 2002; Sneddon et al., 2006; Kates et al., 2016). There is little doubt that the relationship between humans and the environment has since the advent of the industrial revolution morphed from one that that was mutually beneficial or symbiotic into one that remains exploitative and destructive. In many ways this has been legitimated by invoking human advancement and progress. This is seemingly the prevalent ethos even though there is growing acknowledgement of the destructive nature of unfettered industrialization. One of the key implications of the WCED Report was that the environment-development relationship became politicized. The Brundtland Report and subsequent initiatives demanded some sort of action on the part of governments and their political principals. The irony is that those who have benefited from the over-exploitation of the environment have shown a decided lack of political will to address the challenges of sustainable development. A case in point is the American president Donald J. Trump withdrawing unilaterally from the 2015 Paris Agreement on climate change.

During the late 1960s and early 1970s in light of the global turmoil and economic crisis there were initiatives such as the Club of Rome that began to look at questions pertaining to the relationship between world resources and population (Mebratu, 1998). The outcome was the Malthusian conclusion – population growth and resource consumption were in opposition to each other. In other words the levels of consumption could not continue unabated or indefinitely. Growth was therefore not limitless, but limited. These questions were almost always asked in relation to the existing paradigm based on what would evolve into an intense, almost religious focus on economic growth, in a phrase, "growth fundamentalism". During the 1970s with the onset of the neoliberal era while there was an increasing awareness of the importance of environmental issues, the global movement was bent towards increased environmental despoliation and exploitation. The social impacts were being largely ignored with the focus on sustainable development and its relationship to business receiving the bulk of attention (Barkemeyer et al., 2014). Even in the face of changing environmental conditions and the increasing influence of phenomena such El Niño and La Niña, there was little political will to address, progressively, the relationships between the environment, economics, and society.

Subsequent to the Brundtland Report, in 1992, the United Nations Conference on Environment and Development (UNCED), popularly known as the Rio Earth Summit, was held in Rio de Janeiro, Brazil. One of the key outcomes of the conference was the Agenda 21 document which spoke overtly to sustainable development. What was apparent from this document was that sustainable development was a shorthand for the sustainability and proliferation of global capitalism. It also acted as a homogenization device transforming nature into environment. In 2002, the World Summit on Sustainable Development (WSSD) was held in Johannesburg, South Africa. Just prior to this summit, in 2000, the Millennium Development Goals (MDGs) were promulgated. Among the goals were the eradication of extreme poverty and hunger by the year 2015. There were eight goals in all and 21 targets. By 2015 the Millennium Development Goals had been abandoned and replaced by the Sustainable Development Goals (SDGs) which had expanded to 17 goals and 169 targets to be achieved by 2030. It is unclear how 17 goals with 169 targets could be met when eight goals with 21 targets could not be met (cf. Parris and Kates, 2003; Nilsson, 2016). The barometers indicate uneven progress. While there seems to be some optimism around the achievement of the SDGs in Africa by 2030 as evidenced from the following assertion,

> Across the board, African countries perform comparatively well in terms of sustainable production and consumption as well as in climate action (SDGs 12 and 13) but perform poorly in goals related to human welfare (SDGs 1 to 7 and 11),
>
> *(Africa SDG Index and Dashboards Report, 2019: vii)*

in general though, Africa, and sub-Saharan Africa in particular, lags behind in terms of progress in achieving goals and targets. It would be a nigh impossible task to examine all the sustainable development goals and targets in this chapter. It would be expedient, however, to examine issues such as governance and public participation which we argue are indispensable to sustainable development in general and the SDGs in particular.

Governance and public participation

During the past 30 odd years, the concept of government has increasingly given way to governance. The concept of government which has a long provenance simply refers to the exercise of rule over a population. For critics, government signifies domination and control. But, for others it implies orderly rule, in a phrase the rule of law, authorized by a population through mechanisms such as regular elections and parliament. Furthermore, government assumes a relationship between governing – those individuals, parties, or groups who have been given the responsibility for this task – and the governed – those over whom rule is exercised. In effect government acts as both producer and supplier of services while individuals, groups, and communities, in a word citizens, are consumers of those goods and services which government provides. In effect there exists a patron-client relationship. Specifically, the patron is the provider/producer of services (government) and the client is the consumer/citizen. The movement to governance has resulted in a reconfiguration of the relationship between government and citizens. This movement is a dialectical one. It is characterized by a reconfiguration if not transformation of not only government and citizens as separate entities, but additionally to the relationship between the two.

In the first instance, the nature of government has changed. Government is no longer the sole producer and supplier of services. Indeed, in many cases government neither produces nor supplies services. There is a third party which provides these services to citizens. Furthermore, the citizens themselves may be responsible for the provision of services. The era of governance has meant therefore that the unilineal relationship between government and citizens had given way to multi-stakeholder or even multilateral relationships. Furthermore, control by politicians, while democratically elected but seldom accountable had given way to increased power and control being vested in managers. Managers are responsible for decision-making, the allocation of resources, implementation, and so on. The performance of managers is monitored and they are remunerated accordingly. There are also various very lucrative incentives available to managers. A key characteristic here is that managers are unelected and may not be necessarily be accountable to the broader public. Accountability derives from the rules, procedures, and regulations which govern the institution(s). This is one of the features of what has become known as the new public management. Under the new public management regime there are multiple scales at which governance and governing takes place. This is particularly relevant to the nation state which has also undergone transformation. Some have referred to this as the "hollowing out" of the state where its powers and functions are either devolved to intra-state scales or given up to supranational regional, continental or global entities (Rhodes, 1994; Jessop, 1996). While some may argue that this has resulted in the weakening of the nation-state, it does not mean that the nation state has disappeared or is obsolete (see Dicken, 2007). This reconfiguration has meant that unelected individuals, groups, constituencies have an increasing influence on the governance of populations. The advent of governance has meant that the hierarchical arrangement that was characteristic of rule under government has given way to a network arrangement. While the hierarchical arrangement signified a flow of power facilitating control by a minority, the network arrangement is characterized by a horizontal flow of power. Power is a product of multiple relationships among, some would argue, self-interested actors. Sustainable development initiatives therefore take place within this governance context. Later on we will look at an example of the failure of governance of this type.

The patron–client relationship described earlier was very much *de rigueur* during the 1960s and 1970s. The publication of Arnstein's seminal paper in 1969 on citizen participation encapsulated the ethos of the 1960s were modernist planning was being increasingly challenged. Since then there has been a proliferation in the public participation literature. There is no doubt that public/citizen participation as a concept has influenced the practice of politics. Participatory democracy and deliberative democracy are but two examples of the influence of public participation. According to Carole Pateman (2012) there has been a strong movement to deliberative democracy, particularly during the period from the late 1990s. Deliberative democracy presumes citizen participation. During the 1980s citizen participation and indeed public participation in governance began to increase. According to Pateman (2012: 7)

> During the 1980s "participation" began to become part of mainstream development practice – one commentator has stated that participation has "become an act of faith in development" – promoted by nongovernmental organizations (NGOs) and international agencies, not least by the World Bank.

It is now taken as a given that government and governing, in a word governance, requires some form of citizen participation. Indeed citizen participation is integral to governance. Public/citizen participation is a vital component of sustainable development. The rationale here is that there is a need for transparency and accountability.

Yet, there is hardly any consensus about what constitutes public participation and its variants. Simply put public or citizen participation is about the involvement of "the public" in the process of governing. What is often contested is the nature of involvement. Ideally each governance model strives for citizen empowerment and citizen control of decision-making. Some examples of these approaches are participatory budgeting and the panchayat system in the Asian sub-continent. However, in many instances, vested interests are served when information or placation acts as substitutes for the involvement of the public in governance. Furthermore, the public are required to rubberstamp decisions already taken outside of the public gaze. In its more sophisticated incarnation, the public are invited to be complicit in their own domination. An example of this approach is the active citizenry approaches characteristic of Blair's Third Way. Here if a citizen needed to access certain welfare benefits for example, one had to become an "active citizen", that is, engage with the government on their own terms. Additionally, those who did not comply would be actively excluded. Here the strategy is more discursive rather than overtly punitive or penal.

Governance and public participation cannot be considered apart from the public sphere. Perhaps one of the mostly influential writers on this issue is Jurgen Habermas who identifies the public sphere as a space between the state and society. It is also a space characterized by free speech (a space where citizens can engage and speak freely without fear). It is within this context that something approaching public opinion may be formed. For Habermas, this is a non-conflictual space where consensus may be reached through the application of reason and rationality. The media plays an important role here in terms of shaping public opinion. Therefore a free press is essential.

Yet, one may argue, provocatively perhaps, that consensus in the form of the application of global projects such as the MDGs and the SDGs have yielded very little in terms of substantive progressive tangible benefits to the majority of the marginalized which it seemingly targeted. The data on the many and varied goals and their targets have been largely unmet. The spatial relationship between the developed and the underdeveloped/developing world remains intact. Africa remains the producer of raw materials for the consumption of the developed core areas – both traditional and incipient – western Europe/USA/Japan as traditional cores and now the entry of China. The dynamic of capitalist exploitation described earlier in the theoretical statement remains intact. The plunder of the continent takes place within this context. Africa and sub-Saharan Africa in particular remains the most unequal and underdeveloped in the world. The levels of poverty on the continent remain stubbornly high. Here (un)sustainable development has wreaked havoc in the form of its consequences – extreme climatic events – floods of biblical proportions, droughts of equal intensity, famine, pestilence, and conflict. For example Akinyemi (2016) has shown how climate change has led to an escalation of conflict between nomadic herders and crop farmers in Nigeria. In the northern districts of KwaZulu-Natal Province in South Africa there is a severe water insecurity challenge due to climate change (Patrick, 2019). These examples illustrate the scale of the challenge that climate change poses to Africa. Its effects range from Nigeria in the north to South Africa in the south. There are numerous other examples of course. The argument here is that sustainable development is capitalism in its ecological phase. Nature itself is reincarnated as environment and packaged to generate capital. Africa has been the testing ground for some of these projects. There are major solar and wind energy projects underway. Not unsurprisingly these projects are private-sector driven and are initial movements for the privatization of energy on the continent. This movement towards clean energy notwithstanding, fossil fuels still remain dominant on a global scale.

Sustainable development and the failure of governance – the case of South Africa

Under apartheid, the National Party (NP) acted in the interests of Afrikaners in particular and whites in general. The apartheid state was solely responsible for the facilitation of Afrikaners as a fraction of capital in South Africa. This fraction became dominant during the 1960s onwards. Since the 1980s, capital was in crisis in SA. There are a couple of studies which have looked at this issue, primarily from the regulation theory perspective (see for example Gelb, 1991). During the late 1970s and into the 1980s big capital came together to assemble a new accumulation strategy. The new accumulation strategy revolved around the crisis of capital in SA and its resolution through broaching what some writers have referred to as "racial fordism" (mass consumption/mass production). Key to this was of course a post-apartheid dispensation based on universal franchise. The culmination of this initiative was among other key conjunctures, the February 2, 1990 announcement by Mr. F.W. de Klerk unbanning the liberation movement; the negotiations between 1990 and 1993; and the first democratic election held on April 27, 1994. The government of national unity was installed after the elections. The African National Congress (ANC) was the majority party. It remains the majority party 25 years later. The first government faced the enormous challenge of re-creating an inclusive state and economy. By 1996 SA had made several strides politically and economically. Economically, the ANC had effectively jettisoned its faux socialism and substituted it with neoliberalism. It was clear, early on, that the economy was going to be a key battleground, the outcome of which would determine the development trajectory of the nation.

Twenty-five years into democracy, the development trajectory of South Africa is one characterized by a population where over 50% live in poverty (broad definition around $75 per month); 27% are officially unemployed (this rises closer to 40% when those who have given up looking for employment are included) – the unemployment among the youth is well over 50% (54%); inequality is around 0.63, up from 0.59 around 1993/1994. South Africa's economy has dipped in and out of recession and growth rates remain stubbornly low – the GDP grew by 0.8% in 2018 and by 1.3% in 2017. These challenges are stated time and again; however, the indicators show little if any improvement. On the contrary things seem to be getting worse. Perception studies among South Africans indicate that people are increasingly pessimistic about conditions both locally and globally.

Corruption in post-apartheid South Africa

There was a specter which haunted the South African state when it entered the democratic era in 1994. It was one which had decimated several post-liberation countries and many in Africa. And that was the specter of corruption. Sadly, corruption in South Africa has reached embarrassingly high levels. It is corruption which has clearly acted as a brake on development in South Africa. The fruits of liberation are being greedily devoured by a tiny minority – the predatory elite. And, it is being done in the name of the oppressed people of the country. It is an act of unconscionable immorality; however, the predatory elite continue to act with incorrigible impunity/immunity. For some analysts, corruption has been identified as systemic. For others corruption is the shadow state or is a parallel state.

The sheer scope and scale of corruption is being laid bare, quite literally, on a daily basis. Commissions of inquiry, such as the Zondo Commission are daily receiving information about corrupt activities. The Zondo Commission into state capture is among a number of commissions of inquiry which have been proclaimed by the president of South Africa and his predecessor to assist with the uncovering of acts of corruption. The commissions are expected to provide remedial measures for what is ultimately a failure of governance. Ironically, the "failure of governance" is a euphemism for corruption. In South Africa there certainly has been a failure of governance with regard to corruption which is seemingly rampant.

Recently, there was a Commission headed by Judge Nugent into the affairs of the South African Revenue Services (SARS). Judge Nugent found that there were egregious lapses in governance which had led to the virtual implosion of the institution. Indeed a chapter in the final report was titled "A Massive Failure of Governance and Integrity – Never Again". Not so long ago, a period prior to the "dark ages", SARS was hailed internationally as an institution of high repute when it came to the governance of taxation and related affairs in South Africa. And, while it had taken over a decade to build SARS into an outstanding model of tax governance – there were awards which the agency had won – it took a far shorter period for the institution to be systematically emasculated. Under the stewardship of Mr. Pravin Gordhan the rates of tax compliance increased. In consequence so did revenue collection. So much so that there was scarcely a financial year when tax revenue targets were not exceeded. Those responsible for governance at SARS ensured that a singular approach was adopted for individuals regardless of status. Indeed even the former president of South Africa, Mr. Zuma, did not escape scrutiny. Furthermore, those individuals and groups responsible for illegal activity (i.e. tax evasion), were targeted and pursued with dogged commitment by a team of intrepid, enthusiastic, highly competent, qualified staff. There were a number of high profile cases which SARS pursued, among them cases involving the former president, his family, friends, and associates. There were specialized units within SARS who were responsible for pursuing tax delinquents. In his book *Rogue*, SARS investigator and longtime SARS staffer Johann Van Loggerenberg (2016) details the institution in its heyday and its subsequent dismantling – or to use the parlance of authors of the book *The Shadow State* (Peter et al., 2018), how the institution was repurposed. Institutions are only as effective as those who are willing to ensure that the rules, regulations, ethos, and morality which govern them are implemented and respected. Institutional design serves the purpose of ensuring that if rules, procedures, etc. are followed, an expected outcome, beneficial to the institutions and those its serves alike, would be obtained. Moreover, if these procedures are not followed, there is a distinct probability that the expected outcome would not be achieved. In the case of SARS, the primary objective is the collection of taxes. Without certainty in tax collection, the country's fiscal policy is adversely affected, indeed service delivery to the poor in particular is compromised. It is certainly the most vulnerable in society who would be the first to feel the impact of lower tax income. The failure of governance at SARS has further fiscal and monetary implications. Money would have to be borrowed from private banks at premium rates. This has an impact on government's ability to service debt, in turn this may affect the investment ratings of the country leading to ratings downgrades, high borrowing costs, and the flight of capital. This adds to the instability of the South African financial market which at the best of times is volatile, given that there are no strict regulations governing the inflow and outflow of capital, particularly over the short term. This brief overview of SARS indicates that governance of key institutions has enormous implications for development.

Following exposes by, among others, Jacques Pauw (*The President's Men*), Johann Van Loggerenberg (*Rogue*), Crispin Olver (*How to Steal a City*), Louis Myburg (*Gangster State*), and the findings of key inquiries such as the Public Protectors Report and the Nugent Commission, if these were true, it would be very difficult not to conclude that key state institutions such as SARs were consciously repurposed to exhibit the characteristics of the failure of governance – they were consciously corrupted. This coupled with breath-taking incompetence made for a toxic, if not explosive cocktail. The implications are far reaching as briefly elucidated earlier, but what is of greater concern is that it sets South Africa on the path of becoming a failed state. This begs the question why? Given the extended struggle against oppression and underdevelopment in South Africa it is of concern that the erstwhile liberators have been keenly implicated in the project, which has come to be known as state capture. The term derives from the final report of the former Public Protector titled "state of capture" where she detailed in a few instances how the state had been captured by the former president and his cronies. Her concerns were of such a serious nature that she indicated that the President could not play a determining role in the process given that he was implicated in serious malfeasance. The president's response was predictable in that he rejected the findings of the Public Protectors report and launched a court challenge. True to form, he withdrew his application a year later in October 2017. The president was sanctioned for his frivolous application and a costs order was granted against him. Unsurprisingly, he appealed against this as well, and predictably lost again.

The former president has always denied wrongdoing of any kind. Even with evidence mounting up against him at the state capture inquiry, his response has been that "he is satisfied that he has not been implicated in state capture". These denials are difficult to sustain given that the responsibilities of executive authority were seemingly outsourced to a third, unelected party. The third party here is a collective consisting of a rapacious immigrant family; a group of acolytes and hangers on; businessmen and those who cannot be labeled as anything else but thugs. This group has reaped rich rewards all at the expense of the majority. It is a network of patronage – crass, crude, and rapacious – intent on imagining and realizing plunder on an obscene and vulgar scale.

The predatory elite have not had it all their own way though. More and more of the corrupt and the corruptors are being exposed. And, while there is much consternation that those implicated have not been jailed, there is a sense that the wheels of justice while not turning as fast as may be expected, the wheels are turning nonetheless.

There are those who argue that corruption preceded the Zuma era, the period from around 2009 when Zuma was elected president of the Republic. While this is true, it does not explain the unprecedented levels of corruption which have beset the country at present. One should be cognizant of the fact that Zuma was part of the ANC executive until his second term as ANC president ended in 2018. During this period a web of accumulation has been systematically spun with a single center of power. Some analysts have labeled this neopatrimonialism. The structure of the ruling party suits this conceptual framework in terms of the structures of governance which exists in the party. It is a hierarchical structure.

It is a structure which Zuma is intimately familiar with. Indeed he used it to his advantage to defeat his predecessor Mr. Thabo Mbeki. Zuma is also familiar with the ways in which government institutions work. What Zuma was adept at during the years in which he was president was to re-engineer the state apparatus and in so doing create a system of accumulation which can only be described as parasitic.

To achieve this objective, it was necessary to deploy party cadre who in the final instance were loyal to him personally rather than the party itself. In effect this was taken as a given since he had remade the ANC in his own image by controlling the key levers of power within the party itself, namely the National Executive Committee (NEC). The NEC was staffed with members loyal to Zuma. This was an outcome of the slate politics evident at the Polokwane Conference in 2007. Control of the party meant control of the state and state resources. This is a consequence of what Butler (2009) has argued is the party-state. There is no separation between the party and the state, so much so that the ANC resolutions at conferences are expected to be implemented as government policy. ANC policy is a product of dominant factions within the party itself. Taken to its logical conclusion therefore, government policy is a product of a small group of people who have not been elected by the broader populace.

Zuma created a network of patronage, rewarding those loyal to him. The cabinet was expanded substantially. The cabinet of South Africa is among the biggest in the world. It is bigger than the USA and China. Loyalists are also deployed to various State Owned Enterprises (SOEs) and government in general. The network is vast and extends through to the provincial, district, and local municipal levels. There are also the major metropolitan areas which are staffed by party cadre. This is the spatial framework within which corruption operates. In and of itself the spatial framework does not facilitate corruption, but it is used to facilitate and enable corruption. The key reason for this is because there are numerous opportunities for rent seeking and extraction. Central to this are those involved in decisions relating to procurement. Studies have revealed that government in South Africa is outsourced. In the main, the government of South Africa does not produce goods and services. It procures goods and services from service providers. There are indeed instances where procurement of goods and services are outsourced.

The components to enable accumulation or the plunder of resources were assembled over a period of time. This entailed the systematic dismantling of key agencies such as the Scorpions and its replacement with the Hawks. The Hawks came into existence in 2008 when Zuma was president of the ANC. The Scorpions proved to be a very effective crime fighting unit. And, they posed a serious threat to Zuma and his acolytes. Zuma had the Hawks under his effective control and his appointments to head up the unit ensured that key investigations into corruption were quietly shelved. Indeed the unit was utilized to attack Zuma's opponents. Likewise the intelligence services were repurposed to serve the political purposes of the president. The police services were similarly politicized. Those who raised questions were simply replaced. The strategy was clear – agencies which had the power to investigate malfeasance/corruption were under political control. The civil service as a whole was politicized.

With agencies responsible for the investigation of corruption firmly under control, SOEs were targeted for unfettered plunder. Among the SOEs which have the highest procurement spend is Eskom, Transnet, and SAA. These SOEs were targeted by a group centered around the Indian immigrant Gupta family. This family had ingratiated themselves with the president and his family. The president's son works for the Gupta family and is a director in a number of their companies.

Initially, the family started up a media company that would effectively act as a mouthpiece of the ANC and the government. The company was given preferential treatment in terms of advertising and broadcasting arrangements with the SABC – all these were beneficial to the

Gupta media company. Their staff was made up of a large contingent of Indian expatriates. To facilitate the employment of this large contingent of staff, it was necessary for the allocation of visas and related work permits. These were issued by the Department of Home Affairs which was under the political control of Mr. Malusi Gigaba, a trusted and loyal lieutenant of Zuma. Subsequent investigations revealed that Mr. Gigaba had acted in favor of the Gupta family by granting them citizenship when they were not eligible.

When Mr. Gigaba was moved to the portfolio of Public Enterprises, it became clear that he was not acting impartially. At Transnet he attempted to install a Gupta-linked businessman as Chairperson of the tender committee. He failed in this regard when the cabinet refused to approve the appointment. He then created another structure which gave effective control over tenders and installed his preferred candidate. The preferred candidate then acted in the interests of his group. There were two key players installed at Transnet at the time – Mr. Brian Molefe and Mr. Anoj Singh. These two people would later move to Eskom where they presided over deals which favored the Guptas. They facilitated the purchase of a coal mine and ensured that the company received huge amounts for their substandard coal. The impact of this has had a devastating impact on South Africa. They have subsequently fled the country. Eskom was at one stage one of the world's leading power utilities. During the early 2000s the utility was running a surplus. Today, it is highly indebted to the tune of over R400 billion. There are estimates that as much as 1.4 trillion rand or around 100 billion dollars at today's rates was looted from Eskom alone over the past decade or so. Although South Africa's economy is not growing, Eskom is unable to provide a stable power supply. The utility has been devastated by corruption. The impact on the economy is equally devastating.

I have alluded to the fact that corruption is pervasive – it exists at both the macro-national scale and the micro-local level. Crispin Olver (2017) in his book *How to Steal a City* reveals how corruption has occurred at the local level. It is articulated through political control of the Council. Furthermore, there is serious contestation within the ruling party between contending factions – all these factions are intent on looting the local state through control of tenders. Interestingly Olver suggests that communities are mobilized to facilitate plunder. For example protests are organized around various aspects of services delivery. The municipality is forced to provide these services. The service provider is linked to the protest action and benefits through huge tenders. It is a bizarre situation which forces those of us thinking about the way in which social movements develop to consider an alternative theorization to that of Castells et al.

Investigations into corruption in SA have provided rich empirical material on how the various schemes operated. Corruption has been characterized as systemic. What is apparent is that there is a lack of oversight when it comes to procurement. Additionally, government has effectively outsourced procurement. The whole system of procurement needs to be addressed. The politicization of the civil service also needs to be addressed. The civil service needs to be accountable to the public rather than politicians. Public participation in governance needs attention. Several court judgments have ruled against the government for not facilitating public participation in various governmental processes. Lastly, there needs to be a clear distinction between the party and the state. These are but a few issues which need to be addressed, without it, "The executive of the post-apartheid state is little more than a committee for managing the affairs of the predatory capitalist elite" (cf. Marx and Engels).

The case study on corruption in South Africa gives essence to the challenges facing the development on the African continent. As argued earlier governance and public participation is essential to a sustainable development that is inclusive, democratic, and indeed substantive. Yet, as the example indicates, systematic erosion of institutions' aims, goals, and objectives and their effective dysfunctionality militates against the sustainable development. According to

Transparency International, African countries are amongst the most corrupt in the world, with Somalia ranked the most corrupt and the region as a whole being ranked the most corrupt (Transparency International, 2019). As Dasgupta (2007a: 7) argues,

> In the opposite vein, [total factor productivity] TFP could decline over a period. Increased government corruption or civil strife could cause this decline, which not only destroys capital assets, but also damages a country's institutions, both public or civic. When institutions deteriorate, assets are used even more inefficiently than previously, and TFP declines.

Neoliberalism and sustainable development

Over the past four decades or so, neoliberalism has proven itself to be the dominant political-economic discourse on a global scale. First seen as a viable alternative in the early 1970s to the faltering state intervention, welfarist, Keynesian approaches of the post-World War Two era, neoliberalism has succeeded in capturing the development imaginary. Development and by implication sustainable development is conceptualized (imagined) in neoliberal terms. In other words development can only be viable if it is focused on economic growth; adopts free markets; is export orientated; adopts private sector business practice in government; regulates in favor of capital, little or no regulation of capital flow, privatization of basic services, user pays principle, removal of subsidies and welfare support for the poor, transfer of assets to the private sector. These are some of the policy measures adopted by governments following a neoliberal growth path. Since the early 2000s there has been a veritable explosion of literature on neoliberalism, to such an extent that there is hardly consensus on what neoliberalism means anymore. Our perspective here is that neoliberalism is an advanced form of capitalism. Neoliberalism is an homogenizing device in the form of neoliberal globalization, but is hardly homogenous, that is, occurring at the same time, in the same way, everywhere. There are spatial variations in this regard dependent on the local context. In other words neoliberalism has a geography and that geography is a product of the processes of formal and real subsumption described earlier. Therefore, sustainable development is the vehicle through which neoliberalism acts to colonize spaces hitherto unconquered.

In the MDGs and the SDGs there is hardly an attempt to address the root causes of poverty and inequality, for example. The root cause is ironically identified – the focus on growth-oriented development. Yet, the remedy is simply more growth. Some lip service is paid to consumption, but little by way of curbing the gluttony of the wealthy is adopted. Piketty has detailed how the concentration of wealth among the global elite has increased somewhat spectacularly during the last four decades or so, rivaling if not surpassing the levels of wealth in the hands of the minority of the pre-World War One era (cf. Oxfam, 2020). This was the period during which neoliberalism has been in the ascendancy. In consequence there seems little chance of inequality and poverty being substantively addressed in places like sub-Saharan Africa for example.

> While the richest Africans get ever richer, extreme poverty in the continent is rising. Africa is the second most unequal continent in the world, and home to seven of the most unequal countries. The richest 0.0001% own 40% of the wealth of the entire continent. Africa's three richest billionaire men have more wealth than the bottom 50% of the population of Africa, approximately 650 million people. Meanwhile, Africa is rapidly becoming the epicentre of global extreme poverty. While massive

reductions in the numbers living on less than $1.90 a day have been achieved in Asia, these numbers are rising in Africa. The World Bank estimates that 87% of the world's extreme poor will be in Africa by 2030, if current trends continue.

(Oxfam, 2019: 3)

The MDGs and the SDGs do not target systemic change. There is no mention of neoliberalism in the SDGs either in the goals or targets. Yet, it is apparent that during the period of neoliberal hegemony, inequality has been increasing and there has been a concentration of wealth in the hands of a minority. Not only does this minority wield significant economic power, but they also have great political power (cf. Dos Santos in Angola; Kenyatta in Kenya; Ramaphosa in South Africa). It is little wonder that business-friendly policies are adopted. Furthermore, as has been demonstrated in the case study on South Africa, there also exists a predatory elite whose main source of wealth is the plunder of the state. There are of course numerous examples in both a bygone and contemporary era: Mobutu Seseko, Sani Abacha, Dos Santos, Guptas – to name those who gained notoriety.

Development schemes such Structural Adjustment Programmes (SAPs) and the assistance given to Highly Indebted Poor Countries (HIPC) played pivotal roles in deepening misery of sub-Saharan countries in particular. SAPs were applied in a number of countries across Africa. The SAPs were stringent conditions attached to loans given to the most impoverished countries by the International Monetary Fund (IMF) and the World Bank. Far from addressing issues such as high debt, these SAPs led to an increase in indebtedness. The policies which these countries had to adopt were the textbook neoliberal policies that promote the entry and proliferation of the private sector, usually the big business from the Western hemisphere. The largest banks in the world lent money to these countries. Export-oriented development was promoted and worked in fits and starts, but was rarely sustainable. The economies of sub-Saharan Africa is primarily based on the export of raw minerals and cash crops. These commodities are subject to the caprices of the global markets. There have been numerous occasions when hardship has been compounded by market crashes. The focus on the export of raw materials is a cause of underdevelopment. There has been a steady decline in the capital assets and the erosion of institutions of governance which are linked to the management of these assets. The failure to replace capital assets and the virtual implosion of institutions are key reasons for Africa's decline (see Dasgupta, 2007a).

The power of the structural adjustment discourse cannot be overstated. Its legacy still remains. And, so do its policies. Even though the nomenclature may have changed, the key policies which inform economic practice remain intact. The focus is still on private sector led growth and the creation of conditions to facilitate such growth. Indebtedness now includes China. Oxfam (2019: 3) indicates

unsustainable debt has returned to squeeze the populations of Africa. While there have been some exceptional individual cases, there is great risk of a continental crisis that will undermine development for the next decade, while population growth increases. Reckless lending by European financial markets and Chinese banks, combined with reckless borrowing by a number of African governments, is already severely constricting spending on public services.

What has occurred, as in the case of South Africa for example is that countries are disciplining themselves. They are imposing structural adjustment policies on themselves. But, sub-Saharan Africa does not attract the volumes of capital needed to address issues of poverty and inequality. Though it cannot be denied that there are countries in Africa which are among the fastest growing in the world, like Ethiopia, Rwanda, Senegal, Tanzania, which are among those countries exhibiting high levels of growth. It remains to be seen though whether these growth rates can be sustained and whether they will make a substantive difference to poverty and inequality. The biggest and most developed economies on the continent such as South Africa have shown negligible growth. Indeed, Africa's most advanced economy dips in and out of recession. Economist Dasgupta (2007b: 5) has argued that a focus on growth alone is misleading: "the moral is telling: GDP (or for that matter, HDI) is not a measure of long-run human well-being, and that movements in GDP or HDI are a poor basis for judging economic progress".

One of the key difficulties of neoliberalism is that it is a discursive strategy working insidiously, seeping into the interstices of everyday life. So much so that populations view it as normal, even common sense. In other words as a tool to produce consensus, it is very effective. It is therefore hegemonic and has maintained that dominant position, in the face of alternatives. Thus, it is an effective tool of governance. Neoliberal governance has effectively drawn in populations through strategies such as sustainable development. However, not everyone enjoys the same status. There is an internal process of differentiation or reclassification which divides populations into smaller groups of rich people and masses of poor people. In the main the strategy is spatial. Rich people are grouped together in highly resourced exclusive spaces while the masses wallow in spaces of squalor and underdevelopment (cf. Fuchs, 2017). The rich have access to high quality services while the poor have to contend with inefficient, low quality services, if indeed they receive any at all (see for example Woldemariam and Narsiah, 2014; Narsiah, 2017). The reason for this is unambiguous – high quality services are reserved for those who can afford to pay for them. Services to the poor are provided as an afterthought. The removal or reduction of subsidies render the situation of the poor even more tenuous. It is unlikely therefore that issues of poverty, unemployment, inequality, and inequity will be substantively addressed under a neoliberal regime of governance.

Conclusion

The ideals which characterized sustainable development have morphed into its opposite. By linking environment and development explicitly for the first time as Brundtland did in 1987, the WCED succeeded in achieving what Truman had almost 40 years earlier – setting out a roadmap for the continued exploitation of the world's resources. The irony, as the evidence has shown is that development has enabled new relations of exploitation at hitherto unprecedented levels of squalor, underdevelopment, and suffering. As more and more are thrown into poverty, so too wealth becomes concentrated in the hands of a small minority. We have shown how components integral to sustainable development like governance and public participation have failed in the substantive project of human upliftment. So too have attempts to address poverty and inequality. The argument in this chapter is that there is little or no effort to address the systemic issue of neoliberal sustainable development. Without this change the onward march of global capitalism towards an environmentally destructive, unsustainable future will continue apace. While corruption is a global phenomenon, it has affected Africa particularly severely, impoverishing already poor populations. The example of South Africa was used to demonstrate how corruption has led to the virtual destruction of state institutions within a very short space of time.

References

Akinyemi, T.E., 2016. *Climate change, migration and resource contestations: A case study of North-South migration in Nigeria* (Doctoral dissertation, University of KwaZulu-Natal, Pietermaritzburg).

Barkemeyer, R., Holt, D., Preuss, L. and Tsang, S., 2014. What happened to the 'development' in sustainable development? Business guidelines two decades after Brundtland. *Sustainable Development, 22*(1), pp. 15–32.

Baumgartner, R.J. and Korhonen, J., 2010. Strategic thinking for sustainable development. *Sustainable Development, 18*(2), pp. 71–75.

Butler, A., 2009. Considerations on the erosion of one-party dominance. *Representation, 45*(2), pp. 159–171.

Dasgupta, P., 2007a. Measuring sustainable development: Theory and application. *Asian Development Review*, 24(1), pp. 1–10.

Dasgupta, P., 2007b. The idea of sustainable development. *Sustainability Science, 2*(1), pp. 5–11.

Dicken, P., 2007. *Global shift: Mapping the changing contours of the world economy*. London: SAGE.

Escobar, A., 2011. *Encountering development: The making and unmaking of the third world* (Vol. 1). Princeton: Princeton University Press.

Fernando, J., 2003. The power of unsustainable development: What is to be done? *Annals of the American Association of Political and Social Sciences* 590, pp. 6–34.

Fuchs, C., 2017. Critical social theory and sustainable development: The role of class, capitalism and domination in a dialectical analysis of un/sustainability. *Sustainable Development* 25, pp. 443–458.

Gelb, S. (ed.), 1991. *South Africa's economic crisis*. David Philip, Cape Town.

Holden, E., Linnerud, K., Banister, D., Schwanitz, V.J. and Wierling, A., 2017. *The imperatives of sustainable development: Needs, justice, limits*. London: Routledge.

Jabareen, Y., 2008. A new conceptual framework for sustainable development. *Environment, development and sustainability, 10*(2), pp. 179–192.

Jessop, B., 1996. Post-fordism and the state. In *Comparative welfare systems,* pp. 165–183. London: Palgrave Macmillan.

Kates, R.W., Parris, T.M. and Leiserowitz, A.A., 2016. Editorial–What is sustainable development? Goals, indicators, values and practice. *Environment Science and Policy for Sustainable Development*. www.environmentmagazine.org/Editorials/Kates-apr05-full.html

Koester, V., 1990. From Stockholm to Brundtland. *Environmental Policy & Law, 20*, pp. 14–19.

Lélé, S.M., 1991. Sustainable development: A critical review. *World Development, 19*(6), pp. 607–621.

Liverman, D.M., 2018. Geographic perspectives on development goals: Constructive engagements and critical perspectives on the MDGs and the SDGs. *Dialogues in Human Geography, 8*(2), pp. 168–185.

Marx, K. (ed.), 1977. *Capital: Vol. 1*. New York: Vintage Books.

Mebratu, D., 1998. Sustainability and sustainable development: Historical and conceptual review. *Environmental Impact Assessment Review, 18*(6), pp. 493–520.

Narsiah, S., 2002. Neoliberalism and privatisation in South Africa. *GeoJournal, 57*(1–2), pp. 3–13.

Narsiah, S., 2017. The spatiality of citizenship and governance: The case of water and sanitation in Durban, South Africa. In Reilly, T. (Ed.). *The governance of local communities: Global perspectives and challenges,* pp. 183–198. New York: Nova Science Publishers.

Nilsson, M., Griggs, D. and Visbeck, M., 2016. Policy: map the interactions between Sustainable Development Goals. *Nature, 534*(7607), pp. 320–322.

Olver, C., 2017. *How to steal a city*. Cape Town: Jonathan Ball Publishers.

Oxfam, 2019. https://www-cdn.oxfam.org/s3fs-public/file_attachments/bp-tale-of-two-continents-fighting-inequality-africa-030919-en.pdf

Oxfam, 2020. https://oxfamilibrary.openrepository.com/bitstream/handle/10546/620928/bp-time-to-care-inequality-200120-en.pdf

Parris, T.M. and Kates, R.W., 2003. Characterizing and measuring sustainable development. *Annual Review of environment and resources, 28*(1), pp. 559–586.

Pateman, C., 2012. Participatory democracy revisited. *Perspectives on Politics, 10*(1), pp. 7–19.

Patrick, HO. 2019. *Climate change and water security in South Africa: Assessing conflict and coping strategies in KwaZulu-Natal Province* (Doctoral dissertation, University of KwaZulu-Natal, Howard College).

Pearce, D. and Atkinson, G., 1998. The concept of sustainable development: An evaluation of its usefulness ten years after Brundtland. *Revue Suisse d Economie Politique et de Statistique, 134*, pp. 251–270

Peter, C., Friedenstein, H., Bhorat, H., Chipkin, I., Mondi, L., Swilling, M., Buthelezi, M., Qobo, M., Prins, N. and Duma, S., 2018. *Shadow state: The politics of state capture*. New York: NYU Press.
Pezzoli, K., 1997. Sustainable development: A transdisciplinary overview of the literature. *Journal of Environmental Planning and management, 40*(5), pp. 549–574.
Redclift, M., 2002. *Sustainable development: Exploring the contradictions*. London: Routledge.
Rhodes, R.A.W., 1994. The hollowing out of the state: The changing nature of the public service in Britain. *The Political Quarterly, 65*(2), pp. 138–151.
SDG Center for Africa and Sustainable Development Solutions Network, 2019. Africa SDG Index and Kigali and New York: SDG Center for Africa and Sustainable Development Solutions Network.
Smith, N., 2010. *Uneven development: Nature, capital, and the production of space*. Athens: University of Georgia Press.
Sneddon, C., Howarth, R.B. and Norgaard, R.B., 2006. Sustainable development in a post-Brundtland world. *Ecological Economics, 57*(2), pp. 253–268.
Transparency International, 2019. www.transparency.org/news/feature/cpi_2019_Sub_Saharan_Africa
Van Loggerenberg, J., 2016. *Rogue: The inside story of SARS's elite crime-busting unit*. Cape Town: Jonathan Ball Publishers.
Watts, M. and Peet, R., 2004. Liberating political ecology. *Liberation Ecologies: Environment, Development, Social Movements, 2*, pp. 3–43.
WCED, 1987. *Our common future*. World Commission on Environment and Development. Oxford: Oxford University Press.
Woldemariam, B. and Narsiah, S., 2014. The poor and differential access to water in Addis Ababa, Ethiopia. In Asuelime, L., Yaro, J. and Francis, S. (Eds.). *Selected themes in African studies: Contemporary African development issues* (Vol. II, pp. 27–38). London: Springer.

15

CONTRASTING CLIMATE CHANGE KNOWLEDGES IN COLOMBIA

C. Nancy Aguirre

Since the late 20th century, both worldwide policy and action have increasingly focused on global environmental problems mainly based on scientific assessments at global, continental, or relatively small scales (Adger et al., 2001). This has led to paramount interests on pan-disciplinary themes such as environment and development, global environmental change, and sustainability (Turner II and Robins, 2008). Consequently, expanding knowledges on issues such as deforestation, biodiversity loss, and climate change, among others, have arisen. By 2001, Adger et al. argued that climate change was at the forefront of global environmental concerns, albeit the production of many unreadable discourses at local scales that may actually overlook the social processes of adaptation.

Indeed, coordinated international governmental action through the UN Framework Convention on Climate Change (UNFCCC) has been generally agreed upon at the United Nations Conference on Environment and Development (UNCED) since 1992, while the acceptance of climate change as a reality and a problem seemingly has resulted in "both scientific advances and the institutionalization of the issue" (Adger et al., 2001, p. 698).

In 1994, the UNFCCC established aims at preventing "dangerous" human interference with the climate system and now has near-universal membership with 197 countries ratifying it (Parties to the Convention):

> The ultimate objective of the Convention is to stabilize greenhouse gas concentrations at a level that would prevent dangerous anthropogenic (human induced) interference with the climate system. . . . Such a level should be achieved within a time-frame sufficient to allow ecosystems to adapt naturally to climate change, to ensure that food production is not threatened, and to enable economic development to proceed in a sustainable manner. Industrialized countries are expected to do the most to cut emissions on home ground. . . . Industrialized nations agree under the Convention to support climate change activities in developing countries by providing financial support for action on climate change. . . . Industrialized countries also agree to share technology with less-advanced nations. . . . The Adaptation Committee, which Parties agreed to set up under the Cancun Adaptation Framework as part of the Cancun Agreements, is a major step towards a cohesive, Convention-based approach to adaptation.[1]

 DOI: 10.4324/9780429450310-18

After UNFCCC, myriad international agreements – global assessments such as the Intergovernmental Panel on Climate Change (IPCC), and especially since 1998, the Conference of the Parties to the Climate Change Convention – have been established to answer questions on climate change and variability, adaptation, and mitigation (Adger et al., 2005), impacts, costs, resilience (Broto and Bulkeley, 2013; Thornton et al., 2018; Conevska et al., 2019; Wall et al., 2019), social vulnerability (Bohle et al., 1994), rehabilitation, and forecasting modeling (Adger et al., 2001; García et al., 2012), to name a few. To be sure, the IPCC is the United Nations' body for assessing the science related to climate change – restricting themselves to peer-reviewed science (Adger et al., 2013, p. 116) – and was created to provide policymakers with regular scientific assessments on climate change, its implications, and potential future risks, as well as to put forward adaptation and mitigation options.[2]

Relevant scientific definitions of the IPCC (2001) include that of climate change mitigation as involving actions to reducing the rate of climate change by limiting or preventing greenhouse gas emissions and by enhancing activities that remove these gases from the atmosphere; it also includes that of adaptation as "the adjustments in human and natural systems, in response to actual or expected climate stimuli or their effects, that moderate harm or exploit beneficial opportunities: Whereas mitigation deals with the causes of climate change, adaptation tackles the consequences" (IPCC, 2001). Adger et al. (2005) contend that adaptation is translated into actions undertaken in a hierarchical structure beyond the nation state (e.g. UNFCCC) by all actors from individuals to governments, and at diverse scales including municipalities, cities, and markets, while circumscribed by climate change knowledges, although is difficult to distinct climate change adaptation actions from actions prompted by other socioeconomic events.

But it also has been noted that, in turn, society's responses to climate change are twofold, and mediated by culture: while climate change affects cultural dimensions of lives and livelihoods (i.e. identity, community cohesion, and sense of place), the society's cultural dimensions (i.e. production, consumption, lifestyles, and social organization) mediate responses and adaptation to climate-related risks (Adger et al., 2013).

Moreover, research and policy on adaptation and mitigation devoted mainly to their material and quantifiable aspects of climate change, are more often included in policy analyses, notwithstanding the importance of cultural dimensions of climate change, for which research methods are mainly qualitative (i.e. ethnography and participant observation) and thus not easily integrated with the quantitative approaches that are dominant in climate change science (Adger et al., 2013):

> Most areas of public policy seek to promote societal goals through efficient policy mechanisms. . . . *Given imperfect information on risks*, much government action has been focused on reducing uncertainty, *increasing information and protecting productive assets*, for example by *providing climate risk information and reducing risk through regulation and planning*. Yet if the cultural dimensions of climate change are ignored, it is likely that both adaptation and mitigation responses will fail to be effective because they simply do not connect with what matters to individuals and communities.
>
> *(Adger et al., 2013, p. 115–116, emphasis added)*

Conversely, as Boykoff (2008, p. 13) explains, even if climate-change science, policy, and related events have influenced "media reporting and public understanding, journalism and public concern have also shaped climate science and policy decisions" for example by including uncertainty in anthropogenic climate change to nullifying public distress facing it as an environmental *and* social problem.

According to Yusoff and Gabrys (2011, p. 517), climate change, which is regarded by its relationship to futures,

> is as much a political, social, and cultural event, as it is a scientific one. . . . While a majority of work on the social life of climate change has been marked by a distinct concentration on scientific, technological and economic approaches to climate, it is being reimagined as an ethical, societal, and cultural problem that poses new questions and reconfigures the geographic imaginaries of the world.

Moreover,

> In this respect, climate science arguably requires even more sophisticated integration with complex cultural and political systems not just to anticipate and mitigate future impacts, but also to reconsider the political and temporal logics that underpin current scenario trajectories, and examine the descriptive crafts that produce them as spaces for the imagination.
>
> *(p. 518)*

Hence, adaptation practices seem to be practical alternatives through iterative and local everyday practices. (Yusoff and Gabrys, 2011, p. 522).

In this chapter I will address ecological and socioeconomic aspects of climate change in Colombia, focused on the political and institutional apparatus, to note how it consistently blurs when reaching more detailed scales of policy and action from global discourse to local level actions.

The Colombian climate-change structure

Colombia is a multiethnic and pluricultural nation (1991 National Constitution), with around 48 million inhabitants according to the 2018 population census, ranking second in highest biodiversity worldwide, with coastlines in the Caribbean Sea and the Pacific Ocean. The Andes splits in three mountain ranges as it passes in from the southern part into the country, leading to very diverse ecological settings. According to the Colombian Institute of Hydrology, Meteorology and Environmental Studies[3] (IDEAM, 2018), Colombia will be one of most affected countries by climate change in the region, with direct impacts on the environmental goods and services that are fundamental for national productivity and for reducing the causes of poverty. Likewise, IDEAM warns that by year 2050 climate change in Colombia will lead to an increased average temperature of 1 to 2 °C, an increasing sea level between 8 and 12 centimeters, the disappearance of 78% of glaciers and 56% of the paramos, a considerable loss of species in parallel to the effects of development, and a direct upsurge of dengue, malaria, and yellow fever epidemics, among others, over a vulnerable population of at least 8 million people, albeit the more immediate escalation of floods and landslides, and of agricultural soil loss.

The Third National Communication of Colombia to UNFCCC and its biannual update (IDEAM, PNUD, MADS, DNP, CANCILLERÍA, 2017, 2018) shows factors contributing to the country's vulnerability, including paramount informality and illegality, excessive centrality of the Colombian Estate, rural violence accelerating migration processes, historical limited coordination between sectoral policy, scant land use policy, traditional land grabbing as a mechanism of wealth accumulation in opposition to value added activities, and limited coordination and participation of actors involved in climate change policy, science, and action.

Based on a 2016 survey on climate change, results showed in this third communication depicts that 98.3% of the Colombian population thinks climate change is a reality, whereas 87.5% thinks that their municipality or city will not be prepared to cope with adverse climate change conditions if they worsen in a five-year time-frame; 91% of respondents obtain climate change information via television channels, and 80% of participant subjects (in cities) and 87% (in rural areas) do not understand the concept of greenhouse gases.

Also according to this communication, Colombia shares 0.422% of worldwide greenhouse gases emission (with 36% and 26% shares from the forestry and agricultural sectors respectively) thus ranking fifth among 32 Latin American and Caribbean countries. Finally, the country's mean annual temperature is expected to rise 0.9 °C between 2011 and 2040 (0.5 °C in the Caribbean Sea and 0.7 °C in the Pacific Ocean), and will likely increase 1.6 °C between 2041 and 2070 (0.85 °C in the Caribbean Sea and 1.2 °C in the Pacific Ocean), and 2.14 °C between 2071 and 2100 (1.6 °C in the Caribbean Sea and 2.0 °C in the Pacific Ocean); whereas rainfall is projected to decrease between 10% and 40% in 32%, 29.6%, and 30% of the country's continental surface in the three considered periods respectively. These predicted changes in temperature are expected to result in sea-level rise affecting the Caribbean and Pacific Ocean coastlines and settlements; glacier melting affecting paramos and water provision of main cities largely in the Andean region of the country; increasing extreme climatic events; decreasing productivity of agricultural soils; increasing desertification processes; and increasing heat waves in cities. Likewise, foreseen changes in rainfall are expected to cause acceleration and intensification of desertification processes, increased landslides and flooding; loss of water bodies; negative effects on health, agriculture, forestry; and the sub-regional economy; and increased infrastructure damage particularly in mountainous areas.

In contrast to these figures, Pabón (2012, p. 261) suggests a widespread warming of 0.1–0.2 °C per decade and an increasing of the maximum temperature frequency, while scenarios for 21st century show gradual increments of average air temperature of 2–3 °C between 2011–2014 and 3 to 4 °C towards the end of century; although in the valleys of Magdalena (where the Municipality of Ricaurte, Cundinamarca is located) and Cauca the warming would be greater than in the remainder regions; the precipitation towards the end of 21st century would register reductions of up to 30% in the inter-Andean region, and in the Caribbean region.

Disaster risk management and climate change and variability adaptation in Colombia

Stemming from UNFCCC definitions, in Colombia as in many parties to this Convention in the Americas, Risk Management and Adaptation to Climate Change (GRACC) is framed by the National Disaster Risk Management System (SGRD)[4] created by Law 1523 of 2012, which pivots the institutionalized global climate change issues by the United Nations Organization, with some adjustments consistent with the national institutional culture.

The SGRD is structured from the national to the sub-national and local levels, by way of national, departmental, and municipal councils and funds. It performs functions with respect to three processes that are formalized through the Knowledge, Risk Reduction, and Disaster Management Committees. The Presidency of the Republic and the National Council, at the top level, guide this structure, and in turn are represented by the National Unit for Disaster Risk Management (UNGRD), with coordination, direction, and implementation functions. In the subsequent hierarchical level, the department whose highest authority is the governor, this same structure and its competences are replicated, with intervention of authorities of this political-administrative level. A similar organization is reproduced for the third and last level of the political-administrative system of the country, the municipality, whose mayor is the visible head (supported by the council, the committees, and the respective fund), with the purpose of deploying locally the social processes of risk management.

In turn, risk management and adaptation to climate change are broken down by national policy guiding mitigation and adaptation actions, through linkages of three main instances: the National Plan for Adaptation to Climate Change (PNACC), the strategy of low carbon development and REDD (reduction of emissions of greenhouse gases caused by deforestation and forest degradation), and the sustainable development policies. Of these, the PNACC, which is part of the country's political and institutional strategies since 2010, focuses on reducing socioeconomic and ecosystem risk and impacts associated with the country's variability and climate change.

Through the PNACC, two cardinal definitions are assumed for the country (DNP, 2012):

> Adaptive capacity is defined as the capacity of a system and its parts to anticipate, absorb, accommodate or recover from the effects of a disturbance in a timely and efficient manner.
>
> Mitigation is the set of strategies, policies and actions aimed at limiting or reducing greenhouse gas emissions and improving carbon sinks according to what was agreed in the UNFCCC.

In effect, three strategies of adaptation to climate change (knowledge, planning, and development transformation), frame the main objectives of the PNACC: generating better knowledge about potential risks and opportunities (to strengthen management of climatic, hydrological, and oceanographic knowledge, as well as institutional capacities for adaptation to climate change); incorporating climate risk management in sectoral and territorial development planning (to include climate change variables into state planning instruments and to develop resilient investment projects); and reducing socioeconomic and ecological systems' vulnerability to climate events (to manage climate change impacts on biodiversity, agricultural production, and basic infrastructure, based on green growth concepts). These imply, on one hand, the obligation of national public organizations of incorporating adaptation to Climate Change strategies in their Sectorial and Territorial Plans. On the other hand, the PNACC Coordinating Committee should technically support Climate Change Sectorial and Territorial Nodes (NSTCC), and therefore produce and socialize guiding documents while resolving queries and channeling adaptation information for these purposes.

Given the complexity of climate change management, the National Council for Economic and Social Policy (CONPES) in 2011 voiced the need for creation of the National Climate Change System (Sisclima). Nonetheless, it was not until 2016 that its operation began with a main objective of "coordinating, articulating, formulating, monitoring and evaluating policies, standards, strategies, plans, programs, projects, and actions for climate change adaptation and mitigation of greenhouse gases".[5]

The Sisclima seeks facilitating effort-coordination at the national, regional, local, and international levels, and articulating climate change plans and strategies with those of economic, social, and environmental development, as well as articulating public and private initiatives on climate change while encouraging citizen participation in related decisions. The Ministry of Environment and Sustainable Development, the Ministry of the Interior, the Ministry of Finance and Public Credit, the Ministry of Agriculture and Rural Development, the Ministry of Mines and Energy, the Ministry of Transportation, the Ministry of Foreign Affairs, and the National Planning Department integrate the Inter-sectorial Commission for Climate Change

(CICC). Year 2016 was indeed a milestone for Climate Change governmental provisions in Colombia: the Paris Agreement was signed by the country; the National Plan for Disaster Risk Management was launched; and a novel National Policy for Climate Change was established (MADS, 2017). After all this, in year 2018, Law 1931 readjusted guidelines for Climate Change Management in the country, and redress the articulation and complementarity between the processes of adaptation to climate change and disaster risk management that apply for its incorporation in Climate Change Management Plans, and other planning instruments of the Law 1523 of 2012.

Although there is a significant political and regulatory structure for climate change risk management and adaptation in Colombia, due to the relatively short lapse of these achievements, the implementation of the proposed strategies and objectives remains a challenge for the various sectors and territorial levels.

More general, Climate Change Management in Colombia, which seeks to articulate the various geographical levels through territorial development and planning, is also problematic due in part to the limited availability of detailed information and data. At the municipal level, for example, actions must be undertaken (according to the law) based on information and data at scale of 1:25,000, while most available climate and ecological data is at scales of 1:100,000, and in the departmental level (second administrative boundary) territorial management related to climate change is still incipient. This implies at least expanding efforts within the framework of the knowledge strategy of the National Plan for Adaptation to Climate Change (PNACC). Likewise, the available information is not always free-access and interoperable; therefore, the organizations producing such information face challenges to leverage what is prescribed in relevant policies and legislation for profound climate change action. Detailed and oriented evaluation of available Adaptation Plans in the country could allow refocusing efforts, and re-harmonizing strategies and objectives of the PNACC and the Sisclima, and thus the Climate Change Management provisions with the National Disaster Risk Management System.

Adaptation to climate change in Tropical Dry Forest Ecosystems

As to tackling the aforementioned, I address a practical case at the Tropical Dry Forest (TDF) ecosystem in Colombia, based on a land parcel restoration process aimed at adapting to climate change over five years. I detail how these practices (global to national to local) may be examined using methods that include genealogies and participant observation.

I describe how the municipality's official climate change knowledges inside the TDF Ecosystem in Colombia radically blur as soon as pertinent action focuses on local scales. Then I deepen attention to a land parcel restoration process to point out the nuances of everyday life climate variations, in contrast to the actual available climate change information needed for adaptation and resilience processes at this scale. This allows juxtaposition of relevant issues and contradictions between global and local scale information, land use planning policy for climate change adaptation and mitigation, biodiversity, and agriculture, among others.

The land parcel under study is located inside the Colombian Tropical Dry Forest Ecosystem (TDF), more precisely in the municipality of Ricaurte, Department of Cundinamarca. Pizano and García (2014) explain TDF is found in lowlands with a strong rain seasonality (less than 100 mm during four to six months annually) with only 8% of originally nine million hectares of TDF remaining, which makes it one of the most threatened ecosystems in the country. Here, 65% of deforested lands are now undergoing desertification processes and only 0.4% of the original ecosystem area is under the Protected Areas National System. The Ministry of Environment recently declared the TDF as an strategic ecosystem for conservation and currently the TDF is now mainly located on private property so that the engagement of all actors in conservation initiatives is paramount. Pizano and García (2014) also have noted that due to a unique biodiversity adapted to hydric stress, the TDF has a high endemism of species that have adapted to a long dry seasonality with ecosystem services including carbon sequestration that regulates climate. Notwithstanding the previously mentioned, limited biological, ecological, and social data exist, making it difficult to propose integrated and feasible actions for TDF integrated management (Pizano and García, 2014, p. 9).

The municipality of Ricaurte is situated in the Alto Magdalena Province at 142 km to the south from Bogota, and at 284 meters above sea level, in the eastern Andean mountain range with an extension of 130 km^2; mean annual temperature is 29°C, with and around 11,000 inhabitants by 2018. Large urban projects are currently being developed, for tourism purposes mainly, temporarily linking floating population either because of work or tourism coming from different places (Alcaldía de Ricaurte, 2016).

By year 2016, the municipality of Ricaurte had planning instruments for risk management, such as the Municipal Disaster Risk Management Plan, the Municipal Emergency Response Strategy (EMRE) and the Forest Fire Contingency Plan (Alcaldía de Ricaurte, 2016); however these documents need periodical update and are unavailable to the general public. The Institution in charge of emergency and disaster attention has been supported in recent years by the Municipality of Girardot (a neighbor municipality) Official Fire Department, considering that only in April 2015 did the volunteer fire department of Ricaurte come into operation (Alcaldía de Ricaurte, 2016). The municipality has established a Municipal Council of Risk Management; however, their activities are more focused on emergency attention to flood and mass movement events, while climate change plans and actions are not significantly contemplated.

In the words of the Alcaldía de Ricaurte (2016, pp. 129–130), the Autonomous Regional Corporation (CAR) conducted a research study with the National University of Colombia, thus modeling the regional climate to develop future scenarios of climate conditions (at the scale of 1:100.000), which allows forecasting climate trends for 2011–2040, 2041–2070, and 2071–2100, compared with the reference climate from 1971–2000. In relation to climatic threats, the results of the study for the period 2011–2040, compared with 1971–2000, allows recognizing in the CAR area the probable climate variations, and to identify possible climatic threats. In general terms, the climate of the region for the period 2011–2040 in the territory of the CAR, in comparison with the conditions observed in 1971–2000, would be warmer by at least 1 °C, air humidity would be greater by 1–3%, and annual precipitation would increase in the Magdalena valley and throughout the eastern part of the municipality, and decrease in the highland sector, with main effects of climate change in Ricaurte being forest fires and flooding. This is almost all that is found concerning climate change in the municipality of Ricaurte. Interestingly, the Municipal agreement for territorial ordering (land use planning) of 2000 does not mention any aspects of provision for climate change adaptation, resilience, or mitigation (Alcaldía Municipal de Ricaurte Cundinamarca, 2000), although the updating process of this plan begun in 2021[6], which now must include Climate Change Management provisions according to the novel national territorial ordering Decree 1232 of 2020. Notwithstanding, recent Municipal Development Plan for 2020–2023, promotes a "greater awareness of climate change for the prevention and mitigation of emergencies and disasters" (Alcaldía Municipal de Ricaurte, 2020, p. 6).

Climate change local action

Since 2014 a restoration process in a land parcel of 1300 m^2 located inside the municipality of Ricaurte has taken place by private landowners with scant help of local municipal governmental actors. As noted, only general and very limited information on climate change data, policy, and action was available for Ricaurte, considering the project needed detailed scale data. Almost a third of the surface extension of the land lot was left to natural ecological restoration while the remaining area was devoted to aided ecological restoration. During this process, a learning participant observation process took place. None of the available climate change information and data were of help at the local scale for the project's purposes. Beginning the project, the lot was almost completely characterized by bare soil and was used for trash dumping by the neighboring community. Once the restoration process began, a local peasant was engaged in the project as an expert.

First actions included informal conversations with the local population thus noting that only elders would have direct empirical knowledge on climate and variability change, agriculture, local plants and animals. Younger people are increasingly less interested in undertaking ecological restoration nor do they expose pre-existing interest or acquired knowledge on land use management and traditional livelihoods including agriculture.

The general local imaginary was that now this TDF ecosystem was not feasible for agricultural practices, the same as the scientific statement of Pizano and García (2014) for much degraded areas inside this strategic ecosystem. On the other hand, locals do not value endemic tree species but in contrast value more foreign food crops, which have resulted in increased deforestation and loss of local plant species in favor of non-adapted species. Deforestation indeed has been increasing in the last decade. Local experts in the project used their own knowledges for aided restoration in which a number of food trees would be planted and after a five year implementation, the aided restoration process had succeeded: crops began being productive although water provision had to be secured.

The lot area left to natural restoration showed after the same period a wealthy secondary dry forest based on the seed bank of soils and the provision of seeds by birds and other animals. It was evidenced that the TDF has natural restoration potential in the municipality if degradation practices are stopped. Shadow provided by the secondary forest is helping other restoration practices to prosper, while dramatically increasing climate comfort inside this land lot and for nearby neighborhoods.

However, field data on temperature and rainfall (qualitatively stated) was not the basis for understanding climate variability at the site. It was found that seasonality resembled two spring periods, and that temperature and rainfall may change significantly in a 12 km^2 radius. Strong rainfall may be found in one site while not in others at the same time. One-year climate variability is followed by contrasting differential variability in the following year, making action-planning problematic. Thus ecological care should be put forth on a daily basis and the climate variability recorded in the short run (five-year period) may seem unsuitable for many local owners with similar interests.

The above analysis proved to be useful for enhancing knowledge on climate variability and change. However, over the long run, it would not be sufficient for explaining and coping with climate variability and adaptation, as detailed variability seems to be now undetermined for the site, thus additional studies and governmental support are needed.

Flaws, controversies, and opportunities for the country's climate change

Based on the previous case study, we may consider that in Colombia, critical perspectives associated with climate change are hardly found within the bulk of contributions on the broader societal impacts of the relation between development and the environment. Notwithstanding, most contributions (either scientific, policy, or action works) are at global, national, or sub-regional scales, making it problematic to succeed with actions at local levels to adapting or mitigating climate change (Espitia-Falla, 2020).

Indeed, most efforts address the international/global directives as adapted to the country, the political (Lau, Jarvis, and Ramírez, 2011) and normative/legislative issues regarding sectorial and territorial inclusion of climate change dimensions (and particularly in land use planning and disaster risk management), or the physical-biotic data that could inform their causes and consequences (e.g. Andrade and M, 2011 on biodiversity knowledge in Colombia; the CDKM-IDEAM, 2013 Inter-institutional and Multi-sectoral Analysis of Vulnerability and Adaptation to Climate Change for the Agricultural Sector of the Upper Cauca River Basin, Impacting Adaptation Policies; the DNP, 2016 Colombian National Plan for Adaptation to Climate Change; the IDEAM, PNUD, MADS, DNP, CANCILLERÍA, 2016–3rd Communication on the Inventory of Greenhouse Gases for Colombia; or Ferrer-Paris et al., 2019 on the Red List of Ecosystems).

Fewer offerings are devoted to analyzing climate change from contrasting but complementary – and more critical – subfields of investigation that are related to economic inequalities, marginalization of traditional communities to climate change knowledges, or the novel territorial settings that emerged after the national peace treaty was signed in 2017 with armed guerilla groups addressing conflict over 50 years (i.e. see also Echeverri, 2009 on Indigenous peoples and climate change in the Colombian Amazon; Swyngedouw, 2009 on the Hydro-Social Cycle; Lampis, 2013 on vulnerability and adaptation to climate change; Ryan and Bustos, 2019; Béné et al., 2019; Barragán et al., 2018; Lamprea-Quiroga and Sanabria-Marin, 2020). Debates about the contrasting concepts of vulnerability and its measurement are largely absent in this space, which implies that the use of one or the other involves radically different public policies (Lampis, 2013).

As soon as climate adaptation action focuses on more detailed levels – the local – the cultural – then climate change scientific information and data, as well as policy instruments and action, dramatically fade out. Not only do these instruments become unavailable or inexistent at such scales, but also the sociocultural connection or involvement to these efforts are at best, incipient. Certainly, all of the global and national climate change paraphernalia critically blurs as soon as pertinent initiatives are aimed at tangible, concrete, local levels. Indeed, Comprehensive Territorial Climate Change Management Plans (PIGCCT) today reaches sub-regional levels for 59.1% of the country (second level administrative boundaries of 32 departments in total in Colombia), and almost 12% of a total of 1,123 municipalities and districts, which result in limited progress from discourse to action[7].

Conclusions and future knowledges

In more than 150 majoritarian countries, including Colombia, climate change (and variability) policy, science, and action have cascaded from global structures, particularly the United Nations Framework for Climate Change Convention (UNFCCC). Considering climate change, whether anthropogenic and/or natural, as a reality, most contributions have focused on global to national and sub-regional scales, thus leaving the local scales largely unassessed. Likewise, provisions to systematically incorporating sociocultural uniqueness to these knowledges are still incipient but more importantly, increased participation of local actors at these instances as well as the visibility of everyday life nuances, as related to global change, has been limited. This is problematic to decreasing uneven development futures.

Colombia is scientifically projected to be one of most affected countries by climate change in the Americas thus impacting national productivity and significantly, poverty reduction. Mean annual temperature is expected to rise between 2 °C and 3 °C (even 4 °C) by the end of this century, whereas rainfall is projected to decrease between 30% and 40% (almost 50%) during the same period! These worrying figures pose much uncertainty to relations between development and environment. Even more so, the future development-environment context in this country must grapple with its history and a recent peace treaty with a large guerrilla group operating for approximately 60 years. Acute adaptation action at the country level seems feasible as based on prolific and detailed unfolding of scientific and policy prerogatives for climate change adaptation, resilience and mitigation at both national and sub-regional levels. Yet, seen in the case of a small local five-year restoration project developed for climate change adaptation in a tropical dry forest ecosystem, it becomes evident that climate change apparatus at these scales abruptly blurs and fades when pertinent data, information, knowledge, and action is needed from and for local scales.

This analysis helps uncover broader flaws and contradictions of the climate change structure put in place for adaptation and mitigation at global to national levels. But it also poses opportunities for local scale initiatives to increase adaptation and resilience to climate change. On one hand, we witness the unfolding of mounting interest of local communities engaged with similar projects for learning and sharing diverse knowledges on climate change with broader actors, including scientific and governmental entities. On the other hand, we see a rising awareness of both communities and decision-makers for hands-on and down-top initiatives that converge and leverage critical knowledges production and good practices at local scales for better-envisioned development in a context of climate change.

Notes

1 https://unfccc.int/bigpicture accessed on January 18, 2019.
2 www.ipcc.ch/ accessed on January 20, 2019.
3 http://www.ideam.gov.co/web/tiempo-y-clima/escenarios-cambio-climatico accessed on July 31, 2021.
4 http://portal.gestiondelriesgo.gov.co/Paginas/Estructura.aspx accessed December 15, 2018.
5 www.minambiente.gov.co/index.php/politica-nacional-de-cambio-climatico-2/sisclima accessed on December 15, 2018.
6 https://www.ricaurte-cundinamarca.gov.co/NuestraAlcaldia/SaladePrensa/Paginas/INICIA-LA-ETAPA-DE-REVISI%C3%93N-Y-AJUSTE-DEL-ESQUEMA-DE-ORDENAMIENTO-TERRITORIAL-DE-RICAURTE.aspx accessed on July 27, 2021.
7 https://www.minambiente.gov.co/index.php/gestion-territorial-de-cambio-climatico/aproximacion-a-territorio-planes-territoriales-de-cambio-climatico

References

Adger, N., Arnell, N., & Tomkins, E. (2005). Successful adaptation to climate change across scales. *Global Environmental Change*, (15), 77–86.

Adger, N., Barnett, J., Brown, K., Marshall, N., & O'Brien, K. (2013). Cultural dimensions of climate change impacts and adaptation. *Nature Climate Change*, *3*(2), 112–117.

Adger, N., Benjaminsen, T., Brown, K., & Svarstad, H. (2001). Advancing a political ecology of global environmental discourses. *Development and Change*, *32*(4), 681–715.

Alcaldía Municipal de Ricaurte, Cundinamarca. (2000). Acuerdo Numero (036) del 29 de diciembre/ 2000 por el cual se adopta el Esquema de Ordenamiento Territorial del Municipio de Ricaurte 2001–2009.

Alcaldía Municipal de Ricaurte, Cundinamarca. (2016). Plan de Desarrollo Territorial, Ricaurte, Cundinamarca 2016–2019.

Alcaldía Municipal de Ricaurte, Cundinamarca. (2020). Plan de Desarrollo Territorial, Ricaurte, Cundinamarca 2020–2023.

Andrade, C., & M, G. (2011). Estado del conocimiento de la biodiversidad en Colombia y sus amenazas. Consideraciones para fortalecer la interacción ciencia-política. *Revista de la Academia Colombiana de Ciencias Exactas, Físicas y Naturales, 35*(137), 491–507.

Barragán, A., Ramírez, J., Mikán, J., Marín, M., & Torres, G. (2018). *Crónicas de la resiliencia.* Bogotá: Instituto de Investigación de Recursos Biológicos Alexander von Humboldt.

Béné, C., Oosterveer, P., Lamotte, L., Brouwer, I. D., de Haan, S., Prager, S. D., . . . Khoury, C. K. (2019). When food systems meet sustainability – current narratives and implications for actions. *World Development, 113*, 116–130.

Bohle, H. G., Downing, T. E., & Watts, M. J. (1994). Climate change and social vulnerability: Toward a sociology and geography of food insecurity. *Global Environmental Change, 4*(1), 37–48. https://doi.org/10.1016/0959-3780(94)90020-5

Boykoff, M. T. (2008). Media and scientific communication: A case of climate change. In D. G. E. Liverman, C. P. G. Pereira, & B. Marker (Eds.), *Communicating Environmental Geoscience* (pp. 11–18). London: Geological Society, London, Special Publications.

Broto, V. C., & Bulkeley, H. (2013). Maintaining climate change experiments: urban political ecology and the everyday reconfiguration of urban infrastructure. *International Journal of Urban and Regional Research, 37*(6), 1934–1948. https://doi.org/10.1111/1468-2427.12050

CDKN-IDEAM. (2013). Análisis interinstitucional y multisectorial de vulnerabilidad y adaptación al cambio climático para el sector agrícola de la Cuenca Alta del Río Cauca. Impactando políticas de adaptación. Metodología de análisis de vulnerabilidad para la Cuenca Alta del Río Cauca. Reporte Técnico. Retrieved from https://cdkn.org/project/agriculture-adaptation-plan/?loclang=en_gb:

Conevska, A., Ford, J., Lesnikowski, A., & Harper, S. (2019). Adaptation financing for projects focused on food systems through the UNFCCC. *Climate Policy, 19*(1), 43–58. https://doi.org/10.1080/14693062.2018.1466682

DNP. (2012). *Plan Nacional de Adaptación al Cambio Climático. ACB: Adaptación Bases Conceptuales, Marco Conceptual y Lineamientos.* Bogota, Colombia: Departamento Nacional de Planeación.

DNP. (2016). www.dnp.gov.co/programas/ambiente/CambioClimatico/Paginas/Cambio-Climatico.aspx

Espitia-Falla, M. M. (2020). *Guatavita, Cundinamarca. Ficha para la inclusión del cambio climático y la variabilidad climática en el ordenamiento territorial.* Bogota: Ministerio de Ambiente y Desarrollo Sostenible.

Ferrer-Paris, J. R., Zager, I., Keith, D. A., Oliveira-Miranda, M. A., Rodríguez, J. P., Josse, C., . . . Barrow, E. (2019). An ecosystem risk assessment of temperate and tropical forests of the Americas with an outlook on future conservation strategies. *Conservation Letters, 12*(2), e12623. https://doi.org/10.1111/conl.12623

IAvH. Colombia Anfibia: Un pais de humedales. Volumen 2. Bogotá: IAVH.

Echeverri, J. Á. (2009). Pueblos indígenas y cambio climático: el caso de la Amazonía colombiana. *Bulletin de l'Institut français d'études andines, 38*(1), 13–28.

García, M. C., Piñeros Botero, A., Bernal Quiroga, F. A., & Ardila Robles, E. (2012). Variabilidad climática, cambio climático y el recurso hídrico en Colombia. *Revista de Ingeniería, 36*, 60–64.

IDEAM, PNUD, MADS, DNP, CANCILLERÍA. (2016). Inventario nacional y departamental de Gases Efecto Invernadero – Colombia. Tercera Comunicación Nacional de Cambio Climático. IDEAM, PNUD, MADS, DNP, CANCILLERÍA, FMAM. Bogotá D.C., Colombia.

IDEAM, PNUD, MADS, DNP, CANCILLERÍA. (2017). Resumen ejecutivo Tercera Comunicación Nacional de Colombia a La Convención Marco de Las Naciones Unidas Sobre Cambio Climático (CMNUCC). Tercera Comunicación Nacional de Cambio Climático. IDEAM, PNUD, MADS, DNP, CANCILLERÍA, FMAM. Bogotá D.C., Colombia.

IDEAM, PNUD, MADS, DNP, CANCILLERIA. (2018). Segundo Informe Bienal de Actualización de Colombia a la Convención Marco de las Naciones Unidas para el Cambio Climático (CMNUCC). Bogota D.C, Colombia: IDEAM-PNUD-MADS-DNP-CANCILLERIA Retrieved from http://www.ideam.gov.co/documents/24277/77448440/PNUD-IDEAM_2RBA.pdf

Intergovernmental Panel on Climate Change (IPCC). (2001). *Climate Change 2001: Impacts, Adaptation and Vulnerability. Summary for Policy Makers.* Geneva: World Meteorological Organization.

Lampis. (2013). Vulnerability and adaptation to climate change: Debates about the concept of vulnerability and its measurement. *Cuadernos de Geografía Revista Colombiana de Geografía, 22*(2), 17–33.

Lamprea-Quiroga, P., & Sanabria-Marin, R. (2020). Teoría general de sistemas en el diálogo del conocimiento campesino del altiplano cundiboyacense colombiano con las ciencias edáfica y climática. *Perspectiva Geográfica, 25*(2), 34–55. https://doi.org/10.19053/01233769.9283

Lau, C., Jarvis, A., & Ramírez, J. (2011). Agricultura colombiana: Adaptación al cambio climático. CIAT Políticas en Síntesis no. 1. Centro Internacional de Agricultura Tropical (CIAT), Cali, Colombia. 4 p.
MADS. (2017). *Política nacional de cambio climático*. Bogota, Colombia: Ministerio de Medio Ambiente y Desarrollo Sostenible (MADS).
Pabón, J. D. (2012). Cambio climático en Colombia: tendencias en la segunda mitad del siglo XX y escenarios posibles para el siglo XXI. *Revista de la Academia Colombiana de Ciencias*, XXXVI, *139*, 261–278.
Pizano, C. y H. García (Editores). 2014. *El Bosque Seco Tropical en Colombia*. Bogotá, DC, Colombia: Instituto de Investigación de Recursos Biológicos Alexander von Humboldt (IAvH).
Ryan, D., & Bustos, E. (2019). Knowledge gaps and climate adaptation policy: A comparative analysis of six Latin American countries. *Climate Policy, 19*(10), 1297–1309. https://doi.org/10.1080/14693062.2019.1661819
Swyngedouw, E. (2009). The political economy and political ecology of the hydro-social cycle. *Journal of Contemporary Water Research & Education, 142*(1), 56–60. https://doi.org/10.1111/j.1936-704X.2009.00054.x
Thornton, P. K., Whitbread, A., Baedeker, T., Cairns, J., Claessens, L., Baethgen, W., . . . Keating, B. (2018). A framework for priority-setting in climate smart agriculture research. *Agricultural Systems, 167*, 161–175. https://doi.org/10.1016/j.agsy.2018.09.009
Turner II, B. L., & Robbins, P. (2008). Land-change science and political ecology: Similarities, differences, and implications for sustainability science. *Annual Review of Environment and Resources, 33*(1), 295–316. https://doi.org/10.1146/annurev.environ.33.022207.104943
Wall, R., Grafakos, S., Gianoli, A., & Stavropoulos, S. (2019). Which policy instruments attract foreign direct investments in renewable energy? *Climate Policy, 19*(1), 59–72. https://doi.org/10.1080/14693062.2018.1467826
Yusoff, K., & Gabrys, J. (2011), Climate change and the imagination. *WIREs Climate Change*, 2, 516–534. https://doi.org/10.1002/wcc.117

16

SPACES OF ENVIRONMENTAL (IN)JUSTICE AND ACCUMULATION BY DISPOSSESSION IN INDIA

B.S. Butola

Environmental justice: its genesis and trajectory

Justice as a matter of right is always contingent upon citizenship. Subjects were/are neither entitled nor do they demand justice. Modernity distinguished itself from all other previous systems by raising relentless struggles against all forms of powers, particularly the religious, the imperial, and the despotic and ensuring equality in terms of rights to every individual (citizen). It advocated that "Rights" is a substantive and inclusive praxis and there is no need to fight and ensure justice separately. These claims were based on certain assumptions about the "success" of modernity, not only the way it solved two fundamental injustices (contradictions) of human history – non-recognition and maldistribution (Frazer, 1995, pp. 68–93) – but also the way it apparently claimed that freedom, rights, and citizenship are the three non-negotiable conditions for the existence in/of modern society, but as a matter of praxis only at the cost of justice: social as well as environmental.

The so-called three achievements, as well as fighting against injustice, were not the results of the work and efforts of few individuals at certain points of time in a particular region or country. On the contrary, these were the results of the persistent attempts made by various individuals, communities, along with the historical struggles raised by millions of people over a long period of time in different parts of the world. However, the contributions of the four historical figures in articulating and giving concrete shapes to the Spirit of the Age (Weber, 1958) are worth mentioning here.

The first and foremost important contributions were by Rousseau. Rousseau, with his propositions, advocated that the supremacy of "social contract" and "general will" demolish the "divine covenant of predestination" and "atavistic bases of imperial legitimacy". He advocated that the social contract is primarily concerned with the legitimacy of the state or authority over the individual (Saxe and Linscott, 1953, p. 283). The social contract assumes that the individuals have consented either tacitly or explicitly to surrender certain rights and freedoms to the state in exchange of the protection of other rights. This was different from the postulations made by Locke and Hobbes. For Locke the surrender of certain rights to the sovereign is for the protection of private property, to which Rousseau objected and proposed that the invention of private property constituted humanity's fall from grace (Rousseau, 2017, pp. 1–73) and out of the state of first nature and for which an individual surrenders certain rights to the sovereign (Esposito,

DOI: 10.4324/9780429450310-19

2006, pp. 86–111). The second major contribution came from Thomas Paine through his famous declaration on the "Rights of Man". Paine justified the "government which arise out of society" as rational against the previous "government of the priest craft" based on superstition and the "government of the conquerors" founded on naked use of power and by uniting the scepter (superstition) and sword and naming it as "divine right" (Paine, 1792, pp. 3–4). He too presumed that "men are born, and continue, free and equal in respect of their rights" (Paine, 1792, pp. 3–4) and once preservation of natural and imprescriptible rights of man – including right to equality, right to freedom, and right to property as inviolable and sacred etc. – have been achieved, there is no need of fighting right to Justice separately. Similar concerns were expressed by Victor Hugo after he advocated the supremacy of the idea of freedom in such words: "Nothing is more powerful than an 'Idea' whose time has come". And finally, it was John Stuart Mill in his famous essay "On Liberty" who categorically proclaimed Liberty – not in terms of "liberty of the Will", but in terms of civil and social liberty – as "the nature and limits of the power which can be legitimately exercised by society over the individual" (Mill, 1859, p. 6). According to him, "the struggle between Liberty and Authority is the most conspicuous feature in human history" (Mill, 1859, p. 6). Consequently, humans have always fought to limit the powers of the society (authority), which can be legitimately exercised over individual, because authority has always been antagonistic to people. Thus, liberty stands for protection of individuals against the tyranny of authority (political rulers). So, effective Liberty is inclusive of all that is necessary for human welfare and there is no need of justice as a standalone agenda.

For Marx, rights and justice are interlinked to each other only through the social network of private property and it is only in the absence of their embedding on private property that it is possible to achieve both, but in the context of private property both are antagonistic to each other. To him, private property is the root cause of all forms of injustices and one thing that he wanted forthwith even in the communist manifesto was the "abolition of the private property" (Marx and Engels, 1975, p. 63). This is also the summary of the communist manifesto in one sentence, and that the basis of bourgeois individualism is detrimental to the formation of an egalitarian society. For him private property is the basis of social classes and also the state. In their turn as long as private property, classes, and state exist, a tiny minority will enjoy the rights at the cost of the majority and justice will always elude the majority. Therefore, the fight for justice is a fight against private property, individuality, and the state.

Anarchists, particularly Pierre Joseph Proudhon, saw justice antithetical to the state as the central authority. According to him, to govern, which is the main function of the state, quintessentially means

> to be watched, inspected, spied upon, directed, law-driven, numbered, regulated, enrolled, indoctrinated, preached at, controlled, checked, estimated, valued, censured, commanded, by creatures who have neither the right nor the wisdom nor the virtue to do so. To be governed is to be at every operation, at every transaction noted, registered, counted, taxed, stamped, measured, numbered, assessed, licensed, authorized, admonished, prevented, forbidden, reformed, corrected, punished. It is, under pretext of public utility, and in the name of the general interest, placed under contribution, drilled, fleeced, exploited, monopolized, extorted from, squeezed, hoaxed, robbed; then, at the slightest resistance, the first word of complaint, to be repressed, fined, vilified, harassed, hunted down, abused, clubbed, disarmed, bound, choked, imprisoned, judged, condemned, shot, deported, sacrificed, sold, betrayed; and to crown

> all, mocked, ridiculed, derided, outraged, dishonoured. That is government; that is its justice; that is its morality.
>
> *(Proudhon, 1923, pp. 293–294)*

So, for the anarchists, justice is antithetical to state. In place of a state=injustice, they offered a variety of options to bring about justice: mutualism (spontaneous orders without a central authority: Proudhon), community ownership (social anarchists), collectivization by the revolutionaries (Bakunin), direct democracy (Peter Kropotkin), developed individualism (William Godwin), and workers' government (Anarcho Syndicalism), etc.

For the radicals, the rise of the bourgeois system, particularly capitalism based on universal and generalized commodity production through the dominant labor process comprising of wage labor, (in)justice was seen purely in terms of profit maximization, generation of surplus; its appropriation and accumulation by the capitalist class at the cost of the labor is the principal cause of all forms of injustice. It also ensembles the totality of all social relations: "a solid foundation on which there arises a legal and political superstructure" (Marx, 1976, p. 3). The bourgeoisie has become so emphatic that "the executive of the modern state is but a committee for managing the common affairs of the whole bourgeois" (Marx and Engels, 1975, p. 44). So, the ruling class does not only control the material resources of the society but they also control the ideas in such a way that:

> The ideas of the ruling class are in every epoch the ruling ideas, i.e. the class which is the ruling material force of society, is at the same time its ruling intellectual force. The class which has the means of material production at its disposal has control at the same time over the means of mental production, so that thereby, generally speaking, the ideas of those who lack the means of mental production are subject to it. The ruling ideas are nothing more than the ideal expression of the dominant material relationships, the dominant material relationships grasped as ideas.

Therefore, the autonomy and freedom of the individual(ity) and its defense as a sovereign right has been the life blood of the bourgeoisie in its fight against the feudal. The clergy took a complete "U" turn once they got hold of the material bases of the society. Once in power and in control of the economic structure, "in bourgeois society capital is independent and has individuality, while the living person is dependent and has no individuality" (Marx and Engels, 1975, p. 49). Moreover, the creator of all the social wealth (i.e. the workers) gets caught up in web of unchanging contradictions.

> The rising wages gives rise to over work among the workers. The more they wish to earn, the more must they sacrifice their time and carry out slave-labour, completely losing all their freedom, in the service of greed. . . . This class has always to sacrifice a part of itself in order to be wholly destroyed.
>
> *(Marx, 1977, pp. 23–24)*

Thus, it is workers who are not only perpetual sufferers in terms of their material gains but also their ability to lead a human life. Ultimately workers, while working tirelessly for satisfying the greed of the bourgeoisie, sink to the level of the machine, and completely dehumanized, they confront the machine as a competitor. They remain losers under all conditions of production, that is, they suffer with the decline in production, they also suffer when production remains stagnant and continue to suffer even when production increases. Finally,

> the worker becomes all the poorer the more wealth he produces, the more production increases the power and size. The worker becomes an even cheaper commodity the more commodities he creates. The devaluation of the world of men is in direct proportion to the increasing value of the world of things. Labour produces not only commodities: it produces itself and the worker as a commodity.
>
> *(Marx, 1977, pp. 67–68)*

Therefore, from the point of view of the radical critique of the political economy, the rule of private property is the root cause of all forms of social injustice and with the abolition of private property, along with its various manifestations of injustice, will also be abolished particularly the family, individuality, classes, nation states, traditions, and eternal truth.

Overemphasis on the economic injustices by the Marxists was criticized by many. But the most significant ones were the "radicals" who accused such propositions being over deterministic of economic relations and basic economic structure. Included among the critiques were the advocates of anti-colonial and national liberation groups and most of these were from the colonies and their sympathizers in the metropolitan centers. They were of the opinion that everything about colonialism is not about economic exploitation. On the contrary, there were times when colonial powers had to spend far more in terms of economic, political, and human resources in the colonies than what they were able to extract. The American occupation of Vietnam and South Korea and the Russian occupation of the Afghanistan are some of the glaring examples of colonialism, which only showed that "colonialism could be characterized by the search for economic and political advantages without concomitant real economic or political gains, and sometimes even with economic or political losses" (Nandi, 1999, p. 10).

(In)justice and the colonial episteme

In response to the over-determinism of the economic structure there were scholars who opined out that equating colonialism to economic exploitation and drain of wealth is gross injustice to the agonies and pains of the victims. Colonialism is far more serious and devastating than what the theory of economic injustice comprehends. To essentialize colonialism to economic exploitation and political domination is to replace a tragedy with a farce. Colonialism is about the historic as well as civilizational injustices. It is true of all the colonizers and particularly the Western colonizers because their persistence and capacity to stop at nothing (Koning, 1972, p. 116) sets them apart from the rest, which can be verified by the fact that within the first four years of the discovery of the New World by Christopher Columbus and "two years of the administration of the brothers Columbus, an estimated one half of the entire population of Hispaniola was killed or killed themselves. The estimates run from 125,000 to one half million" (Koning, 1972, p. 82). However, the killing continued at no less speed. In 1515 there were not more than ten thousand Indians left alive; 25 years later, "the entire nation had vanished from the Earth. Not one Indian on the island had ever been converted to what Columbus called 'our Holy Faith'" (Koning, 1972, p. 85). The killing was not only done using guns, gunpowder, and steel but also the bacteria, germs, viruses, and microbes, etc. were the most effective allies in mass extinction (Diamond, 2017). Europeans brought with them, like biblical plague, smallpox and tetanus; various lung, intestinal, and venereal diseases; trachoma; typhus; leprosy; yellow fever; and teeth-rotting caries (Galeano, 1997, p. 18). It was much worst in case of the

Spanish-Aztec war of 1519 A D. Cortes landed with 600 Spaniards on the cost of Mexico. They were up against the strong army of Aztec running into the millions. The Spaniards were no match for the Aztec but for the decisive advantage they had their immunity to smallpox, which the Spaniards exported with their infected slave, resulting in an epidemic killing nearly half of the Aztec, including their Emperor Cuitlahuac. The impact of smallpox was so devastating that far more Native Americans died in bed from European germs than on the battlefield from European guns and swords (Diamond, 2017, pp. 200–201).

Consequently, the population of the Aztec was reduced to 1.6 million in 1618 from 20 million in 1519 (Diamond, 2017, pp. 201–202), within a span of 100 years.

The reach of colonialism was not only restricted to the liquidation of the population but it was far more than that. It was also about establishing their cultural, moral, political, epistemic, and civilizational supremacy. "Colonialism minus civilisational mission is no colonialism at all" (Nandi, 1999, p. 11). In case of the British conquest of India,

> colonial exploitation was an incidental and regrettable by-product of a philosophy of life that was in harmony with superior forms of political and economic organisation. This was the consensus the rulers of India sought, consciously or unconsciously. They could not successfully rule a continent-sized polity while believing themselves to be moral cripples. They had to build bulwark against possible sense of guilt produced by a disjuncture between their actions and what were till then, in terms of important norms of their own culture, "true" values.
>
> *(Nandi, 1999, p. 10)*

To accomplish this task they put a whole battery of philosophers, writers, bureaucrats, clergies, politicians, and philanthropists, etc. to construct the Indians as "crypto-barbarians who needed to further civilise themselves" (Nandi, 1999, p. 7). The Indians according to Kipling are "effeminate, passive-aggressive, and half-savage-half-child" (Nandi, 1999, p. 38). For Orwell, the Indians are weak: " 'the weak in a world governed by the strong' must 'break the rules or perish' " (Nandi, 1999, p. 42). Similarly, Macaulay not only undermined the intellectual self-confidence of the Indian but also cast aspersions even on their cultural achievements: "a single shelf of a good European library was worth the whole native literature of India and Arabia" (Sen, 2005, p. 77). He was adequately complemented by James Mill who not only comprehensibly denied intellectual originality of the Indian but also accused the Indian scholars for "fabricating things . . . under the glossing exterior of the Hindu, lies a general disposition to deceit and perfidy" (Sen, 2005, p. 79). He accused that the Indian "pundits had become acquainted with the ideas of European philosophers respecting the system of the universe, and had then proceeded to claim that those ideas were contained in their own books" (Sen, 2005, p. 78). Thus, he concluded that "Indian civilization was on par with other inferior ones known" (Sen, 2005, p. 147) and recommended that India is a "great scene of British action" (Sen, 2005, p. 142). Accordingly, McCauley divided India's history on communal lines: the Hindu Period, Muslim Period, and British Raj and recommended that the country (which, according to the Bishop of Oxford Richard Congreve, "God has entrusted . . . to us to hold it for Him, and we have no right to give it up"[Nandi, 1999, p. 34]) must be constructed as per the design of the Raj. Correspondingly the Imperial Survey of India and the Census of India, etc. were entrusted the task of constructing the geography and people (population) of India, and Macaulay was entrusted the task of constructing the minds of India through educational reforms. He wanted to create the new subject in the colonies: those were born with a different color of their skin pigmentation but always White in terms of their thinking. Though this remained the dominant view and also the view of the imperial power for all practical purposes, yet it was neither the total nor even uncontested notion. Warren Hasting, the Governor of the Presidency of Fort William and head of the Supreme Council, Bengal, was particularly perturbed by the policies of Macaulay and James Mill etc. and proclaimed that the British "had more to learn from the civilisation they ruled than they had to teach"(Nandi, 1999, p. 72). Similarly, Max Muller had revered the great Indian civilization for its great achievements in the field of mathematics, astrology, philosophy, culture, sciences, and literature, etc. but lamented its decline under the colonial rule and "forbade his students to visit India; to him, the India that was living was not the true India and the India that was true had to be but dead" (Nandi, 1999, p. 17).

Thus, the call to fight economic injustice that became the immediate flash point of anti-colonial struggle during the 19th and first half of the 20th century ultimately culminated into a fight against cultural injustices and a fight for identity and national liberation. With the success of national liberation and freedom struggles in the post-colonial countries after the World War II, it was expected that the world and the majority of its population have either gotten or are at the verge of living in a relatively less unjust society – until it encountered yet another major injustice: environmental or ecological injustice.

Environmental injustice

Though the concerns about environmental injustice had been articulated by many at different points of time in different contexts, the voice raised by Martin Luther King Jr. particularly, linking environmental injustice to racial discrimination broughtin new and the most startling facts to light. His support to the striking black garbage workers in Memphis in 1968 was a landmark contribution, though his untimely assassination was a big, yet temporary setback to the movement, which got fresh momentum through the garbage dispute with African American homeowners in Houston, USA in 1979. The dispute was about keeping the garbage dumping away from the White suburban middle-income neighborhood. Within a span of three years the movement snowballed into a full-fledged environmental justice movement, highlighting yet another racial dimension of environmental injustice. This time the site was in African American Warren County, North Carolina. This protest was so strong that the US General Accounting Office was motivated to conduct a study on "Siting of Hazardous Waste Landfills and Their Correlation with Racial and Economic Status of Surrounding Communities" (General Accounting Office, US, 1983). The revelations of the study were not only startling but also highlighted one of the darkest sides of modernity in the capitalist world, particularly the US. The study revealed that

> three out of four of the off-site, commercial hazardous waste landfills in Region-4 (which comprises eight states in the South) happen to be located in predominantly African-American communities, although African Americans made up only 20 % of the region's population.

Most importantly, the movement and the protesters succeeded in putting "environmental racism on the map" (Bullard, 2018, pp. 151–171). Following these developments, the publication of *The Limits to Growth: A Report for the Club of Rome's Project on the Predicament of Mankind* (Meadows et al., 1972) and *Our Common Future* (UNEP, 1987); the First National People of Color Environmental Leadership Summit in 1991; the listing of the Seventeen Principles of Environmental Justice (NPCELS, 1991),[1] Agenda-21 in 1992 (UNDP, 2002), the Million

Development Goals in 2000 (UN, 2002), and the Sustainable Development Goals in 2016 (UNDP, 2015), etc. have brought the issue to the center stage.

The issue became more complex when the idea of justice became more inclusive, particularly after environmental justice embraced racial discrimination within its ambit, vociferously advocated by the Environmental Protection Agency (EPA) (Bullard, 2018, pp. 151–171), which was striving to achieve environmental equity: procedural equity, geographical equity, social equity, distributional equity, generational equity, and outcome equity. The EPA advocated that it is possible by clearly defining environment justice and spelling out its goals:

> to insure that all people, regardless of race, national origin or income are protected from disproportionate impacts of environmental hazards. To be classified as environmental justice committee residents must be a minority and/or low income group, excluded from the environmental policy setting and/or decision making process; subject to a disproportionate impact from one or more environmental hazards and experience a disparate implementation of environmental regulation requirement practices and activities in their community.
>
> *(US Environmental Protection Agency, 2000)*

But above all these were also the concerns of environmental experts from Oxford University and many others who highlighted that neglecting ecological crises and environmental justice is one of the two major insurmountable crises that nation states are facing today. According to John Dunn (Dunn, 1994, p. 4), the ecological crises due to global warming, acid rain, depletion of the ozone layer, emission of greenhouse gases, production and accumulation of radioactive and toxic materials, etc. have totally undermined the sovereignty of the national states, violated their boundaries, and have affected the quality of life of every citizen irrespective of their nationalities. The nation states are not in position to control the same. It is not only the real crises of the nation states (Butola, 1998, p. 141) but also the epitomes of environmental injustice.

The threats of environmental injustice reached new heights with the rise and growth of the "praxis of terrorism" in the 20th century, particularly since April 22, 1915 (Sloterdijk, 2009, pp. 41–47) during the war between Franco-Canadian infantry led by French General Jean-Jules Henry Mordacq and the German Gas Regiment led by Colonel Max Paterson at Ypes. It was a watershed in history as it added a new dimension to environmental justice. The German army, visualizing their defeat at the hand of the formidable Franco-Canadian infantry, introduced a new concept in war and released 150 tons of toxic chlorine gas over an area of 6 km wide and 600 to 900 m deep (Sloterdijk, 2009, p. 42). Prior to this in war one man killing another and one army overpowering the other remained the motive of arms conflict. Hereafter "targeting not the body of the enemy, but his environment" (Sloterdijk, 2009, p. 43) was crucial in determining the outcome of the war. It is a chilling reminder of Shylock's lines in the Shakespearian play *Merchant of Venice*: "You take my house when you do take the prop. / That doth sustain my house; you take my life. / When you do take the means whereby I live" (Shakespeare, 1623, Act-IV, Scene-I, p. 245).

Environmental justice in India

India has been the victim of environmental injustice right from the days of colonial rule in the form of forced migration of the people from Central Indian Tribal Regions and their resettlement in the tea plantations of the North-Eastern Regions, particularly Assam and West Bengal.

Destruction of the natural habitats of the indigenous population in these regions led to the establishing of large tea plantations and the displacement of local populations. The local ecology was destroyed in order to establish hill stations, cantonment towns across the Himalayas, port towns along the environmentally fragile costal zones. Mining and mining towns in the mineral resource rich regions of India have subjected millions of vulnerable people to environmental injustice.

It was expected that the independence of India from colonial rule would not only remove the yolk of foreign rule but also transform the colonial subjects into citizens of Independent India. Citizens that would participate in framing the rules of governance, would legitimize the new rulers though a democratic process of elections and seek informed consent. But it is an irony that the intensity as well as extent of environmental injustices have increased in leaps and bounds in nearly every region, adversely affecting social groups, communities, and sectors of the economy. Some of the most important areas and population groups suffering from the environmental injustices in the post-independence India are the Adivasi or tribal communities, Dalits (classified as scheduled caste), and the poor.

Environmental injustice and rise of Naxalism

> The Naxal movement had its origin in the Naxalbari village, West Bengal in 1967, led by the Communist Party of India Marxist Leninist (CPI ML) leaders Charu Mazumdar, Kanu Sanyal and Jangal Santhal. The movement was influenced by the political philosophy of Mao Zedong. Initially it started a class struggle between sharecropper and landlord, but raised serious objections about the failure of the state government to protect the rights of the tribal and other indigenous communities living in 5th and 6th Scheduled Areas demarcated by the Indian constitution. In case of the West Bengal it was about the rights of the Santhal Scheduled Tribal Population and their rights. Over the years, the extent of the peasant and indigenous peoples' struggle spread across India, right from West Bengal, Odisha, Bihar, Jharkhand, Chattishgarh, Madhya Pradesh, Telangana, Uttar Pradesh, Maharashtra and Andhra Pradesh and have become an arena of Naxalite movement. Initially, the movement was confined to struggle against land alienation, mining and land acquisition leading to displacement of indigenous population due to mining, illegal acquisition of tribal land, national parks and bio-reserves etc. but over the years it has taken over issues related to displacement, ecological degradation, corporate loot, human trafficking, violations of civil liberties and human rights, custodial deaths and ultimately cultural genocide.
>
> *(Padel, 2012)*

National parks and biodiversity (biosphere) reserves

Hailey National Park was the first in India to be commissioned in 1936 in the Terai Region of Uttarakhand (previously United Province, and Uttar Pradesh) and it was subsequently rechristened as Jim Corbett National Park. By the year 2018 there were as many as 104 National Parks accounting for approximately 40,501 Sq. Km. (i.e. 1.23% of the total surface area of the country). Similarly, there are 18 bio-reserves in the country. The Nilgiri Biosphere Reserve was the first one to be established in India in 1986 and at present the 18 biosphere reserves cover around 89,516 Sq. Km. It is an irony that most of these national parks and biosphere reserves are in the areas that have high concentrations of indigenous population, rich in ethnic diversity and biodiversity.

National parks and biosphere reserves are indicators of the ecological health of the country. But these have also become important sources of tourist attraction, livelihood, and employment. Since most of these are located in the remote and ecologically sensitive regions of the country, these directly infringe upon the traditional rights and livelihood of the indigenous communities. Restrictions imposed on the movement as well as collection of the minor forest products; attack by wild animals; high-handedness and persistent harassment, torture, and penalties imposed by the forest, revenue, and police officials have contributed to disrupting the social fabric of indigenous communities. Consequently, most of the people are in direct confrontation with the government. They have been supported by civil rights activists, human rights organizations, environmental protection groups, etc.

Dams and reservoirs

India claims to be a developing country and it has emerged as the fifth largest economy in the world. Agriculture and associated activities still continue to the backbone of its economy, yet rapid urbanization, industrialization, development of transport and communication networks as well as modern IT industries, hotel, and tourism industries and the modernization of agriculture have created an unprecedented high demand for energy. Since India has very poor reserves of hydrocarbons and most of its coal reserves are not of good quality, it loses huge amounts of foreign reserves on the import of petroleum and other related products. Moreover, fast depleting non-renewable resources and increasing environmental pollution due to overuse of carbon, hydrocarbon, and nuclear energy have compelled many countries including India to search for alternative sources of energy. So, there is always a demand to diversify the preferable energy resources in favor of renewable energy. Solar energy, wind energy, tidal energy, geothermal energy are some alternatives, but most of these are either in the nascent stage for want of proper technological knowhow, or they possess limited possibilities of shouldering the burden of the massive energy requirements of a giant sized economy such as India's. Consequently, the most obvious choice is the use of water and hydroelectricity. Construction of dams and reservoirs constitute essential components of these energy sources, which in turn displace the peasant, rural populations in the hills, mountains, and plateau regions of India. Once again most of these are ethnic groups and communities. Submergence of the vegetation, forest areas, agricultural land, pastures, wetlands, and grazing lands increases the vulnerabilities of these people manifold, with successive constructions of dams.

Degradation of landscape, deforestation, and destruction of the natural habitat of wildlife, displacement of people due to dams and reservoirs, loss of livelihood and traditional knowledge and social cultural networking, etc. have created psychological anxieties among the people. Most of these have either become environmental refugees or cultural nomads in their own homeland.

Industries

Industries and industrialization were once seen as the backbone of capitalist development, but they are visualized by the marginalized groups as a menace to their economy, culture, livelihood, and ecology. Industries are seen as consumers of raw materials, destroyers of ecosystems, sources of noise pollution, causes of social conflicts, crimes, diseases, and above all negations of cultural diversity and social plurality. It is a well-known fact that, similar to traumatic experiences of India under colonial rule, the rural and tribal areas of the country are experiencing indiscriminate exploitation of their natural resources, acquisition of their land, and displacement of people due to industrial development. Industries are not only exploiting the natural resources and land in these areas but adding large volumes of toxic wastes too and exposing lives and ecosystems to stresses on their metabolism. Consequently, occupational hazards and life style diseases like mental stress, depression, suicide, alcoholism, accidents, surveillance and interference of police in public and private lives, lockouts, and social tensions, etc. have serious adverse impacts on the life of women, children, and elderly people.

Mining

Mining is one of the oldest human activities, existing long before animal husbandry and agriculture. But, the nature and extent of mining have undergone a sea change under industrialization. Destruction of the environment, ecology; damage to biodiversity; contamination of land, water, and air; noise pollution; land subsidence, etc. have become synonymous to mining activities. Moreover, large-scale migration of workers, contractors, agents, and mafia in mining and related activities has resulted in the breakdown of traditional social networking without replacement with a better alternative. Consequently, damage to surface cover and vegetation, soil erosion, waste overflow, eco-services, large-scale disturbance of hydro-geological systems, reduced ecological-hydrological connectivity, mine trailing spills, loss of biodiversity, depletion of the aquifers, lowering of ground water tables, subsidence, the unprecedented rise in organized crime, drug abuse and diseases, alcoholism, loss of livelihood, immoral human trafficking, etc. have directly impacted the lives and rights of the people.

Infrastructure

Demand for infrastructure is derived from the other sectors of the economy. Moreover, the quality, efficiency, and affordability of infrastructure activities contributes significantly not only in the (under)development of the region but it also contributes to inter- and intra-regional disparities. However, construction of modern infrastructure in a relatively isolated and less dynamic region like the tribal, hill, and desert areas of India is indicative of serious environmental injustices. Construction of massive infrastructure across the length and width of India for the rapid and smooth movement of raw materials, goods, commodities and services, people, etc. have resulted in massive land acquisition, leading to loss of agricultural land, decrease in vegetation cover, increase in mud flow and floods, soil degradation, loss of biodiversity, loss of livelihood, socio-economic displacement of the people, loss of traditional knowledge, and pollution and diseases.

Toxic waste disposal

Toxicity is the introduction of contaminated substances in a natural environment that destroys its capacity to regain its original status, which results in serious metabolic rifts in various natural cycles, including the livelihood and traditional knowledge systems of the inhabitants. Indiscriminate destruction of forests; poaching of wildlife; extraction of plant species; and introduction of radioactive wastes, chemical, and other industrial toxins; hotel and hospital wastes are some examples of toxic waste dumping that have become almost like routine features of development in India. These have caused serious concerns about environmental injustice. The victims of toxic waste disposal are the neighboring communities, biodiversity, and eco-services. It has been a common experience that most often the toxic and solid waste disposal sites are located in the vicinity of the marginalized communities. Consequently, contaminated air, water, soil, and ground water not only disrupt eco-services but also expose these people to different diseases and ailments.

Tourism

Tourism is emerging as one of the most flourishing and sustainable economic activities in India. There are different types of tourist activities in operation: heritage tourists, eco-tourists, adventure tourists, health tourists, medical tourists, rural tourists, slum tourists, and cultural tourists. And even though several researchers and policy makers tout certain forms of tourism as sustainable, lack of sensitivity to the culture and environment can result in disastrous consequences for the local population and the environment. Commoditization of culture and destruction of the environment due to overcrowding and pollution have disrupted the delicate balance between the environment and society and has added stress on biodiversity. Moreover, most of the tourists are also responsible for intellectual property right violation and the destruction of traditional knowledge systems. These adverse impacts of tourism have generated strong resentments among the indigenous communities, particularly women. Losses of landscape, aesthetic degradation, deforestation, loss of biodiversity, and air, water, and noise pollution have adversely affected the livelihood of the indigenous communities. Moreover, increase in the flow of tourists, event managers, and services providers have also resulted in the rise of anti-social, illegal activities: drug abuse, alcoholism, crime, and human trafficking.

Environmental justice and thermal and nuclear power plants

It has been mentioned before that India has emerged as one of the five largest economies in the world, which means it is one of the largest energy consumers in the world. Since India had limited hydrocarbon fuel available within the country, it relies heavily on the import of petroleum products, putting strong stress on its foreign exchanges reserves. Consequently, India has been diversifying its energy options. Generation of thermal and nuclear energy are the most sought-after options but this is not without enormous environmental justice consequences, in addition to other side effects. Air, soil, and water can be polluted, including ground water. Loss of biodiversity, stress on livelihood, and vulnerability to diseases are some of the important aspects of environment injustice.

These are common concerns with nuclear power plants, but there are some additional and more serious aspects associated with nuclear power, including genetic contamination, exposure to radioactive substances endangering not only human health and quality of life but also organisms in general. Moreover, the establishment of nuclear power plants in the backdrop of communities with high illiteracy and poor scientific understanding directly infringes upon the human rights of the inhabitants due to lack of informed consent.

Development of airports and environmental justice

Development in the modes of communication and transport infrastructure is one of the essential parts of globalization. Smooth and speedy movement of goods, commodities, services, and human resources across geographical regions have played significant roles in integrating the world into a single economic unit. Land acquisition, displacement of people, and noise pollution are some of the important reasons behind resentment against the development of airports in most parts of India.

Special economic zones and environmental justice

In a globalized economy but divided political space among sovereign nations, there are bound to be contradictions between economic, environmental, and political interests. But in all eventualities it is usually economic interests that succeed and dominate (Marx, 1976, p. 3)[2] but without apparent display of the defeat of the political interests represented by the stakeholders of the nation states. This has been possible by provisioning "demarcation of areas within the territorial boundaries of the sovereign nation states having special business and trade regulations different from the rest of the country". Apparently, the aim of establishing such zones includes aspects like "trade balance, employment, attract[ing] foreign investment, and efficient administration". These policies also include change in the monetary, tax, employment, patents, copyrights and labor regulations. Theoretically an SEZ should be restricted to a geographically delimited area, usually physically secured (fenced-in), single management/administration; eligibility for benefits based upon physical location within the zone; separate customs area (duty-free benefits) and streamlined procedures (World Bank, 2008). But it is viewed by the local people as enclavization in the post-colonial countries. It is a new form of spatial exclusion at the cost of biodiversity, plurality of cultures, and livelihoods. Moreover, it will increase redundancy and obsoletism among rampant unemployment, which has serious implications for women in the workforce. Moreover, multiple working shifts leads to disorientation and lifestyle changes among the working people. Overstress at the work place and racing against time have been identified as the main causes of death in road accidents and suicides, etc.

These are only miniscule representations of the issues related to the environmental injustice in India. There are conflicts over depleting water resources and destruction of the biomass; there is global warming due to the built-up environment, particularly in the urban areas; there are health hazards and contamination of food items due to excessive use of chemical fertilizers, pesticides, and preservatives, all of which need in depth research to highlight the nature of environmental injustice in India. However, it is pertinent to ponder on the political responses these injustices have received across the political and intellectual spectrum.

Politics of environmental justice

It was mentioned in the beginning of this chapter that justice is contingent upon the idea of citizenship, which in turn is dependent on the nation states which had been the natural ally to modernity. Since modernity itself is a highly differentiated process spatially and historically, it is but natural that nation states, justice, and citizenship too underwent numerous metamorphoses along with modernity. We are citizens under multiple modernities (Chatterjee, 2013, p. 12): commercial society, civic associations or civil society, rational bureaucracy, industrial society, democratic societies based on universal suffrage and the welfare state (Chatterjee, 2013, p. 12). Consequently, there are multiple ways of defining access to justice. John Rawls viewed it as "new form of social contract in terms of maximisation of liberty, equality and opportunity as the central theme seeing justice in the light of fairness". His postulations about justice are based on three cardinal principles (Stanford Encyclopedia, 2017):

1 Greatest Equal Liberty Principle: According to this first principle "each person is to have an equal right to the most extensive total system of equal basic liberties compatible with a similar type of liberty for all".
2 Social and Economic Inequalities Principle: This encompasses both:
   a The Fair Equal Opportunity Principle: society should pursue the greatest good for the greatest number.

   b The Equal Opportunity Principle: it provides equal opportunities to all, particularly "offices and positions open to all under conditions of fair equality of opportunity".

3 The Difference Principle: it is aimed to provide the greatest benefits for the least advantaged so that no person should be blocked from occupying any position.

According to Amartya Sen (Sen, 2009), such an idea of justice suffers from the notion of "perfect arrangement": society is a system of cooperation for mutual advantage between individuals, which is practically impossible when evaluated in the backdrop of multiple modernities, citizenships, democracies, modes of governance, and opinions, etc. It amounts to giving tacit approval to the tyranny of the majority over the minority and, finally, the infallibility of some innate human moral sense, intuitions to judge, and determine right and wrong.

Sen opined that the principle of social justice should define the appropriate institutional mechanism for the distribution of benefits as well as burdens of social cooperation. For this to happen, people should liberate the idea of justice from the narrow confines of "Niti" and embrace the idea of "Nyaya" within it (i.e. correct principles as well as the desired just consequences of the same on individual and society). Failing which, environmental justice will continue to languish in the cross firing of the following.

1 Utilitarian, pleading for maximization of use value;
2 Egalitarian, advocating for equitable distribution and access;
3 Libertarians, seeking
   a Higher returns, greater autonomy to individual entrepreneurial skills (conservative environmentalists),
   b Traditional values and religious values, community life, non-materialistic virtues, prudence, piety, self-restraints in personal life and individual autonomy (traditional environmentalists),
   c Rewards for ownership of private property as self-interested rationalists, inconsonant with equally rational market sensitivity to ecology and guarantees of private property (free-market conservative environmentalists);
4 Natural capitalists (Paul Hawken), believing
   a The unregulated market is responsible for ecological crises because of its urge to externalize the costs,
   b In proper valuation of eco-services,
   c In minimum possible administrative control over market (market socialism),
   d The government should only make the market more responsible to ecology through green taxes (Pigovian taxes) by replacing other taxes,
   e In an intelligent production system – products should be reusable, recycled, and remanufactured,
   f In the public-private-partnership model (PPT);
5 Liberal environmentalists, registering a strong reaction to the natural capitalists, particularly Paul Hawken, insisting
   a Restrictions on freedom of the market will impinge on human freedom but strong welfare measures by the state will help the consumers,
   b The state should create a public good to protect the environment,

c Pigovian taxes will transfer taxes onto consumers,
d Environmental deliberative democracies:

 i Are real democracies that are always pro-environment,
 ii Hold the key to ecological sustainability with the reduction of inequalities and marginalization,
 iii Provide citizens access to environmental data and awareness.
 iv Replace instrumental rationality with human rationality.
 v Have civil society involved in ecological issues and the replacement of administrative and bureaucratic control;

6 Eco-socialists, believing that environmental injustice is the second contradiction of capitalism,

 a Externalize the production cost deterioration of society and environment,
 b Insist ecological problems are global in nature,
 c Claim there is need of an ecological manifesto,
 d Contend social justice and ecological balance are inseparable,
 e Break John Bellamy Foster's suggested four laws of capitalism (quintessentially the root cause of environmental injustice):

  i The only important connection between things is market value,
  ii It is not important where things go after use, as long as they do not interfere with capital accumulation,
  iii The market rather than nature knows best,
  iv The goods of nature are a free gift to the capitalist;

7 Ecofeminists, advocating

 a The process of "othering" is the root cause of injustice: nature, culture, communities, and women, in that order,
 b Development is mal-development,
 c That development marginalizes all (in traditional societies men-women relationships are of interdependence and complementary),
 d In Ariel Salleh's concept of "global debt",
 c That capitalism upsets the balance between society and nature, resulting in ecological crises through three types of debts:

  i Social debt: exploitation of labor,
  ii Ecological debt: exploitation of natural resources,
  iii Embodied debt: exploitation of the reproductive power of women.

Concluding remarks

Justice is a latecomer in the fight for rights. Initially, it was assumed that the fight for citizenship and citizen's rights was substantive and there was no need for raising separate demands for justice. But the developments under capitalism proved that successfully obtaining rights in the developed countries was possible, but at the cost of justice to the marginalized sections within the developed countries and large populations in the colonies.

The success of anti-colonial movements and the rise of nation states mostly in the former colonies gave hope that there would be justice across social, cultural, economic, and political spectrums. But the annals of history prove that there was no change in injustice; on the contrary these societies are exposed to new kinds of injustices.

India, being a post-colonial country, has succeeded to emerge as one of the fifth largest economies of the world but at the huge cost to environmental justice. These injustices are directly related to the promotion of a neoliberal agenda of accumulation by dispossession. Since most of the resources are located in the regions inhabited by indigenous communities, they are the main target of accumulation and dispossession and injustice. Since the ethnic communities have maintained symbiotic relationships with the environment, injustice to one component has cascading and multiplying impacts of the other.

Notes

1 National People of Color Environmental Leadership Summit (1991): It outlined Seventeen Principles: (1) Sacredness of Mother Earth, (2) Mutual respect and justice for all people without discrimination, (3) Right to ethical, balanced, responsible use of land and renewable resources, (4) Universal protection from nuclear testing, extraction, production and disposal of toxic wastes, (5) Right to political, economic, cultural and environmental self determination to all people, (6) Cessation of all the toxins, hazardous wastes, and radioactive materials, (7) Right to participate as equal partners in decision making at every level, (8) Right to safety and healthy work environment to all, (9) Full compensation to the victims of environmental injustice, (10) Government should invoke stringent international laws such as human rights and the United Nations Convention on Genocides against the violators, (11) Rights of the Native people must be recognized and given precedence over others, (12) The rural and urban ecological policies of clean and healthy environment must be upheld, (13) Mandatory informed consent and the halting of the testing of experimental reproductive and medical procedures, (14) Stopping the destructive operations by the multi-national corporations, (15) Opposition to military occupation and repression, (16) Education of the present generations about the social and environmental issues, and (17) Restraint of consumption by individuals of the earth's resources.

2 Marx Karl (1976): Without falling into the fallacy of economic determinism, Marx himself mentioned "production constitutes the economic structure of society, the real foundation on which there correspond definite forms of social consciousness. The mode of production of material life conditions the real social, political and intellectual life-process in general".

References

Bullard Robert D (2018): *Environmental Justice in the 21st Century: Race Still Matters*, Environmental Justice Resource Centre, Clark Atlanta University, Phylon, GA.

Butola BS (1998): The Paradoxes of National Integration, in *Nationalism, Regionalism and Philosophy of National Integration*, (ed.) Malla N. Regency Publications, New Delhi, p. 141.

Chatterjee Partha (2013): *Lineages of Political Society: Studies in Postcolonial Democracy*, Permanent Black, New Delhi.

Diamond J (2017): *Guns, Germs & Steel*, Vintage, London.

Dunn John (1994): Introduction: Crisis of the Nation state? *Political Studies*, Vol. 42, Special Issue, 3–15.

Esposito Roberto (2006): *Communitas: The Origin and destiny of Community*, Stanford University Press, California.

Frazer Nancy (1995): From Redistribution to Recognition? Dilemmas of Justice in Post-Socialist Age. *New Left Review*, No. 212.

Galeano Eduardo (1997): *Open Veins of Latin America: Five Centuries of Pillage of a Continent*, Monthly Review Press, New York.

Koning Hans (1972): *Columbus: His Enterprise: Exploding the Myth*, Monthly Review Press, New York.

Marx K (1976): *Preface and Introduction to a Contribution to the Critique of Political Economy*, Foreign Language Press, Peking, Beijing.

Marx K (1977): *Economic and Philosophical Manuscripts of 1844*, Progress Publishers, Moscow.

Marx K & Engels F (1975): *Manifesto of the Communist Party*, Progress Publishers, Moscow.
Marx K & Engels F (1977): *Feuerbach: Opposition of the Materialistic and Idealistic Outlook*, Selected Works V0l.-I, Progress Publishers, Moscow.
Meadows DH et al. (1972): *The Limits to Growth: a Report for the Club of Rome's Project on the Predicament of Mankind*, Universe Books, New York,
Mill John Stuart Mill (1859): *On Liberty*, Batoche Books, Kitchener, Ontario, 2001.
Nandi Ashis (1999): *The Intimate Enemy: Loss and Recovery of Self Under Colonialism*, Oxford University Press, New Delhi.
National People of Color Environmental Leadership Summit (1991): *Office of Environmental Justice (2000)*, Environmental Protection Agency, Washington DC.
Padel Felix (2012): *Out of This Earth*, Orient BlackSwan Private Limited, New Delhi.
Paine Thomas (1792): *Rights of Man*, J.S. Jordan, London.
Proudhon Pierre Joseph (1923): *General Idea of the Revolution in the Nineteenth Century*, (Translated by Robinson J.B.), Free Press, London.
Rousseau JJ (2017): *The Social Contract*, Cambridge University Press, Cambridge.
Saxe Commins & Robert N Linscott (1953): *Man and the State: The Political Philosophers*, Rousseau J J From the Social Contract, Modern Pocket Library, New York.
Sen A (2005): The Argumentative India: Writings on Indian History, Culture and Identity, Allen Lane, Penguin Books, New Delhi.
Sen A (2009): *The Idea of Justice*, Allen Lane, London.
Shakespeare William (1623): *The Merchant of Venice, The Complete Works of William Shakespeare*, Collins Clear-Type Press, London, p. 245.
Sloterdijk Peter (2009): Airquakes, *Environment and Planning D: Society and Space*, Vol. 27.
Stanford Encyclopedia of Philosophy (2017): *John Rawls, Idea of Justice,* Stanford, U.S.A.
UNDP (2002): *Agenda-21, United Nations Conference on Environment and Development*, United Nations, New York.
UNDP (2015): *Sustainable Development Goals*, United Nations, New York.
UNEP (1987): *Our Common Future: Report of the World Commission on Environment and Development*, Oxford University Press, New Delhi.
United Nations (2002): *Millennium Development Goals*, United Nations Millennium Declaration, New York.
US Environmental Protection Agency (2000) Office of Environmental Justice. https://www.epa.gov/environmentaljustice. Accessed Aug. 24, 2021.
US General Accounting Office (1983): *Siting of Hazardous Waste Landfills and Their Correlation with Racial and Economic Status of Surrounding Communities*, D.C. Government Printing Office, Washington, DC.
Weber Max (1958): *The Protestant Ethics and the Spirit of Capitalism*, Scribner, New York.
World Bank (2008): *Special Economic Zone: Performance, Lesson Learned, and Implication for Zone Development*, World Bank, Washington, DC.

17
NO LIFEBOATS AVAILABLE
Hurricane Harvey and emergency management

Samantha Espinoza-Villejo

The perfect storm

In August 2017, Hurricane Harvey flooded the Houston region with record-breaking rainfall. In this chapter, I uncover the relationship between neoliberal emergency management and the marginalized communities that go unprotected. I outline the basic social structure of Houston and how that spatially translated to being affected by Hurricane Harvey. I then discuss the interviews I gathered from two Houston residents that were affected by Hurricane Harvey in different ways: one affected by flooding, another affected by industrial complex failures. Then, I deconstruct Harris County Office of Homeland Security and Emergency Management (HCOHSEM) disaster preparation documents. Finally I synthesize the information from the interviews and deconstruction to showcase where local emergency management practices fail to protect citizens in low-income communities. This chapter is divided into the following sections:

- Inverting tradition: Houston's social spatiality
- Voices of the bayou: photovoice method and interviews
- Taxes or theft: deconstructing HCOHSEM rhetoric
- Wading through murky waters: engaging emergency management

Inverting tradition: Houston's social spatiality

Houston benefited from the industrial boom of the 1980s. Along with the fortress of skyscrapers and their custodians came the influx of people looking for opportunity. Of the 2.3 million people living within Houston proper, 22%, or around 500,000 people, live below the poverty line (United States Census Bureau, 2017). Low-income residents run the engines of Houston, but they are forgotten by the influence of industrial barons and politicians. The silence of vulnerable communities in the discussion of Hurricane Harvey recovery is deafening and is the heart of this study. I will amplify voices from those who will experience the long-term effects of disaster more intensely and deconstruct emergency management language to show how marginalized communities are left out of modern disaster science.

 DOI: 10.4324/9780429450310-20

Houston is a neoliberal success story, taking seemingly inhospitable, savage land and turning it into a global powerhouse. Where Houston throws neoliberal tradition to the wind is its floodable space. In many societies, there exists an Other that is societally outcast. According to Laura Pulido (2017), the Other finds themselves devalued in both social and economic aspects. The value of the Other, especially in neoliberal societies, is based on the amount of economic power an individual brings to the table. Since Others in neoliberal societies tend to work in low-income jobs, their power is negligible, and their human value disappears into itemized categories that are considered replaceable (Pulido, 2017). When constructing a community, societies tend to reserve the best land for those with the most social and economic power. Those who classify as Other are left with less desirable land that is generally more at risk from environmental hazards (Pulido, 2017). These places are environmental "sinks", places where societally undesirable people and economic sectors pool together (Pulido, 2017). In Houston's case, however, upper-class populations live in the areas at higher risk for flooding. The reason for this is commodification of Houston's waterways and coastlines.

The Buffalo Bayou, Houston's founding waterway, has been a site of capital-genesis since the 1980s, and "parking" Houston waterways has been discussed since 1913 (Urban Land Institute, 2018). Beautifying and gentrifying waterways to attract new and expensive land tenants has extended to the coastlines along Trinity and Galveston Bays as Houston expanded; therefore, those who could afford the beautified water-adjacent properties moved into these areas. Commodification of the coastlines has been a major factor in tropical system-related destruction of coastal communities along the Gulf Coast. The Gulf Coast is home to diverse ecological communities like coastal prairies, Black Mangrove swamps, and Bald Cypress swamps. The plants in these communities evolved to experience intense flooding and funnel water into the bays or into root and soil systems. These ecosystems also act as barriers against erosion, storm surge, and high winds that carry ocean debris. Prairie grasses hold the soil in place and break storm surge. Black Mangrove roots grow into the sand and back up to the surface, ensuring the plants' security against storm surge and high winds. Likewise, coastal trees like Bald Cypress have stabilizing root systems through giant taproots and knobby knees.

Expensive housing developers commodify the coasts and remove the protective ecosystems for unobstructed ocean views and access. Without the protective ecosystems, the expensive coastal neighborhoods leave the coasts undefended against damage. In addition, these communities occupy space that was evolved to accommodate intense weather events, bringing disaster further inland. Commodification of Houston coastlines and waterways caused many expensive neighborhoods to flood beyond repair. The paving of native flood infrastructure allowed flooding to occur in spaces where flooding has historically never happened. For marginalized communities outside of flood risk zones, residents may have experienced flooding, but structurally, the homes remained intact (Boyer & Vardy, 2019).

Low-income housing surviving Hurricane Harvey is extremely important to notice. The term "flood disadvantage" is more applicable than "flood risk" when assessing the impacts of flooding on vulnerable communities. The reason flood disadvantage is more appropriate is those who are closer to water may be more at risk of flooding, but if those residents have more money, they can more easily afford insurance, rebuilding, or relocating (O'Hare & White, 2018). Those at flood disadvantage may not flood as badly, but they lack the resources to start or continue the recovery process, putting these residents at higher risk of not recovering (O'Hare & White, 2018). In affluent neighborhoods, flood insurance was available, along with lawyers to sue on behalf of whole neighborhoods and financial capital to move away or privately rebuild (Boyer & Vardy, 2019). Low-income residents did not have those options. According to Pierre Bourdieu, there are several kinds of capital that are important in societies. In this chapter, I focus on two:

financial capital and social capital. Financial capital is money and its transactions (Bourdieu, 1986). Social capital consists of a community's resources and knowledge of the innerworkings of their society (Bourdieu, 1986). In the neoliberal disaster complex, recovery success is highly dependent on a community's social capital. This requires a community to be aware of what they have communally – maybe one neighbor's generator allows a few neighbors to stay in the same house where they can find some comfort and security. In addition, social capital indicates understanding of the local bureaucracy. Bureaucratic know-how is exceptionally powerful because it allows a group to move through social landscapes to find resources they may not be able to afford or find on their own (Boyer & Vardy, 2019).

Social capital is abundant in affluent, at-risk communities during disaster. The power of social capital for affluent communities comes from government-sanctioned planning. When affluent neighborhoods are built, local governments take stock in how much the city can gain through high property taxes and high-paying jobs; therefore, affluent communities are, from the get-go, afforded luxuries, government protections, and positive government interactions that low-income communities can only dream of.

The power of abundant social capital was evident in communities directly adjacent to the Addicks and Barker Reservoirs on the west side of the Houston metro area, near Katy, Texas. In the 1940s, city engineers built the two reservoirs to ensure flood control of Buffalo Bayou (Wallace et al., 2018). Most of the reservoirs' land is government-owned, but the outskirts of these reservoirs were left as private land, which developers acquired as Houston grew (Wallace et al., 2018). During Hurricane Harvey, the city of Houston evacuated the neighborhoods within the reservoirs and allowed the neighborhoods to flood, resulting in total loss of entire neighborhoods.

While devastating for the residents that lost homes, Addicks and Barker Reservoir residents had social capital to start their journeys to recovery. Whole neighborhoods banded together in class-action lawsuits, requiring the know-how to get a powerful attorney; used their insurance policies, requiring the know-how to work through the complex web of coverage and exceptions in insurance policy; moved, requiring financial capital from the social system that afforded residents a high-paying job; or were bought out, requiring the know-how to negotiate with a local government to ensure an acceptable price for the destroyed home. The means to recovery these reservoir communities used are not available to most low-income residents, especially those residents who do not speak English as a first language.

Unlike the more affluent homes that experienced total destruction, low-income communities may have flooded, but they did not flood enough to lose their homes, requiring them to stay in place. Low-income residents who had some money to spend on rebuilding but could not move were swindled out of their money (Boyer & Vardy, 2019). To profit off Hurricane Harvey, dishonest contractors swarmed the city, promising a repairs at affordable costs (Boyer & Vardy, 2019). Once money was exchanged, contractors disappeared, leaving residents with limited money stuck with unrepaired flood damage (Boyer & Vardy, 2019).

Being stuck in place brings on a host of interlaced frustrations and terrors (Boyer & Vardy, 2019). Low-income residents live in fear every time it rains because of the traumatic stress they endured as Hurricane Harvey forced water down streams that could not compensate (Boyer & Vardy, 2019). If low-income residents attempt to move away like their more affluent neighbors, they risk selling their water-damaged homes at below pre-damage market value, and falling into debt (Boyer & Vardy, 2019).

As was shown in the recovery of the Addicks and Barker reservoir neighborhoods, the city can acquire damaged property, like buyouts, but for low-income residents, buyouts never came (Boyer & Vardy, 2019). The reason is buyouts in low-income areas are financially risky for cities.

The city must maintain bought-out properties, which costs money. In addition, the land in low-income areas cannot be sold at prices the city can profit from (Boyer & Vardy, 2019). Lack of buyouts and worsening home conditions left some families with no choice but to abandon their homes, leaving ghost neighborhoods with little traffic and acknowledgment from Houston (Boyer & Vardy, 2019).

Some low-income areas are coming back, like Greater Greenspoint, which was heavily inundated by the swelling of Greens Bayou (Boyer & Vardy, 2019). Several flooded apartment buildings have sprung back to life, replacing damaged first floors and reopening them for rent (Boyer & Vardy, 2019). Low-income residents that can move to another low-income area are moving into these once-flooded apartments because they are affordable, and the new residents may believe they moved out of a flood zone (Boyer & Vardy, 2019). In these low-income situations, the social capital power imbalance lies with landlords. Landlords know the laws and regulations of operating apartments and rental spaces. This also means property owners have the ability to find loopholes. In Greater Greenspoint, apartment property owners are not telling new tenants that the first floors were completely underwater; therefore, current tenants will try to alert prospective tenants (Boyer & Vardy, 2019).

Like Greater Greenspoint, Harrisburg/Manchester has largely been ignored in its recovery from Hurricane Harvey. Harrisburg/Manchester is low-income, is a predominantly Hispanic community, and, unlike Greater Greenspoint, is surrounded by chemical plants (Friedrich, 2017). There are around 500 hazardous material plants in the Houston area (Friedrich, 2017). During Hurricane Harvey, many of them flooded and released toxic chemicals and metals into the flood waters. Those who came into contact with these chemicals became at risk for more than home damage. Toxicologists found Harrisburg/Manchester was exposed to benzene and dioxins, which are well-known carcinogens (Friedrich, 2017). In addition, these chemicals can cause acute illnesses like headaches, dizziness, and tremors (Friedrich, 2017).

Many low-income residents cannot afford health insurance, never mind home and flood insurance, which are two separate expenses in the state of Texas. When vulnerable populations get sick, they often do not risk an expensive doctor visit and will stay home to fight off illness. There should have been safeguards against disasters like toxic floodwaters, especially with around 500 chemical plants in almost 10,000 square miles; however, Texas has avidly fought to keep federal regulations at bay and was successful (Friedrich, 2017). Two years after Hurricane Harvey, Tropical Storm Imelda stalled over Houston. Governor Greg Abbott removed environmental protections to allow over 100,000 pounds of toxins to be released into the air to protect chemical companies (Trevizo, 2019). This fact is concerning, especially with the knowledge that the Texas government did not know the locations of chemical stores in the city, and Occupational Safety and Health Administration (OSHA) is understaffed, letting many violations go unchecked (Collette & Dempsey, 2016). Chemical companies get passes for polluting bodies and the environment quite often. According to the *Houston Chronicle*, Beaumont, Texas residents who live near the Exxon-Mobil plant experience respiratory illnesses and headaches regularly (Trevizo, 2019). Without insurance, going to the doctor to treat these symptoms is not affordable. Meanwhile, the cause of the symptoms rages on for capital accumulation.

Voices of the bayou

To amplify the voices of those living in marginalized communities, I used the photovoice method. In this section, I will discuss what photovoice is and why I chose this method. Then, I will discuss the interviews and what I learned from my co-researchers. I use the term *co-researcher* to reference those who lent their time and lived experience to participate in this research. This terminology honors both those who participated and the philosophy of photovoice, which recenters research away from the academic researcher.

Photovoice method

Photovoice dates back to the 1990s, when researchers Wang and Burris (1997) used the photovoice method to create community engagement and discussion surrounding public policy and health (Schumann et al., 2018). This method is a feminist approach that removes the researcher from a position of power and gives that power to those participating in the study (Schumann et al., 2018). The methodology revolves around gaining a database of co-researchers who carry a camera to photograph their lives. After photos are taken, co-researchers and the researcher interview in either a group setting or an individual setting.

In the interviews, co-researchers discuss the photos to reflect on their lived experiences and the intricacies of their community. The information is analyzed and processed during the interview, during which the researcher records what the co-researcher synthesizes. After co-researchers share their ideas, common threads can be discussed to find out what the community needs in order to enact social change from within the community. In this study, both co-researchers were given pseudonyms to protect their identity.

Ernesto was the first co-researcher. We met through SER-Jobs for Progress Houston, a non-profit organization that helps train retired, low-income seniors to return to the workforce. From Ernesto, I attempted to use a snowball method to find more co-researchers. Houston's COVID-19 crisis proved too challenging, and the leads disappeared as people fell ill. The second co-researcher is an acquaintance from social media, named Helene. She and I had a few interactions prior to this research, so I reached out to her, and she agreed to be a co-researcher.

After the initial contact, the co-researchers and I discussed gathering photos from Hurricane Harvey and scheduled interviews. Because of the COVID-19 chaos that overtook Houston, I decided it would be safest for co-researchers to gather photos they already had. I also suggested using news clippings and news photos. In each interview, I asked a set of demographic questions that pertained to income, race and gender identification, and disaster experience. Then, I asked each co-researcher to share their photos and discuss what feelings the photos brought up. Given the freedom photovoice provides, the co-researchers and I went off topic several times, which added more context to their feelings towards the provided photos.

Interviews

The first co-researcher I interviewed is Ernesto. Ernesto is from the Greater Third Ward in Houston, a community that experienced flooding during Hurricane Harvey. When I asked Ernesto if he had any pictures from Hurricane Harvey, he told me he lost those photos when he lost his phone during a near-drowning experience. Since Ernesto lost his photos, I asked him to gather some pictures from local news articles. During his interview, I asked questions about social demographics. Ernesto identifies as Hispanic, male, gay, and low-income. His income is $20,000–$25,000 per year. He rents his home in the Greater Third Ward, and he has experienced multiple hurricanes and tropical storms, having grown up south of Houston, closer to the coast.

Prior to Hurricane Harvey, Ernesto said he heard the storm was coming, but he said he's "always an optimist" and thought, "well, it's not going to do that much". As the storm came in and stalled over Houston, Ernesto was trapped by flood waters that isolated his community. Ernesto described Hurricane Harvey as, "It wasn't wind. It was just rain . . . you couldn't go anywhere". To understand how Hurricane Harvey flooding compared to past flooding in Ernesto's community, I asked, "you said that . . . your area is pretty . . . susceptible to flooding . . . so what kinds of disasters have you experienced while living in your home?" Ernesto replied he's experienced "four or five floodings". He then told two stories of how a simple trip can turn into a perilous event:

> Like I was telling you the other day I was going to . . . the Beltway and somebody had bought some food for me since I've been quarantined. And . . . I just got a new car, and . . . it was maybe raining like 15 or 20 minutes, and it was already flooding! And I just got to a really high point and just stood there in the middle of the street because there's no in and out. It floods like *that* here. Just the city needs a huge infrastructure flooding plan, but you know, they just, you know, cities are, they don't have the money to do it. So it, you know, you really feel threatened in Houston. You just do. Houston is a very . . . because the weather just pops out of nowhere. You know, I was by Barnaby's over on West Gray one day, years ago, and I had a little Honda Prelude, and it just started raining real quick. And so I pulled over, and all these trucks came through there and just pushed the water and flooded my engine. And I lost that car in like 20 minutes. I mean, Houston, the flood, it happens so quickly. So . . . you feel very unsafe.

Ernesto's most recent stories are close to Hurricane Harvey in terms of timeline, but in 2001, when Tropical Storm Allison came into Houston, Ernesto lived near the White Oak Bayou. In his recount of the experience he said, "they came and told us that if we didn't get out by boat, right then . . . they were not going to . . . save us, and . . . it was a two story". By Ernesto's account, 16 years before Hurricane Harvey, Houston neighborhoods were already going under water due to the high rainfall, and Tropical Storm Allison's rainfall maximum in Harris County was 11.01 inches less than Hurricane Harvey's rainfall maximum.

After Hurricane Harvey, low-income residents' ontological security (feeling protected by infrastructure) was destroyed (Boyer & Vardy, 2019). Through the interview with Ernesto, it is clear that ontological security was shot long before Hurricane Harvey ever came in. By speaking with Ernesto, it is apparent low-income communities have learned to live with flooding as an occurrence that happens often in Houston; meanwhile, places like the West Buffalo Bayou Park are designed to flood with minimal damage and clean-up because of the money poured into those spaces (Urban Land Institute, 2018).

At the end of the interview, the final question I asked Ernesto, based on his stories, was, "Would you ever consider moving out of Houston?" to which Ernesto vehemently replied, "I would *love* to! I would *love* to! . . . I get it. You know, I could be ignored in another [place] – [where] it's not so hot, and you know, doesn't have hurricanes".

The word "ignored" is something I want to highlight in this particular part of the conversation. Prior to this moment, Ernesto and I deviated from the direct topic of Hurricane Harvey and discussed his personal history in Houston. Ernesto identifies as a gay man and has felt the pressure and ostracizing that many gay men experience if they do not fit the mold of having light skin, abs, and a hyper-masculine presentation. I will not go further on this subject in this chapter, but being ignored by his gay male-identifying peers for not being sexy enough has played a large part in Ernesto's life, and it does not end at feeling excluded by the LGBT+ community. During Hurricane Harvey, Ernesto felt ignored by the city. About his experience with Hurricane Harvey, he said, "I live by myself with my dogs, and I was just sunk here, you know. Nobody could come in or out to help me or anything". Being ignored by a city clearly has ties to affluence of a neighborhood and a person's social status as shown in this research and others mentioned (Boyer &Vardy, 2019) (Friedrich, 2017) (Pulido, 2017).

Ignoring marginalized communities and their infrastructure has historically led to low-income communities being affected by property loss, loss of health, and damages more than their more affluent neighbors. In this interview with Ernesto, he felt the sting of not only social marginalization within an already marginalized group, but he also experienced his city letting him down time after time with each flood event he experienced.

The second co-researcher I interviewed is Helene. She is a 26-year-old white-Hispanic female who makes around $34,000 per year. She and her boyfriend own their home without home or flood insurance, and they do not have health insurance. She lives in Pecan Park in East Houston. Like, Ernesto, prior to Hurricane Harvey, Helene used Tropical Storm Allison as a benchmark for extreme rainfall in Houston. Based on this benchmark, Helene did not evacuate. She said she went to the store to purchase water and non-perishable food, and hunkered down. After the storm, Helene said she had some roof damage that her family financially helped her fix. Helene said an insurance adjuster came out to see the damage but determined the roof damage could not be definitively traced to Hurricane Harvey.

Helene expressed being grateful that her home has two sources of income and was grateful for her family's financial and emotional support. She made note that her community in general is not so lucky. When I asked her if her community could afford something like a first aid kit, she said no. About evacuating, Helene said for her community, evacuation "is a luxury". The process is too expensive to afford the gas it takes to sit on the highway and a hotel to stay in once a destination is reached. When talking about why she did not evacuate prior to Hurricane Harvey, aside from saying she historically hunkered down, Helene said, professionally, a lot of people would not have had time. The night Hurricane Harvey moved into Houston as a tropical storm, Helene, among many others, had to work during the day. Helene said she and her co-workers were released at noon, leaving no time to evacuate, even if they wanted to.

At first, Helene was not sure she would be a good fit for this study because she did not experience flooding; however, Helene was subjected to chemical fumes that "burned" her lungs when she breathed. Helene said she and others in her community noticed something that "smelled like gasoline . . . but then [she] noticed [her] lungs started to get irritated, like really, really irritated, and [she] . . . just got a. . . [face cover] and just said, 'Fuck, I don't know what it is, but don't breathe it'". On Facebook, Helene noticed others in her community were noticing the smell as well. Several people called the non-emergency city number to report the smell. "Weeks later" they heard a nearby refinery had leaks. When it comes to chemical plant accountability, Helene said, "We have TCEQ [Texas Commission on Environmental Quality]; they have no backbone . . . they kind of just turn their backs away. They don't really do anything". In total, this concerning moment of chemical fumes lasted for around 15 to 20 minutes. Afterward, Helene never went to a doctor because she has never had chronic respiratory issues.

In the post-Hurricane Harvey toxicology study of the nearby Harrisburg/Manchester neighborhood (Friedrich, 2017), the damage in low-income neighborhoods was found to be similar to both Boyer and Vardy's study (2019) and this study: most low-income communities were not totaled by floodwaters; therefore, residents stayed and tried to rebuild. Carcinogens and other toxins were found in low-income homes after seeping into dry wall via floodwaters from chemical plants upstream (Friedrich, 2017). Acutely, these toxins cause headaches, shortness of breath, and heart irregularities (Friedrich, 2017). Long term, these toxins can cause cancer and other chronic health problems.

A major difference between Helene and Ernesto is that Helene considers herself recovered from Hurricane Harvey. For her, Hurricane Harvey was a record event, but it did not carry the isolation that Ernesto experienced. Helene differs from Ernesto in that she has a partner, and her family helped her financially when Hurricane Harvey put a hole in her roof. In addition, Helene's social network provided her with support through the storm, and she was offered places to stay outside of Hurricane Harvey's path. Helene's experience of Hurricane Harvey showcases the significant importance of social capital in practice during disastrous events. In future studies, social capital must be considered as a significant recovery factor to better understand how to help those with and without significant social capital.

Taxes or theft: deconstructing Harris County Emergency Management rhetoric

With the voices of Ernesto and Helene in mind, I deconstructed the HCOHSEM disaster prep rhetoric. The documents I analyzed are:

- The *All Hazards Disaster Guide*
- The *2017 Hurricane Brochure*
- The *Emergency Essentials List*

The last two items in the list are available in both Spanish and English. The *All Hazards Disaster Guide*, which includes information on hazardous material-related disasters, winter weather disasters, and what to do if you have children in school during a disaster is only available in English. Overall, the website is simple, and the preparation guides are located in the "Welcome" section.

I will discuss each document in its own section to fully describe the assumptions and responsibilities these documents place upon each resident. The reader should also note that HCOHSEM's assumptions and perceived responsibilities are neither pulled from the ether, nor are they influenced by any natural law of existence. The assumptions and perceived responsibilities placed on residents through these disaster-related documents and their counterparts across the United States are influenced through neoliberalism and shirking of public duty for all from rich politicians and their friends.

The All Hazards Disaster Guide

The *All Hazards Disaster Guide* (HCOHSEM, 2017a) is an eight-page document that includes a small blurb from the County Judge; a disaster supply kit checklist; and a brief overview of planning for hurricanes, tropical storms and floods, evacuation, local city contacts, preparing your pets, hazardous material incidents, and winter weather. It also includes a small list of resources and county contacts and websites. To begin deconstruction, I start at this document's publication date: 2014. As I previously stated, Hurricane Harvey hit Houston in 2017, and Houston has since experienced Tropical Storm Imelda, which stalled over Southeast Texas in September of 2019. To give Tropical Storm Imelda a quantitative relation to Hurricane Harvey, much of Harris County received 10–12 inches of rain in a five-day period during Tropical Storm Imelda (Landreneau et al., 2019) (Figure 17.1). Chambers County, just east of Harris County, received 40–45 inches of rain (Landreneau et al., 2019) (Figure 17.1). During Hurricane Harvey, Harris County received anywhere from 20–50 inches of rain (Blake & Zelinsky, 2018) (Figure 17.2). Given the *All Hazards Guide* is six years old, it is significantly unprepared for the possibility of record-breaking rainfall amounts.

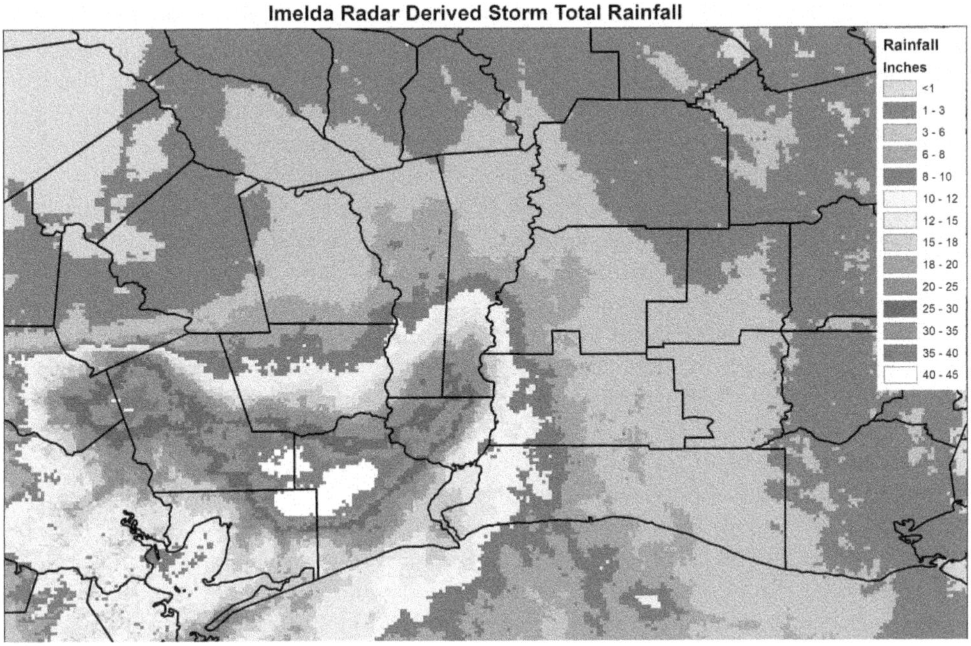

Figure 17.1 Tropical Storm Imelda rainfall totals from September 16, 2019–September 21, 2019

Source: Landreneau et al. (2019).

The first page of the *All Hazards Guide* includes a word from the Harris County Judge. Since the guide is outdated, the words come from former Harris County Judge Ed Emmett, who was unseated in 2018 by Judge Lina Hidalgo. What immediately stood out were the following words, "There are four basic components to being prepared: get a kit, make a plan, stay informed, and get involved" (HCOHSEM, 2017a:1). The idea that a resident can simply "get a kit" (HCOHSEM, 2017a:1) is out of touch with the income diversity in a community as large as Houston, as low-income residents may not be able to outright afford to buy a disaster kit. Since disaster kits are not used every day, other necessities, such as food and rent payments will take precedence. "Make a plan" (HCOHSEM, 2017a:1) is more attainable, but a successful plan depends significantly on HCOHSEM recommendations. Finally, "stay informed, and get involved" (HCOHSEM, 2017a:1) requires fiscal and social capital, and bureaucratic innerworkings (Boyer & Vardy, 2019), which can be a problem for low-income residents who may not have adequate money to spend on necessities for reliable technology and may not have access to powerful people, groups, or resources.

Continuing Emmett's words, "your ability to survive a disaster depends on you doing your part to prepare for the unexpected" (HCOHSEM, 2017a:1). With 17 words, Emmett removed responsibility from HCOHSEM to protect its constituents. Emmett continues, "your emergency plan remains the most dependable tool to keep you safe" (HCOHSEM, 2017a:1). The idea that everyone has the individual responsibility to take care of themselves renders the very idea of emergency management as moot. If HCOHSEM is not willing to take responsibility for ensuring safety of their constituents, HCOHSEM should not receive the tax money that supports them. I will finish deconstructing Emmett's words with this last quote, "Relief workers can not [sic] reach everyone right away" (HCOHSEM, 2017a:1). Remembering Ernesto's words from his interview, no one could reach his community because the flooding was so bad, including relief workers. The words "right away" (HCOHSEM, 2017a:1) instill a false sense of ontological security – the idea that the government will eventually rescue those in peril (Boyer & Vardy, 2019). This false sense of security can lead to disastrous results, as people like Ernesto go without their medications and cannot leave their homes.

The rest of the *All Hazards Guide* consists of important numbers to county offices and tips for preparing for multiple hazards, including "Hazardous Material Incidents" (HCOHSEM, 2017a:6). In this section, the HCOHSEM states, "Staying indoors until the emergency is over may be much safer than going outside where the air may be unsafe to breathe" (HCOHSEM, 2017a:6). Considering that much of Houston's low-income communities are situated in chemically risky places, like Harrisburg/Manchester and Pecan Park in East Houston, low-income residents are constantly inhaling potentially dangerous fumes from neighboring industrial complexes, which are allowed to build right next to family homes (Figure 17.2).

The constant chemicals low-income Houston residents live with seep into their soils and have effects on their bodies, manifesting in ways such as heart issues and breathing troubles (Friedrich, 2017). To write that residents may have to "shelter indoors" during a hazardous material incident (HCOHSEM, 2017a:6) seems ignorant of the toxicological dangers already at play. A chemical spill or other related event is disastrous. But in terms of risk, responding to a disaster while ignoring the near constant inhalation of hazardous chemicals seems ill-prioritized. In toxicological terms, toxic risk is determined by not only toxicity but exposure duration. By this assessment of risk, HCOHSEM should be working with other agencies to address the constant presence of toxic chemicals already in the air in addition to the less frequent fires and spills.

Figure 17.2 Houston family home directly next to an industrial complex

Source: Photo by Paul Hester (2017)

A big issue in this toxicological disaster zone that is exacerbated by disasters like hurricanes is the zoning in Houston. A large part of Houston's uniqueness is its zoning policies, of which there are few. As shown in Figure 17.2, it is possible to zone for residences in one parcel of land while having dirty industry right next door. The idea of no-zoning comes with the idea that Houston is a unique mosaic, where everyone and everything is intermingled with one another, creating an inclusive city culture. In reality, this no-zoning has sacrificed the health and quality of environment of low-income communities. In the next section, I will deconstruct the *2017 Hurricane Brochure*. Both the *All Hazards Guide* and *2017 Hurricane Brochure* contain at-a-glance disaster prep kit lists. Since both of the lists are very similar save a few items, I added the lists together. As I deconstruct the *2017 Hurricane Brochure*, I include the combined at-a-glance lists to uncover the real cost of what it takes to materially prepare for a disaster like a hurricane.

The 2017 Hurricane Brochure

The *2017 Hurricane Brochure* (HCOHSEM, 2017c) is a short two-page brochure containing information about hurricane preparations and tips, and an evacuation zone map. The brochure contains a truncated disaster prep kit. Because the disaster prep kit in the brochure is very similar to the *All Hazards Guide*, I combined the lists to create a very basic disaster prep kit based on the quick-read literature from HCOHSEM (Table 17.1).

The disaster prep kits suggested by HCOHSEM are just that – suggestions – but the power in these lists comes from the official government office producing them. Combining the lists with former County Judge Ed Emmett's words is significant. The responsibility of being prepared lies solely on the individual rather than the local government. Because the individual must assume responsibility, the cost of what it takes to be considered prepared must be exposed. Table 17.1 contains an itemized list of every item suggested by HCOHSEM in the at-a-glance lists. To favor HCOHSEM's insinuation that everyone can be prepared, I chose to use Walmart.com as the retailer – Walmart is ubiquitous in the United States and known for being affordable. The items I chose from Walmart.com are the cheapest available items, to assume the list is made by an imaginary resident who is low-income. To make the imaginary resident more realistic, I gave the resident a single dog. The dog in this scenario is healthy with no special dietary needs. Due to the extremely variable cost of children's diapers and needs, the imaginary resident has no child. The imaginary resident also is single with no second income. Under "Personal Hygiene", I included pads to represent residents who experience periods, as bodies that experience periods are subjected to the added cost of necessary tampons and pads. To compensate for a diversity of bodies and their processes, I chose pad packs that had the most variety in size and menstrual flow.

The cost of the list (Table 17.1), $144.99, may be an affordable bill for middle-income to wealthy residents, but for low-income communities, $144.99 can come down to preparing for a hurricane, or keeping the utilities on or getting groceries. To assume a resident has income to spend on disaster kits, which are used significantly less than utilities, medications, cars, etc., comes from a position of privilege. When a person has a salary of $140,000/year like former County Judge Ed Emmett (Texas Judicial Branch, 2013), the idea of spending $145.99 on a quick kit is as easy as working less than one eight-hour day.

Understanding the cost it takes to be prepared can drive home what it realistically takes to prepare funds to get remotely close to becoming prepared. Being ready for a disaster in the context of a list does not take into account *where* a resident lives, *what* kind of embedded disasters a major disaster can have, *how much* a resident earns per year, *who* lives with a resident, and so on.

Table 17.1 Deconstructed budget of the cost to prepare for a disaster (at a glance)

At-a-glance disaster prep kit cost		
Item	*Cost/unit*	*7-day cost*
Food		
Great Value (GV) canned chicken (2 pack)	$7.72	$30.88
Bush's Pinto Beans (3 pack)	$4.08	$12.24
Water		
GV gallon of water (one gallon/person/day)	$0.80	$5.60
First aid		
Equate All-Purpose First Aid Kit (85 items)	$9.66	$9.66
Backup prescriptions (assume the resident has medical insurance that covers prescriptions)	$0	$0
GV acetaminophen GV (200 ct bottle)	$3.22	$3.22
Personal hygiene		
Equate Moisturizing Hand Sanitizer (34 fl oz)	$3.97	$3.97
Angel Soft Toilet Paper (12 pack)	$5.97	$5.97
Scott Paper Towels double roll (6 ct)	$5.98	$5.98
Dental and vision (assume the resident has this)	$0	$0.00
Change of clothes, shoes, and blanket (assume the resident has these items)	$0	$0.00
Stayfree Ultra-Thin Pads, Regular Wingless Unscented (44 ct)	$4.74	$4.74
Zest Coco Butter & Shea 8 pack bar soap	$2.98	$2.98
Lighting		
Rayovac Mini Flashlight, batteries included	$9.98	$9.98
Duracell AAA Pack of 4 (x 3) (one pack initially for the LED lantern and two packs for various replacement needs)	$7.47	22.41
Communication		
Stansport FM/Weather Band Dynamo Radio with LED Light	$17.75	$17.75
Extra radio batteries (none; the weather radio has a hand crank)	$0	$0
Tools		
Stalwart Household Hand Tools, Tool Set (6 piece set)	$11.70	$11.70
Bulldog Hardware 3/4 in. 18 Gauge Stainless Steel Nails (1.5 oz)	$1.97	$1.97
Cash		
(assume negligible in this list)		
Pet items		
Vibrant Life 15 lb Chicken and Rice recipe	$12.54	$12.54
GV Gallon of water	$0.80	$4

At-a-glance disaster prep kit cost		
Item	*Cost/unit*	*7-day cost*
MidWest iCrate Metal Dog Crate (assume the resident has a leash, bowls, pet bedding, and toys)	$16.99	$16.99
Totals		
Total without tax		$133.86
Total with tax		$144.99

Reducing a resident to a list and leaving the disaster to take its course is a poor way to govern in general, but as I will show in the next section, it gets worse for low-income residents.

The Emergency Essentials List

The *Emergency Essentials List* (HCOHSEM, 2017b) is a one-page PDF published in 2017 and is endorsed by former County Judge Ed Emmett. It is a checklist, and the design is clean and simple. The instructions on the *Emergency Essentials List* say, "Print a copy of the emergency essentials list [,] and take it with you to the store". The instructions assume every resident has access to a printer, which can be considered nit-picky, but for those at or below the poverty line, accessing printers can be challenging. In addition, if they must print the document at a FedEx store or library, low-income residents may have to work out that errand through a public transportation schedule, which takes more time than using a personal car.

The idea that each resident can make their own decisions is understood; if a resident does not have access to a printer or smart phone, they may choose to write the list on their own accord. The language assumes all residents have access to expensive things like printers and smart phones. When planning for disasters at a massive scale, local governments must understand their audience/constituents, or they risk wasting time creating a checklist that does not serve their most vulnerable populations. Understanding where social capital is most important is crucial to planning for a disaster, and without understanding privilege and community, communities with less social capital are put at a higher, unnecessary risk of disastrous effects.

Like in Table 17.1, I used the same imaginary resident to construct Table 17.2: someone of low-income, someone who is single, someone with a single healthy dog that has no dietary restrictions, and someone whose body experiences periods. The list in Table 17.2 is much more comprehensive, including things like home safety measures, insurance, pet micro-chipping, and evacuation items.

Preparing for disaster based on HCOHSEM's unabridged suggestions is a whopping $646.71. For former County Judge Ed Emmett, that is less than two days of work in the office (Texas Judicial Branch, 2013). For low-income residents, that may be a monthly rent payment. According to the postings on Realtor.com (2020), the Greater Third Ward of Houston rent prices are around $750/month for a one-bedroom apartment if the property has not been gentrified. Disaster prep kits are never a one-time purchase; you must replace items as they age out. Canned foods need to be replaced at some point, batteries can deteriorate or come out of the package defective, and critters can get into items if they chew a hole in packaging.

Table 17.2 Deconstructing a disaster prep budget (full kit)

Full disaster prep kit cost		
Item	*Cost/unit*	*7-day cost*
Food		
Great Value (GV) canned chicken (2 pack)	$7.72	$30.88
Bush's Pinto Beans (3 pack)	$4.08	$12.24
goodcook Can Opener (hand-operated)	$3.87	$3.87
GV Plastic Plates (50 ct)	$4.48	$4.48
GV Plastic Party Cups (50 ct)	$2.98	$2.98
GV Everyday White Cutlery (120 ct)	$3.64	$3.64
Water		
GV gallon of Water (one gallon/person/day)	$0.80	$5.60
First aid		
Equate All-Purpose First Aid Kit (85 items)	$9.66	$9.66
Backup prescriptions (assume the resident has medical insurance that pays for their prescriptions)		
Personal hygiene		
Equate Moisturizing Hand Sanitizer (34 fl oz)	$3.97	$3.97
Angel Soft Toilet Paper (12 pack)	$5.97	$5.97
Scott Paper Towels double roll (6 ct)	$5.98	$5.98
GV Small Kitchen Trash Bags 4 gal (40 ct)	$3.34	$3.34
Dental and vision (assume the resident has insurance)	$0	$0.00
Change of clothes, shoes, and blanket (assume the resident has these items)	$0	$0.00
Clean air items		
3M™ Aura™ 9211+ N95 Particulate Respirator (Box of 10)	$22.49 (special price on www.envirosafetyproducts.com/3m-9211plus-aura-n95-respirator.html)	$22.49
Husky 10x25 Clear Sheeting, 3.5mil	$9.27	$9.27
Utility Duck Tape Brand Duct Tape 1.88 in x 55 yd	$2.97 (was $4.71)	$2.97 (was $4.71)
Baby items (The imaginary resident has no child)	$0	$0.00
Lighting		
Rayovac Mini Flashlight, batteries included	$9.98	$9.98
Duracell AAA Pack of 4 (x 3) one pack initially for the LED lantern and two replacement packs for various needs)	$7.47	22.41

Full disaster prep kit cost		
Item	*Cost/unit*	*7-day cost*
Ozark Trail LED Camping Light 100 Lumens. Camping Lantern. (Model 4878)	$3.46	$3.46
BIC Multi-Purpose Lighters	$4.97	$4.97
Communication		
Stansport FM/Weather Band Dynamo Radio with LED Light	$17.75	$17.75
Landline phone (omit; phone services are first to go down during disasters, and many people do not have landlines anymore)	$0	$0
Extra radio batteries (none; the weather radio has a hand crank)	$0	$0
Extra cellphone battery (omitted due to newer phone models)		
Belkin 24W Dual-Port USB-A Car Charger + 5ft. USB-C to USB-A Cable, Black	$19.94	$19.94
Ozark Trail Safety Whistle	$2.56	$2.56
Family safety items		
Kiddie Fire Sentry Micro Profile 3-Year Smoke Alarm, 9-volt battery	$4.44	$4.44
Duracell Coppertop Alkaline 9-Volt Batteries (2 pack for extra smoke alarm batteries)	$7.98	$7.98
Kidde AA Battery Operated Basic Carbon Monoxide Alarm	$16.84	$16.84
Everready AA batteries (4 pack)	$2.18	$2.18
Kidde 1A10BC Basic Use Fire Extinguisher (2.5 lb)	$17.44	$17.44
Clorox Disinfecting Bleach (30 oz)	$3.08	$3.08
Neutrogena Pure & Free Baby Mineral Sunscreen with SPF 50 (3 fl oz) (chosen because many regular sunscreens contain oxybenzone, a known irritant)	$10.97	$10.97
Cutter Unscented Insect Repellant Aerosol (6 oz)	$2.84	$2.84
Ozark Trail Folding Shovel	$10.44	$10.44
Stalwart Household Hand Tools, Tool Set (6 Piece set)	$11.70	$11.70
Bulldog Hardware 3/4 in. 18 Gauge Stainless Steel Nails (1.5 oz)	$1.97	$1.97
Transportation items		
Rand McNally 2021 Road Atlas Paperback	$11.40 (was $14.95)	$11.40
Stalwart Emergency Roadside Automobile Assistance Kit (30 pc)	$12.99	$12.99
Slime 11-piece Deluxe Tire Plug Kit W/ glue	$5.88	$5.88

(*Continued*)

Table 17.2 (Continued)

Full disaster prep kit cost		
Item	*Cost/unit*	*7-day cost*
Briggs & Stratton 4-Cycle Engine Oil (48 oz)	$7.64	$7.64
Games, books, puzzles (assume the imaginary resident has these items)	$0	$0
Pet needs		
Vibrant Life 15 lb Chicken and Rice recipe	$12.54	$12.54
MidWest iCrate Metal Dog Crate	$16.99	$16.99
(assume the imaginary resident has a leash, bowls, pet bedding, and toys)		
GV gallon of water	$0.80	4
Pet First Aid Kit from Walmart.com	$14.99	$14.99
(Free delivery; not available in stores)		
Full disaster prep kit total without a generator		333.03
Full disaster prep kit total with a generator		
Solar Power Station With 110V AC Outlet, DC12V, USB Output For Camping Travel CPAP Emergency		$150.12
Full disaster prep kit total with generator		483.15
Full disaster prep kit total with miscellaneous items from the truncated kit lists, flood insurance, and micro-chipping		
Flood insurance (TDI.Texas.gov)	58.33	$58.33
Avg cost to get a pet microchipped (without the appointment fees) (Petfinder.com)	$45	$45
GV Acetaminophen (200 ct bottle)	$3.22	$3.22
Stayfree Ultra-Thin Pads, Regular Wingless Unscented 44 ct	$4.74	$4.74
Zest Coco Butter & Shea 8 pack bar soap	$2.98	$2.98
Full disaster prep kit total without tax		597.42
Full disaster prep kit total with tax		646.71

There are some items on the comprehensive budget that are more expensive than normal. For example, disposable masks are at high prices due to the current COVID-19 pandemic. In addition, any other item listed that did not come from Walmart.com is due to the fact Walmart does not carry that particular item. To ensure I selected the cheapest items, I searched to find what was readily available with the cheapest price and shipping cost.

When pricing out a generator, I chose the cheapest generator, which can power a small fridge or CPAP machine. When pricing flood insurance, I went to TDI.Texas.gov. The Texas government understands the average cost of flood insurance to cost about $700/year (Texas Department of Insurance, 2020); however due to the neoliberal nature of insurance, prices can vary from individual to individual. To get the monthly cost of flood insurance, I divided $700 by 12 months.

When pricing out pet needs, I followed the suggestions given by the *Emergency Essentials List*; however most pet lovers understand that carrying a picture of your pet as suggested is good, but micro-chipping is much more efficient and permanent. I used PetFinder's average cost of $45 for a dog's procedure (n.d.). The $45 is generous to the budget in that it does not take into account the cost of initial intake – if the pet has not been to the vet performing the procedure – and does not take into account the cost of the appointment to see the pet at the time of the procedure.

Wading through murky waters: engaging emergency management

In this chapter, I discussed:

- The spatial structure of Houston's communities
- The photovoice methodology
- The voices from co-researchers
- The HCOHSEM publications surrounding disasters

The spatial nature of Houston's social structure is unique in that it does not necessarily place low-income neighborhoods in areas that experience total flood destruction. In fact, the city's vulnerable neighborhoods are generally more mid-elevation and experience flooding that mostly results in partial damage. This is significant in that these communities do not receive help from the city through buyouts, enabling them to move out of the area (Boyer & Vardy, 2019), and these communities must live with toxic industry that affects residents not only during flood events but in their everyday lives through the air residents breathe (Friedrich, 2017). Through talking to my co-researchers about their experiences, Boyer and Vardy (2019) and Friedrich (2017) did not discover communities of anomalies in their studies. Their findings along with findings synthesized by my co-researchers and me show that low-income residents in Houston are at less risk of flooding; however, in terms of flood disadvantage, low-income residents are more likely to be at risk of living in dangerous conditions than their more affluent neighbors (O'Hare & White, 2018).

The burden of living in dangerous conditions is the reason I used a photovoice method in this study. I wanted to refrain from inserting my own thoughts and lived experiences into my co-researchers' experiences of disaster. By stepping back and handing the study to my co-researchers, their voices came through untainted and raw. This study is only an amplifier to help create social change from within a resident's community. The challenges with photovoice during this time come from not only the regular issue of not hearing back from potential co-researchers but from the COVID-19 pandemic. This pandemic has hit Houston and its massive healthcare center, while state officials strip city and county power away from Mayors and County Judges who wish to protect their residents. By stripping power away from Mayors and County Judges who want to protect residents, Texas has granted powers to corrupt government officials who would rather sacrifice their constituents to this pandemic than see their paychecks cut. Through these violent acts, low-income residents are at risk of more than flooding. They risk their lives going to work to keep an oppressive economy alive for the rich who view low-income bodies as replaceable.

In the best interest of my co-researchers' safety, I asked them to provide pictures they already had from Hurricane Harvey rather than take pictures of places during a pandemic. I also adapted to each co-researcher to ensure this study was not a burden on their mental health in a time where, in addition to the COVID-19 pandemic, anti-science and white-supremacist rhetoric is falling from the skies in a torrent.

In sharing their experiences, the co-researchers and I discovered several conclusions: their experiences of being marginalized and left to their own devices is not limited to disastrous events. Much of their daily lives revolve around living in communities that are constantly exposed to flooding (Ernesto) or living in communities that are constantly exposed to chemically toxic air (Helene). These forgotten populations are not forgotten by accident. The West Buffalo Bayou area in Houston is in a highly flood-prone area (Urban Land Institute, 2018). The difference between Ernesto's and Helene's communities is the fact the West Buffalo Bayou area was redesigned to flood frequently and with high volumes of water, so the residents there stay relatively safe. Ernesto and Helene, however, must live with constant community flooding outside of disaster and toxic air because their communities do not have trendy bars or magnolia-crowned plantation-style manors that encourage massive city revenue.

To further investigate Ernesto's and Helene's stories of bearing the city's burden of disastrous events, I looked into HCOHSEM's publications on disaster preparation. I analyzed the disaster preparation documents on Prepare.ReadyHarris.org, the site used by HCOHSEM for public disaster information. I deconstructed:

- The *All Hazards Disaster Guide*
- The *2017 Hurricane Brochure*
- The *Emergency Essentials List*

By deconstructing material intended for all residents of Harris County, I discovered the language used in current Harris County emergency management is unaware of the diverse population they serve in most aspects. The documents do not consider the social capital of marginalized Houston communities, and the documents certainly do not consider the financial burden it is to prepare for a disaster. Based on former County Judge Ed Emmett's words, HCOHSEM follows an individualistic approach to preparation, which is common in neoliberal settings; however, the idea that it is up to the individual to prepare for disaster, when a government entity is specifically created for disaster, is ludicrous and theft of tax dollars. I write that former sentence with a hint of hope and a possible change in direction for low-income Harris County residents. Since Ed Emmett lost the election in 2018, County Judge Lina Hidalgo has been overseeing HCOHSEM. Based on the current COVID-19 safety measures she has taken the liberty to challenge the Texas Governor's office through mask mandates. At the time of this writing, Harris County's disaster adaptation seems to be moving towards a more inclusive philosophy, wherein the communities are responsible for their members and the members of other communities in their areas. However, until real change ripples through into the policy documents and publications, residents are still left on their own.

When intertwining the rhetoric from HCOHSEM and the stories from Ernesto and Helene, I found there are gaping holes in the preparation documents for those in low-income communities. The cost to prepare a small disaster prep kit is \$144.99, while a fully prepped kit costs \$646.71. While the numbers are hypothetical, I used Walmart.com almost exclusively to build the deconstructed lists. The exceptions to using Walmart.com came when I could not find an item, such as face masks. To play favorites with HCOHSEM suggestions – to pretend a kit is attainable for everyone – I chose the cheapest items possible or the items that had the most bang for your buck. In addition, I considered the hypothetical situation to apply to a single person, who lives alone, and who experiences periods. By creating an imaginary resident who has periods, I calculated in extra items for those who must buy tampons/pads. I also gave the imaginary resident a healthy dog, since many people have pets, and dogs are easier to calculate for since they do not require litter box care. Based on my deconstruction of the lists provided, creating a disaster prep kit, even a short one, is unattainable for many low-income residents who would be forced to decide between paying the bills or creating a disaster prep kit that may be used once or twice per year.

Ultimately, the publications presented by HCOHSEM are not designed for vulnerable communities. The publications are designed for the former County Judge Ed Emmetts of the world who can afford to buy quality items that will last both in storage and during use. Based on the stories from Ernesto and Helene, HCOHSEM and other government entities must listen to the stories of those most impacted by disaster and disaster recovery. Without listening, governments fall short of preparing for anyone outside of middle-income and wealthy residents. Vulnerable populations need more help and more transparency from their governments because like their more affluent neighbors, vulnerable populations pay taxes too. Without receiving resources, low-income taxes become monetary theft, when these communities have already had their health, minds, and bodies stolen for capital accumulation.

References

Blake, E.S., & Zelinsky, D.A. (2018). *National Hurricane Center Tropical Cyclone Report Hurricane Harvey* [PDF file], 63. Miami, FL: National Hurricane Center. Retrieved from www.nhc.noaa.gov/data/tcr/AL092017_Harvey.pdf

Bourdieu, P. (1986). The Forms of Capital. In *Handbook of Theory and Research for the Sociology of Education*, edited by J. G. Richardson, 241–258. New York: Greenwood Press.

Boyer, D., & Vardy, M. (2019, March 1). *Recovery, Relocation, and Alluvial Awareness in Post-Harvey Houston*. Crosscurrents Spring 2019. Denton, TX, United States: University of North Texas.

Collette, M., & Dempsey, M. (2016). Chemical Breakdown Part 2: An Industry Left to Police Itself. *Houston Chronicle*. www.houstonchronicle.com/news/investigations/article/Chemical-Breakdown-Part-2-7925960.php

Friedrich, M.J. (2017). Determining Health Effects of Hazardous Materials Released During Hurricane Harvey. *American Medical Association*, 318 (23): 2283–2285.

Harris County Office of Homeland Security and Emergency Management (HCOHSEM). (2017a). *All Hazards Guide* [PDF file]. Houston, TX: Harris County Office of Homeland Security and Emergency Management. Retrieved from https://prepare.readyharris.org/Portals/24/Documents/2014_Disaster%20Guide_r10.pdf?ver=2016-03-04-101440-240

Harris County Office of Homeland Security and Emergency Management (HCOHSEM). (2017b). *Emergency Essentials List* [PDF file]. Houston, TX: Harris County Office of Homeland Security and Emergency Management. Retrieved from https://prepare.readyharris.org/Portals/24/Documents/2017_HCOEM_EssentialsKit_Eng_web.pdf?ver=2017-05-09-100459-613

Harris County Office of Homeland Security and Emergency Management (HCOHSEM). (2017c). *Hurricane Brochure* [PDF file]. Houston, TX: Harris County Office of Homeland Security and Emergency Management. Retrieved from https://prepare.readyharris.org/Portals/24/Documents/2017_Hurricane%20Brochure_web.pdf?ver=2017-05-09-101425-630

Hester, P. (2017). ClimateChange.ie. www.climatechange.ie/the-newer-greener-houston-isnt-for-everybody/

Landreneau, D., Brazzell, J., Mogged, T., Erickson, R., & Warthen, S. (2019). *Tropical Storm Imelda*. Lake Charles, LA: National Weather Service. www.weather.gov/lch/2019Imelda

O'Hare, P., & White, I. (2018). Beyond 'Just' Flood Risk Management: The Potential for – And Limits to – Alleviating Flood Disadvantage. *Regional Environmental Change*, 18: 385–396. DOI: 10.1007/s10113-017-1216-3.

PetFinder. (n.d.). *Pet Microchip FAQs*. PetFinder.com. www.petfinder.com/dogs/lost-and-found-dogs/microchip-faqs/

Pulido, L. (2017). Geographies of Race and Ethnicity II: Environmental Racism, Racial Capitalism and State-Sanctioned Violence. *Progress in Human Geography*, 41 (4): 524–533. DOI: 10.1177/0209132516656495

Realtor.com. (2020). https://www.realtor.com/rentals/Greater-Third-Ward_Houston_TX. Date Accessed: July 21, 2020.

Schumann, R.L., Brokopp Binder, S., & Greer, A. (2018). Unseen Potential: Photovoice Methods in Hazard and Disaster Science. *GeoJournal*, 84 (1): 273–289. DOI: 10.1007/s10708-017-9825-4

Texas Department of Insurance. (2020). *Flood Insurance*. Texas Department of Insurance. www.tdi.texas.gov/consumer/storms/flood-insurance.html#:~:text=What%20it%20costs,your%20deductible%2C%20among%20other%20factors.

Texas Judicial Branch. (2013). *Salaries of Elected State Judges* [PDF file]. Texas Judicial Branch. Retrieved from www.txcourts.gov/media/10689/5-judge-salary-2013.pdf

Trevizo, P. (2019, September 25). Imelda Cited in Release of Almost 100,000 Pounds of Air Pollutants. *Houston Chronicle*. www.houstonchronicle.com/news/houston-texas/houston/article/Imelda-cited-in-release-of-almost-100-000-pounds-14465369.php

United States Census Bureau. (2017). *QuickFacts Harris County, Texas*. United States Census Bureau. www.census.gov/quickfacts/fact/table/harriscountytexas/LFE305217#LFE305217

Urban Land Institute. (2018). *Buffalo Bayou Park Houston, Texas*. https://developingresilience.uli.org/case/buffalo-bayou-park/

Wallace, T., Watkins, D., Park, H., Singhvi, A., & Williams, J. (2018, March 22). How One Houston Suburb Ended Up in a Reservoir. *New York Times*. www.nytimes.com/interactive/2018/03/22/us/houston-harvey-flooding-reservoir.html

Wang, C., & Burris, M.A. (1997). Photovoice: Concept, methodology, and use for participatory needs assessment. *Health Education and Behavior*, 24 (3): 369–387.

PART 3

People and communities

18

CHALLENGES OF THE ANTHROPOCENE FOR PROTECTED AREAS AND CONSERVATION IN COSTA RICA

Carlos Morera Beita, Daniel Avendaño Leadem, and Luis Fernando Sandoval Murillo

The current era – the Anthropocene – is a new geological epoch based on the observation that human impacts on planetary processes have driven the Earth out of the Holocene period in which agriculture and sedentary communities prevailed, into a period in which socially and technologically complex human societies developed (Steffen et al., 2018). These conditions have triggered the necessity to define crucial strategies regarding ecological conservation as one of the initial steps towards confronting natural resource depletion.

Under these conditions, protected areas are considered an indispensable strategy required for biodiversity conservation (Saura et al., 2017; Jepson et al., 2017). In this current Anthropocene period, such areas have been reconsidered due to their importance towards mitigating the effects of climate change, as they allow places for species movement, carbon fixation, and regulation of atmospheric temperature, among other benefits. Nevertheless, there is little contemplation of climate change issues within the strategies of protected area systems in tropical countries (Fung et al., 2017).

Costa Rica is a country in which conservation has been a key issue for over a century, as is the case of many Latin-American countries. However, conservation initiatives have been founded on the "myth of the untouched nature", as affirmed by Diegues (2005). This allegory sustains that natural ecosystems are commonly considered as strict spaces absent of sociocultural footprint, rather than spaces that include some sort of human dimension. This understanding of protected areas was replicated from the ecological conservation principles implemented in the United States (Boza, 1993).

Due to its biodiversity conditions, Costa Rica has been the object of research for natural specialists since the 18th century during which it has stood out amongst other tropical countries for its efforts and commitment towards environmental and biodiversity conservation (Steinberg, 2001). In the Central American region, other countries such as Belize (37.7%) and Nicaragua (37.2%) have more surface protected under a conservation category (Morera & Jover, 2018), whereas the budget that Costa Rica assigns for protected areas is higher than all of the other Central American countries combined. In this sense, Costa Rica has become very well known

DOI: 10.4324/9780429450310-22

in the world not only for its ecological resources, but also for the actions targeting their preservation. It has progressed in biodiversity conservation despite the fact that current conditions create new challenges that fall in line with the national reality – in an era of the Anthropocene. These challenges impact implementing actions towards ensuring the preservation of marine resources, strengthening natural ecosystem connectivity, and considering local participation, as well as ensuring well-distributed economic benefits.

Moreover, the last decade has seen an awakening in indigenous and Afro-American rights movements, led by actors who have organized themselves in order to recover their autochthonous vision regarding the relationship between nature and society. A factor in this is that most of the indigenous territories have recovered a large percentage of their natural vegetation for which they receive some economic retribution. In accordance with the country's multiethnicity and the challenges related with conservation, national environmental policies still need to incorporate this vision.

Costa Rica has based its tourism development on protected areas and has emerged as one of the leading eco-tourist destinations. However, recently this nature tourism archetype has been shifting strongly towards a sun-and-beach model of eco-tourism/eco-development (Morera & Nel-lo, 2017). The present chapter discusses the establishment of protected areas in Costa Rica during different periods in which they were created, their actual uses as well as their current conditions in order to analyze the main social and environmental challenges surrounding their management. Doing so, we offer a notable case of novel development and environment contradictions embedded in protected areas and conservation in the era of the Anthropocene.

Protected areas epochs in Costa Rica

Due to its geological origin as a space of connectivity between Nearctic and Neotropical biota (Kappelle, 2016), besides its location between the Pacific Ocean and the Caribbean Sea, Costa Rica, as well as other countries of the Central American isthmus, have been renowned for their high presence of biodiversity. Because of these conditions, many well-known scientists have visited the country during the last two centuries, including Mortz Wagner, Kart Scherzer, Karl Hoffman, Herman Wenland, Felix Bally, Henry Pittier, Adolfo Tonduz, and Pablo Bioley (Hilje, 2013).

These visits generated many articles that promoted Costa Rica's biodiversity within the scientific community. Based on this knowledge of Costa Rica's ecological conditions and their importance, a system to develop protected areas was established at first as the Servicio Nacional de Parques (National Parks Services), which later evolved and transformed into the SINAC (National System of Conservation Areas). During the '90s, SINAC implemented a regional approach in which protected areas were organized within Conservation Areas. This was then considered as an innovative and unique approach because it was the first time that ecological preservation addressed management as a territory in which the whole area was protected, despite the fact that there was little dialogue between them (Morera & Alvarado, 2018). Currently, 27.5% of Costa Rica's terrestrial land is within a protected area, of which 11.4% corresponds to government owned lands, such as biological reserves and national parks (Table 18.1).

Costa Rica's approach to conservation management is a result of decision-making processes and practical implementation from which five stages can be defined (see Figure 18.1). The first one corresponds to the period before 1970 within which the establishment of protected areas was characterized by isolated initiatives with strong personal leadership. During these periods, the national parks protecting the Turrialba and Irazú volcanoes were created, along with Cahuita and the Cabo Blanco Absolute Reserve.

Table 18.1 Protected area categories in Costa Rica, 2018

Categories SINAC	*Landowner*	*Quantity*	*Surface (ha.)*	*Percentage of national territory*	*Management*
Biological reserves	Government	8	21.432	0.40	SINAC
National parks	lands	25	567.800	11.00	
Subtotal		33	589232	11.4	
Wildlife refuges	Government-	47	175.524	3.40	Private- SINAC
Forest reserves	private lands	11	284.133	5.50	
Subtotal		58	459657	8,90	
Reserves for water	Private lands	31	163.714	3.00	Private- SINAC
Protection		14	92.771	1.70	
Wetland		13	17.392	0.03	
Others					
Subtotal		58	273877	4,73	
TOTAL		148	1,322.818	25.03	
Biological corridors	Private lands	44 (30)	1,902.574	36	Private
Total		192	3,225.392	61.01	

Source: SINAC (2018); Pfaff and Sanchez-Azofeifa (2004) and Morera (2011)

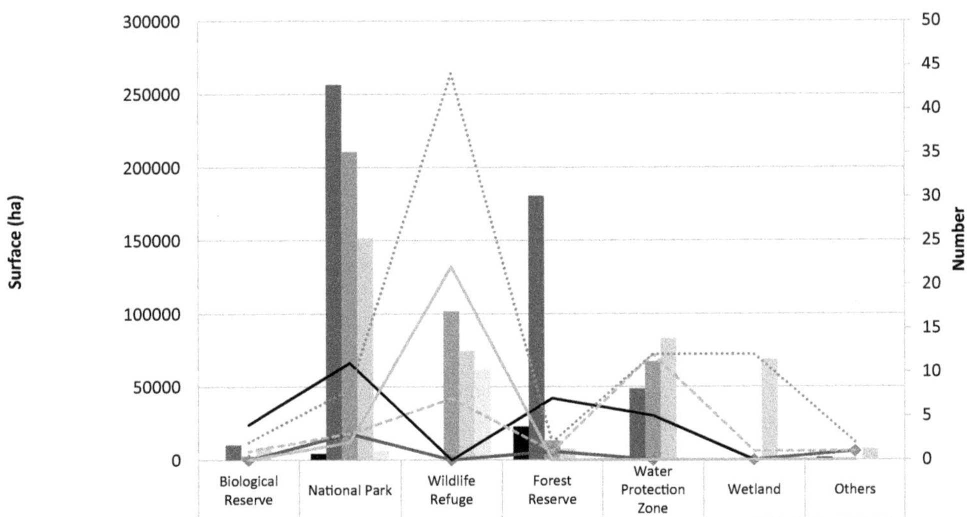

	Biological Reserve	National Park	Wildlife Refuge	Forest Reserve	Water Protection Zone	Wetland	Others
Less 1970 (Surface)	0	4355	0	22625	0	0	1296
1970-1980 (Surface)	10280	256396	0	180512	48401	0	0
1981-1990 (Surface)	2612	210406	101503	13140	66956	611	230
1991-2000 (Surface)	8601	151723	74246	7733	83170	68641	7342
2000-2017 (Surface)	0	6619	61565	0	0	8	0
Less 1970 (#)	0	3	0	1	0	0	1
1971-1980 (#)	4	11	0	7	5	0	0
1981-1990 (#)	1	3	7	1	12	1	1
1991-2000 (#)	2	8	44	2	12	12	2
2000-2017 (#)	0	2	22	0	0	0	0

Figure 18.1 Protected area categories in Costa Rica, 1970–2017

The second period occurred soon after during the '70s and is recognized as the period in which most of the current protected areas of the country were established. Most of these protected areas were characterized by not being state-owned, a condition that required the expropriation of significant portions of these lands. As a result, conflicts emerged between local communities historically inhabiting such spaces and lacking legal ownership, and the government. Emblematic national parks such as Santa Rosa, Manuel Antonio, Rincón de la Vieja, Barra Honda, Chirripó, Corcovado, Tortuguero, Coco Island, Palo Verde, Carara, and Braulio Carrillo were stablished during this stage. These important preserved areas correspond to 39% of the country's total protected areas, and 52% of the surface corresponding to the national park category, which is defined officially as singular character areas with national or international interest, which must include representative samples of significance ecosystems and have little evidence of human activity. In addition, they offer important attractions for visitors and have capacity for recreational and educational use. Likewise, four biological reserves were established during these years, corresponding to 50% of the surface area within this category. This period is the most significant in terms of protected area establishment, considering quality and quantity of ecological conditions in the areas being protected. Recreational activity within these spaces was then incipient and basically corresponded to domestic visitation.

During the 1980s (stage 3), the country's economic crisis affected the establishment of additional conservation categories, even though a new category was promoted in which the corresponding lands did not necessarily have to be state-owned (Evans, 2010). In most cases, these areas were areas located adjacent to larger patches of natural ecosystems, which were already preserved. In this sense, the progression of protected area establishment was being determined by emerging categories in which these spaces were privately owned, for instance wildlife refuges and water zone protection areas. Additionally, the largest national park was established during this phase; La Amistad (Friendship) National Park (198336 ha) extending between Costa Rica and Panama. This stage coincides with the country's tourism boom.

During the 1990s (stage 4), protected area consolidation was dominated by spaces with reduced surface and less relevant ecological value, corresponding to private properties in which tourism activities were allowed despite having few facilities for their realization. Some national parks were declared, such as: Guanacaste, Diria, Piedras Blancas, Arenas, Juan Castro Blanco, Volcán Tenorio, and Marino Las Baulas.

During the present century (stage 5), conservation efforts have been focused on the establishment of connectivity spaces, such as biological corridors, reaching a total of 44 in 2018, that cover 38% of the country's surface. However, only 30 biological corridors have been established under the official terms that SINAC requires (local council, technical profile, and a strategic plan). Biological corridors have a lot of heterogeneity according to area, landscape form, and connectivity-fragmentation level as a result SINAC is making some polities in order to define priorities to ensure those areas as ecological connectivity spaces that have a land use according to preservation goals.

Protected area uses

Protected areas were traditionally established for ecological conservation reasons with the possibility of implementing touristic and recreational uses. However, this trend is currently changing as innovative approaches are emerging in which new uses and possibilities of added value are being considered for these lands. Amongst these alternative uses are tourism, bioprospecting, and payment for ecosystem services.

Tourism

Protected areas have been a key element for the development of tourism in Costa Rica. Based on the country's ecological fame within the scientific community, Costa Rica was one of the first countries in the world to promote ecotourism (Morera & Nel-lo, 2017). Nevertheless, most of the protected areas, with regards to surface area and quantity, were established before the tourism boom of the 1990s. Tourists that visited protected areas in Costa Rica favored national park visitation not only because of their attractiveness, but also the presence of required facilities as restrooms, lunch areas, and visitor centers. The most visited national parks were Irazú Volcano, Poás Volcano, and Manuel Antonio (Morera & Alvarado, 2018).

Due to their level of ecological fragility, visitation to biological reserves was restricted and overnight stays were not allowed, with only certain exceptions due to scientific reasons. During the last years, private wildlife refuges have become important attractions; however, they are still far from the appeal of national parks. The former is due to the growing interests in conservation and natural attractions on behalf of tourists, which has led to an increase in the number of private protected areas, such as the Monteverde Reserve, that in many cases complement public efforts (Nel-lo, 2008). Conversely, the actual tourism model of Costa Rica is based on a sun-and-beach model which has triggered the construction of many hotels linked with international corporations, which have furthermore established facilities that in fact compete with the government-owned/managed protected areas. Due to this reason, the number of tourists that visit Costa Rica have increased every year since 1990 with the exception of 2009, while visitation to protected areas does not have the same growth (Figure 18.2). The gap between these figures is increasing each year, revealing a weak relationship between conservation efforts and the tourism industry in Costa Rica. In other words, this disparity increases as the tourist model favors the sun-and-beach tendency and retires from ecotourism model.

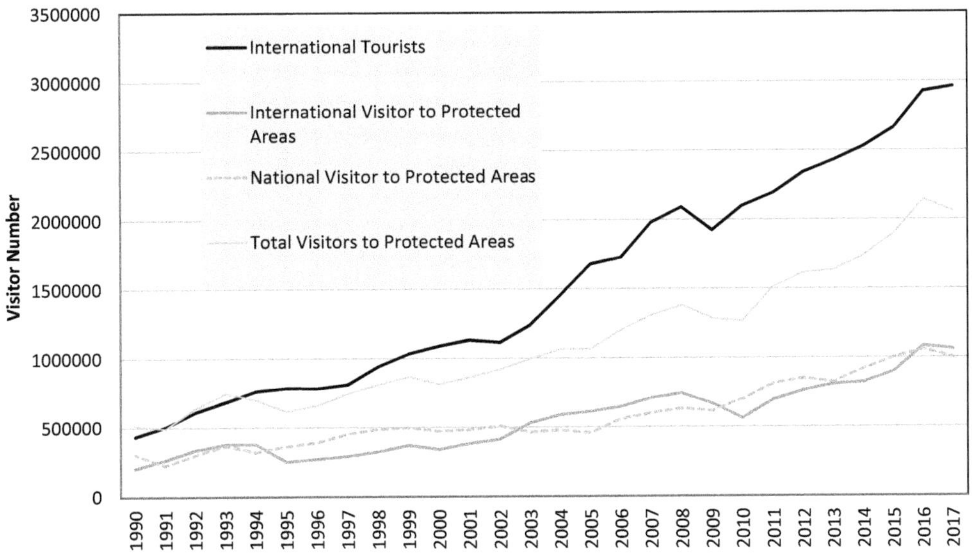

Figure 18.2 Comparison of tourist visitation to the country and protected areas in Costa Rica, 1990–2017

Along the same line, the tourism model has been shifting from small to large-scale businesses linked with international capital. Thus, many lodging companies have developed their own natural attractions in order to reduce the expenses of protected area visitation, hence increasing their profit. This factor has triggered the current economic crisis of protected areas. The income generated by the entrance fee to protected areas is recollected and administrated by the government and subsequently distributed to each particular area, not before leaving a considerable portion in the government's pocket. Another essential factor that adds to this problem has been the differentiated protected area entrance rate which has been higher for tourists than for locals since 2003. Nevertheless, there are no studies that evaluate the actual impact of this differentiation, and some parks, such as Poás, Irazú, and Manuel Antonio, display a steady increase in their visitation.

Even though biodiversity is still Costa Rica's main tourist attraction, it is not necessarily carried out exclusively as an activity held in protected areas, becoming a tendency that reflects the fact that the many tourism companies do not contribute economically towards conservation despite receiving tax exemptions. For example, the main attractions in the Arenal Volcano National Park are the geomorphological structures, formations that most tourists can access without paying a fee. Another similar case occurs in Tortuguero National Park where tourists visit with the main objective of watching turtles even though only 10% of the nesting area is actually located within the park (Jacobson & Lopez, 1994). This situation has forced some actions such as determining by regulation that the park can only be accessed through the Tortuguero channels in order to control and obligate visitors to pay an entrance fee.

In addition, some conflicts between tourism businesses and local communities have been emerging during the last years. Issues surrounding water availability and supply in the province of Guanacaste are an example in which the actual tourism model favors the interests of larger companies by compromising the access of local communities to natural resources and exhausting them without guaranteeing their future preservation (Silva, 2019). Such companies are scattering all over developing countries in such a way that if some country decides to change the procedures, they would simply move elsewhere. In consequence, during recent years there has been an idea to generate a special tourism tax for supporting protected areas in Costa Rica.

Bioprospecting

The development of informatics and molecular biology technologies during the last century opened up the possibility for new opportunities of bioprospecting, which is considered as an activity of exploratory non-destructive research with the purpose of commercialization of new products based on biological resources (Nielsen et al., 2004). Aware of these novel possibilities, Costa Rica sought to take advantage of these new prospects by establishing the National Institute for Biodiversity (INBIO), which was an NGO established in 1989 precisely to pursue this objective. This marketable use of biodiversity seeks to enhance the sustainable use of natural resources for human benefit, not only implying economic and environmental benefits, but also promoting the improvement of capacities for local community development and the management of nature-friendly technologies (Murillo & Ramírez, 2008).

For this innovative bio-commercial activity it is necessary to conduct inventories regarding insects, plants, and other species that are dispersed throughout Costa Rica's protected areas. Rodríguez (2013) discusses that this model is inadequate because it empowers a private company with the responsibility for marketing genetic resources which are state property. Even though INBIO was a model that many countries tried to replicate, this private organization has practically disappeared due to financial difficulties, reflecting the fact that biodiversity uses for pharmaceutical and biotechnology purposes has not been a successful viable economic opportunity. In addition, the main lesson learned from this experience is that private organizations should not have complete control over biodiversity management and the state should play a leading role.

Payment for environmental services

The development of Costa Rica's Payment for Environmental Services (PES) scheme is a pioneering experience of environmental services developed by means of a specific policy instrument with national coverage and articulated by state management. It was institutionalized by the 4th Forestry Law (#7575) as early as 1996, prior to popularization of the environmental services concept and the first definition of PES schemes (Le Coq et al., 2015). The country's environmental policies directed towards strengthening the forestry and environmental protection sector commenced as mere forest incentives over four decades ago and eventually evolved into today's PES scheme.

The PES program is a financial recognition granted by the Costa Rican State to the owners of farms that establish forest plantations, agroforestry systems, and/or undertake natural regeneration, forest protection, and management projects. It is a compensation to private landowners, indigenous communities, organizations of agricultural or forestry producers and private landowners for the ecosystem services generated by plantations and forests, which affect climate change mitigation, biodiversity protection and recovery, water source protection, and the scenic beauty (FONAFIFO, 2007; Pagiola, 2008).

The National Forestry Finance Fund (FONAFIFO) manages Costa Rica's PES program and is in charge of allocating the financial resources in the projects, as well as the operational tasks related to the control and monitoring of these. Its main source of financing comes from taxing fossil fuel consumption and allocating 3.5% of this tax specifically to the program. In addition, current legal framework (Forestry Law) establishes other potential sources of financial resources to strengthen the program, such as contributions received from the State; regular and extraordinary public budgets; donations or credits from national and international bodies, as well as resources collected through the issuance and placement of credit titles (FONAFIFO, 2019).

Costa Rica's PES scheme is intended to stimulate carbon market development within specific sectoral strengthening perspectives and as part of strategies integrated into national development plans (IICA, 2010). However, the development of this market has been slower than expected and resources available for investment have so far not been sufficient to supply the growing demand (McAfee, 2016). Herein, FONAFIFO has developed mechanisms and agreements with local private companies to generate alternative sources of financing for the PES program, allowing more producers to benefit from the resources mobilized through the application of this instrument.

This PES experience is not envisioned by the country as an objective *per se*, but rather as an instrument for influencing land use toward the continued or enhanced provision of forest-derived ecosystem services (Daniels et al., 2010). It is destined to forest owners as a form of compensation and less for the opportunity costs of alternate land use voluntarily foregone. Despite Costa Rica's PES scheme being depicted locally and internationally as an exemplary market mechanism to promote natural cover protection and restoration, it is contended by some to basically portray a disguised administration of subsidies (Fletcher & Breitling, 2012).

In general terms, PES have been positive for conservation since deforested areas have been recovered, sustainable management of forest areas has been promoted, and natural forests have been protected, as well as conservation measures been considered in valuable ecosystems lacking

State protection (located in private areas). In addition, among the indirect consequences of forest conservation, the program has generated jobs and increased the number of people who have been favored with payments (Daniels et al., 2010; CGR, 2011; Allen & Vásquez, 2017).

On the downside, the National System of Conservation Areas (SINAC) and FONAFIFO have failed in defining PES policies to determine priorities, objectives, goals, activities, methods, and indicators to control and evaluate the program. The lack of strategic planning has favored short-term decision making and institutional actions that, whilst strengthening the forestry sector, move away from the perspective of environmental services. According to the country's National Regulator, operational planning lacks indicators to adequately measure the actual effects of the PES program in crucial areas, such as the amount of resources invested in the program, percentage of contracts formalized by women, number of approved credit operations, indigenous population participation, as well as amount of proposals specified at national and international levels for resource recruitment (CGR, 2011). Moreover, it has been demonstrated that Costa Rica's PES program has resulted in payments generally favoring larger landowners, with a tendency towards excluding certain kinds of smallholders despite concerted state efforts to include the rural poor (Lansing, 2014).

Beyond these achievements and limitations, an important aspect of its evolution is an approach that has been enhanced in the last decade with an increasing number of studies to improve performance and functioning. Flaws within Costa Rica's PES experience have demonstrated that there are several essential aspects that have been overviewed and are required to promote an environmental policy ensuring stable outcomes for direct beneficiaries and communities in the long run. These aspects should include promoting active participation and access to information (within the process of implementing, monitoring, and readjusting the mechanism), devolving decision legitimacy through consultations and negotiations on distribution, as well as defining the governance shaping the scheme's procedural aspects (Pagiola, 2008; Vignola & Morales, 2011; Fletcher & Breitling, 2012).

It is also important to consider the opinion of different sectors indicating relevant actors' procedural and distributional concerns. For example, small agricultural producer associations have expressed concerns about exclusion of their members from ongoing PES schemes. They claim for a series of improvements including greater flexibility of the land property registry, adequate payment for recognized native species, increased flexibility in authorizing the use of dead wood, as well as contemplating community lands that are not provided with tenure title (Daniels et al., 2010; Börner et al., 2017). The same concerns are expressed by indigenous groups and local communities with respect to the current system. These groups perceive the necessity to consider several requirements such as considering customary and cultural rights in redesigning the PES scheme, establishing community-owned monitoring systems, and providing local communities with more power to define use rights on forest resources in their territories (Vignola & Morales, 2011).

With regards to measuring the actual impact of the evolution of Costa Rica's PES scheme, it is important to consider three interrelated themes: the gap between vision and execution in the program's administration, spatial data considerations, and the actual spatial-temporal dynamics within the country's landscape. The latter refers to forest cover regrowth since the 1980s and its steady ongoing trajectory of recovery ever since (Daniels et al., 2010; Allen & Padgett, 2017).

The explicit criteria and technical procedures regarding spatial data required under the PES scheme have generally been ambiguous for regents (third-party forestry consultants charged with PES monitoring) and experts to follow. Crucial aspects such as data projection units and information formats have resulted in unnecessary spatial errors when changing the datum and projection. This factor makes satellite-based monitoring for early years of PES a challenge, particularly if it is not combined with extensive field mapping and verification (Daniels et al., 2010).

With regards to spatial-temporal dynamics within the country's landscape, Morera et al.'s (2017) research regarding the spatial distribution of the PES in Costa Rica, identifies certain asymmetries on a regional scale for the period between 2011 and 2015. For instance, there are regions with an average PES submitted farm area of 30 ha (e.g. Norte region) while others exceed 91 ha (e.g. Palmar region). This reflects the country's significant socioeconomic differences. According to FONAFIFO's management priorities, PES should be located within biological corridors (BC), however only 52% meet this condition.

The fact that Costa Rica's PES scheme is an innovative environmental policy applied in a developing country that has been running for over 20 years becomes an interesting opportunity to point out certain lessons for the implementation of payment for ecosystem services. However, it is not only important to learn from its successful implementation and results, but also to look into mistakes and the perception of differently scaled stakeholders.

Conservation challenges in Costa Rica

During recent years, SINAC and other environmental organizations have considered ecological connectivity as one of the main goals to improve conservation in Costa Rica. Although official protected categories exclude biological corridors as such, the government established the National Biological Corridor Program (NBCP) by means of Executive Order 33106 of the Ministry of Environment, Energy, and Telecommunications (MINAET) in 2006 (Villate et al., 2009), with the intention of expanding the PES program. In reality, around 38% of the country's terrestrial area is located within a biological corridor. In addition, SINAC has established the first international marine corridors in cooperation with other countries, for instance with Galápagos (Ecuador), Coco Island (Costa Rica), Coiba (Panama), as well as the Gorgona and Malpelo islands (Colombia) that are an initiative supported by UNESCO. Similarly, the municipalities of the GAM (Greater Metropolitan Area) have been establishing urban corridors connecting and preserving fragments of riparian vegetation (Gastezzi-Arias et al., 2017), including the cases of the María Aguilar, Torres, Reventazón-Agua Caliente, Pará, Tibas, and Garci-Muñoz rivers.

SINAC's current economic crisis reflects fewer economic and human resources which have affected the management of biological corridors directly. Essentially, there are few personnel available to attend to the requirements of each biological corridor, and as a result, some of them are only recognized properly on paper or, what is more, there is only one person in charge of managing BCs in each Conservation Area. In other words, BC supervision depends basically on the organization of local committees. For this reason, there is no intent of assessing the actual connectivity degree on behalf of SINAC, but rather an evaluation of the respective management organization. Thus, most of the exceptional BCs are managed by NGOs and not by SINAC, such as Talamanca-Caribe, Paso La Danta, and San Juan-La Selva.

Presently, Costa Rica has numerous BCs with different surface areas, landscape structure, and ecological conditions. For example, there are BCs ranging from a micro-local scale (0–25,000 ha) to a macro-regional scale (+ 100,001 ha), and even some with the scale so large that there are no guidelines regarding their management and size regulation. In addition, there are no regulations to determine the suitable size of a BC founded on key species or socioeconomic conditions. With the transformation of Costa Rica's economy, in which rural areas are dominated by monoculture production (pineapple, palm oil, and banana) and unfair PES distribution, there are few triggering factors for farmers to maintain and protect the natural cover of their land.

On the other hand, Costa Rica has made progress in the establishment of protected areas, particularly terrestrial areas, though little has been achieved regarding the preservation of marine resources. For example, El Salvador and Costa Rica have less than 8% of marine resources protected under some conservation category, which is low compared to countries like Belize that have 10.1%. SINAC is required to take action in order to preserve marine resources that have been internationally questioned due to cases of shark finning (Paniagua, 2007). In addition, Costa Rica is one of the highest consumers of agrochemicals in the world, which due to narrow geomorphological form, many of these substances drain straight into the oceans.

In recent times, SINAC has extended the establishment of coastal national parks in marine areas such as the cases of Santa Rosa, Marino Las Baulas, Manuel Antonio, Corcovado, Piedras Blancas, Tortuguero, Cahuita and Coco Island, Caño Island, Ostional, Caletas, Cueva del Murciélago, Playa Hermosa-Punta Mala, Río Oro, and Gandoca Manzanillo protected areas. Nevertheless, many of these marine parks only exist formally on paper, and/or lack resources and personnel (Chacon, 2016).

SINAC has been generating policies for the participation of communities as in the case of Cahuita National Park in which a local organization co-manages the protected area. In general terms, it becomes imperative to articulate conservation interests with the surrounding communities, especially in relation to the generation of economic benefits, which in many cases focus on prioritizing the transfer of services from protected areas to organized local groups. This is a way to privatize government functions, yet it is also a tool to empower local development.

It is important to highlight the fact that conservation is not only limited to protected areas. Other elements such as spatial planning, waste management, agricultural and industrial policies also have an important influence over biodiversity conditions.

In the new Anthropocene, conservation efforts need to be embedded into current, actual local dynamics in such a way that they generate income and communities become motivated to participate in this process. This chapter reveals the evolution that has shaped serious contradictions of development and environment, but also offers a possible way forward for the future. Under this conception, grassroot tourism, payment for environmental services, and bioprospection could become key integrated actions in order for conservation to become an effective instrument of socioeconomic improvement for developing countries.

References

Allen, K., & Padgett Vásquez, S. (2017). Forest cover, development, and sustainability in Costa Rica: can one policy fit all? *Land Use Policy*, *67*, 212–221. DOI: 10.1016/j.landusepol.2017.05.008

Börner, J., Baylis, K., Corbera, E., Ezzine-de-Blas, D., Honey-Rosés, J., Persson, M., & Wunder, S. (2017). The effectiveness of payments for environmental services. *World Development*, *96*, 359–374. DOI:10.1016/j.worlddev.2017.03.020

Boza, M. A. (1993). Conservation in action: past, present, and future of the national park system of Costa Rica. *Conservation Biology*, 7(2), 239–247.

Chacon, V. (2016). *Costa Rica está en deuda con sus mares: Protección de la riqueza marina en Costa Rica.* Semanario Universidad 25-8-2016.

Contraloría General de la República (CGR). (2011). Informe acerca de los efectos del Programa Pago por Servicios Ambientales (PSA) implementado por el Estado Costarricense. San José, Costa Rica: CGR – División de Fiscalización Operativa y Evaluativa del área de Servicios Ambientales y de Energía.

Daniels, A. E., Bagstad, K., Esposito, V., Moulaert, A., & Rodriguez, C. M. (2010). Understanding the impacts of Costa Rica's PES: are we asking the right questions? *Ecological Economics*, *69*(11), 2116–2126. DOI: 10.1016/j.ecolecon.2010.06.011

Diegues, A. C. (2005). *El mito de la naturaleza intocada.* São Paulo: NUPAUB.

Evans, S. (2010). *The green republic: a conservation history of Costa Rica.* Austin, TX: University of Texas Press.

Fletcher, R., & Breitling, J. (2012). Market mechanism or subsidy in disguise? Governing payment for environmental services in Costa Rica. *Geoforum, 43*(3), 402–411. DOI: 10.1016/j.geoforum.2011.11.008
FONAFIFO. (2007). *Programa de pago por servicios por servicios ambientales (PPSA) en Costa Rica*. Retrieved from www.fonafifo.go.cr/
Fonafifo. (2019) *Visión de futuro 2040 y Plan Estratégico Institucional 2020–2025*. San José, Costa Rica: MINAE.
Fung, E., Imbach, P., Corrales, L., Vilchez, S., Zamora, N., Argotty, F., & Ramos, Z. (2017). Mapping conservation priorities and connectivity pathways under climate change for tropical ecosystems. *Climatic Change, 141*(1), 77–92.
Gastezzi-Arias, P., Alvarado-García, V., & Pérez-Gómez, G. (2017). La importancia de los ríos como corredores interurbanos. *Biocenosis, 31*(1–2).
Hilje, L. (2013). *Trópico agreste: Las huellas de los naturalistas alemanes en la Costa Rica del Siglo XIX*. Cartago, Costa Rica: Editorial Tecnológica. Instituto Tecnológico de Costa Rica.
IICA. (2010). *Un nodo de cooperación sobre los servicios ambientales en Costa Rica*. San José, CR: Instituto Interamericano de Cooperación para la Agricultura.
Jacobson, S. K., & Lopez, A. F. (1994). Biological impacts of ecotourism: tourists and nesting turtles in Tortuguero National Park, Costa Rica. *Wildlife Society Bulletin (1973–2006), 22*(3), 414–419.
Jepson, P. R., Caldecott, B., Schmitt, S. F., Carvalho, S. H., Correia, R. A., Gamarra, N., & Ladle, R. J. (2017). Protected area asset stewardship. *Biological Conservation, 212*, 183–190.
Kappelle, M. (2016). *Costa Rican Ecosystems*. Chicago, IL: University of Chicago Press.
Lansing, D. M. (2014). Unequal access to payments for ecosystem services: the case of Costa Rica. *Development and Change, 45*(6), 1310–1331. DOI: 10.1111/dech.12134
Le Coq, J. L., Froger, G., Pesche, D., Legrand, T., & Saenz, F. (2015). Understanding the governance of the payment for environmental services programme in Costa Rica: a policy process perspective. *Ecosystem Services, 16*, 253–265. DOI: 10.1016/j.ecoser.2015.10.003
Mcafee, K. (2016). Green economy and carbon markets for conservation and development: a critical view. *International Environmental Agreements: Politics, Law and Economics, 16*(3), 333–353. DOI: 10.1007/s10784-015-9295-4
Morera, C. (2011). Sinergias del Modelo Turístico con los espacios protegidos estatales en Costa Rica. En: Las vías del desarrollo sustentable en el medio rural: Naturaleza, sociedad rural y turismo en América Latina. Memorias de Seminario Internacional. Centro Interdisciplinario de Investigaciones y Estudios sobre Medio Ambiente y Desarrollo, Instituto Politécnico Nacional, México.
Morera, C., & Alvarado, M. (2018). Parques Nacionales y Paisajes en Costa Rica: Un análisis de la visitación. In: Suarez, R., & Pinto, J. (Eds.), *Análisis del Paisaje; Gestión turística en espacios naturales; Sustentabilidad*. España: Universidad de Girona.
Morera, C., & Jover, J. (2018). Turismo y áreas protegidas en Centroamérica: una sinergia urgente. En: Abellán, F. C., Martí, F. J. J., & González, R. C. L. (Eds.), *América Latina en las últimas décadas: Procesos y Retos* (Vol. 17). España: Ediciones Universidad De Castilla La Mancha.
Morera, C., & Nel-lo, M. (2017). Local level policies for tourism management in protected areas: experiences from Costa Rica (Chapter 6). In: Mukul, S. A., & Manzoor Rashid, A. Z. M. (Eds.), *Protected Areas: Policies, Management and Future Directions*. New York, USA: Nova Science Publishers.
Morera, C., Sandoval, L., & Loría, A. (2017, April 26–29). *Dinámica espacio-temporales de los Servicios Ambientales (PSA) en Costa Rica 2011–2015*. Lecture presented at the XVI Encuentro de Geógrafos de América Latina EGAL 2017 in La Paz, Bolivia.
Murillo, C., & Ramírez, R. A. (2008). Biocomercio: una alternativa para el desarrollo sostenible. *Revista de Ciencias Económicas, 26*(1).
Nel, M. (2008). La intensidad de la gestión, la clave para un desarrollo turístico sostenible en áreas protegidas: diagnóstico de las áreas protegidas de América Central. *Boletín de la Asociación de Geógrafos Españoles*, (47), 333–354.
Nielsen, V., Hurtado, P., Janzen, D. H., Tamayo, G., & Sittenfeld, A. (2004). Recolecta de artrópodos para prospección de la biodiversidad en el Área de Conservación Guanacaste, Costa Rica. *Revista de biología tropical, 52*(1), 119–132.
Pagiola, S. (2008). Payments for environmental services in Costa Rica. *Ecological Economics, 65*(4), 712–724. DOI: 10.1016/j.ecolecon.2007.07.033
Paniagua, F. (2007). *Caracterización de los conflictos socioambientales en Costa Rica: 2006*. San José, Costa Rica: Programa del Estado de la Nación.

Pfaff, A. S., & Sanchez-Azofeifa, G. A. (2004). Deforestation pressure and biological reserve planning: a conceptual approach and an illustrative application for Costa Rica. *Resource and Energy Economics*, *26*(2), 237–254.

Rodríguez, S. (2013). *El despojo de la riqueza biológica: de patrimonio de la humanidad a recurso bajo soberanía del Estado* (No. 333.95 R696d). Heredia, Costa Rica: EUNA.

Saura, S., Bastin, L., Battistella, L., Mandrici, A., & Dubois, G. (2017). Protected areas in the world's ecoregions: how well connected are they? *Ecological Indicators*, *76*, 144–158.

Silva, A. (2019). *Lorena (Guanacaste): Agua, territorio y espacios de representación*. www.albasud.org/noticia/es/1076/lorena-guanacaste-agua-territorio-y-espacios-de-representaci-n

SINAC. (2018, December 15). *Sistema Nacional de Áreas de Conservación (SINAC)*. Retrieved from www.sinac.go.cr

Steffen, W., Rockström, J., Richardson, K., Lenton, T. M., Folke, C., Liverman, D., & Donges, J. F. (2018). Trajectories of the earth system in the Anthropocene. *Proceedings of the National Academy of Sciences*, *115*(33), 8252–8259.

Steinberg, P. F. (2001). *Environmental leadership in developing countries: transnational relations and biodiversity policy in Costa Rica and Bolivia*. Cambridge, MA: MIT Press.

Vignola, R., & Morales, J. P. (2011). Equity in the Costa Rican PES scheme: lessons for distributional and procedural justice in REDD +. *REDD-net Programme*.

Villate, L., Canet-Desanti, O., Chassot, O., & Monge-Arias, G. (2009). *The San Juan-La selva biological corridor: a successful conservation strategy* (pp. 1–96). Costa Rica: SINAC.

19

ARCHAEOLOGY AND TOURISM AT MESA VERDE NATIONAL PARK

An environmental justice heritage

Steve Wolverton, Robert Melchior Figueroa, C. Melinda Levin, and Porter Swentzell

Introduction

Archaeology, the study of the human past via remains of material culture, is a field steeped in a colonial disciplinary history, which includes roots in classical archaeology and anthropology (McIntosh et al., 1996; Ndoro and Wijesuriya, 2015). Archaeologists concern themselves with meeting the sophistication needed for their best capacity for writing the stories of the unwritten past by discovering, excavating, and analyzing sites and artifacts. Put succinctly by archaeologist Stephen Lekson (2018: 158), "my job is to figure out the past, and to write pre-history". Thus, archaeology explicitly concerns people in places, at any given time. Upon closer examination, the past is valued and connected to contemporary people who have some heritage connection even if it is remote and indirect. Only within recent decades have archaeologists commonly come to acknowledge the impacts of their research on local peoples, many of whom have ancestral ties to artifacts and sites (McIntosh et al., 1996). This recognition includes focusing on heritage ethics inclusive of field approaches that engage local people, termed community archaeology (Marshall, 2002; Tully, 2007; see recent discussions by Armstrong and Brown, 2019; May et al., 2017; Goh et al., 2019). Despite these positive directions, which are significant and growing, much of archaeology resides in the relative realm of tourism of preserved places, such as national parks, monuments, and other state-protected areas.

In managed areas, such as parks, the stories of the past as written by archaeologists are distilled to their basic core for the benefit of visitors. Goals of such distillation include ensuring an appreciation for the past, which archaeologists view as an irreplaceable record of preserved material culture. Given the colonial and then settler and modern development histories of areas such as the American West, it is less common (indeed rare) that Indigenous perspectives about cultural keystone places, such as ancestral homes and meaningful landscape features that hold deep cultural meaning, are offered to tourists as stories of the past (Butler, 2002). Those stories were marginalized, eroded, and compartmentalized as Indigenous groups were pushed off their native lands (McIntosh et al., 1996). Thus, there may be multiple stories – including those of Indigenous protest and the continuous heritage of environmental trauma upon cultures of Indigenous peoples – that are left out of the tourism package. Correspondingly, archaeologists

 DOI: 10.4324/9780429450310-23

and parks often become stewards of a marginalized past (knowingly or unknowingly) trying to craft a narrative where recentering these stories means presenting stories of places in romantic terms of fascination and awe for the benefit of tourists (sensu Laing and Crouch, 2009). Descendant communities receive little real benefit from this exchange and may actually experience an additional level of exploitation as they add value to the ancestral places tourists consume (Whyte, 2010). Our research has humbly recognized that contemporary peoples who identify such places as their ancestral homes have long-existing accounts and very common restrictions that vehemently resist the excavation, cultural objectification, or occupation by contemporary state and industrial tourist development (Taylor and Wolverton, 2015; Wolverton et al., 2016). In this vein, following Macbeth (2005: 963) in the context of tourism development, the archaeologist (amongst others) "should engage with and understand the ethics of their positions" in order to "foster research, planning, and development". That is, "ethically reflexive scholarship is required".

In this chapter we present a discussion of how displaced stories of the past represent environmental injustices upon heritage, particularly the heritage of people who recognize protected archaeological places as ancestral homes. A conceptual framework from environmental justice provides a set of critical guideposts for asserting the socio-environmental importance of Indigenous narratives over time. We center our discussion on Mesa Verde National Park (MVNP) in the American Southwest of the United States where we made site visits between 2010 and 2016 as an interdisciplinary team that included scholars from the sciences, arts, and humanities, including ones who claim the park as an ancestral home. Descendants with direct heritage connections to MVNP include the Pueblo nations of Taos, Picuris, Ohkay Owingeh, Santa Clara, San Ildefonso, Nambe, Pojoaque, Tesuque, Cochiti, Santo Domingo, San Felipe, Santa Ana, Zia, Jemez, Sandia, Isleta, Laguna, Ysleta del Sur, Acoma, Zuni, and the Hopi tribe.

We begin this exploration of MVNP with a brief introduction to Environmental Justice Studies (EJS) as a transdisciplinary framework for understanding heritage ethics at MVNP. Then we introduce archaeological and Pueblo stories of the past, which is followed by examples of how worlds collided during site tours and visits from our own fieldwork. We use these vignettes to highlight how an environmental justice framework may dramatically shift the visitor experience towards greater critical reflection of the power that narratives can invoke upon socio-ecological relations. MVNP narratives (as experienced by visitors, rangers, and Indigenous peoples) provide an opportunity to discuss critical components of an environmental justice framework that provides a pathway toward recognition thereby opening possibilities for reconciling the worlds of MVNP. Undoubtedly, should reconciliation happen at this coarse intercultural scale, new challenges may emerge within Pueblo societies.

Tourism, archaeology, and environmental justice at MVNP

MVNP is a touristic, scientific, and heritage landscape that is a product of development. The park was founded in 1906 by the United States Congress and President Theodore Roosevelt. Its status as a National Park is part of the larger history of federal land management during the height of the industrial revolution. Specifically, the creation and continuation of the US Forest Service and the US National Park Service testify to the symbolic significance of authority and federal oversight in land use, natural resource use, and philosophies of preservation. However, such institutions are also the ongoing product of settler colonialism carrying out the expansionist ethics of manifest destiny and frontier breaking (Cronon et al., 1992:17; Macbeth, 2005). Understanding the park as a junction of development provides a better understanding of how industrialization sustains settler colonial ethics through state institutions, such as the National

Park Service. Likewise, the birth and growth of the park's tourism have served as an economic lever and scientific laboratory that has generated opportunities for archaeology's investigative practices and positivist principles. Contemporary tourism combines with science to shape the industrial scale of the park's cultural legitimacy and land management, which is an illustration of developing heritage around place through modern industrial material culture.

MVNP is located in what is today the southwestern corner of the State of Colorado. It represents one of the earliest national parks preserving cultural heritage, and in 1978 the UNESCO World Heritage Commission recognized the universal value of its cultural significance for connecting ancient and contemporary Pueblo people through its outstanding, well-preserved cliff dwellings, such as Cliff Palace, Balcony House, Spruce Tree House, and many other places in the park. MVNP is famous as an international and national tourist destination. It embodies a complex history that represents a national pride in the park system (imagine the condition of the ancient sites if they were not within a national park), awe, and intrigue about the deeper past of Indigenous Americans (including evidence of prior appropriation of lands, and pre-colonial arts, science, politics, economics, ecology, and spiritual values). Most commonly the park embodies a tourist experience into the archaeological acquisition of heritage, which is something far enough in the past to be part of universal human history and yet maintained and passed on to present and future individuals using the collective material means of development to capture the economic, scientific, and political arena of development and distribution. In many ways, it represents the park ideal that was established in the US and exported to other places (Timothy, 2000). What tourists encounter at MVNP is a series of constructed narratives that combine the histories of their lives, the places they have been to and come from, and the stories they encounter – MVNP is socially reproduced in the context of colliding geographies of economy, politics, knowledge, and communities. That is, the "slippery, mercurial nature of heritage [in places like MVNP] allows it to be endlessly reinterpreted for personal and political possibility. There is no unmediated past" (Meskell, 2015: 2).

As a UNESCO World Heritage Site and US National Park, and being adjacent to the Ute Mountain Park, which is operated by the Ute Mountain Ute Tribe, there are several broad heritage paradigms that influence how tourists encounter Ancestral Pueblo villages at MVNP (Figueroa, 2015). Similarly, archaeologists have conducted research in the park for over a century, adding and altering the narratives of the Pueblo sites through the cultural narrative of Western science. Particularly, the strong current of scientific positivism in archaeology that has dominated research in the discipline since the 1960s (O'Brien et al., 2005; see Watson et al., 1971) is responsible for a large part of contemporary narratives about how Ancestral Pueblo people lived in the famous cliff dwellings and villages throughout the park. Although these currents continue to flow through the discourses of US political and academic structures, they also intersect and at times collide with the history of contemporary Pueblo people who hold long-living connections to those places (Naranjo, 1995, 2006; Ortman, 2012; Swentzell, 2015; Lekson, 2018).

Tourism at the park represents an ongoing interaction of global, national, and Indigenous worlds (sensu Butler, 2002), specifically a leading global state in which the traditional needs of contemporary Pueblo people have been marginalized, disrespected, and excluded for future heritage. Environmental justice refers to the conceptual connections and causal relationships between environmental issues and social justice. The relationship between critical tourism studies and environmental justice has been established as has the relationship between archaeology and environmental justice (Whyte, 2010; Figueroa, 2015; Figueroa and Waitt, 2010; Wolverton et al., 2016). This is because the transdisciplinary nature of environmental justice studies focuses on case-based and grassroots interpretations of ecological, political, scientific, and historical experiences. Environmental justice studies (EJS) provides a transdisciplinary arena where methodologies cut across knowledge paradigms for coping with questions of justice, ecological adaptation, social capacities, and multiple standpoints to determine harms against a heritage; cultural harms that are to be addressed whether or not they were intended. Transformative justice, or the capacity to move institutions and collective actions towards systemic change, is the ultimate goal for ameliorating environmental traumas faced by communities. However, such communities, inclusive of human and nonhuman beings, must be able to address dimensions of recognition justice; that is, a community must be able to enact self-determination concerning its history, identity, heritage, and ability to direct how transformations should be adopted.

Tourism as a cultural and material exchange among communities denotes distributive forms of environmental justice, particularly related to locations and systems that determine environmental benefits and burdens. These cultural exchanges are denoted by recognition forms of environmental justice that are central for addressing how power, especially pertaining to collective identity, heritage, and self-determination, needs to be navigated in terms of agency and respect (Figueroa, 2006; Waitt et al., 2007; Whyte, 2010). We focus on four concepts within recognition dimensions of environmental justice: *environment heritage*, which expresses a community's *environmental identity* (self-perceptions related to one's material environment) over time; *moral terrains*, which represent webs of heritage practices and values (e.g. norms or lenses through which people encounter places); and *collective continuance*, or the self-determination of a collective to employ adaptive capacities (strategies of resilience) for ongoing social-ecological changes that impact values, practices, and institutions. These concepts highlight perspectives and practices that may reveal opportunities for tourists, archaeologists, and other visitors to recognize the importance of places in the park for Pueblo people. Before delving deeper into EJS, we first share different stories of the Mesa Verde region and the interpretations of the past we bring to the present discourse.

Stories of the Mesa Verde past

How archaeologists tell the story of Mesa Verde archaeology varies, particularly regarding what happened in Ancestral Pueblo culture leading up to depopulation of the region at roughly AD 1300. Prior to AD 700, people lived in small pithouse villages throughout much of the region, and by AD 750 were reorienting into small farming villages that incorporated room blocks of above-ground architecture. There are perturbations in population growth and decline, but by AD 900 village size increased with growing expression of architectural characteristics of places to the south, particularly Chaco Canyon in northwestern New Mexico. Archaeologists demarcate a second period of demographic expansion after AD 1150, and by the early to mid-1200s site locations shifted dramatically to what are generally interpreted as defensive locations on the landscape, such as canyon heads and balcony alcoves that include the famous cliff dwellings of MVNP. Archaeologists describe the decades leading up to AD 1300 and depopulation as a time in which animal resources declined, droughts occurred, and intervillage socio-political strife increased. Archaeologists emphasize a number of themes in terms of determining what caused regional depopulation, which include socio-political changes that varied by scale across the cultural landscape (Glowacki, 2015), environmental change (Benson et al., 2007), and/or shifts in subsistence (Kohler et al., 2008; Ellyson et al., 2019). There were likely "push factors" that led to migration from the region and "pull factors" that drew people to other areas (Cameron, 1995, 2006; Ortman, 2012). Tourists may or may not be aware of this archaeological version of the story of the past when they arrive at MVNP, but those who come to the park are likely to hear or read about this version at the visitor center, on tours, or in books that they purchase at the gift shop (e.g. references in Noble, 2006). If visitors research Mesa Verde archaeology prior to their visit, they are likely to find resources that summarize the story of the Mesa Verde past in a generally similar manner.

The story of the Mesa Verde past as told from the point of view of a contemporary Pueblo person is likely quite different than that summarized in the preceding paragraph (e.g. Naranjo, 1995; Suina, 2002; Swentzell, 2015). One key difference relates to the fact that many contemporary Pueblo people from the Rio Grande Valley, as well as communities in other parts of New Mexico (e.g. Zuni) and parts of Arizona (Hopi), hold that ancient sites in the Mesa Verde region represent their ancestral homes; Mesa Verde was not abandoned. Central to their heritage is that ancestors still live in these places, which should be respected and left undisturbed. The future well-being of Pueblo people is dependent upon such cross-cultural respect (see collective continuance in the Interpreting MVNP within environmental justice section). Thus, the fact that MVNP is a significant tourist destination and that the archaeological story of the past is pervasive and dominant, makes contemporary Pueblo people vulnerable to harm (intended or unintended). This is so because the very tone and nature of, as well as public fascination with, the archaeology of Mesa Verde has left little room for the Pueblo narrative – a situation that does appear, however, to be shifting in the last two decades within the US Park Service.

Consultation with Pueblo communities was an important element of the construction of a new interpretive center at the entrance of MVNP, which more overtly recognizes the ongoing relationship between contemporary communities and their ancestral homes. While this is an important step, there are issues with exploitation of Pueblo peoples. For example, government research on economic impacts of national parks estimated that visitation to MVNP brought in excess of $70 million to the local economy of the area surrounding the park (Brown, 2018). At a coarse social scale, in what way does this contribute to Pueblo peoples? In essence, Pueblo peoples' history, culture, and sacred places are commercialized within the context of a park for the benefit of a relatively large geographic region. However, Pueblo peoples see no tangible benefit to such exploitation as versions of stories of their past are shared with hundreds of thousands of visitors per year. At a finer scale, should Pueblo people recover additional tangible benefits, who then within Pueblo communities would benefit and how so? The answer will, of course, be determined within Pueblo communities. Nonetheless, the story of Mesa Verde as told by the archaeologist and operationalized by the park operates at a coarse, intercultural scale, which is our focus in this chapter.

Visiting Mesa Verde National Park

Authors of this chapter made multiple visits to MVNP together and independently during the last decade. In this section, we portray experiences from site visits. These include observations of how archaeologists, rangers, and Pueblo people engage with and/or portray Mesa Verde, including differences among tours, shifts in perspective when contemporary Pueblo people are present, how the presence of archaeologists impacts narratives and related experiences. Though our experiences represent but a very small sample of park visits, we are able to demonstrate that what a visitor experiences during their time at MVNP varies by context.

Boundaries at Spruce Tree House

During our fieldwork, we made several visits to Spruce Tree House (STH), which at the time was open to the public. Currently, STH is closed due to the potential for rock fall from the cliff above the alcove, but prior to 2018, it was the most accessible cliff dwelling in MVNP. STH also boasts some of the best-preserved architecture in the park. In terms of archaeology, STH is a classic, large cliff dwelling that was occupied up to regional depopulation just prior to AD 1300 (more can be learned about the archaeology of STH in Glowacki [2015]). Prior to 2018 when the site was still open, visitors could walk through the pueblo on a self-guided tour, they could hire a local tour guide, and/or they could visit with on-site rangers. Very rarely, tourists might interact with archaeologists doing research at the site or with contemporary Pueblo people visiting the pueblo. Members of our interdisciplinary team visited STH under each of these conditions; in this section, we describe interactions with archaeologists, rangers, and Pueblo people, emphasizing how physical boundaries and storytelling varied.

Visitors to STH (when open) can walk on a self-guided tour that presents the basics of the archaeological rendition of culture history in the area and at the site. Signs and pamphlets also incorporate basic information on how and why the site and the park are important to contemporary Pueblo people. Interactions with rangers during such visits are generally informal and reflect either: (1) topics such as how rooms of various types were used, how people lived in the past, how archaeologists do research at the site, and other information that can broadly be characterized as "archaeological"; or (2) prohibitions of use of space as there are areas of STH (e.g. rooms) that are off limits to tourists (Figure 19.1). Thus, the ranger is part storyteller and part guardian of STH. For the vast majority of tourists, the ranger is the primary expert about STH and other places in MVNP. In the following section, we portray how the story of the past can vary among rangers (see section Visiting Cliff Palace and Balcony House).

Steve Wolverton and a group of environmental science students visited STH with archaeologist Donna Glowacki in 2011 during a period in which she was finishing detailed architectural documentation of the site (see Glowacki, 2015: 206–207). Students were able to see a few parts of STH that are typically off-limits to tourists during self-guided tours. In addition, Donna Glowacki is an expert on Mesa Verde archaeology and provided specifics about construction, archaeological documentation, room use, and various other details that are not typically available to visitors. Although members of the group were allowed closer access to some areas of the site, there was substantial emphasis on leaving as little impact on the architecture (e.g. walls, floors, etc.) as possible, and some areas of the site were considered off-limits. Thus, compared to a self-guided tour, visiting STH with an archaeologist who is an expert on Mesa Verde emphasized the details of archaeology, respect for the physical space of the site, and provided access to some rarely visited rooms and areas.

The power of story and the content of narrative shifts considerably if contemporary Pueblo people visit STH and interact with rangers. Porter Swentzell is a recognized Pueblo historian and cultural conduit for archaeologists working in the region. His namesake carries significance for experienced rangers at MVNP because of his recent matriarchal lineage, including his mother who is a well-known artist who has revitalized Pueblo art and his grandmother, who is a scholar from Santa Clara Pueblo known partly for her book on the Pueblo city structure of Santa Fe. Further, his aunt, Tessie Naranjo, is an important holder of traditional knowledge and a scholar of Ancestral Pueblo history. When we visited STH, rangers recognized his name and his status as a Pueblo person. The positionality of storytelling completely shifted, as did physical boundaries at the site. Rangers became the audience and Porter was recognized as an expert on STH. He spoke of the meaning of different forms of architecture and spaces in the pueblo and rangers asked a number of questions to learn as much as possible from him. In a later conversation between archaeologists Steve Wolverton and Donna Glowacki, it became clear that when contemporary Pueblo people visit the site, rangers open the space out of respect for their heritage. Donna mentioned a time when she was documenting architecture at the village when a group of young dancers from Jemez Pueblo visited. She remarked in awe her experience

Figure 19.1 Porter Swentzell at Spruce Tree House with MVNP rangers

Source: Taylor and Wolverton (2015, p. 79); reprinted with permission from University of Utah Press

of witnessing them hold traditional dances in this place that is deeply important to their heritage. In terms of tourism, archaeology, and heritage, then, the question becomes, "how do we [scholars, tourists, archaeologists] engage in places in ways that more commonly bring Indigenous narratives to the forefront?" We address this question in the final section of the chapter when we discuss important framing concepts from environmental justice studies. First, we share an additional series of experiences from tours in MVNP that highlight how the story of the past can vary within short periods of time, even in the same place (e.g. within the span of a couple of hours at Cliff Palace).

Visiting Cliff Palace and Balcony House

The first part of this section is a presentation of Melinda Levin's experience during a reconnaissance trip to MVNP in preparation for filmmaking related to our interdisciplinary project. Written in the first person, her vignette portrays how tourist experiences can vary, even within the same day at the park. The second section is a brief vignette from Steve Wolverton about how a ranger introduced a guided tour of Balcony House, which is another popular tour at MVNP.

Cliff Palace – One summer, late in the season but while it was still warm and busy in Southwestern Colorado, I took what I call a "reconnaissance" trip. As a documentary film director, I find it difficult and assumptive to arrive at a place I've never been, with people I've never met, and start filming. If at all possible, I take one or more journeys in advance to settle in a bit, to begin to steep in the location, get a feel for personalities, observe the light, and listen to the natural ebb and flow of sounds. Before any of these scouting trips, I will have accomplished dozens of hours of literary and other research, so I tend to go in with some certainty that I have prepared off-site as deeply as possible.

Prior to one such documentary film project, I took such an exploration of MVNP, and in particular, to the Cliff Palace location. It was mid-morning, and I had driven up to the park from my home in northern New Mexico. For anyone interested in Southwestern archaeology, indigenous culture, and/or "road trips for the family", the Cliff Palace dwelling is well known, and highly photographed. As I arrived, MVNP was saturated with tourists, and the traffic prompted me to park quite a distance from the actual site. As I walked back through the long, bustling parking lot, I passed cars and trucks from dozens of U.S. states, and even a few from Mexico and Canada. There were also tourist buses, and I heard fellow visitors speaking in Chinese, French, German, and Japanese. The opportunity to purchase a ticket, earn with that ticket an official "guide", and be allowed to scale down a cliff ladder from the mesa above to see where, and be offered a glimpse of "how", people from long ago lived, obviously was worth long travel for me and many of my fellow travelers.

I purchased a ticket for the 10:00 am tour, and my group consisted of approximately 25 people. I noted that we were mostly American tourists, with about 8 or so Japanese visitors. Many of the visitors were families with children. The mid-morning slot allowed for moments of pause in the shade, while we waited for the 9:00 am tour to complete their visit just ahead of us. During one of these pauses, while we were partially down in the canyon but not quite yet to the coveted Cliff Palace location, our group's female National Park Ranger pointed to the shadows under the trees below us, and told us that was where the children of the ancient Cliff Palace villagers would play with the community dogs, and take care of the small group of village turkeys. In answer to a question from one of the Japanese men (who spoke fluent English), she expressly denied one of the prominent theories that cultural strife and warfare were to blame for the decline of Mesa Verde culture. The visitors around me smiled, felt a sense of positive human connection, and expressed awe at this link through time and space. Soon, we were led through the actual archaeological site, though kept at a safe distance by small barriers and "do not touch" signs.

After the 10:00 am tour (which was touted as a one-hour visit, but clocked-in at 48 minutes), I made a quick decision to make a second visit. My intent with this re-do was to observe changes in sunlight for future filming purposes. I also had spied a specific spot I might want to do an interview in, and I hoped to analyze the location better. Truth be told, I was also simply enamored with the mystical place I had just been hustled through, and wanted another chance to spend time there. This second tour was a smaller group, again dominated by American visitors, but with substantially fewer children. We also welcomed about half a dozen tourists from China, along with their translator.

This group, with the shadows now a bit thinner, and the temperature about 5 degrees F hotter, was led by a middle-aged National Park Ranger. Mustachioed and muscular, I immediately decided he had been in the military, and was taking this job in early retirement. When we arrived at the spot previously noted, the one that allowed us to see the Cliff Palace dwellings from a distance and the trees below us, but not quite interface with the Cliff Palace ruins yet, the guide asked everyone to be quiet so he could give us some information. He was very serious, and indicated to us that what he was about to share was official, and endorsed. I remember thinking that his ego was a bit too involved in this story.

What he proceeded to outline was a story of aggression, violence and hate. He told us of cannibalism, and warfare. As our summer trip in the American Southwest became quieter, he vividly described cannibalism, infanticide due to drought, and the tools of violence. By the time we arrived at the beautiful Cliff Dwelling structure, I had an entirely different feeling about the physical space I was standing in, and had mostly forgotten about the light and sound elements I was wanting to evaluate.

It is clear from Melinda Levin's experiences that the story of the past at Cliff Palace, the most famous of sites at MVNP, can vary considerably between tours (even ones that are back to back). At Balcony House co-authors experienced another example of how rangers talk about places in the park. Whereas Melinda's experience at Cliff Palace highlights the story of the past, the experience at Balcony House highlights who gets to claim ownership of the park as a national park and world heritage site. The following vignette is portrayed in the first person by Steve Wolverton:

Balcony House – Members of our interdisciplinary team are joining a site tour at Balcony House with a group of students from Crow Canyon Archaeological Center, a non-profit public education and research institute in nearby Cortez, Colorado. Co-authors and students are members of a week-long workshop entitled Transformations during which the group visits sites throughout southwestern Colorado, encountering those places through the lenses of Pueblo heritage, the arts, humanities, and archaeology. The group amasses near the parking lot during the late morning, co-mingling with a large group of tourists in order to await the ranger who would guide the tour. The group is approximately 35 people with international membership including a large contingent of tourists from northern Europe. The ranger, an amicable mid-twenties-aged male, meets us at the trailhead; as is typical for guided tours, he introduces himself and begins to set up the narrative for the site visit. The visit to the site is not memorable, as I have been on a number of these tours over the years.

Given our interdisciplinary work centering on environmental justice and Pueblo heritage, what stands out to me is how the ranger introduces visitors to the site. He expresses great concern that tourists show respect for the site, emphasizing that touching the sandstone masonry and wooden timbers can damage the site's integrity. He

> stresses not veering from the prescribed public spaces, also to protect the physical condition of the place. I am paraphrasing, but what really gets my attention is a statement that "MVNP is a world heritage site. *It belongs to all of us; think of it as your home and my home*". The intent of this statement is clearly to encourage tourists to show respect for the place as a physical, archaeological space. I hear no mention of Balcony House as the ancestral home of contemporary Pueblo people. Indeed, what strikes me is the opposite claim, which by default marginalizes the Pueblo claim of heritage – that MVNP belongs to the public. It is not what is intended that disturbed me, it is what is missing, the already marginalized narrative of Pueblo people. This represents the very nature of marginalization, the complete absence of status.

Summary

At STH members of our research team were able to witness how details of the story of the past as well as the fluidity of spatial boundaries shift depending on who is being regarded as an expert. In our experiences at STH, rangers tended to stick to the basics of the archaeological story of the past by relaying information about subsistence, functions of different types of rooms, and the fundamentals of how archaeologists do research. Their descriptions of Ancestral Pueblo culture, however, vary greatly. Visits with archaeologists delved into finer details of architectural documentation and contemporary questions in archaeological research. The presence of Indigenous Pueblo people shifts the dynamics of storytelling and the use of space at STH. Porter Swentzell, for example, would express the details of how the site layout and room orientations do or do not relate to contemporary Pueblo worldviews. Further, he was able to access areas of STH that are commonly off limits to visitors. Melinda Levin's experience at Cliff Palace highlights that the story of the past can shift by storyteller, and renditions of the Pueblo past can vary dramatically among rangers leading tours of the same village on back-to-back tours. Steve Wolverton's recollection of the visit to Balcony House highlights that not only is the Pueblo narrative missing from many site tours, but it may be replaced by one of public ownership of MVNP. The ranger on the site tour even referred to Balcony House as the "home" of the tourists. None of these experiences – with the exception of the presence of Pueblo people during visits (which is a rare event) – highlight MVNP as *the ancestral home of contemporary Pueblo people with implications for their communal well-being*, which relates to the park being a tourism destination in the context of a history of Western development. Adopting concepts from EJS can aid in restoring Pueblo narratives.

Interpreting MVNP within environmental justice

Expanding upon our earlier introduction of EJS with an emphasis upon recognition justice, this section illuminates three concepts – environmental identity (and heritage), moral terrains, and collective continuance – in order to help clarify the critical importance that Pueblo environmental heritage has for archaeologists, scholars in the arts and humanities, and the visitors and tourists of MVNP.

Environmental identity is an amalgamation of cultural identities, ways of life, and self-perceptions that are connected to a given group's material environment (Figueroa, 2006). It concerns the elements of place, heritage, and community that shape a person's sense of self according to these environmental relations (Figueroa, 2001; Clayton, 2003). Environmental heritage unfolds from environmental identity, where the meanings and symbols of the past frame values, practices, and places preserved for members of a community. In other words, environmental heritage is environmental identity in relation to a community viewed over time (Figueroa, 2006). As the classifications of World Heritage and National Park signify, a majority of tourists arrive from afar with environmental identities shaped by heritages and places that are *not here*, and yet the tourist desires to bring this place into the scope of their own environmental identity and graft it onto the broader environmental heritage that shapes future imaginary values and relations. Environmental heritage is a forward-looking direction of environmental justice, as an active collective process occurs to address a host of challenges such as conflicting heritage between settler colonial (US and international tourists, scientists) and Indigenous self-determination. The environmental heritage of tourists could take multiple trajectories, but the alignment of tourism with Western science puts the contemporary Pueblo heritage at several possible disadvantages. Who speaks for the Pueblo heritage is of a different kind of significance than who speaks for archaeology. Whereas the former may be much fewer individuals (community representatives of Pueblo peoples), the latter is a profession of experts whose skills include responsibility to maintain the integrity of the profession, convey facts to other experts, and to the general public (including tourists). Archaeologists speak *for* their profession and ought to be speaking *with* Pueblo people. Tourism speaks into a web of environmental identities, histories, and institutional systems.

These webs are moral terrains representing varying narratives, including the affective domain of the feeling world of material existence that invokes meaning, structure, and collective values over a place. Respecting the environmental heritage of Indigenous communities requires considering the moral terrains beyond those of one's own environmental identity. For instance, contemporary Pueblo people may hold archaeological interpretations as co-significant, or secondary to their environmental heritage that sustains Pueblo environmental identity; meanwhile, archaeologists invoke their own environmental identities that move along the moral terrains of Western sciences (physical and social) interacting with experts in the field and invoking public understanding. It is inaccurate to compare the Pueblo heritage and the heritage of archaeology as *merely* the competition between two different narratives (Gill and Lantz, 2014; Hopkins et al., 2019; McGhee, 2008; Reid et al., 2014; Reo, 2019; see Watkins, 2005). Both narratives infer the co-modern discourse that has been inherited from interactions of colonialism, development, tourism, park management, sovereign recognition, and a myriad of local to global scale relations that can be multi-systemic including racism (Kendi, 2016, 2019; see Byrne and Wolch, 2009 for discussion on race, parks, and geography; see Blakey, 2020; Carey, 2019; Franklin et al., 2020; White and Draycott, 2020 for current perspectives on racism in archaeology).

Of course, moral terrains are projected onto places as people encounter them; a prominent moral terrain for the archaeologist approaching MVNP is Western science – the park represents a place of study under the archaeologist's lens. Mesa Verde archaeology is a study subject, and sites and artifacts, indeed landscapes, are viewed as objects (of study) (sensu Bhabha, 1990: 72–75). The moral terrain of a tourist is threaded into the dominant terrain of science, via objectivism that may relate to consuming the park as a learning and/or voyeuristic experience. But, an experience into what? What environmental identity, environmental heritage, and intermeshing of moral terrains does the tourist project into the park, and what will graft onto her/his environmental identity afterwards? These questions are intended to bring the issue back around to what voices, interpretations, and narratives the tourists are able to sustain while visiting the park as well as the environmental heritage take-away from their experience at MVNP. The scientist and tourist share MVNP along an entwining set of narratives and performative actions demonstrating behavior, feelings, and embodied practices, often obviated on the landscape by guided walkways, trails, and visitor assimilation to establish the boundaries of occupying a space and place in the park. These are normative structures of who goes where in the park, who knows what, and who determines where others go and what others know about the park. Consider three environmental identities along one general dimension of *visiting MVNP*: tourist, scientist, and interpretive ranger likely do not hold places in the park as their "ancestral home"; they would have to undergo a transgression of their environmental identities to do so. Only Pueblo peoples can truly claim MVNP as their ancestral home.

The moral terrain of objective experience is an attitude to study *it* and/or consume experiences of *it*, and then to leave *it*. At MVNP this attitude is generally shared by both tourist and scientist. The ability to achieve the physical distance of coming to and departing imitates the behavior of an observer who can "step back" from this place, despite experiential attachments *to it*. Tourists and scientists return to other places, where their environmental identities have to be *reconciled* with their place-based experiences from MVNP. Reconciliation is a moral affect, a feeling of resolution that ought to be achieved with as much recognition of the other as possible; the more recognition justice the greater the collective capacity for recognition justice because inclusion, respect, robust participation, and self-determination are frontloaded into the discourses. Place-based experiences of MVNP graft onto experiences that anchor one's environmental identity, and these are reconciled along the moral terrains that extend into and from MVNP. The moral terrains of tourist or scientist lead away from MVNP and align with the moral terrains of experiences outside of the park, those experiences that are normative and perhaps even reexamined after experiencing the park.

Stepping further into EJS, there is a restorative justice dimension that undergirds reconciliation, to restore the dignity of inclusion to bring forth new responsibilities – those that feed into a collective capacity to continue lifeways in the face of dramatic challenges. For the tourist, such a challenge might be reconciling (1) that contemporary Pueblo people want villages in MVNP to be left undisturbed because they are ancestral homes with (2) the expectation that encountering such sites provides a sense of awe and discovery because they are tourist destinations. The tourist might have to face that their fascination comes at the expense of contemporary Pueblo people who experience the park within a separate moral terrain (ancestral home). The tourist, thus, may recognize that something has been lost and Pueblo people are harmed because their ancestral homes are disturbed, which is an imbalance to be restored. Restorative justice, thus, includes elements ranging from restoring something to what it was before impacts (e.g. those of settler colonialism), to compensation for some material loss, and/or a variety of strategies for direct recognition allowing for interactions between the agent-of-harm and the subject-of-harm. The reconciliation process of restorative justice is always intended as an alternative to retributive justice whereby punishment is the goal. As a result, reconciliation of one's individual experience can include resolving cognitive-affective dissonance of moral terrains, including those that may be restricted to or beyond the epistemological horizon of one's particular environmental heritage. For example, a tourist *learning* about someone else's ancestral home while standing in it during a ranger-guided tour implies that their environmental identity lacks a version of heritage that would address the collective capacity that Pueblo peoples have to access that place as their ancestral home. Reconciling these disparate moral terrains illuminates a pathway for the tourist's environmental heritage – new knowledge and experiences that can be collectively passed forward, including one's limitations.

Along the same trajectory of objective experience, the National Parks Interpretive Ranger provides another account to bridge the tourist and scientist environmental identities. Interpretive rangers are encouraged and required in their employment to invoke their own research and generate their version of the site, its history, archaeology, management, and heritage message, all if not in part, from their personal and professionally accumulated knowledge. The extent to which site tours are scripted by the National Parks Service or to which individual interpretive rangers are censored from providing critical information on controversial injustices against Indigenous environmental heritage only serves to make our argument more salient that rangers deliver accounts of sites that are privileged with epistemic authority of expertise and objectivity.

A portion of this interpretive authority of rangers transects another moral terrain broadly characterized as enforcement. The extent to which an individual ranger has the enforcement authority of restraining someone is determined by their pay grade, but every ranger's first priority is to ensure park regulations are obeyed. The effect delivered by the state authority in the badge, the uniform, and the ranger namesake of the American West's elite law enforcement official delivers an authoritative narrative for her/him to provide their expert version of the place (sensu Althusser, 2001). Thus, the objectivist and voyeuristic moral terrains entwine with authoritative structures of touristic consumer, science, and state enforcement. Similar to archaeologists and tourists, the ranger has an objectivist project based in the enforcement of universals – equality and protection under the law. However, enforcement entails many behaviors and roles. It may mean safety, access, or physical protection of persons and/or sites, or preservation of the sites and artifacts of MVNP in perpetuity so that it is available for experiential consumption by future visitors, and (thus) maybe (also) protecting Pueblo heritage at this particular ancestral home.

The concept of collective continuance is wedded to environmental heritage and collective self-determination. As Kyle Powys Whyte clarifies, collective self-determination is more than recovering and/or maintaining decision-making authority but, "rather it comes to mean *a society's overall capacity to adapt to social and environmental changes, or resilience*" (Whyte, 2018: 351, emphasis in original). Thus, environmental heritage, the capacity of a collective to interpret its environmental identity over time and to continue in forward looking justice, becomes a "capacity of a social-ecological system to absorb or withstand perturbations and other stressors such that the system remains within the same regime, essentially maintaining its structure and functions" (Resilience Alliance, 2002; Whyte, 2018: 354). In Whyte's model, environmental heritage becomes a vital adaptive capacity of "a society's persistence in the face of environmental viability" (Whyte, 2018: 351). The adaptive capacity of environmental heritage, is therefore the linchpin for maintaining ancestral connections, even when temporal and geographical remoteness may be apparent to people of another environmental heritage (Figueroa and Waitt, 2010). For example, the Pueblo ancestral home is connected to living people by the self-determination of collective continuance because the environmental heritage has worked to adapt systems, beliefs, practices, and cosmological vision that ultimately led Pueblo People to the Rio Grande region. The model here suggests that by the deeply adaptive capacity to maintain narratives of environmental heritage, Pueblo ancestors were able to migrate to ensure collective continuance for both the "old world" of the Mesa Verde region and in the new world of the Rio Grande. The adaptive capacity of larger Pueblo narratives of environmental heritage foster self-determined collective continuance concerning MVNP. Indeed, the overall collective capacity of the Pueblo People is recognized by the multiple adaptive capacities their ancestors were able to actualize in order to overcome the social-environmental perturbations (including mega-stressors such as drought, population crisis, and adaptive reevaluations for future peoples) by extensive migration that sustained ancestral systems through self-determination.

That is, heritage describes ways the present agents interpret the past to frame environmental identity for present and future collectives. Thus, the moral terrain of ancestral home is significant for contemporary Pueblo people because it clarifies that the future well-being of members of their communities depends upon respect for and stewardship of ancestral villages in MVNP and surrounding areas. These places include the moral terrains where environmental heritage maintains the world of the ancestors in the MV region, especially the park and surrounds.

Environmental heritage holds a vital point for connecting narratives across geography and history (between ancestral lands and current Pueblo communities). Maintaining a myriad of adaptive capacities within an environmental heritage provides a broader narrative of values, systems, lifeways, and especially self-determination of cultural continuance in the face of environmental viability (Whyte, 2018)

Adaptive capacities may be broad. Environmental heritage, for instance, provides a narrative of reciprocity within a social-ecological relational ethic where more specific adaptive capacities such as customs, roles, responsibilities, stories, and individual obligations are bound together to read environmental perturbations. These nested capacities then function to evaluate and pass forward the reciprocal adjustments necessary for collective continuance (Whyte, 2018; Figueroa and Waitt, 2010). Each of the adaptive capacities (from broad to specific) are webbed by moral terrains, especially when obligations to ancestors and settler colonial interactions have centuries to entwine. Experiencing the paths of moral terrains at MVNP means reconciling moral and epistemological gateways that open to us, as well as those closed off from us precisely due to the limited capacities of *our particular collective environmental heritage as visitors*. To come back around to the visitor's moral terrain, there is a distinction between intentionally stepping back to experience the virtue of objectivity and a lack of collective capacity to step back from the moral terrain of MVNP as one's ancestral home. Visitors, meaning tourists, employees, and scientists, lack the narrative of moral terrains that build upon the adaptive capacity of environmental heritage to ensure Pueblo collective continuance because only the Pueblo people have moral terrains open for combining their adaptive capacities (environmental heritage as one of the broader systems of adaptive capacities) into self-determined collective continuance. Indigenous peoples in particular face intense pressures from our current social-ecological systems of development (e.g. Blair, 2019; Caron-Beaudoin and Armstrong, 2019), especially state enforced tourism that promotes economic and historical regimes that include implicit systems expressed in the MVNP interpretative tours that may be taken for granted, such as gender, warfare, and the legacy of manifest destiny.

Discussion

Our discussion of environmental identity, moral terrains, and collective continuance is readily integrated within contemporary environmental justice scholarship and commonly implied in grassroots environmental justice discourses. Upon introducing recognition justice into the environmental justice paradigm, Figueroa included a relational conceptual framework that would address explicitly leveraging environmental identities, moral terrains, and the adaptive capacity of environmental heritage (Figueroa, 1999; Whyte, 2010, 2011). Whyte's significant contributions to recognition environmental justice pulls the collective continuance of Indigenous peoples into policies and programs (Whyte, 2013, 2014). When considering Indigenous tourism and tourism of Indigenous places, recognition justice is challenged in two co-imbricating ways. First, Whyte notes the injustice of direct derogation of people because of their identity: [it is a] "recognition injustice, to have to live in a society where members of a majority get to impose . . . their conceptions" of identity (e.g. stereotypes of Indigenous peoples [Whyte, 2010: 79). Second, and at the same time there is denial of the importance of their heritage in broader society, which includes dominant and marginalized communities: when social circumstances and cultural terms "are not duly weighed or appropriately emphasized or deemphasized within institutional proceedings and transactions among members of a society" (Whyte, 2010: 79).

Archaeologists face Whyte's two forms of recognition injustice because Indigenous perspectives were marginalized and neglected during the colonial history of the discipline. This injustice, however, extends beyond the disciplinary boundaries of archaeology because of the key status of the field in the public imagination, one of awe and discovery that "allows [stories of the past] to be endlessly reinterpreted for personal and political possibility" in places such as MVNP (Meskell, 2015: 2). At the same time, archaeologists are a direct influence on the public impression of MVNP and thus have obligations to the second feature stated by Whyte, the importance of how and whose heritage is dominant or diminished for social understanding and institutional practices.

To assist archaeologists, an understanding of environmental justice for recognition minimally requires concepts and interpretations of environmental identity, moral terrains, and collective continuance as guideposts for conversations *at MVNP*. Although this chapter is largely an academic exercise, the transdisciplinary nature of EJS entails that our audience includes *contemporary visitors* of the park; thus, we propose a transformative ethic in which rangers and tourists in particular think more carefully of the meaning of villages in the park for members of contemporary Pueblo communities. We surmise that for most visitors of the park, in contrast to our proposed ethic, the experience is one on a bucket list of places to visit – remnants of an exotic, past cultural landscape imbued with awe and fascination. An important step in raising consciousness is to recognize histories that converge in terms of environmental justice, beginning with the material conditions for the development of the park in the first place, while simultaneously expressing an informed tacit understanding that "visits" represent entry into someone else's ancestral home for which MVNP collects an entry fee without compensation to contemporary Pueblo people. The National Park System stems from settler colonial disruptions forming extreme power imbalances and injustices experienced by Indigenous peoples up to the present point in global development. MVNP was set aside for protection and the opportunity to use the park as an expression of a frontier ethic. The same heritage ethics of settler colonialism became a global paradigm for land preservation and the development of national park systems beyond the United States (e.g. Ahmed, 2008; Butler and Boyd, 2000; Cock and Fig, 2002). Thus, MVNP for contemporary Pueblo people embodies these much larger narratives of power, injustice, and history. Our chapter raises and leaves open the question: "in what real and tangible ways does the park contribute to Pueblo communities whose ancestral homes are being discovered, visited, explored, and exploited?"

Conclusion

An important arena in which development and archaeology intersect is within tourism, particularly in reference to heritage sites within state, national, and international parks. Other arenas include cultural resource management in reference to urban, energy, and transportation development as well as museums where people come into some form of contact with archaeological materials. These represent "contact zones" that are

> social spaces where cultures meet, clash, and grapple with each other, often in *contexts* of highly asymmetrical relations of power, such as the legacies of colonialism, slavery, or their aftermaths as they are lived out in many parts of the world today.
>
> *(Pratt, 2002: 4 emphasis added; see Kersel and Lake, 2015: 80)*

Our visits to MVNP represent a series of interactions in these contact zones, interactions during which asymmetrical power relations were visible. However, one reason that we keyed upon those asymmetries is that members of the team were increasingly cognizant of an EJS framework in our interdisciplinary research.

It is time for an ontological turn related to the visitor experience at MVNP (sensu Armstrong and McAlvay, 2019). In terms of development theory, the park shares much with other places. As development represents "social reproduction within environment" (Peet and Hartwick, 2009: 288), it is critical to ask what is being reproduced at MVNP? For the bucket-list tourist, encountering a history of injustice is likely to be a letdown. For the park, which relies on tourism for sustenance, messages of historic injustices contrast starkly with the expected awe and fascination with what is to the tourist a foreign, even exotic, past. Despite considerable recognition by members of the park staff of the history of Pueblo peoples and their ties to the Mesa Verde landscape, there is an inherent structural inertia to a discourse that fully embraces a history of injustice. We see our academic discussion as essential for framing new reasoning.

We argue in this chapter that concepts from EJS provide a framework for bringing forth Indigenous or local narratives that may be on the disempowered side of asymmetrical relations. In contact zones, such as Cliff Palace or other Pueblo villages at MVNP where people-to-place-to-people relations occur, there is an opportunity to recognize that environmental identities converge. The concept of moral terrains asserts that there are webs of values projected upon a place; such terrains represent a person's (or group of people's) rendition of what such a place represents as informed by their environmental identity.

Of the EJS concepts we employ, collective continuance holds particular significance for places such as MVNP because it conceptually embodies multiple adaptive capacities working together. Collective continuance links the *future* well-being of communities to self-determination experienced over time, which may be represented in a place. One's environmental identity matters and informs the moral terrain a person occupies in a place; recognizing diversity and power relations using those concepts is a first step. A second step is that one's environmental identity is informed through one's heritage, a critical adaptive capacity for the community's future well-being under socio-environmental challenges – self-determined collective continuance. We argue that these concepts offer guideposts for engaging conversations in contact zones that respect and recognize Pueblo heritage at MVNP. Recognition justice, however, is only a first step; as one of the co-authors (Porter Swentzell) states, it represents "the battle for lip service" and does not represent a fuller shift in norms through policy. Indeed, when it comes to policy, recognition by the state, because of dramatic power asymmetry, often undermines self-determined opportunities for collective continuance.

Going beyond recognition requires that Pueblo communities recover the power to enrich and preserve the collective continuance of their heritage. At a finer intracommunity scale, recognition justice may introduce new challenges for Pueblo peoples concerning who within Pueblo societies hold stories of places, such as those for Mesa Verde. Further, should Pueblo people receive economic resources from places such as MVNP, there may be new challenges concerning who benefits. Those conversations, of course, will belong to Pueblo peoples. At the coarser, intercultural scale between the moral terrains for archaeological science and Pueblo narratives, restorative environmental justice promotes interactions of recognition that entail a Pueblo community-determined course of action, one that is enacted through community leaders voicing concerns about tourism in places such as MVNP (recognition) and that combines voices to assess shifting norms and practices of tourism that may compound impacts from development regimes that problematize adaptive capacities for social-ecological resilience. These types of restorative justice outcomes have occurred in some places (e.g. Figueroa and Waitt, 2008, 2010), but the battle for "lip service" continues in places such as MVNP.

References

Ahmed, W. 2008. Indigenous communities and their marginalized space in India. In *Studies in Indian economy*, ed. K. Gupta, 188–218. New Delhi, India: Atlantic Publishers.

Althusser, L. 2001. Ideology and ideological state apparatuses. In *Lenin and philosophy and other essays*, 85–126. New York, NY: Monthly Review Press.

Armstrong, C. G., and C. Brown. 2019. Frontiers are frontlines: Ethnobiological science against ongoing colonialism. *Journal of Ethnobiology* 39(1):14–31.

Armstrong, C. G., and A. C. McAlvay. 2019. Introduction to special section on action ethnobiology. *Journal of Ethnobiology* 39(1):3–13.

Benson, L., K. Petersen, and J. Stein. 2007. Anasazi (pre-Columbian native-American) migrations during the middle-12th and late-13th centuries – were they drought induced? *Climatic Change* 83(1–2):187–213.

Bhabha, H. K. 1990. The other question: Difference, discrimination and the discourse of colonialism. In *Out there: Marginalization and contemporary cultures*, eds. R. Fergusson, M. Gever, T. T. Minh-ha, and C. West, 71–89. New York: The New Museum of Contemporary Art.

Blair, J. J. A. 2019. Splintered hinterlands: Public anthropology, environmental advocacy, and Indigenous sovereignty. *Journal of Ethnobiology* 39(1):32–49.

Blakey, M. L. 2020. Archaeology under the blinding light of race. *Current Anthropology* 61s22:183–197.

Brown, C. 2018. *2018 Tourism to Mesa Verde creates $72.4 million in local economic benefits* [Press Release]. Retrieved from www.nps.gov/meve/learn/news/19-07_2018-mevetourism.htm

Butler, R., and S. W. Boyd, eds. 2000. *Tourism and national parks: Issues and implications*. New York, NY: Wiley.

Butler, R. W. 2002. The development of tourism in frontier regions: Issues and approaches. In *Tourism in frontier areas*, eds. S. Krakover and Y. Gradus, 3–20. New York, NY: Lexington Books.

Byrne, J., and J. Wolch. 2009. Nature, race and parks: Past research and future directions for geographic research. *Progress in Human Geography* 33(6):743–765.

Cameron, C. M. 1995. Migration and the movement of southwestern peoples. *Journal of Anthropological Archaeology* 14(2):104–124.

Cameron, C. M. 2006. Leaving Mesa Verde. In *The Mesa Verde world*, ed. D. G. Noble, 139–147. Santa Fe, NM: School of American Research Press.

Carey, M. L. 2019. Toward and antiracist archeology. *The Activist History Review*. Retrieved from https://activisthistory.com/2019/09/27/toward-an-antiracist-archeology/

Caron-Beaudoin, E., and C. G. Armstrong. 2019. Biomonitoring and ethnobiology: Approaches to fill gaps in Indigenous public and environmental health. *Journal of Ethnobiology* 39(1):50–64.

Clayton, S. 2003. Environmental identity: A conceptual and an operational definition. In *Identity and the natural environment: The psychological significance of nature*, eds. S. Clayton and S. Opotow, 45–65. Cambridge, MA: MIT Press.

Cock, J., and D. Fig. 2002. From colonial to community-based conservation: Environmental justice and the transformation of national parks (1994–1998). In *Environmental justice in South Africa*, ed. D. A. McDonald, 131–155. Athens, OH: University of Ohio Press.

Cronon, W., G. Miles, and J. Gitlin. 1992. Becoming west: Toward a new meaning. In *Under and open sky: Rethinking America's western past*, eds. W. Cronon, G. Miles, and J. Gitlin, 3–27. New York, NY: W.W. Norton.

Ellyson, L. J., L. Nagaoka, and S. Wolverton. 2019. Animal resource use related to socioenvironmental change among Mesa Verde farmers. *Journal of Anthropological Research* 75(3):000–000.

Figueroa, R. M. 1999. *Debating the paradigms of justice: The bivalence of environmental justice*. Boulder, CO: University of Colorado Press.

Figueroa, R. M. 2001. Other faces: Latinos and environmental justice. In *Faces of environmental racism: Confronting issues of global justice*, ed. B. E. Lawson and L. Westra, 167–186. Boston, MA: Rowman and Littlefield Publishers.

Figueroa, R. M. 2006. Evaluating environmental justice claims. In *Forging environmentalism: Justice, livelihood, and contested environments*, ed. J. Bauer, 360–376. New York, NY: M.E. Sharpe.

Figueroa, R. M. 2015. Fire in the rain: Exploring the moral terrains of Mesa Verde. In *Sushi in Cortez: Interdisciplinary essays on Mesa Verde*, eds. D. Taylor and S. Wolverton, 118–147. Salt Lake City, UT: University of Utah Press.

Figueroa, R. M., and G. Waitt. 2008. Cracks in the mirror: (Un)covering the moral terrains of environmental justice at Uluru-Kata Tjuta National Park. *Ethics Place and Environment* 11(3):327–349.

Figueroa, R. M., and G. Waitt. 2010. Climb: Restorative justice, environmental heritage, and the moral terrains of Uluṟu-Kata Tjuṯa National Park. *Environmental Philosophy* 7(2):135–164.
Franklin, M., J. P. Dunnavant, A. Omilade Flewellen, and A. Odewale. 2020. The future is now: Archaeology and the eradication of anti-blackness. *International Journal of Historical Archaeology* 24:753–766.
Gill, H., and T. Lantz. 2014. A community-based approach to mapping Gwich'in observations of environmental changes in the Lower Peel River watershed, NT. *Journal of Ethnobiology* 34(3):294–314.
Glowacki, D. M. 2015. *Living and leaving: A social history of regional depopulation in thirteenth-century Mesa Verde*. Tucson, AZ: University of Arizona Press.
Goh, H. M., C. Y. Saw, S. Shahidan, M. Saidin, and D. Curnoe. 2019. Community heritage engagement in Malaysian archaeology: A case from the prehistoric rock art site of Tambun. *Journal of Community Archaeology & Heritage* 6(2):110–121.
Hopkins, D., T. L. Joly, H. Sykes, A. Waniandy, J. Grant, L. Gallagher, L. Hansen, K. Wall, P. Fortna, and M. Bailey. 2019. "Learning together": Braiding Indigenous and Western knowledge systems to understand freshwater mussel health in the Lower Athabasca region of Alberta, Canada. *Journal of Ethnobiology* 39(2):315–336.
Kendi, I. X. 2016. *Stamped from the beginning: The definitive history of racist ideas in America*. New York, NY: Nation Books.
Kendi, I. X. 2019. *How to be antiracist*. New York: One World.
Kersel, M. M., and C. Lake. 2015. Civil societies? Heritage diplomacy and neo-imperialism. In *Global heritage: A reader*, ed. L. Meskell, 70–93. Malden, MA: Wiley Blackwell.
Kohler, T. A., M. D. Varien, A. Wright, and K. A. Kuckelman. 2008. Mesa Verde migrations new archaeological research and computer simulation suggest why ancestral Puebloans deserted the northern southwest United States. *American Scientist* 96(2):146–153.
Laing, J. H., and G. I. Crouch. 2009. Myth, adventure and fantasy at the frontier: Metaphors and imagery behind an extraordinary travel experience. *International Journal of Tourism Research* 11(2):127–141.
Lekson, S. H. 2018. *A study of southwestern archaeology*. Salt Lake City, UT: University of Utah Press.
Macbeth, J. 2005. Towards an ethics platform for tourism. *Annals of Tourism Research* 32(4):962–984.
Marshall, Y. 2002. What is community archaeology? *World Archaeology* 34(2):211–219.
May, S. K., M. Marshall, I. Domingo Sanz, and C. Smith. 2017. Reflections on the pedagogy of archaeological field schools within Indigenous community archaeology programmes in Australia. *Public Archaeology* 16(3–4):172–190.
McGhee, R. 2008. Aboriginalism and the problems of Indigenous archaeology. *American Antiquity* 73(4):579–597.
McIntosh, R. J., S. K. McIntosh, and T. Togola. 1996. People without history. In *Archaeological Ethics*, eds. K. D. Vitelli, 185–197. Walnut Creek, CA: Altamira Press.
Meskell, L. 2015. Introduction: Globalizing heritage. In *Global heritage: A reader*, ed. L. Meskell, 1–21. Malden, MA: Wiley Blackwell.
Naranjo, T. 1995. Thoughts on migration by Santa Clara Pueblo. *Journal of Anthropological Archaeology* 14(2):247–250.
Naranjo, T. 2006. We came from the south, we came from the north: Some Tewa origin stories. In *The Mesa Verde world*, ed. D. G. Noble, 49–57. Santa Fe, NM: School of American Research Press.
Ndoro, W., and G. Wijesuriya. 2015. Heritage management and conservation: From colonization to globalization. In *Global heritage: A reader*, ed. L. Meskell, 131–149. Malden, MA: Wiley Blackwell.
Noble, D. G., ed. 2006. *The Mesa Verde world: Explorations in Ancestral Pueblo archaeology*. Santa Fe, NM: School of American Research Press.
O'Brien, M. J., R. L. Lyman, and M. B. Schiffer. 2005. *Archaeology as a process: Processualism and its progeny*. Salt Lake City: University of Utah Press.
Ortman, S. G. 2012. *Winds from the north: Tewa origins and historical anthropology*. Salt Lake City, UT: University of Utah Press.
Peet, R., and E. Hartwick. 2009. *Theories of development: Contentions, arguments, alternatives*. New York, NY: Guilford Press.
Pratt, M. L. 2002. Arts of the contact zone. In *Professing in the contact zone: Bringing theory and practice together*, ed. J. M. Wolff, 1–18. Urbana, IL: National Council of Teachers of English.
Reid, M. G., C. Hamilton, S. K. Reid, W. Trousdale, C. Hill, N. Turner, C. R. Picard, C. Lamontagne, and H. D. Matthews. 2014. Indigenous climate change adaptation planning using a values-focused approach: A case study with the Gitga'at Nation. *Journal of Ethnobiology* 34(3):401–424.

Reo, N. 2019. *Inawendiwin* and relational accountability in Anishnaabeg studies: The crux of biscuit. *Journal of Ethnobiology* 39(1):65–75.
Resilience Alliance. 2002. *A consortium linking ecological, economic and social insights for sustainability*. www.resalliance.org
Suina, J. H. 2002. The persistence of the corn mothers. In *Archaeologies of the Pueblo revolt: Identity, meaning, and renewal in the Pueblo world*, ed. R. W. Preucel, 212–216. Albuquerque, NM: University of New Mexico Press.
Swentzell, P. 2015. Irrigating astrofalfa. In *Sushi in Cortez: Interdisciplinary essays on Mesa Verde*, eds. D. Taylor and S. Wolverton, 73–89. Salt Lake City: University of Utah Press.
Taylor, D., and S. Wolverton, eds. 2015. *Sushi in Cortez: Interdisciplinary essays on Mesa Verde*. Salt Lake City, UT: University of Utah Press.
Timothy, D. 2000. Tourism and international parks. In *Tourism and national parks: Issues and implications*, eds. R. Butler and S. W. Boyd, 263–282. New York, NY: Wiley.
Tully, G. 2007. Community archaeology: General methods and standards of practice. *Public Archaeology* 6(3):155–187.
Waitt, G., R. M. Figueroa, and L. McGee. 2007. Fissures in the rock: Rethinking pride and shame in the moral terrains of Uluru. *Transactions of the Institute of British Geographers* NS 32(2):249–263.
Watkins, J. 2005. Through wary eyes: Indigenous perspectives on archaeology. *Annual Review of Anthropology* 34:429–449.
Watson, P. J., S. A. LeBlanc, and C. L. Redman. 1971. *Explanation in archaeology: An explicitly scientific approach*. New York, NY: Columbia University Press.
White, W., and C. Draycott. 2020. Why the whiteness of archaeology is a problem. *SAPIENS*. Retrieved from www.sapiens.org/archaeology/archaeology-diversity/.
Whyte, K. P. 2010. An environmental justice framework for Indigenous tourism. *Environmental Philosophy* 7(2):75–92.
Whyte, K. P. 2011. The recognition dimensions of environmental justice in Indian country. *Environmental Justice* 4(4):199–205.
Whyte, K. P. 2013. Justice forward: Tribes, climate adaptation and responsibility. *Climatic Change* 120(3):517–530.
Whyte, K. P. 2014. Indigenous women, climate change impacts, and collective action. *Hypatia* 29(3):599–616.
Whyte, K. P. 2018. Food sovereignty, justice, and indigenous peoples: An essay on settler colonialism and collective continuance. In *The Oxford handbook of food ethics*, eds. A. Barnhill, M. Budolfson, and T. Doggett, 345–366. New York, NY: Oxford University Press.
Wolverton, S., R. M. Figueroa, and P. Swentzell. 2016. Archaeology, heritage, and moral terrains: Two cases from the Mesa Verde region. *Ethnobiology Letters* 7(2):23–31.

20

COMMUNITIES AND CONSERVATION

Between two models of development

Alexandra Areiza-Tapias and Luis Sánchez-Ayala

The strategies for economic development proposed by the Colombian government in the last three decades position as its center pillar the industrial activities of the agricultural and mining-energy sectors. In the context of this governmental agenda for the development of the country, many communities, to protect their livelihoods, have proposed local development alternatives, which in turn are encouraging strategies and practices that benefit the conservation of the environment and the responsible management of natural resources. In this sense, community development strategies and initiatives has positioned itself as an important and, above all, effective approach that needs the attention of the governments when it comes to launching their "engines" of development. In this case, community development strategies and initiatives can be of great benefit to local populations and to society in general.

In this chapter, we present how community development strategies not only represent an alternative for the benefit of communities, but also for the conservation of the environment. From this perspective, local development and environmental protection have been in confrontation with large-scale productive activities, which have traditionally been at the center of the official agendas for development and economic growth. We show these alternatives of local development and their potential effects in terms of conservation of the environment through two specific cases in Colombia. The first case is located in the Pacific region, specifically in the municipalities of Bahía Solano and Nuquí, located in the Department (state) of Chocó. In these two municipalities the large fishing industry saw its activities modified and influenced by different grassroots initiatives; initiatives that had a positive impact on the conservation of fishery resources, as it also translated into an economic benefit for local communities. The second case takes place in the Orinoquia region, particularly in the Department (state) of Casanare. The Colombian government considers Casanare as a department with a great agroindustrial potential; the department is currently experiencing an increase in land transactions and investments for the development of large-scale productive activities, such as the cultivation of rice and palm oil, among other crops. However, some of the inhabitants of the region are formulating strategies aiming for a balance between large-scale agricultural activities, the protection of traditional production methods, and the conservation of nature.

To meet our objectives, we open this chapter with a brief discussion on local and community development. This discussion intends to situate ourselves conceptually, and to offer greater clarity about the dynamics that are taking place in several places of Colombia regarding

DOI: 10.4324/9780429450310-24

the current agendas and strategies for economic development and environmental conservation. After that first part, we continue our discussion by presenting the two case studies. Through these cases, we can account for the motivations, the effects, and the applications related to these community strategies, as they also allow us to demonstrate the implications of such strategies in terms of conservation, biodiversity, and ecosystem services. The chapter closes with some final considerations on the Colombian experience regarding the relationship between development, environment, and society.

Community development

Community development involves different experiences where communities actively formulate alternatives for economic, social, political, and cultural participation, having incidences in the common local welfare. Some authors refer to these community dynamics using concepts such as local development (Núñez et al., 2016; Muñoz, 2012; Coffey and Polèse, 1984), community organization (Ware, 1986), community social action (Marchioni, 1989; Trigueros, 1991), and community development (Porzecanski, 1983; Ander-Egg, 1987). Therefore, community development is a particular grassroots form of local or regional development, in which the internal factors of the community are central to the initiatives that may arise. This, in turn, is a participatory process to address and solve a variety of socio-economic, cultural, and environmental problems with the aim of producing sustainable development and improving the quality of life of the population (Núñez et al., 2016). In this context, for Ware (1986), this form of community organization is a means of promoting general improvements and using the local resources to meet the needs of the population.

This suggests that community development is as a process of social transformation that seeks to improve the living conditions of the people who inhabit a place, through the empowerment of these same people. In more concrete terms, for Coffey and Polèse (1984) local or community development must be understood in stages: (1) the emergence of the local initiative; (2) the "takeoff" of local strategies; (3) the expansion of these strategies beyond the local region; and (4) the achievement of a regional economic structure based on local initiatives and comparative advantages created locally. Therefore, what is central to community development is that the community itself must lead the process of local development. The fundamental factor is that it is not an action on the community, but an action of the community. Therefore, supporting actions between local groups, as well as the creation of small action groups, can be very effective, potentially with a high social impact, in order to face big challenges or big players (government, corporations, institutions, among others).

The social, economic, and political dimensions involved in a community development process are diverse. Its aims can go beyond general purposes such as well-being, autonomy, and subsistence, as these community processes can also influence livelihoods and life projects, means of production, education and health, the environment, the use and access to resources, social services, among others. The basis of community development strategies lie in the fact that local communities know how to respond adequately to their own problems, and the realization of such strategies requires a new way of relating and coordinating the actions of the State, regional and local authorities, and the private sector (Muñoz, 2012).

Development and conservation: contributions from Colombia

In Colombia, there are isolated regions, difficult to access and with little connectivity with the rest of the country. This is largely due to the lack of infrastructure, but above all, to the inoperativeness or lack of will of the government to change that reality. Some of these regions are the Pacific, the Amazon, and the Orinoquia, all of them regions located in the periphery of the Colombian State (strongly centralized in the Andean zone). The isolation of these peripheral areas has become a mechanism for the protection of natural resources, in some cases consciously (by the State), and in other cases, due to illegal actions (the presence and control of irregular armed groups). These two elements are closely related, and not only reflect the little attention given by the State to these regions for many years, but also support the great need for local development strategies.

These peripheral and marginal regions, in turn, correspond to the places with the greatest concentration of natural resources, such as large forest areas, floodplains, fishing resources, the presence of precious minerals such as gold and coltan, and the existence of hydrocarbons (oil and natural gas). These places also congregate large conservation areas (National Parks), indigenous and Afro-descendant communities, peasants, as well as illegal mining activities, deforestation hotspots, illicit crops (coca), and the illegal extraction of fishery resources. All of these must be understood in the context of the absent, or a weak presence, of the state apparatus. Therefore, this isolation has favored an endogenous local development. This scenario is present in the cases of different communities in the Pacific and Orinoquia regions, where local communities are generating synergies between lifestyles and natural resources.

The Pacific region is located on the Colombian west coast, bordering the Pacific Ocean. The region extends to the south to the border with Ecuador, to the east, it borders the Andean region, to the northeast meets with the Caribbean region, while to the northwest, it reaches the Serranía del Darién, which marks the border between this Colombian region and the Republic of Panamá. Internally, in terms of political organization, four departments (states) compose the region: Nariño, Valle del Cauca, Chocó, and Cauca. This is the area of greatest biodiversity and rainfall density in the country. This is true particularly in the department of Chocó, where heavy rains play an important role in contributing to the development of the region's biodiversity. Consequently, the Colombian Pacific has one of the jungle areas with the highest number of species per square kilometer in the world, with a marine area exceeding 300,000 km^2. In this marine area can be identified 17 species of corals, 130 species of mollusks, 267 species of crustaceans, and about 400 species of fish. Among the most important species are the mero (grouper), the tuna, the jurel (horse mackerel), the pargo (red snapper), the hammerhead shark, the galapagos shark, and manta rays. In addition, humpback whales annually visit the waters of the Pacific region to give birth to their young.

This region is also the ancestral territory of native peoples such as the Emberá, the Wounann, the Gunas (Kunas), and the Awá. In addition, the Pacific is the place of concentration of many Afro-Colombian communities. These are mostly communities that historically have depended on fishing resources for their livelihood. Therefore, the great marine wealth of the Pacific region supports the local and regional economy.

However, not only the inhabitants of the Pacific coasts make use of these fishing resources, as some businesses with large-scale industrial activities operate in this area. These businesses practice bottom trawling, and mostly export their product outside Colombia. These extractive dynamics lead to the decrease of fishery resources, and, therefore, to the detriment of the livelihoods of local communities, causing a conflict between the local artisanal fishers and the industrial fishers. Facing this situation, the communities of the Colombian Pacific coast organized themselves to create local development initiatives that guarantee their livelihoods. As one of the community leaders of Bahía Solano tells us:

> We had an objective, to preserve some of our territory, but in the sense of food security. We noticed that with the pass of the time the fishing production decreased, the fishers felt the economic blow.
>
> *(Interviewee 1, Redefrío)*

These local initiatives led to the promotion of "fishing agreements" between the artisanal fishers, the largescale industry, and the State. Thus, after several years of negotiations on the protection of marine resources, in July 2008 the Colombian government, with the participation of artisanal and industrial fishers, declared the Exclusive Zone of Artisanal Fishing (ZEPA, Zona Exclusiva de Pesca Artesanal). This was a great achievement for the community in their struggle for their productive means of subsistence, which in turn had the effect of protecting the natural marine resources of the Colombian Pacific; as Interviewee 1 himself says,

> What we wanted was to get rid of the industrial ships, because we needed the area [where the industrial ships were operating] for our food security. We never imagined that our initiative would lead to the creation of a figure [a protection law]. We were fighting for what is ours, but not knowing well the objective where we were going. Everything fell into place, and today we have the ZEPA.

The ZEPA is a management instrument for the marine zone that covers an area of 2.5 nautical miles, which promotes the use of traditional fishing methods that are friendly with the environment. Therefore, the ZEPA regulates the industrial fishing activities and prohibits bottom trawling, but also demanded that artisanal fishers change their old unsustainable fishing practices (e.g. the use of fishing mesh and trammel nets). Therefore, the ZEPA is committed to the development of fishing activities aimed at preventing over-use of fishery resources, the deterioration of the marine ecosystem, and therefore, promoting the generation of economic development opportunities for local artisanal fishers.

The community development initiatives that led to the creation of the ZEPA also achieved the creation of a Special Zone for Fishery Management (ZEMP, Zona Especial de Manejo Pesquero), which covers a buffer zone of more than 310,000 hectares (where the restrictions and prohibitions apply), from the 2.5 nautical miles included in the ZEPA, up to 12 nautical miles. On this matter, another of the leaders involved in these community processes comments,

> The consolidation of the ZEPA and the ZEMP is the product of a grassroots effort, through which we claim our rights as owners and responsible for the territory. It was a tough process, many years, and many meetings, but we achieved our goals.
>
> *(Interviewee 2, Consejo Comunitario Los Delfines)*

The ZEPA's expee has been successful because of its structure of governance. The center of this structure is the Inter-institutional and Community Artisanal Fishing Group (GICPA, Grupo Interinstitucional y Comunitario de Pesca Artesanal), composed by governmental agencies and community entities related to the management and use of fishery resources.

Consequently, in December 2014, authorities from the Colombian government, with the support of local and community groups, created the protected area "Regional District of Integrated Management (DRMI, Distrito Regional de Manejo Integrado) Gulf of Tribugá – Cabo Corrientes", with an extension of 60,183 hectares in the municipality of Nuquí. Like the ZEPA and the ZEMP, this protected area contributes to the management and protection of fishery resources and the ecosystem that maintains them.

Together, the ZEPA, the ZEMP, and the DRMI make up a group of conservation and biodiversity management figures on the north coast of Chocó, which allow the development of productive activities in a sustainable manner, through co-management strategies. Referring to this topic, one of the community leaders of the process says,

> The process in Nuquí had its roots in the understanding that this type of process can be achieved locally through organizational processes, with the contribution of fishers, women, children, local organizations. We analyzed which was the most appropriate conservation figure for the municipality of Nuquí, ZEPA or DRMI, and we decided that the DRMI was the right one for us.
>
> *(Interviewee 3, Nuquí)*

On the other hand, the Orinoquia region corresponds, largely, to floodplains. It is located in eastern Colombia, extending along the Orinoco River, bordering Brazil to the east, the Amazon region to the south, the Andean region to the west, and Venezuela to the north. It has an area of about 253,000 square kilometers. Four physiographic units compose it: the Piedmont, alluvial plains, wind plains, and the Altillanura. Twenty-two ecosystems are differentiated; ten have different types of forests (piedmont); three are pure or mixed palm tree zones (piedmont, alluvial plain); one corresponds to a low scrub forest (Altillanura) and eight are based on pasture (Altillanura – alluvial plain – wind plain) (Rangel et al., 1995).[1] The departments that are part of this region are Arauca, Meta, Casanare, and Vichada. The region is also known as Llanos Orientales, due to the large plains that characterize the area. The gallery jungles, or riverbank jungle, are the main vegetation of the region.

The agroindustry (rice and oil palm) and the extensive cattle ranching (milk and meat production) are the basis of the economy, thanks to the vast plains that cover the region. The exploitation of oil and natural gas also represent an important economic factor, since it is in this territory where the country's largest oil reserve is located (mainly in Arauca, Meta and Casanare).

Due to its location, the culture of the region differs from the rest of Colombia. The local customs are based on the *llanera* culture; on the ways and traditions characteristic of the Llanos Orientales, which is linked to the life around cattle ranching activities, also called traditional stockbreeding. The main characteristic of the traditional stockbreeding (and the main difference with extensive cattle ranching) is the use of local livestock breeds that are typical of the region, which have adapted to the flooding conditions of the Orinoquia, and as well have adapted to feed with natural pastures. Consequently, these activities generate a high dependence on the natural conditions of the savannah, allowing the conservation not only of the environment but also of the *llanera* culture.

However, the inhabitants of the Colombian Orinoquia have seen their culture and livelihoods affected by different economic activities that have ventured into their territories, such as the oil extraction industry and the rice agroindustry. The department of Casanare is one of the most affected, since it is the place of the greatest concentration of the rice agroindustrial activities. According to the National Administrative Department of Statistics (DANE,

Departamento Administrativo Nacional de Estadística), 188,663 hectares had been planted with rice in Colombia during the first semester of 2018, with a production up to that date of 1,020,738 tons. Likewise, according to the mechanized rice census carried out by the DANE, until July 2018, the department of Casanare represented the 69% of the whole area planted with rice in Colombia, producing, in turn, around 600,000 tons. Therefore, Casanare is currently in the midst of a "rice fever", partly due to the good price of the grain.

Likewise, the massive arrival of rice producers has generated some alerts among the environmental authorities, but, above all, among the peasants who inhabit the region. The peasants consider that these large-scale agricultural practices put the biodiversity of their territory at risk. In addition to the peasants' considerations, the vertiginous increase of the crop (rice) is generating criticism due to the effects it causes to the environment. On this matter, the secretary of environment of the municipality of Trinidad (2017), Carlos Betancur-Amaya, argues, "There is a bombardment to the environment taking place in the region", due to the large-scale use of agrochemicals. Likewise, environmentalists warn that the abuse of chemicals could turn the Llanos Orientales into a vast desert. In the same way, the peasants denounce the impacts to their crops of the aerial fumigations carried out by the rice producers. Interviewee 4, a peasant from the municipality of Paz de Ariporo, argues that livestock also suffers because, "downstream the water comes dirty due to the rice extraction and the cattle go and drink that. . . . I am the one being hurt". Likewise, Interviewee 5, an inhabitant of the municipality of Trinidad, states that they have lived with chemical waste, and, therefore, "We feel this is affecting us because the fumigation affects the livestock, and also their waste [from the rice producers] gets into the stream".

The estimates of the secretary of environment of the municipality of Trinidad are discouraging. This official argues that "it is a soil that can be said that Casanare is going to be the biggest desert in Colombia". The secretary based the estimates on the results of a study by the National University of Colombia led by Professors David Alejandro Muñoz and Adriana Martínez,[2] who found that the large-scale rice production in the area brings major environmental challenges. According to the study, on the one hand, wet rice production implies affectations in the composition of the soil, since having it constantly subjected to flooding, many particles are left in suspension and when the lots are drained to continue the production process, these particles are washed away. On the other hand, the furrows in the farming process of the dry rice obstructs the natural water circulation. Likewise, according to Professor Muñoz, the intensive use of machinery can produce the compaction and sealing of the soil, which in turn would require considerable investments and time to recover.

This panorama, where the peasants not only see the physical landscape of the region threatened, but also their traditional subsistence activities affected, has triggered some local initiatives for the development of community. Therefore, motivated by the protection of their livelihoods associated with traditional stockbreeding, as well by their high dependence on the natural conditions of the savanna, the local community have articulated alternatives that allow their economic sustenance to be in harmony with the protection of the environment. In this context, the initiative to declare a Civil Society Natural Reserve emerged.

A Civil Society Natural Reserve (RNSC, Reserva Natural de la Sociedad Civil) is an estate that, by decision and initiative of its owners, becomes a private protected area under parameters of conservation, restoration, and sustainable production. This private form of conservation was accepted and recognized by the Colombian State, so that a person or group of people interested in establishing an RNSC can initiate the process to register an estate in the National Registry of Protected Areas (RUNAP, Registro Único Nacional de Áreas Protegidas) to be legally recognized. This registry makes the RNSC part of the National System of Protected Areas (SINAP,

Sistema Nacional de Áreas Protegidas). This implies that the owners of the estates that are part of an RNSC are obliged to comply with the regulations regarding the protection, environmental conservation, and management of natural resources, as well as to adopt preventive measures and/or prohibit activities and uses that generate environmental risks or potential of negative impacts to the natural ecosystem.

On March 8, 2013, several RNSC were declared in the village of Altagracia in the municipality of Trinidad, after a community process of eight years. This process involved the peasants who own the land, different levels of the government (Department and Town), and the Regional Office of National Natural Parks of Colombia. As a result, there is now a conservation node of 10,972 hectares, made up of several private estates, whose owners joined forces to preserve their livelihoods while preserving the environment. This initiative of conservation has managed to contain the "advance of rice", as well as hunting activities that in the past were practiced in the area. Therefore, the RNSC of the Altagracia village have not only had a positive effect on the conservation of wildlife, it has also provided the possibility of preserving the traditional modes and practices of subsistence in the region, such as traditional stockbreeding.

This experience has given way to the fact that nowadays more farmers and landowners have decided to join the initiative and initiate other processes for the conformation and declaration of other RNSC, mainly in the municipalities of Trinidad, San Luis Palenque, and Paz de Ariporo in the Department of Casanare. With this, the inhabitants of the area mainly seek to mitigate the negative effects of the rice industry's land hoarding and the oil industry. Their objective is to preserve the floodplains the gallery forests, and the water resource that allows them to develop their traditional stockbreeding and subsistence agriculture activities, while conserving valuable ecosystems and ensuring sustainable management of existing resources. In this regard, a property owner in the municipality of San Luis Palenque mentions,

> We have a farm of around 4,000 hectares, of which we have 800 hectares in conservation. The idea is to declare a natural reserve in order to protect it. Keeping that land protected is a benefit, and we can continue to raise cattle for meat without that [declaring the natural reserve] to be a detriment to our activities.
>
> *(Interviewee 6)*

Another benefit that the local community has seen from the declaration of RNSC in Casanare has been tourism. In this sense, many peasants and landowners have been able to develop ecotourism activities in their natural reserves (RNSC), thus, representing an additional economic income for them. Therefore, in Casanare, the RNSC are not only an instrument of environmental conservation, but they have also become instruments of local development and preservation of local livelihoods.

Final remarks

The local development processes embodied by the Exclusive Zone of Artisanal Fishing (ZEPA) in the northern Pacific coast of Chocó, and the conservation of the floodplains of the Orinoquia in Casanare through the traditional stockbreeding, are generating sustainable practices for the communities. These strategies had captured the government's attention at different levels, therefore achieving few changes in the traditional state-level policies, encouraged by a development plan and agenda, which take little account of the local contexts, as well as the livelihoods of communities and the conservation of natural resources. In this sense, at the regional and local level, the governments had supported these communities' initiatives (recognizing the social movement generated at the local level). While at the national level, the government has demonstrated its support by issuing the regulations that govern the exclusive zones for artisanal fishing, and by creating the integrated management district. Two legal features that promote the conservation of fishing resources without restricting subsistence activities (artisanal fishing).

In the case of the ZEPA, its creation process positively influenced the environmental dimension as well as the social one. One of the main achievements of the initiative was the recovery of fish stocks and other marine species. This in turn had an influence in the recovery and revitalization of the local traditional fishing activities, which guarantees the sustenance of many families of the Colombian Pacific region. Likewise, the large-scale fishing industry has benefited, as the increase in the population of different marine species (especially shrimp and tuna) strengthens the profitability of their activities. Nevertheless, even more importantly, the declaration of the ZEPA, the ZEMP, and the DRMI have strengthened the community organization, since the conformation and continuity of these initiatives require collaborative work between local communities, and between them and the State.

On the other hand, the Civil Society Natural Reserves have proven to be an effective strategy to reconcile the *llanera* culture and livelihoods with the efforts for the conservation of nature. Each natural reserve entails an articulation of work between owners, community, private sectors, and the government to ensure the best possible results from the conservation-production processes. In turn, these articulation processes help to build more resilient production systems. Some of these articulation initiatives have managed to influence the decision-making processes and the structuring and adjustment of territorial planning instruments and development plans, taking into account the environmental dimension and sustainable production ways from a community perspective.

The RNSC of the Department of Casanare are the only conservation strategy that has achieved the status of protected area. This form of private conservation is the only instrument that is protecting the ecosystems of the Orinoquia in the Department of Casanare and is the strategy that contributes the most, in terms of the extension of land conserved. These RNSC have also managed to position themselves as a conservation strategy that not only is preserving biodiversity, but is also protecting the *llanera* livelihoods. Additionally, this conservation effort has generated a local development process that produces economic benefits for the community. These benefits are the result of not only the agriculture and livestock production but also due to the promotion of ecotourism. This emerging ecotourism has generated an economic potential based on, for one part, the preservation of cultural practices of the *llanera* culture, and, for the other part, the incredible biodiversity that characterizes the region.

The experiences of both, the Pacific region and the Orinoquia, demonstrate the power of the communities' agency when confronted with large-scale development plans and initiatives that take little account of the local contexts. These communities have proven their capacity to formulate alternatives for the economic, social, and political participation that affect the common local well-being, and that in turn embody forms for the sustainable use and management of natural resources. Therefore, each of these strategies (ZEPA, ZEMP, DMRI, and RNSC) represent alternatives of community development, where the local and community-specific factors are central. Consequently, the participatory process that characterizes community development is transformed into concrete and effective ways to address and solve a variety of socio-economic, cultural and environmental problems, thus improving the living conditions of people who inhabit a place, through the empowerment of these same people.

Notes

1 Rangel-Ch, J. O., Sánchez-C, H., Lowy-C, P., Aguilar-P, M. and Castillo, A. M. P. A. R. O. (1995). Región de la Orinoquia. Colombia, Diversidad Biótica I. Instituto de Ciencias Naturales. Convenio Inderena-Universidad Nacional de Colombia, 239–254.

2 Muñoz, D. A. and Martínez-Arias, A. (2015). "Evaluación de la calidad de los suelos en áreas cultivadas con arroz riego y secano en los llanos de Casanare". *Guarracuco* 19(30): 33–42.

References

Ander-Egg, Ezequiel. 1987. *La Práctica de la Animación Sociocultural.* Buenos Aires: ICSA/HVMANITAS.

Coffey, William, and Mario Polèse. 1984. "The concept of local development: A stages model of endogenous regional growth". *Papers of the Regional Science Association* 55(1): 1–12.

Marchioni Marco. 1989. *Planificación Social y Organización de la Comunidad.* Madrid: Editorial Popular.

Muñoz, José. 2012. "Intervención comunitaria: concepto. El desarrollo comunitario". *Contribuciones a las Ciencias Sociales.* www.eumed.net/rev/cccss/22/

Núñez, Jorge, Galia Figueroa, Ariamnis Alcázar, and Isvieysys Armas. 2016. "Higher education, technological change, and local development: Experiences and challenges in Cuban context". In Latif Al-Hakim, Xiaobo Wu, and Andy Koronios Yongyi Shou (eds.), *Handbook of Research on Driving Competitive Advantage through Sustainable, Lean, and Disruptive Innovation*, pp. 376–394. Hershey, PA, USA: IGI Global.

Porzecanski, Teresa. 1983. *Desarrollo de comunidad y subculturas.* Buenos Aires: Editorial Humanitas.

Rangel-Ch, J. O., Sánchez-C, H., Lowy-C, P., Aguilar-P, M., and Castillo, A. M. P. A. R. O. (1995). *Región de la Orinoquia. Colombia, Diversidad Biótica I*, pp. 239–254. Bogotá: Instituto de Ciencias Naturales. Convenio Inderena-Universidad Nacional de Colombia.

Trigueros, Isabel. 1991. *Manual de prácticas de trabajo social comunitario en el movimiento ciudadano.* Madrid: Siglo XXI de España.

Ware, Caroline. 1986. *Estudios de la Comunidad.* Buenos Aires: Editorial Humanitas.

21

CIRCUMSCRIBING LOCAL DEVELOPMENT

The role of community-based conservation in Tanzania

Christine Noe and Asubisye Mwamfupe

Introduction

> These conserved territories ought to be taken into consideration as a system, as they are mutually supportive and beneficial. For instance, village forest reserves and community conserved wildlife habitats interact with state protected areas to provide areas for ecological connectivity and wildlife dispersal. While also meeting gaps in protection, these enhanced areas for conservation also provide opportunities for the development of tourism and other livelihoods.
>
> *(IUCN, 2017: 3)*

This passage is part of the report of the IUCN-facilitated workshop for the assessment of actions and governance for conservation and sustainable livelihoods in Tanzania (March 21–22, 2017). It speaks widely to the discussion of this chapter, how conservation actors associate their activities with local development and the importance that vast protected landscapes that are surrounded by community-conserved areas are considered mutually beneficial. It points to the popular narrative that conservation and development (especially through tourism) are highly interdependent.

Tanzania is considered an exemplary performer in biodiversity protection with the largest number of national parks in the world (IUCN, 2017). A total of 19 national parks, 25 game reserves, 15 marine reserves, multiple forest reserves and woodlands as well as three marine parks pull together about 40% of the country's landmass (UNEP-WCMC, 2019; World Bank, 2019). Critics suggest, however, that global sources of data are the most authoritative yet incomplete and inaccurate, since going by calculations, terrestrial area alone under conservation is around 42% and a corroborated estimate of current and planned spatial extent of protected land is 50% of the landmass (Bluwstein et al., 2018). In any case, the proportion is far beyond the Aichi Biodiversity Target 11 of the Convention on Biological Diversity which hopes to see just 17% of terrestrial areas protected by 2020. The country's development depends largely on the natural resources, the most important of which is the agricultural land for that supports 85% of the population through small holder farming. Collectively, however, agriculture, wildlife, forestry, and fisheries contribute only about 30% of gross domestic product (GDP) (National Bureau of

 DOI: 10.4324/9780429450310-25

Statistics [NBS], 2017) with the travel and tourism sector – largely based on biodiversity and wildlife – contributing about 13% (World Travel & Tourism Council (WTTC), 2017).

In the National Development Vision 2025 and also the universal Sustainable Development Goals (SDGs), Tanzania is committed to ending poverty and hunger; achieving good health, well-being, quality education; and, above all, graduating from a least developed to a middle-income country with a high level of human development by 2030. At the moment, the country remains within the category of those with low levels of human development, ranked 151 out of 188 countries globally (URT, 2017). Official statistics indicate that the real GDP growth was 7.0% in 2018, slightly higher than 6.8% in 2017. However, about 26.4% of people live below the national basic needs poverty line (URT, 2019). Both incidences of poverty (share of the population whose income or consumption is below the poverty line) and extreme poverty are higher especially in rural compared to urban areas at the rate of 31.3% and 15.8% respectively (NBS, 2019). Indeed, the recent World Bank's country environmental analysis provides substantial evidence that Tanzania is one of only ten countries in Africa – and of only a few more countries in the rest of the world – where total wealth per capita is declining, mostly due to the increasing dependence on natural resources (World Bank, 2019).

Efforts to balance nature protection and development is evidently contradictory at both policy and practical levels. On the one hand, the country's policies draw heavily from a broader conservation agenda that calls for the integration of local development needs into the planning and implementation of biodiversity protection initiatives. Among other things this call support arguments for community-based conservation through which wildlife corridors and buffer zones have been established in communal lands. Literature supports how communal conservancies, wildlife management areas, and village forest reserves have recently contributed towards protected area expansions and creation of their networks (USAID, 2017; Bluwstein et al., 2018; Ramutsindela and Noe, 2012). In a way, conservation expansion converges with other national policies at poverty alleviation – since more protected areas are associated with tourism income.

On the other hand, protected area expansions go simultaneously with agricultural transformations that aim at improving the sector's contribution to the economy. Tanzania signed the Comprehensive African Agriculture Development Program (CAADP) in 2010, thus agreeing to develop a national framework of implementation of a regional objective (of 6% annual growth in the agricultural sector) in order to reduce poverty and achieve food security (URT, 2011). At the 2011 World Economic Forum in Davos, the then President Jakaya Kikwete promised to provide land enough to invest in modern agriculture.

At the Rio+20, the country's Minister of Environment announced an area amounting to 350,000 hectares of land to develop the Southern Agricultural Growth Corridor of Tanzania (SAGCOT). Backed by World Wide Fund for Nature (WWF) and the Africa Development Bank's side event at the Rio+ conference, SAGCOT was declared a new "laboratory" for testing and implementing the concept of green growth in response to Davos 2011 resolutions to establish agricultural corridors (Bergius and Buseth, 2019). While the SACGOT area covers one-third of mainland Tanzania and it is designed to have aggregate impacts on various strategic areas of agricultural transformations through public-private-partnerships (URT, 2016), it is also a rich biodiversity area with over ten forest reserves, four protected wetlands, six national parks, nine game reserves, and 18 community-based wildlife management areas (WMAs). As such, SAGCOT overlaps with the Greater Selous ecosystem. Nowhere else have agriculture and conservation been promoted than in the SACGOT region.

This chapter seeks to contribute to the understanding of how conservation projects have defined and pursued their development interventions in the country. Using community-based initiatives for wildlife and forest protection, the chapter demonstrates how communal lands that are currently designated as Wildlife Management Areas (WMAs) and Community-Based Forest Management (CBFM) have increased livelihood insecurities that deter local development. The main argument is that the advancement of different conservation regimes continue to divert large parts of Tanzania's land from local control and production. However, local development is still attached strongly to agriculture but constrained by the protected area expansions that foreclose future use of land for non-conservation activities. We bring some insights from the case of wildlife and forest protected areas that involve village lands.

The chapter is organized in six sections. After this introduction, the second section provides the context and methods used to collect and analyze the data. The third section revisits the literature with particular focus on conservation partnerships in Tanzania. The next section will bring insights from two selected sites where different global and local partners have converged to support community participation in conservation while developing mechanisms for local benefits from these resources. The sixth section discusses the data by reflecting on the politics of conservation business, which is the key link to most development claims.

Context and methods

Regardless of specific resource contexts in which community-based conservation has been applied in Tanzania, village lands have successfully established wildlife management areas (WMAs), Community-based Forest Management areas (CBFM), Beach Management Units (BMUs), and similar conservation classifications. WWF reported in 2014 that over 3% of total village lands that lie outside protected areas have been secured for wildlife protection (WWF, 2014). These lands are where most rural Tanzanians depend on for their everyday lives – settlements, food, cash needs, and so on. In whose perspective is this change developmental? We draw from the ongoing collaborative research project on New Partnerships for Sustainability (NEPSUS), which investigates whether conservation partnerships, and putatively more participatory processes in the governance of wildlife, forest and coastal resources, result in more equitable and sustainable livelihoods and environmental outcomes. The project compares "more complex" partnerships to relatively "simpler", more traditional top-down and centralized management systems and to instances where these partnerships have not been established.

For the purpose of this chapter, however, we focus on the wildlife and forest sites that are identified with complex partnerships (in terms of number and type of actors, roles and interests as well as institutional set-up). This way, we narrow down to WMAs and CBFM. Although empirical details of these areas can be limited by the available space for the chapter, more details can be obtained from project publications (Noe et al., 2017; Ponte et al., 2017; Kalumanga et al., 2018).

Tanzania's conservation and development partnership trajectory

Like in other parts of the global south, Tanzania's conservation and development trajectory can be divided into two main regimes. The first is the colonial and postcolonial regime, which is deeply intertwined with efforts to set aside protected areas. During this regime, the popular narrative referred to the lack of consideration for the interaction in natural ecosystems, the role of larger wild mammals in the ecology, the economy of African grasslands as well as the increasing pressure on habitats (owing to rapid human population that depended on these resources) (Curry-Lindahl, 1974). The pressure on habitats was predicted to intensify and threaten to decimate the world heritage that seats on the African continent, especially following the independence of most of the countries. Protected areas were consolidated and shielded by the commitments made by the independent states to embrace partnerships with international actors. This history, as already discussed in the literature (see for example Noe et al., 2017; Neumann, 2001; Ramutsindela, 2004), was responsible for the consolidation and expansion of protected areas across the continent.

The second regime is the neoliberal era beginning from the 1990s that focused on reducing state control while complementing protected fortresses with community-based conservation (Bluwstein et al., 2018). Threats to forest and wildlife have arguably increased over time, owing to the growing human population which is dependent on natural resources. By the 1990s and 2000s, governance reforms in the natural resource sectors had prepared institutional and legal grounds for the formalization of conservation partnerships in view of strengthening biodiversity protection (beyond protected areas), while also seeking the support of the surrounding local communities from which threats originate. Government agencies, non-governmental organizations (NGOs), the private sector, development partners, and local communities were all to join hands in maximizing biodiversity's value for humanity.

Our discussion focuses more on the later regime which is closely related to the rhetoric of a "win-win" – which represents simultaneous achievement of economic growth, environmental protection, and social development – and which is also central to the emergence of community-based conservation initiatives in the country and elsewhere in the continent. As Crawford (2003) argues, the rhetorical and instrumental use of "partnership" by international actors has mainly supported governance reforms to disguise and legitimize the interventions of external agencies in domestic matters which mystify power asymmetry.

Contrary to the official discourse of encouraging locally formulated reform strategies, partnerships reflect conspicuously the power of neoliberalism, its reshaping of landscapes and profound alterations of the lives of rural people in areas targeted for conservation. The agenda of international agencies remains relatively unchallenged, both in terms of what is included and what is excluded (Crawford, 2003). It has become increasingly clear that conservation is an organized political project in which private sector businesses and their networks are dominant, hence transforming conservation in ways that accommodate the interests of global capital (MacDonald, 2010). At the same time, despite the cloying and deceitful rhetoric and the adverse outcomes of new partnerships for some of the local partners, they can still provide institutions and resources for otherwise marginalized rural groups to challenge powerful interests (Wright, 2017). Accordingly, Ponte et al. (2017) argues that the current knowledge base lacks clear criteria for evaluating the impacts of conservation partnerships.

Tanzania's low economic profile on the one hand and its accomplishments in conservation on the other are both the cause and impacts of partnerships that have recently taken different interpretations as *territorialization* (Bluwstein and Lund, 2016), *green squeezing* (Bersaglio and Cleaver, 2018), and *green modernization* (Bergius and Buseth, 2019). The commonality of these rather authoritative and sustained projects is that they entail irreversible changes in local resource use and tenure systems in ways that powerful actors adopt and promote conservation around long-standing narratives of development. These projects have supported expansion of protected areas and their networks while increasingly rendering local livelihoods precarious. Without missing the connection, the fact that most of expansions occur in rural areas is crucial. The rural hosts most protected areas, are a home for the majority of poor population of Tanzanians whose lives are dependent on land and other natural resources. Notably, narratives about threats to wildlife and forest resources have referred to these populations in ways that make their poverty both a cause and opportunity for nature's (in)securing. This becomes a logical entry point into the promotion of conservation partnerships that have recently transformed rural landscapes.

Partnership framework for landscape transformations is dominated by prominent conservation and development partners whose programs are aligned with global agenda for conservation and poverty alleviation, suggesting, for example, that "biodiversity is not just a luxury for the rich, it is a necessity for the poor" (USAID, 2017). For example, IUCN's Global Business and Biodiversity Programme supports the SUSTAIN-Africa initiative to implement the Sustainable Development Goals in growth corridors of Africa with aims of facilitating green growth that is inclusive and climate-resilient. Working in partnership with government and civil society in Tanzania since 2015, SUSTAIN-Africa uses the landscape approach to demonstrate Inclusive Green Growth (IGG) by integrating land and ecosystem management with sustainable business. The primary focus has been the Southern Agriculture Growth Corridor of Tanzania (SAGCOT) where the initiative fosters partnerships between national conservation agencies, local authorities, villagers, business, and agricultural investors whose activities have impacts on biodiversity. Over 700 villages have been targeted with prospects for turning small holders into contract farmers and out growers.

While SAGCOT has pulled together support of both local and international financial institutions and private sector partners, forest and wildlife protection are central to the objectives of the SUSTAIN initiative, especially through the creation and maintenance of ecological corridors within the larger Selous ecosystem. Overall, the targeted ecosystem extends to 90,000 km^2 to include the surrounding national parks of Udzungwa, Ruaha, and Mikumi, several forest reserves, open areas, and WMAs. Connectivity is crucial for the maintenance of this large ecosystem and its link with the Ruvuma landscape that includes 42,000 km^2 of the Niassa game reserve in Mozambique. As such, the African Wildlife Foundation (AWF) partners with SUSTAIN-Africa and SNV Netherlands Development Organisation to facilitate "greening of growth" through smart agriculture. By smart agriculture, AWF implies to influence planning for small- and large-scale farming through agricultural conservation methods across the value chain (www.sustaininitiative.org) in order to secure space for forest and wildlife.

Similarly, USAID's Sustainable Conservation Approaches in Priority Ecosystems (SCAPES) project operates in the country at the landscape scale since 2009. SCAPES had three priorities: promote community-based natural resource management; unite Mozambique and Tanzania on transboundary conservation issues; and help communities and governments adapt to climate change (USAID, 2014). This work was channeled into four wildlife corridors deemed vital to the long-term health of the Ruvuma landscape, covering most parts of southern Tanzania and northern Mozambique. The program activities were implemented by three partners: AWF, Wildlife Conservation Society (WCS), and World Wildlife Fund (WWF) but with their different components. In particular, WWF has been training farmers on conservation agriculture. Block farming, in which villagers farm in a single area rather than scatter their fields throughout the forest, is more consistent with conservation agriculture and allows them to better defend their crops against wildlife (USAID, 2015).

In the meantime, WWF has been involved in establishing forest protection partnerships in Southern Tanzania to address alarming deforestation of forests generally and miombo woodlands in particular (Milledge et al., 2007). This and other international NGOs establish partnerships with locally registered organizations and the private sector to support specific community-based initiatives. Since 2004, the Mpingo Conservation and Development Initiative (MCDI) was locally registered based in Kilwa district. Starting from the same year, the organization received funds from WWF to support participatory forest management and later to be a champion for demonstrating financial benefits of CBFM. This was to be through selling timber from Village Forest Reserves (VLFRs). By 2009, MCDI started receiving support from other partners, including the Norwegian government, to design a REDD+ project that could induce carbon forestry on top of CBFM models (Khatun et al., 2017). What is common in the activities of these conservation and development partners is their focus on accessing and influencing the use of village land in ways that support securing land for their large-scale projects.

Insights from local conservation partnerships in Rufiji and Kilwa districts

The foregoing top-down-oriented projects and funding have depended on partnerships to translate global initiatives into local practices. While invoking notions of development, forest and wildlife protection are central to the formation of these partnerships at the local level. In this section, we bring insights from two cases of conservation partnerships (for wildlife [WMAs] and forest [CBFM]), both of which are situated within the larger Selous ecosystem and Ruvuma landscape. The aim is to demonstrate how these partners have defined and pursued development from their perspectives while giving limited attention to the local needs. We present this in a comparative analysis focusing on *how land is secured* and *the economic benefit streams* in WMAs and CBFM.

The case of Wildlife Management Areas (WMAs) in Rufiji district

Our research categorizes WMAs as a complex form of partnership. The degree of complexity is determined and measured by three aspects: number of actors involved, their composition, and institutional set-up for wildlife-related partnership. A detailed discussion of this complexity can be found in a separate paper. It suffices to point out that WMAs are themselves a form of partnership between villages, protected area management, and private sector actors who are involved in conservation business. The process of setting aside land for WMAs involve two or several villages that agree to let their land be used for conservation. This is made possible through a land use planning process which is largely facilitated by international NGOs concerned with conservation and through the government agencies for wildlife protection (Tanzania National Parks, Tanzania Wildlife Authority and Wildlife Division). Also, the local authorities (District and Village Councils) are partners by virtue of their role as representatives of the central government. While the formation of Community-Based Organizations (CBO) is central to the WMA process, Civil Society Organizations and the private sector actors (in hunting and photographic business) are important partners.

As Rufiji District hosts an important part of the Selous wildlife ecosystem that connects to Mikumi, Udzungwa, and Ruaha national parks, the activities of several international conservation NGOs relate to the establishment of wildlife corridors and buffer zones around these protected areas. It has been argued elsewhere that WMAs are the main way in which these corridors and buffer zones are established, which makes them an important part of the conservation mission in the country (Ramutsindela and Noe, 2012; Noe et al., 2019). On their side, conservation proponents have hailed WMAs as both a conservation tool and a legal mechanism for communities to benefit from their involvement in wildlife protection (USAID, 2014; WWF, 2014). Local communities have therefore found themselves as partners by the sheer presence of protected areas in their ancestral homes and the right to land that is considered valuable for wildlife.

We studied four villages which are part of two different WMAs. These are, namely, Ngarambe, Tapika, Ngorongo, and Mloka. All of these are adjacent to the Selous game reserve with formal partnership configuration built on WMAs. Precisely, the two villages of Ngarambe and Tapika formed among the earliest WMAs in the country by 2007, namely, "*Muungano wa*

Ngarambe na Tapika (MUNGATA)". This is literally the partnership between Ngarambe and Tapika villages. The other two are part of 13 villages that formed "*Jumuiya ya Hifadhi ya Wanyamapori Ngorongo, Utete na Mwaseni* (JUHIWANGUMWA)" recently in 2016. This is literally the Wildlife Conservation Community of Ngorongo, Utete, and Mwaseni.

MUNGATA WMA is a product of the earliest community-based conservation activities supported first by the German Development Agency (GTZ) (now German International Cooperation [GIZ]) under the Selous Conservation Program beginning from 1988, and USAID and WWF in later years. The process for JUHIWANGUMWA was first initiated by the Belgium Technical Cooperation through its project for Eastern Selous Conservation. While these earlier projects established buffer zones around the reserve under the umbrella of community-based conservation, the later projects formalized the conservation status of these areas as WMAs, which share the border with themselves and the reserve. To complete the circle, landscape conservation projects discussed in the previous section connect with Belgium Technical Cooperation's support for JUHIWANGUMWA WMA. This has been through Kilombero and Lower Rufiji Wetlands Ecosystem Management Project (KILORWEMP). The project seeks to create ecological connectivity between the Selous game reserve and Udzungwa, Ruaha, and Mikumi national parks.

While KILORWEMP works up north from the Selous reserve, other conservation partners work down south from Udzungwa and Mikumi national parks. These include AWF through IUCN's SUSTAIN-Africa program and in partnership with the Southern Tanzania Elephant Program (STEP), Reforest Africa, and others to create the Magombera-Mwanihana wildlife corridor (STEP, 2018). Hence, despite their difference in approach, conservation partners have focused on the same thing; securing village land for WMAs. Across the country, two or more WMAs merge to form wildlife corridors that are formally secured through the new law specific for wildlife corridors, dispersal areas, buffer zones, and migratory routes (URT, 2018).

The benefit to communities are projected in the form of revenues, employment, sales of goods, and other activities connected with tourism or trophy hunting. To generate benefits, WMAs that have registered Authorized Associations (AAs) (the governing body for WMAs) enter into joint ventures or concession agreements with photographic tourism or trophy hunting companies. Tour operators usually make an agreement with a CBO that has user rights (through their Authorized Association) to use a portion of village land or a WMA to invest in physical properties that are used for tourism promotion (usually in return for a fee per tourist bed night (in the case of photographic tourism) or hunting fee (in the case of hunting tourism).

Nevertheless, as the government of Tanzania re-regulated activities in WMAs through its revised Wildlife Act of 2009, the government is still in control of most of the revenue streams. Various fees that are charged for photographic tourism activities to generate revenue for WMAs or for the villages that allow private businesses are determined by the government through a revenue sharing formula (see Table 21.1 and 21.2). It is important to note that the use of revenue proportion for AAs or villages (65%) is still subject to government control. AAs are required to allocate at least 15% of their gross revenue for resource protection (including for patrols and paying salaries for village game scouts), 50% for disbursement to WMA member villages, and at least 25% for AA management costs. The AAs can use the remaining 10% as they deem fit. As villages have choices to use their funds, most of them invest in community development projects as opposed to direct distribution to local residents.

We find that the two WMAs of MUNGATA and JUHIWANGUMWA were formalized almost ten years apart. However, they have different contexts that have constrained their operations to the extent that there is no significant difference between them in terms of benefits accrued so far. In the case of MUNGATA, a hunting tour operator Game Frontiers of Tanzania

Table 21.1 Revenue sharing formula for tourist hunting activities in WMAs

s/n	*Fee types*	*Tanzania Wildlife Protection Fund*	*WMA*	*District Council*	*Treasury*
1	Block fee	25%	75%	0	0
2	Game fee	25%	45%	15%	15%
3	Conservation fee	25%	45%	0	30%
4	Observers fee	25%	45%	0	30%
5	Entry permit fee	25%	15%	0	60%

Source: URT (2012) WMAs Regulations; pp. 65

Table 21.2 WMA/village revenue sharing for photographic tourism activities

S/N	*Institution*	*% Share of total revenue*
1	Village or AAs	65
2	District Council	15
3	Wildlife Division	20

Source: United Republic of Tanzania (2014)

(GFT) operated when the government had still full control of hunting blocks. The investor contributed to local development (not legally obligatory) through support for education (school fees), hunting for residents' bush meat, building of mosques, and some food relief during food shortage (which is not unusual). This way, GFT became the favorite, not because of real contributions to development but the level of poverty and basic needs that the villagers experienced at that time. In expectations to renew the contract, GFT influenced significant border changes during the review of the WMA management plan which was due in 2012. Through its local influence and connection to the corrupt officials at the Wildlife Division, the WMA land use was altered in ways that forest harvesting and resident hunting portions of land that had been set aside for use by villagers were both merged with the hunting block. The new MUNGATA plan (2012–2022) lacks these important community areas and with little local understanding of these critical changes. As a result, almost 97% of the two villages of Ngarambe and Tapika are a hunting block for a single private investor (Noe et al., 2019).

The end of the GFT contract with the government in 2013 coincided with legal changes that required the hunting investor to have only one hunting block in WMAs. This disqualified GFT, as the company had other more attractive hunting blocks. The new investor, HK Safaris, entered into contract in 2014. However, the company took three years to set up the business (no fees were paid for these years). The AA filed a court case which made it even easier for the investor (who apparently happens to be a lawyer) to continue his occupancy of the block. Only in 2018 was the block fee USD 55,000 finally paid to the AA and subjected to the formula in Table 21.1.

In the case of JUHIWANGUMWA, the formalization of its process that started way back in 1996 took a long time due to the lack of funds. In the meantime, the targeted land has been and still is a refuge for most animals that escape hunting activities in the reserve and the bordering block (in MUNGATA WMA). After the WMA formalization in 2016, the tender process for the hunting block was blocked by the government claiming that it required to grade the hunting block. This is even though these areas had in the past been the government's hunting blocks with grades already. Until the end of fieldwork for the NEPSUS project and, on claims that the government lacked funds, the hunting block had not been graded. To this end, JUHIWAN-GUMWA has never received any revenues related to its establishment as a WMA.

The case of community-based forest management in Kilwa district

The importance of the forest sector is increasingly echoed in policy and research arenas in relation to the sector's socio-economic and ecological potentials (Trupin et al., 2018; Kalumanga et al., 2018; Horning, 2018). Sustainable management of forest resources is called for in order to deliver improved livelihood outcomes especially for rural communities and the country's GDP. While deforestation is directly associated with poverty of rural communities, its complexity as an environmental threat is also connected to the loss of habitat and climate change (Horning, 2018). Deforestation is therefore a sustained narrative locally (as in other parts of Africa) and it relates directly to local livelihoods through low productivity of agricultural and other forms of rural-based productions.

In a recent environment assessment report by the World Bank, Tanzania ranks fifth for highest annual forest net loss worldwide (World Bank, 2019). It is cautioned that charcoal consumption levels are expected to increase further due to population growth coupled with urbanization and relative price increases of other fossil fuel-based energy sources. The narrative of forest loss drives policies on forest protection and the need for increased participation of local communities. In the process, however, interests of the governments and other actors have been aligned with recent forest governance changes from state-centric to forming partnerships with local communities and non-state actors.

Participatory Forest Management in Kilwa district started in 2001 with the DANIDA funding for the project known as Utunzaji Misitu (UTUMI), or literally forest protection. The Forest Policy (1998) and Forest Act (2002) had shortly been adopted from a decade of governance reform process that opened doors for non-state actors to play roles in supporting participatory forest initiatives. The DANIDA funding was directed towards supporting district councils with financial and technical resources to implement forest decentralization as per the policy and the Act. In 2004, MCDI took up initiatives that were under the UTUMI project and started to consolidate a platform for Community Based Forest Management (CBFM). It supports local communities in Kilwa and nearby districts in, supposedly, understanding their rights, rules of forest access, management skills, and marketing of forest products (Horning, 2018; Corbera et al., 2017; Blomley and Iddi, 2009). In doing so, the focus has been on solving legal, socio-cultural, and technical challenges that hinder the establishment of village forest reserves (VLFRs).

MCDI is currently a well-established local NGO and central to the establishment of CBFM in the district. The organization is also a convener and mediator of other partners and the facilitator of different forest-related initiatives including REDD+ and Forest Certification. Despite the two decades of service, however, local communities have not yet developed technical capacities, marketing skills, and financial power to cover various costs of sustaining CBFM (Kalumanga et al., 2018).

The entry into communal land is given by the country's Village Land Act of 1999 that provides for communal ownership of village lands. However, that does not give communities the common ownership of resources such as forests. Instead, the Forest Act gives villages the powers to own the forest within a specified section of the land where communities can declare and gazette a village land forest reserve (VLFRs) through the process stipulated in the law. As

Table 21.3 Distribution structure of income from forest products sales in CBFM villages

S/N	*Institution*	*% of share of total revenue*
1	Village Council	50%*
2	VNRC	40%*
3	MCDI	5%
4	Kilwa District Council	5%
TOTAL		100

* Village Council and VNRC may vary depending on the specific village

Trupin et al. (2018) argues, communities are not fully aware of their legal rights in the process of establishing VLFRs, which calls for non-state actors such as MCDI to facilitate local involvement in CBFM.

In terms of benefits, it is argued that CBFM has empowered local communities to designate VLFRs and villages have been able to get more money from timber sales and increased transparency (Kalonga et al., 2015). Funds obtained from VLFR are distributed based on the allocations approved by local communities in village assemblies. Increased democracy is also cited often in the sense that the village committees that oversee VLFRs on behalf of villagers are democratically elected. Like it is for WMAs, funds from timber sales are retained to support community development and its distribution is through an agreed formula (Table 21.3). Unlike in the case of WMAs, villages are legally allowed to retain a substantial amount of timber income.

Also, through MCDI, some villages have been supported to acquire a group certificate from the Forest Stewardship Council (FSC), in which emphasis is put on sustainable harvesting and marketing of forest products. FSC uses principles which require that timber buyers purchase from sustainably conserved forests, and that local communities are well involved in the process. Certified forest products earn premium prices which are collected by villagers. CBFM have been hailed for the significant positive impact on forest conditions due to increased patrols that have minimized illicit logging and forests fires. The implementation of forest certification in Kilwa district is done on forests that are under CBFM. As a result, CBFM is considered as a foundation for enabling conservation partners to abide with the FSC standards.

As Sungusia and Lund (2016) finds, the consequences of turning unreserved lands that have open access into VLFRs with controlled access is largely ignored. Despite this trust built on CBFM and MCDI as its facilitator, there are reports that a timber business company called Sound and Fair Company is among the timber buyers in certified forests. However, former board members of MCDI are major shareholders in the company. This means that MCDI is, in a way, also a timber buyer. The fact that MCDI is responsible for monitoring certified forests and products and Sound and Fair the buyer, there is clear conflict of interest. As CBFM villages plan to increase their VLFRs to maximize revenues from selling certified timber, they are also increasingly experiencing challenges of unreliable markets as powerful actors have prepared and taken over the business.

Conclusion

The contemporary politics of conservation and development in Tanzania hinges on radical ways in which large-scale land-based investments have turned onto village lands. The need for conservation expansion and ambitions for agricultural modernization on the one hand and local needs for food and cash crop production on the other are in sharp contrast. Political ecologists are especially critical of livelihood impacts of conservation expansions into village lands as the poor communities have become culprits for the small and/or unequal distribution of revenues accrued from the lands. As critics suggest, investments that are labeled as community-based and fostered within village lands have mainly squeezed local rights to land while constructing modern green territories. Development aspirations of villages remain divorced from the bigger pictures of powerful conservation partners and local elites. Although plans for local development enrich the bigger picture, it is mainly in the context of those whose lives are far from rural settings. It is not surprising that Tanzania has the largest number of protected forests and wildlife areas, and that it is known to be a stable and safe country, but its population remains among the poorest in the world ranks.

Partnerships for wildlife and forest protection are neither entirely voluntary nor can their impacts be assessed directly through improvements in local livelihoods. Rather, partnerships are an entry point into village lands that are most needed for conservation. Regardless of the resource type and context, these partnerships have circumscribed protected area neighbors across the country (Raycraft, 2019). Indeed, the law provides that communities as entities can abandon their commitments to conservation partnerships but the land they allocate for wildlife and forests remain locked into the partnership. These partnerships have therefore served broader interests of securing land for conservation rather than fostering local development. As a result, local communities have lost significant proportions of their land and ability to decide on the use of the remining areas. In both our case studies, crop cultivation remains the main source of food and income for more than 95% of households. We conclude by questioning the development rhetoric, and by observing that partnerships reinforce conservation expansions but that lead to wins for only some actors, thereby indirectly aggravating local power struggles.

References

Bergius, M. & Buseth, J. 2019. Towards a green modernization development discourse: the new green revolution in Africa. *Journal of Political Ecology*, 26, 57–83.

Bersaglio, B. & Cleaver, F. 2018. Artisanal mining in Southern Tanzania: preliminary reflections on a 'green squeeze'. *The Extractive Industries and Society*, 5, 274–277.

Blomley, T. & Iddi, S. 2009. *Participatory Forest Management in Tanzania: 1993–2009 Lessons Learned and Experiences to Date*. Dar es Salaam: Ministry of Natural Resources and Tourism, Forestry and Beekeeping Division.

Bluwstein, J. & Lund, J. 2016. Territoriality by conservation in the Selous – Niassa Corridor in Tanzania. *World Development,* 101 (1), 453–465.

Bluwstein, J., Lund, J., Askew, K., Stein, H., Noe, C., Odgaard, R., Maganga, F. & Engström, L. 2018. Between dependence and deprivation: the interlocking nature of land alienation in Tanzania. *Journal of Agrarian Change*, 18, 806–830.

Corbera, E., Martin, A., Springate-Baginski, O. & Villasenor, A. 2017. Sowing the seeds of sustainable rural livelihoods? An assessment of participatory forest management through REDD+ in Tanzania. *Land Use Policy*, In Press. https://doi.org/10.1016/j.landusepol.2017.09.037.

Crawford, G. 2003. Partnership or power? Deconstructing the 'Partnership for Governance Reform' in Indonesia. *Third World Quarterly,* 24, 139–159.

Curry-Lindahl, K. 1974. The conservation story in Africa during the 1960s. *Biological Conservation*, 6, 170–178.

Horning, N. 2018. *The Politics of Deforestation in Africa: Madagascar, Tanzania, and Uganda*. London: Palgrave.

IUCN. 2017. *Reactive Monitoring Mission. Selous Game Reserve World Heritage site, United Republic of Tanzania*. February 8–15. Dar es Salaam: IUCN/UNESCO.

Kalonga, S., Midtgaard, F. & Eid, T. 2015. Does forest certification enhance forest structure? Empirical evidence from certified community-based forest management in Kilwa District, Tanzania. *International Forestry Review*, 17, 182–194.

Kalumanga, E., Olwig, M., Brockington, D. & Mwamfupe, A. 2018. *Partnerships and Governance in Forest Management in Tanzania.* Working Paper 6. Copenhagen Business School and University of Dar es Salaam, New Partnerships for Sustainability (NEPSUS).

Khatun, K., Corbera, E. & Ball, S. 2017. Fire is REDD+: Offsetting carbon through early burning activities in south-eastern Tanzania. *Oryx*, 5, 43–51.

MacDonald, K. 2010. The devil is in the (Bio)diversity: Private sector 'Engagement' and the restructuring of biodiversity conservation. *Antipode,* 42, 513–550.

Milledge, S., Gelvas, I. & Ahrends, A. 2007. *Forestry, Governance and National Development: Lessons Learned from a Logging Boom in Southern Tanzania.* Dar es Salaam: TRAFFIC East/Southern Africa/Tanzania Development Partners Group/Ministry of Natural Resources of Tourism.

National Bureau of Statistics (NBS). 2017. *National Accounts Statistics of Mainland Tanzania 2017.* Dar es Salaam: National Bureau of Statistics.

NBS. 2019. *Tanzania Mainland Household Budget Survey: Key Indicators Report 2017–18.* Dar es Salaam: Ministry of Finance and Planning.

Neumann, R. 2001. Africa's 'last wilderness': reordering space for political and economic control in colonial Tanzania. *Journal of the International African Institute,* 71, 641–665.

Noe, C., Brockington, D., John, J., Bwagalilo, F. & Kamnde, K. 2019. *Interrogating Wildlife Conservation Partnerships in Rufiji and Kilwa Districts, Tanzania.* NEPSUS Working Paper 2019/3. Copenhagen Business School and University of Dar es Salaam.

Noe, C., Sulle, E. & Brockington, D. 2017. *Tanzania's Partnership Landscape: Convergence and Divergence in the Wildlife Sector.* NEPSUS Working Paper 2017/3. Copenhagen Business School and University of Dar es Salaam.

Ponte, S., Noe, C., Kweka, O., Mshale, B., Sulle, E., Brockington, D., Kalumanga, E., Minja, R., Budeanu, A., Mwamfupe, M., Henriksen, L., Olwig, M., Silvano, S., Namkesa, F., John, R., Katikiro, K. & Mabele, M. 2017. *New Partnerships for Sustainability (NEPSUS): Concepts, Research Design and Methodologies.* NEPSUS Working Paper 1. Copenhagen Business School, CBS.

Ramutsindela, M. 2004. *Parks and People in Postcolonial Societies. Experiences in Southern Africa.* Dordrecht: Kluwer.

Ramutsindela, M. & Noe, C. 2012. Scalar thickening: wildlife management areas and conservation scales in southeast Tanzania. *Singapore Journal of Tropical Geography*, 33, 137–151.

Raycraft, J. 2019. Circumscribing communities: Marine conservation and territorialization in southeastern Tanzania. *Geoforum*, 100, 128–143.

STEP. 2018. *Restoring the Selous to Udzungwa Wildlife Corridor: Last Chance to Maintain Elephant Connectivity in Southern Tanzania.* Southern Tanzania Elephant Program (STEP). www.stzelephants.org, Accessed 03/09/2019.

Sungusia, E. & Lund, J. F., 2016. World development perspectives. *World Development, 3*, 35–37.

Trupin, R., Morgan-Brown, T., Doulton, H. & Nelson, F. 2018. *Making Community Forest Enterprises Deliver for Livelihoods and Conservation in Tanzania.* Dar es Salaam: USAID.

UNEP-WCMC. 2019. *Protected Area Profile for United Republic of Tanzania from the World Database of Protected Areas, August 2019.* www.protectedplanet.net.

United Republic of Tanzania. 2014. *An Overview of the Wildlife Sub Sector: Achievements, Challenges and Priorities for Financial Year 2014/15.* Presented at the Natural Resources Sector Review Meeting on 16th October. The National College of Tourism, Bustani Campus, Dar es Salaam.

URT. 2009. *Wildlife Conservation Act.* Dar es Salaam: Government Printers.

URT. 2011. *Tanzania Agriculture and Food Security Investment Plan (2011-2012-2020-2021).* Dar es Salaam: Ministry of Agriculture, Livestock and Fisheries.

URT. 2016. *Agricultural Sector Development Strategy – II 2015/2016–2024/2025.* Dar es Salaam: Ministry of Agriculture, Livestock and Fisheries Development.

URT. 2017. *Tanzania Human Development: Social Policy in the Context of Economic Transformation.* Dar es Salaam: Economic and Social Research Foundation, United Nations Development Programme, Government of the United Republic of Tanzania.

URT. 2018. *The Wildlife Conservation (Wildlife Corridors, Dispersal Areas, Buffer Zones and Migratory Routes) Regulations.* Dar es Salaam: Government Printers.

URT. 2019. *SGDs Voluntary National Review (VNR): Empowering People and Ensuring Inclusiveness and Equality*. Dar es Salaam: Ministry for Finance and Planning.

USAID. 2014. *SCAPES Landscape Profile: The Ruvuma Landscape.* USAID. www.usaid.gov/biodiversity

USAID. 2015. *Final Evaluation of the Sustainable Conservation Approaches in Priority Ecosystems Program (SCAPES)*. Washington, DC: USAID.

USAID. 2017. *Conserving Biodiversity and Forests*. USAID. www.usaid.gov/biodiversity, Accessed 23/8/2019.

World Bank. 2019. *Tanzania Country Environmental Analysis: Environmental Trends and Threats, and Pathways to Improved Sustainability*. Washington, DC: The World Bank.

World Travel & Tourism Council (WTTC). 2017. *Travel and Tourism: Economic Impact 2017 Tanzania*. London: World Travel and Tourism Council.

Wright, V. C. 2017. Turbulent terrains: The contradictions and politics of decentralised conservation. *Conservation and Society*, 15, 157–167.

WWF. 2014. *Tanzania's Wildlife Management Areas: A 2012 Status Report*. Dar es Salaam: WWF.

22

UNDERSTANDING THE RELATIONSHIP AMONG GENDER, SPACE, AND THE ENVIRONMENT

The case of Waorani Women in Gareno, Ecuador

Ana Gómez Donoso and Paula Soto Villagrán

Between the '80s and '90s, very significant contributions have been made about the interaction between gender and the environment from two different positions: on the one hand ecofeminisms, as an academic-theoretical interpretation; and on the other hand the analysis of social and political mobilization carried out by women as a form of political action in the face of environmental degradation at different geographical scales. These positions have centered the need to incorporate a gender perspective in environmental analysis.

This line of questioning arises from the relationships among the condition of women and measures to confront inequalities, imbalances, and rights in political, social, and economic spaces, which have been acquiring centrality with discussions on development, the growing wear and tear of the ecological system, and the feminization of poverty, among other discourses.

Thus, gender relations have been robustly incorporated in the last three decades in discussions about the environment, as a new look at old problems related to development, poverty, and lately with the notion of sustainability. Indeed, different studies have shown a deep and dynamic understanding of how the environment constitutes urban gender inequality and how this affects the identities of men and women.

Theoretical contexts: gender, development, and environment, specifying the concepts

Paolisso (1996) identifies three currents at the conceptual-theoretical level and at the level of activism of different researchers interested in the relationship among women, gender, and the environment. The first, women and the environment, rescues the role of women as managers of environmental resources, in such a way that any initiative taken in relation to the environment must take into account the contributions of women. The second, ecofeminism, states that the link between women and nature lies in the fact that the domination of women has been parallel to the domination of nature. This is why the participation of women in women's social or environmental

DOI: 10.4324/9780429450310-26

movements occurs from the difficulty they have in obtaining necessary resources to fulfill their role as caregivers of the family, with the deterioration of sanitary conditions or contamination problems (Sabaté, 2000). The third current places gender as a key that mediates male and female relationships with the environment. A concern is distinguished that has to do with how female/male relationships affect or are affected by environmental change and a second concern has to do with how cultural gender roles influence perceptions of the environment and cultural change.

This line of research has introduced different objects of study that link woman-nature, and that show the historical parallels that have occurred in the treatment of both by modern societies. In particular, the interrelation among gender, poverty, and the environment has been fundamental, since it has shown how dynamic processes such as the domestic organization of work, consumption patterns, food issues, pollution, and environmental destruction all affect life but specifically women's lives. At the same time, it places them as key subjects for environmental change.

Some descriptive studies in Latin America show how women's caregiving jobs are closely associated with environmental problems in which they find themselves. There are two interesting phenomena to point out. First, while it is women who manage environmental risk more easily as a result of their domestic tasks, at the same time they have an environmental culture that must be considered (Rico, 1996a, 1996b).

Secondly, linkages to the feminization of poverty have been compelling[1] with respect to territorial and environmental aspects, because the particular use of different urban spatialities, as well as environmental problems that arise in areas of poverty, affect women more severely, whose movements are frequently restricted to where they are housed and their immediate surroundings. This process of spatial confinement is essential to establish social vulnerability and lower levels of health and well-being (Schteingart, 2009; Vega, 1996). From this perspective, it is relevant to consider notions of development, empowerment, and agency, because as has been indicated in other studies, the discourses of empowerment in Latin America have specifically originated in the articulations between gender and development, especially around the limitations of development theories and models when considering different social sectors and a multiplicity of actors that are involved in this process.

The theoretical conception called Women in Development (WID, or MED in its Spanish acronym)[2] articulated diagnostics, proposals, and strategies aimed at highlighting the role of women and promoting their transformation (Portocarrero, 1993). The assumption behind the MED approach is that economic strategies have negatively impacted women and that they should be integrated into development through access to employment, the market. They should be provided with legal and educational resources that allow them to join the productive sector under better conditions, since the origins of female subordination are identified in both the family and in the market. In this line of argument, importance is assigned to economic independence as a necessary prerequisite for equity. However, this would not necessarily exempt them from traditional domestic workloads. One of the main criticisms made of this approach is that women are seen as the ultimate target of actions taken; therefore, they are considered both a cause and a means to overcome their own problems.

For its part, Gender in Development (GED) arises as a result of the actions of grassroots organizations of women and feminist reflections from the third world. This approach recognizes that inequalities between men and women must be considered in development projects, since they affect them differently. The origins of female subordination would be in both the reproductive and productive fields, emphasizing that women are affected by development conditions differently according to race, class, colonial history and current position in the world, international economic order, and where the gaze is set on strategic gender interests.

In this development scheme the notion of autonomy becomes relevant, defined as the power of people – as individuals, groups – to have control over their own lives and the right to determine their options and influence social change (Lycklama, in Ricco, 1993: 14). This approach integrates the importance of power relations, conflict, and gender relations as axes for a better understanding of the subordination of women (Moser, 1991). Indeed, it is under this approach to development that the concept of empowerment begins to become visible and, which will later be recognized as the empowerment approach.

Empowerment in the GED strategy tries to overcome the limitations of the MED; power is re-signified, giving great importance to women increasing their power through access to control over material and non-material resources. According to León (2001), the most relevant element of empowerment is to contain the concept of power itself, where the term empowerment by feminisms has roots in unequal gender relations in society.

The concept of empowerment would be the antithesis of paternalism, the essence of self-management, which builds on the existing forces of a person or social group to grow its capacities, that is, to increase those pre-existing forces (Durston, 2000: 33–34). Therefore, empowerment is articulated with the ability to make decisions, in the perspective of Kabeer (1999), that is, empowerment is expressed directly in decision-making. Thus, it refers to the expansion of people's abilities to make strategic decisions for their lives in conditions in which these abilities have been previously rejected (Kabeer, 1999: 437).

For this last author, the concept of empowerment is defined in relation to that of agency and is defined as what a person has the freedom to do and achieve in search of the goals or values that he or she considers important (Sen, 1985: 203). Thus, there is a close complementarity between individual agency and social institutions (Sen, 2000: 16). The interesting thing here is that agency is fundamental for empowerment, in the perspective that in our case it is women who develop an exercise of contextualized decision-making.

This is why in our analysis of women's empowerment, we must consider spatial relationships and relationships to the environment, as well as experience and significance. We are conceiving space as a fundamental element in the structuring of gender inequalities, and at the same time an axis of analysis to think about how decision-making processes, the exercise of autonomy, and self-management are spatial.

An approach toward collaborative participatory methodology

It is common for the terms empowerment and development to refer to urban populations or groups of people, since development is measured in terms of money and accumulation of capital or goods, making groups of people who do not live in large cities invisible, and who live under another logic; realities that would revolutionize these concepts.

Next we present some findings from a pre-doctoral research effort, a study about and towards the reality of Waorani women within the Ecuadorian Amazon jungle. The field phase was carried out in the Gareno community, with the use of tools typical of qualitative research, and based on respect for the community and its culture. Therefore, this methodological approach is based on a participatory or collaborative methodology, in which the approach is centered on the worldview, a way of thinking, language, etc., where we work through treating participants involved as subjects rather than as objects (Ferreira de Faria, 2016).

In addition, the knowledge discourse proposed by the Boaventura de Sousa Santos school[3] derived from epistemologies of the global south was founded on the urgent need to look for another more globalized social science that pays attention to the cultural diversity that humanity offers (Santos, 2012). This elevates feedback so that such epistemes are not lost, committing *epistemicide*, as he calls it, referring to the destruction of other non-Western knowledge or knowledges not assimilated by Western culture.

Thus, in the first fieldwork phase, we conducted participant observation and in-depth interviews with key actors, using these first results in order to establish a knowledge dialogue. The information was prepared using the tool of participatory mapping, through which we seek to understand women's perception of the use of territory and space, women-nature configuration their way of relating to men and among women, and thereby the accessories of their empowerment.

It should be noted that this chapter, based on preliminary results of the fieldwork conducted, seeks to exemplify the use of a non-extractivist methodology to understand Waorani culture, and the real needs of women within the community of Gareno, and thus to frame the empowerment scheme for development of a vulnerable group in Ecuador.

Gareno and its spatial-territorial context

One of the realities quite distant from the urban one is that of Gareno, a community of the Waorani nation,[4] who live in the Ecuadorian Amazon jungle, Chontapunta parish, Tena canton, Napo province; and form part of the nearly 32 communities of the nation, distributed within the Yasuní National Park between the Tiputini and Curayay rivers, near the border with Peru, and close to the Andean foothills.

Gareno is located in the territory occupied by block 21, in proximity to the activity of the Petroamazonas company, as can be seen on the map in Figure 22.1.

Gareno was created in 2001 after internal separations from the Waorani tribe; the relocation is due to the proximity of markets and roads, in addition to the historical presence of evangelizing missionary groups in the area that have contributed to the constant changes in their way of life since approximately 1969, beginning with the appearance of the Summer Institute of Linguistics. The latter, after obtaining permission from the Ecuadorian government, carried out its protection plan for the Waorani nation, called a protectorate, of approximately 66,600 hectares, which is an area one-tenth the size of the territory traditionally occupied by the nation. The activities they carried out were education, evangelization, and gradual insertion into the national economy, turning communities into sedentary ones and occupying the riverbanks. It is presumed that two families resisted this way of life, the so-called uncontacted: Tagaeri and Taromenane, of whom very little is known, but it is documented that their way of life is still nomadic.

It should be emphasized that despite multiple efforts to civilize them, the Waorani communities maintain their worldview. So being, they do not have deities, but within the jungle there are animals that are respected and interact with certain codes, as explained by the anthropologist Kati Álvarez (2018). The harpy eagle, when flying low means a danger alert, not when flying high, which indicates tranquility in the area. Some of the people who refer to god are the product of the evangelization of missionary groups, in the protectorate stage. Figures 22.2, 22.3, 22.4 are photographs taken in the Gareno community, where evangelization processes are still taking place.

Gareno is made up of 25 families, 45% men and 55% women. The Pekenanis (older adults) speak Waoterero but also Spanish. In the homes where Pekenanis live, their language is transmitted to the children, and they also learn the English language at their school. The clothing, food, and their daily housing could be called Western. Even so, in their festivities they wear their traditional costumes, take the yuca chichi, and their meeting center is a traditional structure called *maloca*.

The activities that the community engages in can be divided into two main branches: activities for self-consumption, and the services they provide to the oil companies. Self-consumption

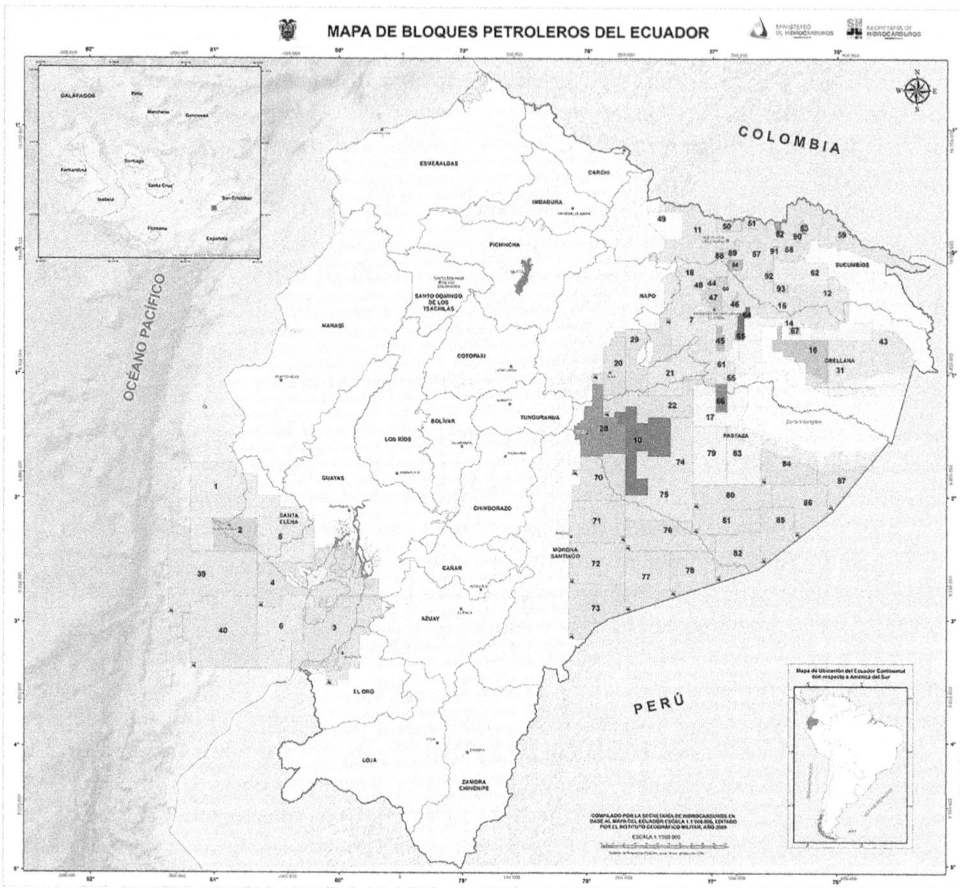

Figure 22.1 Mapa de bloques petroleros del Ecuador (Map of oil blocks in Ecuador)

Source: Ministerio de Hidrocarburos del Ecuador. Retrieved from: www.secretariahidrocarburos.gob.ec/mapa-de-bloques-petroleros/

activities focus on the collection of products for food: hunting animals, fishing, and cultivation in farms. On the other hand, the activities they carry out for the oil companies are road cleaning and forest clearing, among others.

Analyzing the case of the women of Gareno

One aspect that is fascinating when approaching the Gareno community, and Waorani in general, is the understanding of gender roles. In Gareno, both men and women, to a greater or lesser extent, work the farm, build housing, hunt, take care of the little ones, cook, sing as a process of transmitting knowledge, etc., but it is the proximity that the inhabitants have with nature that sheds light on a differentiation of roles between genders.

In this sense, one of the obvious activities that differentiates women and men is giving birth. Women have the task of giving birth and breastfeeding. Even so, there is a myth that Álvarez explains: at first the men also suckled the young, but after a problem with the moon, their

Figure 22.2 Maloca, Gareno, 2018

Figure 22.3 Unidad Educativa Comunitaria Intercultural Bilingüe Gareno

breasts were removed, and that labor became the role of women. Despite this, men can be seen with their young on their chest lying in their hammocks (Álvarez, 2018). Another example of this contact between Waorani people and the natural environment is that provided by Tipa Irumenga (Leader in Gareno). When someone is chosen to sing, a ritual is performed that involves cutting the tongue of a bird and hiding it under the infant's tongue, thus the infant acquires the

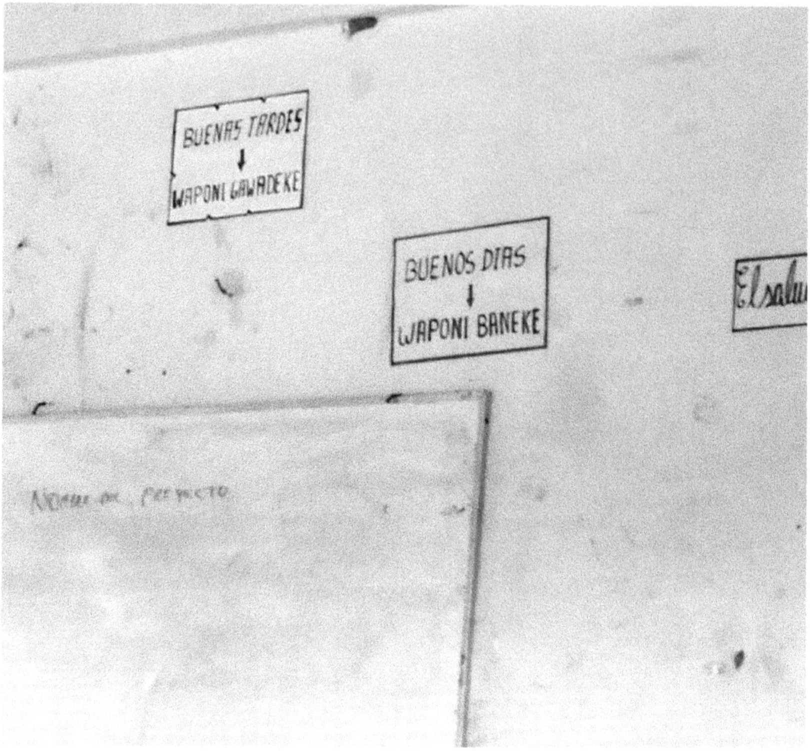

Figure 22.4 "Buenas Tardes, Buenos Dias"

Source: Photos by A. Gómez (2018)

power to emit songs to transmit knowledge, ideas, and life lessons. It should be noted at this point that one of the myths of the Waorani culture tells that men and women were born from the same animal, from the Boa snake; the woman comes out of the head and the man from the back, indicating the parity with which they live in the same social space (Álvarez, 2018). Likewise, the jaguar, in particular, is an animal with which they can feel identified at a cultural level in terms of housing, food, and defense. When they die, they transform into jaguars, where the women become male jaguars and the men become female jaguars.

On the other hand, in Gareno there is a school in which Kichua and Waorani teachers work. Girls and boys learn Waoterero, Spanish, and English, but within the community orality is maintained as a traditional way of transmitting their knowledge, which is carried out through songs in Waoterero. Older people (mothers, grandmothers, fathers, grandfathers) sing their experiences, myths, even the situations they have experienced during the day. The sons and daughters listen and learn life lessons both at home through these songs or through walks in the jungle. In this sense, men and women from zero to three years old have the duty to learn to walk on two legs and eat meat; from three to 18 years they walk through the jungle to learn tasks such as planting, hunting, whistling, knowing about trees, etc. Mothers and fathers take their children to the jungle to learn about these activities. Figures 22.3 and 22.4 are photographs of the Gareno School:

On the other hand, the socio-political and kinship organization is quite complex, but for the purposes of the present characterization of the Gareno community it is summarized as follows:

there are nuclear families made up of a father, mother, and an average of five children per home; few families still have Pekenanis. In addition, as previously stated, Gareno has 25 families which make up the community. There are leaders in the community, and both women and men can be leaders as soon as they show strength; they are legitimized within the family. More than one person can be found in front of families.

There are two organizations with which the Waorani nationality is presented to the Ecuadorian State and worldwide, as a bridge between the communities and the West: The Waorani Women's Association (AMWAE) and the Organization of the Waorani Nationality of the Ecuadorian Amazon (ONHAE), who represent the Waorani Nationality of Ecuador (NAWE). The AMWAE has been dedicated to the organization of Waorani women around the development of handicrafts that are marketed in cities and abroad. Prior to the AMWAE, the ONHAE organization was formed, who, by representing the entire nation in the face of socio-environmental problems, generally with oil companies, experienced the influence of the capitalist, colonial, and patriarchal system. Therefore, their actions were run only by men, giving limited spaces to women.

Despite this organizational reality at the level of the entire Waorani nation, in Gareno, there is a female leader and a male leader, the two offspring of the Pekenani warrior, Monka. Tipa Irumenga, Monka's daughter, is a leader because she has stood out for her vast knowledge of natural medicine, traveled through the Waorani territory around the country, and is a nurse at the Gareno health center.

Some final thoughts

After briefly characterizing the Gareno community, its reality and that of women in particular, we can affirm that the Waorani worldview is in a condition of spatial vulnerability. As has been maintained throughout this work since the actions of the protectorate and the entry of oil companies, changes have been experienced in daily life, showing a certain adaptation and resistance to the Western system, classified as capitalist, colonial, and patriarchal.

By showing that the inhabitants of the Gareno community have begun their gradual entry into the Western system, an inequitable future is envisaged between women and men. Therefore, within the GED trend for the development of women, an empowerment process as explained before is the urgent parallel step that must be carried out together with the AMWAE association, which seeks economic well-being, exploring micro-enterprises as an alternative for their development and improvement of quality of life, without losing sight of the existing and frequent practice of *epistemicide* that has been generated for hundreds of years to indigenous nations.

Thus, we find lines of significance that clearly speak to us of some ruptures aimed at the reorganization of productive and reproductive roles between women and men within the family, which refer us to small agency exercises in the capacity of Waorani women to expand their life choices and make their own decisions. This allows us to glimpse some cultural changes, such as with Tipa Irumenga (a leader in Gareno). These possibilities narrow the empowerment of women in walking within the Western logic, which prepares them to face new challenges such as micro-enterprises or the commercialization of their handicrafts, activities that are being carried out today.

Specifically, it is necessary to accompany the Gareno community, who approached the West protecting their ancestral culture, by keeping several characteristics active such as maintaining a leader, but that are gradually differentiating internal gender roles after the presence of oil companies. That is why it is necessary to locate, within the daily life of women in Gareno, the breaking points from which they are forced to rethink their role in the community with respect to being a couple, at work, a mother, managing resources, due to the high vulnerability to exploitation of both the system and the change of roles with respect to the male gender.

The proposed approach to the Waorani culture, and in general to non-Western nationalities, is for the use of methodologies that are more respectful of culture to venture into new tools to maintain a subject-subject dialogue in which to learn about their daily lives and to understand their real needs in this space of vulnerability within the capitalist, colonial, and patriarchal system.

Finally, it is important to mention that efforts in Ecuador to work on the conservation of ancestral communities intensified in the '90s. Thus, from organizations such as Corporación Grupo Randi, and from academia, the Universidad Andina Simón Bolívar-UASB, Latin American Faculty of Social Sciences-FLACSO Headquarters in Ecuador, several approaches have been made to shed light on the problem of better management to the benefit of indigenous women and Ecuadorian ancestral nationalities.

Translated by: Patricia Solís

Notes

1 The feminization of poverty (Anderson, 2003) is a concept that emerged in the 1980s to designate an increasing trend of the incidence and prevalence of poverty among women. She placed two problems in her sights: women heads of household and early motherhood (frequently linked to being a single mother, head of an incipient household or of a subsidiary unit within another household). This concept led the statistical census to be disaggregated by sex. It created a climate for the subsequent entry of a concern for the voices of victims, the perception of female actors, and the use of qualitative investigative methods to uncover them.

2 The name refers to first world women's groups linked to international development agendas who, based on criticism of the effects of development, studied proposals to integrate women.

3 Boaventura De Sousa Santos, PhD, Portuguese. Doctor in Sociology from Yale University. He works on the Alice: Leading Europe a New Way project. General coordinator of the international course, "Thoughts and situated struggles: for a cartography of the south", Universidad Sur-Sur, Clacso y Ces.

4 Waorani is derived from the term waoterero, Wao, which means "free-people".

References

Álvarez, Kati. (2018) *Territorio y Vida de la Nacionalidad Waorani. Zonificación Etno Territorial de 14 comunidades Waorani.* Quito.

Durston, John. (2000) *¿Qué es el capital social comunitario?* Serie Políticas Sociales No 38. Santiago de Chile, División de Desarrollo Social, CEPAL.

Ferreira de Faria, Ivani. (2016) *Visto de Adentro. Projeto PIBID/Diversidade/CAPES.* Brasil: Universidade Federal do Amazonas/UFAM.

Kabeer, Naila. (1999) Resources, Agency, Achievements: Reflections on the Measurment of Women's Empowerment. *Development and Change*, Vol. 30 (3), pp. 435–464.

León, Magdalena. (2001) El empoderamiento de las mujeres: Encuentro del primer y tercer mundos en los estudios de género. *Revista La Ventana* (13), pp. 94–106.

Moser, C. (1991) *Planificación de Género y Desarrollo.* Lima: Teoría, Práctica y Capacitación, Red Entre Mujeres y Ediciones Flora Tristán, pp. 56–124.

Paolisso, Michael. (1996) *Avances de la investigación sobre género y medioambiente*, en Velázquez, Margarita (Coord.). *Género y ambiente en Latinoamérica.* México: Centro Regional de Investigaciones Multidisciplinarias, UNAM.

Portocarrero, Patricia. (1990) (Ed.) *Mujer en el desarrollo. Balances y propuestas.* Lima: Flora Tristán.

Rico, M. (1996a). *Género, ambiente y pobreza. Un estudio exploratorio en el medio urbano popular de Santiago.* En Velázquez, M. (coord.). Género y ambiente en Latinoamérica. México: Centro Regional de Investigaciones Multidisciplinarias. México: CRIM-UNAM, pp. 185–240.

Rico, M. (1996b). *Asentamientos humanos en América Latina y el Caribe. Una mirada desde la perspectiva de género y desarrollo.* En Asentamientos humanos, pobreza y género, Santiago: Ministerio de Urbanismo/ GTZ-Cooperación Técnica Alemana, pp. 77–96.

Ricco, M. Nieves. (1993). *Desarrollo y equidad de género: una tarea pendiente. Organización de las Naciones Unidas.* Santiago: División de desarrollo social, Unidad Mujer y Desarrollo.

Sabaté, Ana. (2000) *Género, Medio Ambiente y Acción política: un debate pendiente en la Geografía actual, Anales de Geografía de la Universidad Complutense*, núm. 20, pp. 177–191.

Santos, Boaventura de Sousa. (2012) *Introducción: las Epistemologías del Sur, Una epistemología del Sur.* México: Siglo XXI.

Schteingart, Martha. (2009) *Pobreza, condiciones de vida y salud en la ciudad de México.* Mexico: El Colegio de México.

Sen, A. (1985) *Well-Being, Agency and Freedom: The Dewey Lectures* 1984. *The Journal of Philosophy*, Vol. 82 (4), pp. 169–221.

Sen, Amartya. (2000) *Desarrollo y Libertad.* México: Planeta.

Vega, Silvia. (1996) *Hogares urbanos y medioambiente: comportamientos ambientales y salud familiar.* En: Velázquez, Margarita. (Coord.) Género y ambiente en Latinoamérica. México. Centro Regional de Investigaciones Multidisciplinarias, UNAM.

Interviews

Fecha	Nombre
11/7/2018	Tipa Irumenga
29/5/2018	Kati Álvarez

23

UPGRADING THE SHOCK THEORY

Female resilience in reconstructing Puerto Rico after Hurricanes Irma and María

Carlos Jorge Guilbe López

> I hate to tell you, Puerto Rico, but you've thrown our budget a little out of whack because we've spent a lot of money on Puerto Rico.
>
> *(President Donald Trump in Puerto Rico, October 17, 2017)*

> Women are the physical and moral mother of nationality.
>
> *(Pedro Albizu Campos, Puerto Rican nationalist leader [1891–1965])*

The Commonwealth of Puerto Rico confronted a convoluted 2017. This American non-incorporated territory received the brunt of two hurricanes within 20 days. Both natural systems uncovered poverty, marginalization, and political uncertainty for many people in the island and its diaspora. Months before the destruction, the local press was debating on the fiscal policies imposed by the federal government due to the bankruptcy of the Puerto Rican government. This imposition was accompanied by an official ratification of the colonial status by the United States Supreme Court in 2016. The combination of economic problems, lack of political power, and two large-scale hurricanes define the geographical framework of this chapter.

This chapter describes and analyzes the disputes that emerged in Puerto Rico during the recovery and reconstruction processes. Post-hurricanes resilience and sustainability may be defined by two opposite paths: the response to implement neoliberal policies by the Commonwealth's government and the development of self-supporting projects developed by many communities. The hurricanes detonated this distinctive atmosphere. Weeks after the hurricanes, thousands of families emigrated to the United States, dozens of rural communities were isolated and over 2,000 people died because they did not have access to medical and other basic services. Local, state, and federal governments were inefficient in responding to the necessities. Nevertheless, the isolation and "collective sense of powerless" (Llorens, 2018) strengthened community relations across the island. For thousands of families, small-scale projects became the immediate response to mitigate their daily needs. Simultaneously, the Puerto Rican government focused on the recovery and reconstruction proceedings by intensifying the discourse of long-term neoliberal strategies as privatization, public sector cutbacks, selling governmental properties, enabling external capital by abolishing existing legislation, and claiming additional federal assistance.

DOI: 10.4324/9780429450310-27

The alternatives for the recovery and restructuring process and the resilience response have been accompanied by polarized views among public officials and community leaders. Many communities, specifically after the hurricanes, have been headed by women. Females are becoming the frontage of grassroot movements and non-profit organizations. They are leading the debates on the reconstruction from small-scale bases through resilience using self-management, food security, and protection of the island's natural resources.

The governmental position is handling the natural hazards as an opportunity to encourage the integration as the 51st state of the United States of America. The tactics adopted by local public officials are analogous to the main dogmas of the "Shock Theory" exposed by Naomi Klein in 2008. The Puerto Rican experience after the hurricanes served to upgrade her theory under the "disaster capitalism".

Upgrading the shock theory into SASAS doctrine

Naomi Klein is a Canadian writer and environmental activist. She recognizes himself as an active "anti-capitalist" (Ojeda, 2019). Her best-seller book, *The Shock Doctrine* (2008) was an open critique to neoliberal policies worldwide. Under this dogma, Klein introduced the concept of "disaster capitalism" as a governmental practice of privatization and deregulation in response to major events such as natural disasters or armed conflicts. She defined the shock doctrine as:

> the quite brutal tactic of systematically using the public's disorientation following a collective shock – wars, coups, terrorist attacks, market crashes, or natural disasters – to push through radical pro-corporate measures, often called "shock therapy".
>
> *(Klein, 2017, p. 2)*

The shock refers exclusively to the quickness and intensity in the development of projects and implementation of policies in an area after a major destructive event. Worldwide examples of this practice among the past 40 years are:

- The political situation in Chile after the military overthrow of Salvador Allende in 1973
- The reconstruction of Afghanistan after the war against the former Soviet Union in the late 1980s
- Russia and Poland with the fall of the socialist governments in the early 1990s
- The overthrow of Saddam Hussein in Iraq in the early 2000s
- The attacks in the United States after September 11, 2001
- The reconstruction of Sri Lanka after the tsunami in 2004
- New Orleans after Hurricane Katrina in 2005

This process matches with the neoliberal platform in minimizing the government's presence in public economic affairs. The deconstruction of the regulatory state is an essential element in the merging of new market environs. The bold participation of the private sector as prime agent for development defines the goals of revitalization and reconstruction. Klein (2017) identifies the organization of the communities as a method to oppose neoliberal actions after a destructive event:

> these tactics *can* be resisted. To do so, two crucial things have to happen. First, we need a firm grasp on how shock politics work and whose interests they serve. That understanding is how we get out of shock quick and start fighting back. Second, and equally

> important, we have to tell a different story from the one the shock doctors are peddling, a vision of the world compelling enough to compete head-to-head with theirs.
> (*p. 8*)

These ideas serve as the framework for resilience. However, under this scenario, many communities face opposition from lawmaking institutions, corporative resources, local bourgeoisie, and transnational capital. These forces intermingle at a large scale under the official discourse of modernization, globalization, and technology. In order to resist the shocks, communities must act as a single front.

The Puerto Rican experience in 2017 and 2018 has reformulated the shock theory. Klein (2018) describes it as the *Shock-After-Shock-After-Shock doctrine* (SASAS doctrine). The magnitude and nature of the projects beyond the scope of the reconstruction under a colonial and bankruptcy setting provides a new panorama of the theory. According to Klein, this attribute defines the Puerto Rican experience.

A brief historical background of the Puerto Rican experiment

The political condition of Puerto Rico is not a priority in Washington. However, the Caribbean as a backyard still an important issue. Ultramarine expansion, after the Spanish-American War in 1898, left the United States with a set of territories (Smith, 2003). The geographical extension was part of the imperial project under the military doctrine of Alfred Mahan. He hypothesized that a nation that had a good location in the navigation routes and possessed the naval power to dominate enjoyed an excellent position to achieve economic and political dominance on a global scale (Estades Font, 1999, p. 26).

The American government rehearsed with experimental policies in the acquired territories (Burnett, 2005).[1] Under this scheme, Puerto Rico was part of a socio-political laboratory as an overseas colony (McCoy and Scarano, 2009). The United States Congress granted citizenship to Puerto Ricans under the *Acta Jones-Shafroth of 1917*. However, it was a statutory citizenship (Font-Guzmán, 2015). The governors and major public officials were appointed in Washington. By 1930, "Porto Rico" was labeled as "the poorest country in all of America". The island continued to be a strategic naval location during World War II (Rodríguez and Bolivar, 2012, 2015). Through this conflict, Puerto Rico was a site of experiments to reduce birth rate. The high mortality and birthrates turned the island into an "economist nightmare" (Hanson, 1949 in Taylor, 1952, p. 42). This peculiarity made feasible the implementation of birth-control experiments. This Neo-Malthusian perspective led to an unprecedented mass sterilization program (Earnhardt, 1982). American academics such as Hill et al. (1959) also used the island for population control and sexual behavior research. The friction with the local Catholic Church was widely open and became a political issue against the Americanization process (Colón, 2003; Acevedo, 2010). In the end, sterilization was not effective (Taylor, 1952 and Davis, 1953), and emigration became the way to stabilize population.

Puerto Rico continued to be a place for social experiments for the United States after WWII (Lapp, 1995). The Commonwealth (*Estado Libre Asociado*) was another experiment to cover the colonial nature in the political relationship. To support this "self-government" option, an economic model was created. The US Congress conceived Operation Bootstrap (*Operación Manos a la Obra*). This was an industrial prototype to be implemented later in Latin American. The "America's poor house" in the 1930s became "America's showcase" for the "Third World Countries" during the Cold War in the 1950s and 1960s. The implementation of the program and the dissemination of the new ideology emerged from a centralized state planning process

(Picó, 1953). Both initiatives led to the intense suppression and criminalization against the sectors that favored the self-determination and independence of Puerto Rico.[2]

Operation Bootstrap was a comprehensive industrialization program based on inviting foreign capital, receiving corporate tax exemptions and state-funded infrastructure without the application of US federal regulations. The benefits of the program were mixed (White, 1963). Emigration to the "mainland" (United States) continued to be high but was a veiled event during this period (Guilbe, 2016). The investment on heavy industries and mining activities led to the organization of rural communities to oppose large-scale industrial projects. Bergman (1993) and Concepción (1995) argue these conflicts as the launch of community organization in Puerto Rico. Unfortunately, many communities and environmental leaders became labeled as "leftists" or "anti-Americans" (Colón et al., 2014).

The global oil crisis in the mid-1970s slowed down the industrialization program. The US Congress approved Section 936 of the Internal Revenue Code to mitigate the economic collapse in the colony.[3] Under this provision, dozens of pharmaceutical and other high-tech industries with manufacturing plants in the mainland and the island had the advantage to repatriate earnings from the states with no penalty and huge tax exemption. "Las 936" also strengthened local banks and its positive multiplier effect invigorated a confident sense about the colonial experiment. The emerging industrialized landscape was composed of expensive residential projects, a state-of-the-art highway, large-scale shopping centers, luxury hotels, and American industries. Once more, the congressional economic formula was a remarkable achievement and the "commonwealth" continued to be a display example about American economic development in the tropics, especially in Latin America.

Regardless of the economic transformation during the 1970s, poverty continued to be a masked problem. Urban areas were modernized but rural areas remained undeveloped. The US Congress included the island into the federal food stamp program as a strategy to mitigate poverty (Colón, 2013).[4] Furthermore, the popularity of squatter communities ("invasiones de terrenos" or "comunidades espontáneas") turned out to be an important issue. These social movements fueled 186 new settlements where 16,821 homes were built with the participation of 84,104 people (ibid., p. 141). Female leadership was broad and visible within these social crusades.

The accessibility to federal grants and transfers continued to be an engine for economic development in Puerto Rico. Accessibility to programs of the Housing and Urban Development (HUD), Urban Mass Transportation Administration (UMTA), Urban Development Action Grants (UDAG), State Block Grant Program (SBGP), among others, financed much of the infrastructure of road and collective transportation, housing, retail, industries, arts, and education. The economic growth during this period preserved an illusion of development in the Puerto Rican society. This growth was convoyed with a bold increase of the public sector. In 1980, federal transfers were $2,873,000,000 but accompanied by an increase in public debt of 233% during the same period.[5] The dependency of federal assistance as a strategy to mitigate poverty strengthened pro-statehood movements.

The presence of the pro-statehood political party in the Commonwealth's public sector transformed the economic, political, and social structure at the end of the 20th century. The assimilation discourse has been focused exclusively on the transformation of Puerto Rico as an incorporate state. This political alternative gained sympathy in the local electorate. Since the 1990s, the vision of this political movement has been defined by neoliberal policies. The dismissal of federal industrial incentives along with the privatization programs and the creation of an intense campaign to demand a statehood bill were the major initiatives of the colonial administrations in the 1990s and 2000s. The environment in Washington has been to minimize this political option. The position of indifference and "congressional immobility" (Berríos, 2004, p. 417) has created uncertainty in the attitudes of the Puerto Rican electorate. Beyond the minority groups in favor of the self-determination and independence, Ayala and Bernabe (2011) understand that the local political future will be interrelated to the currents of a global capitalism emanating from the United States. However, Grosfoguel (2003) sustains the island will continue to be an "abandoned" territory in the 21st century.

In the new century, Puerto Rico's economic and political structure continued to be battered. The loss of US federal industrial tax breaks, political incapacity to join international trades, public-sector debt, inability to pay in the international bond market, government and corporate corruption, and the end of the political experiment made the island economically unfeasible. Government debt reached $74 billion in 2015 (Pérez, 2016). This setting forced Congress and President Barack Obama to create *Puerto Rico Oversight, Management and Economic Stability Act* (PROMESA).[6] Although Puerto Ricans may elect their governor and their legislature, under this act all local decisions must be reviewed and approved by the board. The bankruptcy of Puerto Rico was treated different from insolvencies that have occurred in other American jurisdictions and territories. PROMESA is the recent federal experiment on the island's economic and political structure.

The natural response to the continuous governmental failures has been emigration. Negrón-Muntaner (2018) described this behavior of "emptying island" as a colonial response to the historical feeling of collective vulnerability, fear, and isolation. Because of this demographic drive, the proportion of female-headed households living in the island with income under the US poverty level is rising (Colón, 2010). This tendency is usual in the Caribbean region (Safa, 1998).

Women as a central axis in the organization of communities

Emigration to the United States is a permanent demographic scenario in Puerto Rico since 1898. The US Census of Population marked a historic demographic point in 2004 when it reported that the number of Puerto Ricans living in the mainland exceeded the population in the island. *The Wall Street Journal* introduced the label of "refugees" to describe the economic crisis and migrations taking place from Puerto Rico to mostly the southern states (Toro-Morn and García, 2013). In 2017, before the hurricanes, the population was 3.34 million (León, 2018). Pérez (2016, p. 197) highlighted a loss of 600,000 people for an average of 164 people per day in 2016. Local projections suggest a population of 2.9 million in 2050. In addition to the population loss, Puerto Rico has one of the lowest sex ratios worldwide. In 2006, it was estimated at 92 and 91 a decade later. In the year 2017, the estimated shrank to 90.[7] This demographic shift is replicated in the political presence of women. Acevedo Gaud (2013) identified a constant, but slow, increase of women in local politics (register voters, participation and elected officials) since the 1990s. This presence includes leadership at the community level.

Historically, community organizations have served as a social trench against powerful interests from the private-corporate sector. They have also been the centers of the environmental debates and education (Concepción, 1995). In a local radical approach, Vázquez-Torres (2015) asserts that Puerto Rican society can be transformed exclusively from the community bases. His political argument emanates from the following principle: "Democracy must not be representative, but participatory and more than that, decisional in the hands and minds of communities" (p. 59, translation from Spanish). Cuadrado et al. (2016) endorsed this grassroots perspective when the Puerto Rican Planning Society (SPP) proposed a new sustainable planning model for

Puerto Rico. García et al. (2018) also validated the approach describing flourishing experiences in several communities before the hurricanes. Both highlight the participation of women as important mediators in community organization and militancy.

The military exercises of the US Navy in the island-town of Vieques is an emblematic event of women leadership in Puerto Rico. For decades, local communities managed to discontinue the use of the island as a bombing zone and military exercises of the armed forces. For 50 years, the US Navy used over 70% of this island-town as a center for military exercises including bombing. Residents, fishermen, environmental groups, and anti-war organizations confronted military leaders and politicians to discontinue military activities. Community leaders were labeled as leftist and jailed by the federal government. With males confined, females became the community leaders in Vieques. Kadalie (2009) claims that much of the outcome was based on the female presence as community leaders. The US navy left Vieques in 2000 and the events served as an aspiration for hundreds of communities with different struggles across the island.

Vieques became a showcase of community organization, empowerment, and resilience education in Puerto Rico. In 2001, the local government recognized this success and implemented a set of legislation oriented to formalize the organization of low-income communities. The Special Communities Office (*Oficina de Comunidades Especiales*) was created as the main project of the Commonwealth's administration.[8] Despite the overwhelming efforts for empowerment since 2001, the basic infrastructure in hundreds of communities continues to be inadequate against major natural hazards such as hurricanes and earthquakes.

The 2017 hurricane season in the Caribbean

Puerto Rico has a "sizable potential for natural disasters" (Palm and Hodgson, 1993, p. 280). The island is located within the natural track of tropical cyclones in the Atlantic Ocean. Tropical depressions, storms, and hurricanes move at will along this path. These systems originate annually in the African tropical western coast between April and November. The Caribbean region has an average of 12.1 cyclones annually since 1982. The island has been impacted by 94 cyclones (including 29 hurricanes) between 1867 and 2017 (Castro and López, 2018).

Hurricane Irma was a Category 5 event with winds up to 185 MPH and an average speed of 16 MPH. Rainfall ranged between 10 and 15 inches. During the first week of September, Irma caused damage to the islands of Barbuda, St. Bart, St. Martin, and St. Thomas, passing 55 miles north of San Juan (see Figure 23.1). Its passage through Puerto Rico left 1,079,465 million residents (33%) without electric power and 100,000 people without water. Over 2,800 people were displaced to shelters. In spite of the damage recorded, local organizations were able to bring aid to the Lesser Antilles, especially in receiving refugees from the American Virgin Islands.

Two weeks later, Hurricane María crossed Puerto Rico with 155 MPH winds from southeast to northwest. Record-breaking rainfall (Kellings and Hernández-Ayala, 2019) ranged up to 37.9 inches in the central mountains (Cordillera Central) during the first 48 hours. The natural landscape, including major rain forests were significantly depleted (Feng et al., 2018). Its trajectory affected the road system, electric grid, telecommunications, agriculture, housing, and other urban activities. Floods, rock falls, landslides, falling trees caused blockages of the roadway system.[9] The Puerto Rico Department of Transportation and Public Works identified 1,500 segments of urban and rural roads with substantial damages. An estimated of 50,000 landslides. Storm rainfall affected 1,220 signalized intersections and 16.6% of the bridges in the island (Ríos et al., 2018). Government officials estimated damages and losses at $94 billion.

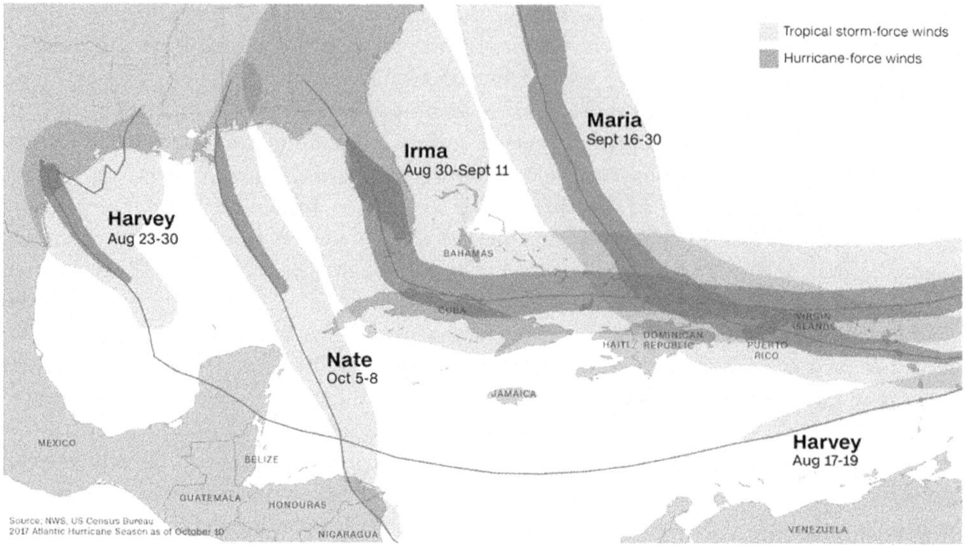

Figure 23.1 Major tropical storms in the Caribbean in 2017

María ranks as the third most expensive hurricane in the United States.[10] According to reporter Nicole Goodkind:

> Puerto Rico has lost 3.4 billion customer-hours of electricity service because of Hurricane María, making it the largest blackout in U.S. history and the second largest in the world, behind only that caused when Typhoon Haiyan hit the Philippines in 2013.
>
> *(Newsweek online, March 18, 2018)*[11]

The blackout defined resilience in the Puerto Rican society. Upgrading and modernizing the electric grid became a priority to public officials and community leaders. No rehabilitation was possible with an obsolete and vulnerable infrastructure. Table 23.1 compares estimated damages and economic losses in Puerto Rico with Florida and Texas in 2017.

Twenty-two (22) *Municipios* were affected by the center of the Hurricane María.[12] Eighteen (18) of these places have a poverty level above the island's average and 12 had a population with over 50% of the families living under the US poverty level (Segarra, 2018). The first official reactions were to request that residents keep calm, stay at home or in shelters, and avoid movement on the affected roadway system. A state of emergency was signed by the governor with a police curfew from 7:00 PM to 5:00 AM. The survival mode of many Puerto Ricans after the hurricanes provided an easy lane to bring "big ideas". During a newspaper interview, Roy Brown, a local songwriter argued:

> What is happening is so outrageous that you would expect everyone to take to the streets to protest. . . . But most of the people do not understand because of the great confusion caused by the blow of María.
>
> *(Carrero Galarza, Hurricane María has made Puerto Rico vulnerable to exploitation, The Morning Call 2018)*

Table 23.1 Estimated damages and economic losses in Texas, Florida, and Puerto Rico after Hurricanes Harvey, Irma, and María, respectively

	TEXAS, H. Harvey issued on August 26, 2017				*FLORIDA H. Irma issued on September 12, 2017*				*PUERTO RICO H. Maria issued on November 27, 2017*			
Area (squares miles)	52,446				56,623				3,452			
Number of census tracts	1,424				4,207				904			
Number of people in the region	7,178,872				18,801,310				3,725,789			
General building stock	Building Count		Dollar Exposure ($M)		Building Count		Dollar Exposure ($M)		Building Count		Dollar Exposure ($M)	
Residential (R)	2,288.678		596,747		6,629,491		1,755,905		1.390,338		230,498	
Commercial (C)	108,746		71,995		430,910		329,804		4,900		2,667	
Other (O)	51,128		37,746		201,832		133,955		1,381		909	
Total	2,448,552		706,487		7,262,161		2,219,664		1,386,619		234,074	
Number of buildings damaged	R	C	O	T	R	C	O	T	R	C	O	T
Minor	21,000	900	400	22,000	290,00	16,000	7,700	310,000	300,000	700	200	300,000
Moderate	9,300	500	300	10,000	58,000	6,000	2,300	66,000	270,000	1,400	400	270,000
Severe	3,000	300	200	3,400	3,800	1,000	600	5,400	160,000	1,200	400	160,000
Destruction	2,200	<10	10	2,300	2,2 00	20	40	2,300	19,000	20	20	19,000
Total	35,000	1,700	900	38,000	350,000	23,000	11,000	380,000	750,000	3,300	1,000	750,000
Displaced households	3,500				12,000				130,00046			
Short-term shelter (#people)	900				2,900				49,000			

Source: HAZUS, Econometrika, 2019

As days passed, the country remained with no communications (phone and Internet), no service from the outdated electric grid, or drinkable water. FEMA satellite phones did not correctly operate in the Caribbean and the agency did not have a proactive plan to address the emergency in the island. The government priorities were hospitals, seaports, and airports. With the arrival of electric power to the airport, emigration to the United States intensified. The country lost 6% of its population after María. This migration was promoted as "humanitarian aid" by federal officials (Criollo Oquero, 2018). For many families, the alternative was to take a free flight to Florida as refugees (Healy and Ferré, 2017; Matthews, 2017; Rojas, 2017; Stone, 2017). This was the first shocking response of thousands of families after the hurricanes.

Puerto Rico became, again, an experimental ground. FEMA implemented a pilot program. After Hurricane María, this agency implemented an untested protocol in coordinating contracts with state and local agencies, processing security, clearances of personnel, and staff recruitment procedures. According to the report *The United States' Actions after Hurricanes Harvey, Irma and María* (2019), this implementation triggered delays and limited the number of participants in the federal response.

The governmental chain of command collapsed (Robles, 2018; Newkirk, 2018; Sullivan, 2018). Commonwealth officials were not aware of the daily reality. Governmental neoliberal plans spawned consecutives shocks to the families and communities not only in the island but also in the United States. The official response became an illusion. Constant interruptions in communications, disagreements between agencies and bureaucratic processes, political controversies and corruption became obstacles during the first six weeks after the hurricanes.

Polarized reactions post-hurricanes

The devastation of the hurricanes led the local government to broadcast an optimistic discourse focusing in building a modern-day infrastructure and developing a new model for economic growth. "*Puerto Rico se levanta*" (Puerto Rico Rise) and "*Yo no me quito*" (I don't quit) were the official mottos to boost recovery and reconstruction efforts. The obsolete and demolished infrastructure were conceived as an opportunity to craft a dream place with newer players. The axis of the recovery was to create a seductive fiscal environment. Conversion as a tax haven and a retirement paradise for opulent sectors were strategies compatible with the conversion into the 51st state. Journalistic reports by Bruno (2018) in Washington and Vizcaino (2018) in Florida cited Governor Ricardo Rosselló describing this opportunity as a *lienzo en blanco* "the blank canvas" to start over. Costa (2018) reported on the potential of converting Puerto Rico into the "Hong Kong of the Caribbean" within the crypto-currency industry (*bitcoin*). This potential would turn the island into "Puerto Crypto" instead Puerto Rico.

Government officials sought to capitalize on the destruction of the hurricanes with speedy legislation under a "state of emergency". This was an historical opportunity to carry out their economic and political agenda: transform the island into a neoliberal paradise politically integrated to the United States of America. The projects included privatization of public services, sales of government assets, changes in the land use and zoning ordinances. The proposed economic environment was designed to be feasible to private energy providers, developers, speculators, and bankers.

The re-introduction of the statehood bill was another shock. The political integration to the "mainland" was re-introduced as part of the governmental discourse for post-hurricanes reconstruction. Dr. Ricardo Rosselló, Governor of Puerto Rico argued that the real problem in the reconstruction of the island was its colonial status. He testified before the US Senate Energy and Natural Resources Committee:

> 17 months after a long process of recovery, and when faced more than $100 billion in damages, it is evident that the lack of equality in treating Puerto Rico, in comparison with other jurisdictions in the United States, threatens our ability to ensure that a recovery and reconstruction is carried out successfully.
>
> *(Hernández, 2019, p. 6)*

Beyond the privatization of the electric system and the proposal for integration to the mainland, the government also announced shutting down public schools; privatizing health systems, nursing homes, and collective transport; permanently closing government offices; and reorganizing the public education system. These decisions included a 50% downsizing of the public university system.

The practices of the government of Puerto Rico and the United States Congress (through PROMESA) illustrates the principles of *Capitalism of the Disaster* argued by Naomi Klein in *The Shock Doctrine: The Rise of Disaster Capitalism* (2008). For this reason, a group of college professors invited the Canadian reporter to Puerto Rico in January 2018. Klein shared her views with environmentalists, community leaders, students, schoolteachers, and public officials. The Puerto Rican experience served to upgrade her premises on the shock doctrine. Irma and María provided the framework for the SASAS (Shock After Shock After Shock) doctrine. Months later she published *The Battle for Paradise; Puerto Rico Takes on the Disaster Capitalists.*

The colonial structure had an imperative function in developing internal polarized reactions. International assistance was controlled by the United States. The aid from neighboring countries, such as medical assistance from Cuba, fuel from Venezuela, and products from Dominican Republic or Colombia were rejected or delayed by the State Department in Washington. This restriction increased the isolation feelings and shock reactions in the island.

Mainstream media played an important role in exporting the shock reactions of Puerto Ricans. Reporters (local and international) projected and constructed an apocalyptic desolation of the areas affected by the hurricanes. The lack of communication between local residents and families in the United States raised the imaginary of a "devasted landscape" occupied by "powerless" residents (Figures 23.2 and 23.3). For the Commonwealth government, these images were effective propaganda to convert the island into a disaster zone and re-orient the redevelopment with a neoliberal compass. With a "disaster zone" declared by the President of the United States, local officials expected more resources for the transformation of Puerto Rico.

Behind this devastation and away from the shocking tragedies exposed by newscasters and political debates, hundreds of communities were in the reconstruction of their environs and basic infrastructure. Llorens (2018) expresses this bias:

> However, rather than depicting a people busy rebuilding their lives in the midst of an unfolding disaster, photojournalists instead brought images to the US mainland that depicted a powerless people. The images that circulated in the aftermath of Hurricane María in mainstream media showed unimaginable suffering and devastation without context of resolution. (p. 137)

The hurricanes unbolted small-scale fronts from communities. The reconstruction begun without a direct support from the Commonwealth and federal governments. In many instances,

Figure 23.2 "Devasted and powerless landscape" in the rural and urban areas of the island projected by mainstream media

Source: Photo by Gabriella N. Báez (2017)

Figure 23.3 "Devasted and powerless landscape" in the rural and urban areas of the island projected by mainstream media

Source: Photo by Gabriella N. Báez (2017)

political and personal issues between public officials (Governor of Puerto Rico, Mayor of San Juan, President of the United States), bureaucracy and institutional inexperience became an obstacle. For the local residents, the governmental agenda was ineffective and unrealistic. The only alternative was to focus on their communities as the path for reconstruction and recovery.

Community resilience as a grassroots response

The concept "resilience" became a cliché after the hurricanes. Public officials, media, politicians, researchers, and academics gave different meanings and interpretations. In 2018, García and Sayers redefined the concepts in order to understand the Puerto Rican context. They delineated community resilience as the "ability of communities to resist and mitigate the stress caused by a disaster" (p. 294). Resilience must be assessable. To measure the degree of community resilience, they adopted the classification from Hall and Zautra (2010) and identified six (6) types of infrastructures: physical (housing, roads, community use), social (organizations and associations of neighbors), civic (committees, agreements of understanding), economic (jobs, salaries), human (education, recreation) and health (medical services, health services).

The first local response after the hurricanes focused on small-scale physical infrastructure. Housing reparation, clearing roadways, removing debris and providing roadway access were the major task. These tasks strengthened neighborhood bonding. In many urban areas, security became a vital issue. Despite the official curfew, nightly lootings became a major problem. Many middle-class communities and exclusive residential areas were forced to organize residents to guard their blocks.

The accessibility to financial assistance, specifically from federal and state agencies became another priority in the communities (Segarra, 2018). Most of the resilience's experience gravitated around this task. Puerto Rico was declared a disaster area in November 1, 2017. From this moment, families were eligible to apply for assistance under the Hazard Mitigation Grant Program. Over $5 billion was granted by the US Congress. Organized communities were able to create informal centers with the Internet as social infrastructure. Residents used these facilities to find assistance, apply to aid programs, and follow up their application status. Communication companies, teachers, social workers, and students volunteered in assisting community residents. Filling out the online applications using limited Wi-Fi spots became an important element in the social and civic infrastructure. Other services such as counseling, workshops, food, and entertainment were provided in these centers.

Nina (2017) described the resilience process within the Dominican community in San Juan Municipio with a singular experience. Dominicans are the largest minority group in Puerto Rico. According to the 2010 census, the population was 63,647 but local leaders estimated 250,000 (Páres, 2018). Nina argued that Dominican residents and retailers were more organized as a community because they did not expect, as immigrants, any government assistance (local and federal). Dealing with no electric power and other limitations was not new in their country. They were prepared for the events. Days after the hurricanes, many Dominican retailers were open using fuel plants to provide ice, water, fuel, and other basic products. These communities became part of the most visited places by Puerto Ricans after Hurricane María.

Lack of electric energy was a major concern in communities and neighborhoods. Electricity returned in many communities up to seven months after the hurricanes. Upgrading and modernizing the electric grid was a priority, not only for the government but to many community leaders in southern Puerto Rico such as in San Sebastián Municipio. Small-scale non-fossil projects such as solar panels and the local management of sedimented old hydroelectric plants were some of the projects proposed by community leaders. Communities organized as

Figure 23.4 Manuel Rojas School transformed into a residential building for families affected by the hurricanes in Lares, Puerto Rico

Source: Photo by Carlos J. Guilbe

cooperatives were promoting this integration (Caquías Cruz, 2019). Contrary to the community efforts, government officials sought to capitalize on the destruction of the hurricanes with speedy legislation under a "state of emergency". Public officials centered in the sale of the Electric Power Authority, the largest public utility in the United States. The privatization of the electric system was a trounced project inserted as part of the effort in reconstructing the country. Similar clashing courses of actions between community initiatives and the governmental discourse occurred in shutting down public schools, health services, transportation, and college education.

In Lares Municipio, female-headed households transformed an abandoned primary school into a residential building for families that lost their residents during the hurricanes (Figure 23.4). They integrated solar plants to provide energy to the facilities. The school lunch room became a community center and the school yard a community-based farm. A cooperative was organized as a coffee shop and catering services. The community project integrated donations from local retailers and also served as an example to other rural communities to use vacant buildings as permanent shelters (Meléndez-García, 2018 and Graulau, 2018). Other small-scale projects such as the community-based agro-ecological farming in Orocovis Municipio were also led by females as community leaders (Gies, 2018).

Similar to the Vieques experience in 2003, female leadership became visible during the clashing, reconstruction, and recovery. Mariel Cruz remarkably summarized this experience in *Vogue Magazine*:

> Five months later, the women of Puerto Rico are moving fast. In the days and weeks and months after Maria, they've waded into flooded neighborhoods to extricate the abandoned, and put together soup kitchens to feed the hungry. They've canvassed their communities in order to diagnose the most critical needs – street by street, mountain by mountain, house by house, family by family – and have returned when they said they would with supplies and support. They've created free legal aid societies to

> help families navigate the confusing and ill-designed processes required to file FEMA claims, and connected Puerto Ricans with aid groups far more active and impactful. They've raised money and rebuilt roads and devised innovative mass communication methods in light of limited or no electricity or internet access.
>
> *(March 8, 2018)*

Women have a leadership presence beyond the community level. Local politicians also were recognized locally and internationally as representatives of the reconstruction of the island. The voices from Puerto Rican mayors, senators, and other public officials are, for the first time in local history, female. This shift was also summarized by Cruz (2018) as follows:

> The women of Puerto Rico have spoken up about subpar leadership and have challenged the inequalities, the broken systems, and have even called out an ignorant, out-of-touch president live on television (as well as firing back at him on *Twitter*). By empowering themselves – and each other – the women of Puerto Rico have empowered the entire island.
>
> *(March 8, 2018)*

The hurricanes were a wake-up call to communities and its leaders. Puerto Ricans understood the importance to organize at a neighborhood scale. In isolated geographical areas, where reconstruction was slow or ineffective, communities organized civic infrastructure to claim attention. This situation spawned an entitlement process and provided the frame for permanent communal organization.

Communities with a well-defined organization before the hurricanes were more successful in mitigating problems than neighborhoods without any organizations. The tropical storms provided the bases for local organizations. In addition to the hurricanes and related atmospheric events, the island is located in a region of intense seismic activity. Food security, green energy, health services for the elderly and handicapped, environmental education, alternative housing, and security are some of the issues to be addressed as part of the community resilience in Puerto Rico.

For many Puerto Ricans, emigration is not an option. Over 2.5 million still live on the island and many in the diaspora expect to return. They recognize that hurricanes (and earthquakes) are a continuous threat. In addition to the issues of energy and food, environmental education will become important in framing community resilience. According to the World Meteorological Organization (WTO):

> on the hurricane season in the Caribbean came to the conclusions that significant changes to early warning systems are necessary in order to strengthen resilience. To this end, a regional strategy must be put in place that addresses the critical challenges and strengthens institutions, increases resources and builds more resilient operational structures.
>
> *(Eckstein et al., 2018, p. 15)*

The agenda for the development of Puerto Rico from a grassroot perspective is not a simple political issue. The physical geography of the island is forcing many Puerto Ricans to develop a holistic perspective in environmental education. The empowerment gained as a result of the hurricanes may play an important role in facing other important issues such as the never-ending debate about the colonial relationship with the United States.

Conclusion

Hurricanes Irma and María unfurnished the Puerto Rican landscape. Beyond the effects on its land cover, the tropical systems unveiled hidden trenches: the government neoliberal agenda for development as a path for statehood on one side and community organizations implementing small-scale projects as a template for empowerment on the other. In principle, both outlooks are compatible. However, the ideological bases behind each frontage are not.

Public officials tried to capitalize the devastation and "powerless landscape" of the colony to promote an integrated future with the United States. The intensity in exposing non-popular transformations after the hurricanes provided an opportunity for Naomi Klein to upgrade her theory about the shock doctrine. However, the local fiscal crisis, the erratic delay of the federal assistance, and performance of federal and local agencies are eroding the short-term neoliberal agenda. These governmental lags are forcing many residents to organize and design their future. The success of female leadership at the community level is generating a trend in neighborhood planning on the island. After the hurricanes, Puerto Rican women are becoming the frontage of the social movements. For many conservative and mainstream sectors, this entitlement is extremely liberal.

Based on the types of infrastructure proposed by Hall and Zautra (2010), the Puerto Rican experience is organized on human infrastructure. Environmental education at the community level is its backbone. Education gravitates around conservation, empowerment, local sustainability, green energy, food security, and health. However, based on the recent natural events, energy production is the core in environmental education. According to Kellings and Hernández-Ayala (2019), the island will face more record-breaking rainfall, unprecedent flooding, and landslides as part of recent climatic changes in the Atlantic Ocean. Understanding this scenario requires an integration of natural hazards, social response, and local development.

The reconstruction and development of the Commonwealth of Puerto Rico is a long-term task. On the one side, community leaders manage the resilience as part of the country's integrated reconstruction process starting from small-scale projects, taking as a true north food security and energy self-sufficiency, citizen participation, gender equity, and environmental protection. On the other side, the construction of an American state-of-the-art tax tropical haven for a wealthy population is the definition of resilience for government officials. It is David's fight against Goliath among those who aspire to live the "American Dream" on the island and those who believe in small-scale community structures to assemble a genuine and "non-experimental" Puerto Rican nation.

Notes

1 Puerto Rico is the largest American colony. Other territories are the American Virgin Islands and Navassa Island in the Caribbean, Northern Mariana Islands, Guam, American Samoa, Wake Island, Howland Island, Baker Island, Jarvis Island, and other atolls in the Pacific Ocean. See Immerwahr (2019).

2 Nelson Dennis (2016) in *War Against All Puerto Ricans; Revolution and Terror in America's Colony* describes the threats and tortures to thousands of "independentistas" (pro-independence supporters) from the United States and the "insular" government during this period in major military installations and *La Princesa* jail in Puerto Rico.

3 Puerto Rico does not have representation in Congress. In the House of Representatives, the island has one seat, but there is no vote and no representation in the US Senate. American citizens residing on the island have the right to vote in the primaries but not in the general US elections.

4 By the end of the decade, 58% of Puerto Rican families received food stamps (Pérez, 2016, p. 140).

5 Many federal programs provided through shared costs (share). Although minimal or in-kind, local input required resources that were affecting government debt.
6 PROMESA translation into Spanish is "Promise". See H.R. 5278-PROMESA, 114th Congress, congress.gov., http://congress.gov/bill/114th-congress/house-bill/5278
7 According to the World Bank and the United Nations Department of Economic and Social Affairs, eight (8) of the top 25 countries with the lowest sex ratio are in the Caribbean Antilles.
8 Law # 1 of 2001 was the first bill endorsed by Sila María Calderón, the first female governor of the Commonwealth of Puerto Rico. According to Dr. Linda Colón, the first director of this office, over 60% of the community leaders were females.
9 Hu and Smith (2018) found a high correlation between topographic elevation and damages on the land cover in the rural areas island.
10 According to the Office for Coastal Management (National Oceanic and Atmospheric Administration), Hurricane Katrina in 2005 was $160 billion and Hurricane Harvey (2017) with $125 billion (https://coast.noaa.gov/states/fast-facts/hurricane-costs.html).
11 Based on Customer Hours of Lost Electricity Services, also known as CHoLES in electrical engineering.
12 This political unit is not equivalent to *municipality* in the United States. The Puerto Rican *Municipio* has features of the county, but they are not similar. The US Census Bureau does not translate and maintain *Municipio*.

References

Acevedo, Juan Antonio (2010). *Sobrepoblación: ¿Pobreza o riqueza de Puerto Rico?* San Juan, PR: Publicaciones Gaviota.

Acevedo Gaud, Luz (2013). Género y procesos electorales en Puerto Rico. *Ambitos de Encuentros*, Vol. 6, No. 1, pp. 269–305.

Ayala, César and Rafael Bernabe (2011). *Puerto Rico in the American Century: A History since 1898.* Chapel Hill, NC: The University of North Carolina Press.

Berman, Déborah (1993). Colonialism, Resistance, & the Search for Alternatives: The Environmental Movement in Puerto Rico. *Race, Poverty & The Environment*, Vol. 4, No. 3, Latinos & the Environment (Fall), pp. 3–5.

Berríos Martínez, Rubén (2004). *Un mapa para la ruta.* Santo Domingo, DR: Editora Corripio.

Bruno, Josian (2018). Rosselló continua agenda de recuperación tras María. *El Vocero* (April 4). www.elvocero.com/rossell-contin-a-agenda-de-recuperaci-n-tras-mar-a/article_95f9c4b6-4a43-11e8-a212-5bc16405d560.html

Burnett, Christina Duffy (2005). Untied States: American Expansion and Territorial Deannexation. *The University of Chicago Law Review*, Vol. 72, No. 3 (Summer), pp. 797–879.

Caquías Cruz, Sandra (2019). Propuesta para energizar tres pueblos de la cordillera central. *EsNoticia* (March 1–14), pp. 4–5.

Carrero Galarza, Milton (2018). Hurricane María has much Puerto Rico Vulnerable for Exploitation. *ProQuest* (May 8). http://biblioteca.uprrp.edu:2087/docview/203633669?accountid=44825

Castro Rivera, Abimael and Tania del Mar López (2018). *Cartilla de los ciclones.* Sea Grant Program. Mayaguez, PR: University of Puerto Rico, Mayaguez Campus.

Colón, Jorge, Féliz Córdoba and José Córdoba (2014). *El Proyecto de explotación minera en Puerto Rico (1962–1968).* San Juan: Ediciones Huracán.

Colón Reyes, Linda (2013). *Sobrevivencia, pobreza y "mantengo"" La política asistencialista estadounidense en Puerto Rico: el PAN y el TANF.* San Juan, PR: Ediciones Callejón.

Colón Warren, Alice (2003). Puerto Rico: Feminism and Feminist Studies. *Gender and Society*, Vol. 17, No. 5, pp. 664–690.

Colón Warren, Alice (2010). Women Employment and Changing Gender Relations in Puerto Rico, *Caribbean Studies*, Vol. 38, No. 2, Dignity and Economic Survival: Women in Latin America and the Caribbean and the Work of Helen I. Safa, pp. 59–91.

Concepción, Carmen (1995). The Origins of Modern Environmental Activism in Puerto Rico in the 1960s. *International Journal of Urban and Regional Research*, Vol. 19, No. 1, pp. 112–128.

Costa, Dennis (2018). San Juan será la sede de varias conferencias sobre el "blockchain". *El Nuevo Día* (March 12). www.elnuevodia.com/tecnologia/tecnologia/nota/sanjuanseralasededevariasconferenciassobreelblockchain-2405690/

Criollo Oquero, Agustin (2018). Puerto Rico lost 6% of its population after Hurricane María. *Caribbean Business* (March 9). https://caribbeanbusiness.com/puerto-rico-lost-6-of-its-population-after-hurricane-maria/
Cruz, Mariel (2018) Puerto Rico Transformation: Five months after Hurricane María, the rebuilding effort in Puerto Rico is powered by women. *Vogue Daily* (March 8). www.vogue.com/projects/13542078/puerto-rico-after-hurricane-maria-2/
Cuadrado, Luz, Lucilla Fuller, Mariecel Maldonado, Evelyn Moreno y Mari Villarini (2016). *Planificación para un Puerto Rico sostenible: Fundamentos del proceso.* San Juan, PR: Sociedad Puertorriqueña de Planificación.
Davis, Kingsley (1953). Puerto Rico: A Crowded Island. *The Annals of the American Academy of Political and Social Science*, Vol. 285, Puerto Rico a Study in Democratic Development (January), pp. 116–122.
Dennis, Nelson (2016). *War Against All Puerto Ricans: Revolution and Terror in America's Colony.* New York: Nation Books.
Earnhardt, Kent (1982). *Development Planning and Population Policy in Puerto Rico: From Historical Evolution Towards a Plan for Population Stabilization.* Rio Piedras, PR: Editorial de la Universidad de Puerto Rico.
Eckstein, David, Marie-Lena Hutfils and Maik Winges (2018). *Global Climate Risk Index 2019: Who Suffers Most from Extreme Weather Events? Weather- Related Loss Events in 2017 and 1998 to 2017* (Briefing Papers). Bonn: Germanwath e.V.
Econometrika Corp (2019). *The United States' Actions After Hurricanes Harvey, Irma and María.* Report. San Juan, Puerto Rico.
Estades Font, María Eugenia (1999). *La presencia militar de Estados Unidos en Puerto Rico 1808–1918: Intereses estratégicos y dominación colonial.* Río Piedras, PR: Editorial Huracán.
Feng, Y., R. Negrón-Juarez, C.M. Patricola, W.D. Collins, M. Uriarte, J.S. Hall and J.Q. Chambers (2018). Rapid Remote Sensing Assessment of Impacts from Hurricane María on Forests of Puerto Rico. *PeerJ PrePrints.* http://biblioteca.uprrp.edu:2062/10.7287/peerj.preprints.2659v1
Font-Guzmán, Jacqueline (2015). *Experiencing Puerto Rican Citizenship and Cultural Nationalism.* New York: Palgrave Macmillan.
García, Gustavo, Carmen Concepción y Alejandro Torres Abreu. Editores (2018). *Ambiente y Democracia; Experiencias de la gestión comunitaria ambiental en Puerto Rico.* Río Piedras, PR: Editorial de la Universidad de Puerto Rico.
García-Robles, Rut y Sean Sayers Montalvo (2018). *Resiliencia; Medición de la fortaleza interna.* Hato Rey, PR: Publicaciones Puertorriqueñas.
Gies, Heather (2018). *Agroecology as a Toll of Sovereignty and Resilience in Puerto Rico after Hurricane María.* https://civileats.com/2018/10/19/agroecology-as-a-tool-of-sovereignty-and-resilience-in-puerto-rico-after-hurricane-maria/
Goodkind, Nicole (2018). Puerto Rico's Hurricane María Power Outrage is now the World's Second Largest Blackout. *Newsweek* (April 12). www.newsweek.com/puerto-rico-power-hurricane-maria-blackout-882549
Graulau, Bianca (2018). Puerto Ricans Who Lost Homes After Hurricane Maria Take Over Abandoned Schools. *WTSP* (May 23). www.wtsp.com/article/news/local/puerto-ricans-who-lost-homes-after-hurricane-maria-take-over-abandoned-school/67-557717893
Grosfoguel, Ramón (2003). *Colonial Subjects: Puerto Ricans in a Global Perspective.* Oakland: University of California Press.
Guilbe, Carlos (2016). Operación Manos A La Obra: A Key for Development in Puerto Rico? In *Race, Ethnicity, and Place in a Changing America.* 3rd edition. John Frazier, Eugene Tettey-Fio and Norah Henry (Editors). Albany, NY: State University of New York Press.
Hall, J.S. y Zautra, A. (2010). Indicators of Community Resilience: What They, Why Bother? In *The Handbook of Adult Resilience* (pp. 350–371). J.W. Reich, A.J. Zautra and J.S. Hall (Editors). New York: The Guilford Press.
Hanson, Earl (1949). *New Worlds Emerging.* New York: Duell, Sloan, Pearce.
Healy, Jack and Luis Ferré (2017). For Many Puerto Rican, the Most Coveted Item is a Plane Ticket Out. *New York Times* (October 5). www.nytimes.com/2017/10/05/us/puerto-rico-exodus-maria-florida.html
Hernández, J.A. (2019). Slow Recovery Due to Colonial Status, Rossello Says, *The San Juan Daily Star*, Wednesday, February 27, page 6.
Hill, Reuben, J. Mayone and Kurt Back (1959). *The Family and Population Control; A Puerto Rican Experiment in Social Change.* Chapel Hill, NC: University of North Carolina.

Hu, Tangao and Ronald Smith (2018). The Impact of Hurricane Maria on the Vegetation of Dominica and Puerto Rico Using Multispectral Remote Sensing. *Remote Sensing*, Vol 10, No. 6, p. 827. http://doi.org/10.3390/rs10060827

Immerwahr, Daniel (2019). How the US has hidden its empire. *The Guardian*. www.theguardian.com/news/2019/feb/15/the-us-hidden-empire-overseas-territories-united-states-guam-puerto-rico-american-samoa?CMP=Share_iOSSpp_Other

Kadalie, Modibo M. (2009). The Emerge of Women in the Leadership of a Struggle for Ecologica, Justice; The Case of Vieques. *Race, Gender & Class*, Vol. 16, No. 3 and 4, pp. 99–108.

Kellings, David and José Hernández Ayala (2019). Extreme Rainfall Associated with Hurricane Maria Over Puerto Rico and Its Connections to Climate Variability and Change. *Geophysical Research Letters*, Vol. 46. https://doi.org/10.1029/2109GL)82077

Klein, Naomi (2008). *The Shock Doctrine: The Rise of Disaster Capitalism*. New York: Picador Publisher.

Klein, Naomi (2017). *No is Not Enough: Resisting Trump's Shock Politics and Winning the World We Need*. Chicago, IL: Haymarket Books.

Klein, Naomi (2018). *The Battle for Paradise; Puerto Rico Takes on the Disaster Capitalists*. Chicago, IL: Haymarket Books.

Lapp, Michael (1995). The Rise and Fall of Puerto Rico as a Social Laboratory, 1945–1965. *Social Science History*, Vol. 19, No. 2 (Summer), pp. 169–199.

León, Lynnette (2018). Rebuilding Puerto Rico: The Aftermath of Hurricanes Irma and María. *Institute of Transportation Engineers. ITE Journal*. Vol. 88, No. 7, pp. 22–27, ProQuest. https://biblioteca.uprrp.edu:2016/docview/2077797217?accountid=44825.

Llorens, Hilda (2018). Imaging Disaster: Puerto Rico Through the Eye of Hurricane María, *Transforming Anthropologies*. Vol. 26, No. 2, pp. 136–156.

Matthews, Dylan (2017). What the Hurricane Maria Migration will Do to Puerto Rico-and the US. *Vox* (October 5). www.vox.com/policy-and-politics/2017/10/5/16403952/hurricane-maria-puerto-rico-migration

McCoy, Alfred, Francisco Scarano and Courtney Johnson (2009). On the Tropic of Cancer: Transitions and Transformations in the U.S. Imperial State. In *Colonial Crucible: Empire in the Making of Modern American State*. McCoy, Alfred and Francisco Scarano (Editors). Madison, WI: The University of Wisconsin Press.

Meléndez-García, Lyanne (2018). Plantel escolar abandonado se convierte en viviendas para familias. *Metro PR* (April 18). www.metro.pr/pr/noticias/2018/04/18/plantel-escolar-abandonado-se-convierte-vivienda-familias.html

Negrón Muntaner, Frances (2018). *The Emptying Island: Puerto Rican Expulsion in Post Maria Time*. http://beta.hemispherinstitute.org.

Newkirk II, Vann R. (2018). The Situation of Puerto Rico in Untenable. *The Atlantic* (September 20). www.theatlantic.com/politics/archive/2018/09/hurricane-maria-anniversary-puerto-rico-trump/570928/

Nina, Daniel (2017). *El gas en fila*. Río Piedras, PR: Puerta del Sur Editores.

Ojeda, Felipe (2019). Naomi Klein: Soy una anticapitalista, este sistema está en guerra con nuestro ecosistema. *Culto* (February 13). http://culto.latercera.com/2019/02/02/naomi-klein-anticapitalista/

Palm, Risa and Michael Hodgson (1993). Natural Hazards in Puerto Rico. *Geographical Review*, Vol. 83, No. 3, pp. 280–289.

Páres Arroyo, Marga (2018). Perfil socioeconómico muestra que la población dominicana enfrenta grandes retos. *El Nuevo Día* (September 24). www.elnuevodia.com/noticias/locales/nota/perfilsocioecono micomuestraquelapoblaciondominicanaenfrentagrandesretos-2448959/

Pérez, Silverio (2016). *La vitrina rota o ¿qué carajo pasó aquí?* San Juan, PR: Ediciones Callejón.

Picó, Rafael (1953). The Role of Planning in Puerto Rico. *The Annals of the American Academy of Political and Social Science*, Vol. 285, Puerto Rico a Study in Democratic Development (January), pp. 70–75.

Ríos, Benjamín Colucci, Alberto Medina and A.M. Santiago (2018). Lessons learned for Puerto Rico's Transportation Infrastructure after Hurricane María. *Institute of Transportation Engineers. ITE Journal*, Vol. 88, No. 10, pp. 38–43. https://biblioteca.upr.edu:2068/docview/2122485801?accountid=44825

Robles, Frances (2018). FEMA was Sorely Unprepared for Puerto Rico Hurricane Report Says. *New York Times* (December 20). www.nytimes.com/2018/07/12/us/fema-puerto-rico-maria.html

Rodríguez Beruff, Jorge and Josè Bolivar Fresneda. Editors (2012). *Puerto Rico en la Segunda Guerra Mundial: Baluarte del Caribe*. San Juan: Ediciones Callejón.

Rodríguez Beruff, Jorge and José Bolivar Fresneda. Editors (2015). *Puerto Rico en la Segunda Guerra Mundial: El Escenario Regional*. San Juan: Ediciones Callejón.

Rojas, Rick (2017). Added to the Storm's Toll, Guilt and Heartache for Puerto Ricans Who Fled. *New York Times* (November 26). www.nytimes.com/2017/11/26/nyregion/puerto-rico-hurricane-maria.html

Safa, Helen (1998). Female-Headed Households in the Caribbean: Sign of Pathology or Alternative Form of Family Organization? *The Brown Journal of World Affairs*, Vol. 5, No. 2, pp. 203–214.

Segarra, Aileen (2018). María y la vulnerabilidad en Puerto Rico. *Revista de Administración Pública*, Vol. 49, pp. 13–38.

Smith, Neil (2003). *American Empire; Roosevelt's Geographer and the Prelude to Globalization.* Berkley: University of California Press.

Stone, Lyman (2017). How Low Will Puerto Rico's Population Go? *Medium* (November 17). https://medium.com/migration-issues/how-low-will-puerto-ricos-population-go-c8d108ac8b3b

Sullivan, Laura (2018). FEMA Reports Acknowledge Failures in Puerto Rico Disaster Response. *National Public Radio* (July 13). www.npr.org/2018/07/13/628861808/fema-report-acknowledges-failures-in-puerto-rico-disaster-response

Taylor, Milton (1952). Neo Malthusianism in Puerto Rico. *Review of Social Economy*, Vol. 10, No. 1 (March), pp. 42–54.

Toro-Morn, Maura and Ivis García Zambrano (2013). Gendered Fault Lines: A Demographic Profile of Puerto Rican Woman in the United States. *Centro Journal*, Vol. XXIX, No. III, pp. 10–37.

Vázques-Torres, Rubén Dario (2015). *Las luchas políticas comunitarias: Comunidades Boricuas/pobres.* Río Piedras, PR: Publicaciones Gaviota.

Vizcaino, Elena (2018). Rick Scott: Puerto Rico tiene una oportunidad "increíble" para reconstruir después de María. *Orlando Sentinel* (June 20). www.orlandosentinel.com/elsentinel/puerto-rico-florida-huracan/os-es-florida-orlando-rick-scott-rossello-maria-summit-cumbre-p3-gobernador-20180619-story.html

White, Byron (1963). Puerto Rico. A Partial Developmental Model. *The American Journal of Economics and Sociology*, Vol. 22, No. 4 (October), pp. 539–542.

24

GENDERED ACCESS TO WETLAND GARDENS (*DIMBA*) IN NORTHERN MALAWI

Rhoda Nyirenda

Introduction

Due to increasing degradation of upland fields and erratic rains, smallholder farmers in Africa have resorted to wetland cultivation for livelihoods security (Mutambikwa et al., 2000). In rural parts of sub-Saharan African countries such as Malawi, Zambia, Mozambique, Kenya, Tanzania, and Ethiopia, smallholder farmers are heavily dependent on wetland cultivation. Wetlands have been cultivated since 300 AD due to their reliable water supply (Kotze, 2011). The practice has varied across countries. A cultivatable wetland area is considered prime land, and a desired opportunity that every rural family needs for the purpose of food production. Research conducted in Simlemba, central Malawi, indicated that wetland cultivation accounts for 37% of domestically consumed food and generates 55% of cash income per household (Msusa, 2011). A similar analysis in Cyabayanga in Rwanda found that the per household contribution of wetland cultivation to gross margin was 74% (Nabahungu and Visser, 2011). In Tanzania, Mahenya (2008) found that the contribution of wetland cultivation to total household cash income was more than 60%. Comparing production of upland fields and wetland gardens Nyamadzawo et al. (2015) observed that smallholder farmers in Zimbabwe harvest more from wetland gardens (i.e. an average of 2–3 t ha^{-1}) than upland fields which yield less than 1 t ha.$^{-1}$ And according to Dixon (2005) dry season cultivation of maize in the wetland has significantly helped smallholder farmers to survive the famous long "hungry" season in Ethiopia. This study used a mixed methods approach to understand gendered access to wetlands in northern Malawi as adaptation to climate change.

In Malawi, wetland agriculture takes place as formal agreements that include government irrigation schemes and informal irrigation arrangements, which include smallholder wetland gardens locally called *dimbas*. Estimates have shown that 123,000ha are under informal wetland cultivation compared to 27,000ha under formal wetland irrigation (Kambewa, 2005). Wetland gardens play a vital role in rural livelihood security and resilience to weather and economic shocks. Through wetland garden cultivation smallholder farmers can grow different crops more than once a year, thereby improving rural income and food availability. In the face of increasing occurrence of droughts and floods and increasing food prices, wetland gardens also act as a means of improving rural livelihoods' resilience. Through wetland gardens, smallholder farmers and communities survive long dry seasons and avoid seasonal hunger. However, wetland

 DOI: 10.4324/9780429450310-28

agriculture in Malawi and most of the sub-Saharan African countries is marred with issues of gendered access to land (Simon, 2014; Kambewa, 2005).

Although women are at the center of smallholder agriculture, particularly wetland cultivation, women's access to wetland gardens in terms of either customary law or formal law are nearly invisible in Malawi and southern Africa in general (Helliker and Murisa, 2011). Existing modes of access, allocation, use, and control over wetland gardens continue to ignore women. This is despite existence of the knowledge that if rural women had the same access to agricultural resources (i.e. land) as men, yield could increase by 20–30% and the total number of hungry people around the world could be reduced by 12–17% (FAO, 2012). It is against this background that I hypothesized that improving information about existing modes of access to wetlands is key to development and improvement of wetland cultivation and production and consequently to reducing rural livelihoods' vulnerability.

Practically, this study was rooted in a mixed method approach to study how access, control, ownership, and use of wetlands as adaptation to climate change in Malawi is influenced/governed by social and power relations. I conducted the study in the rural district of Nkhata Bay in northern Malawi. The study zeroed in on these specific questions: What governs ownership, access and use of wetland garden land at community level? How does wetland garden cultivation contribute to food security and livelihoods' adaptation to climate change? Given that Malawi is exposed to frequent adverse climatic shocks, understanding of gendered access to wetland gardens will contribute to improving wetland cultivation and rural livelihoods.

This chapter is organized as follows: first, theoretical underpinnings that explain land tenure in Malawi in relation to political regimes and gender and wetlands are discussed. Then the methods section follows, which is a description of the study area, data, and the data analysis. Then results and discussions are discussed followed by the conclusion.

Theoretical underpinnings

Situated in a mixed method approach, this study drew on land tenure in Malawi, gender, and wetland agriculture as adaptation to climate change.

Land tenure systems in Malawi: Many Malawians depend on land for their livelihoods, employment, and social security (Chinsinga et al., 2013; Kishindo, 2004). Estimates indicate that 84% Malawians directly depend on land for agriculture which employs about 85% of Malawians (Chinsinga et al., 2013). Access to land in Malawi therefore is vital as it is the source of livelihood of the majority. However, land is unequally distributed. Land tenure impacts land distribution in Malawi. The tenure system in Malawi dates to the British government. Under colonial government, land was categorized as either public, private, or customary land. When Malawi attained independence in July 1964, the same land tenure system was adopted. Even with the political transition from a one-party system to multiparty democracy in 1994, the land tenure system remained the same up until 2016. In 2016 the categories of land were reduced in number two: public and private. The intricacies about the distribution of land among the different social groups (i.e. men and women) were not adequately addressed. Further, wetlands were also not adequately covered. This section briefly reviews the land tenure system in Malawi about the colonial period, post-colonial period, wetlands, and gender.

Colonial land tenure: During the colonial period land was defined as public, private and customary. Private land meant all land that was owned, held or occupied under a freehold, leasehold, or a certificate of claim (Chirwa, 2004). Examples to this category included cash crop estates like tea, sugar, coffee, tobacco estates. It was restricted to white settlers who did estate farming. Public land defined as land occupied, used or acquired by the government and any other not being customary or private (Chirwa, 2004). On the other hand, customary was land for use by indigenous people. However, under British colonial law, indigenous people had no formal title to land because customary land was vested in the Secretary of State who was mandated to administer it to the "natives". The colonial state favored private landowners (white settlers) because they promoted the colonial agenda of creating a "capitalist" economy based on estate (or large scale) agriculture (Deininger and Xia, 2017). This led to conversion of indigenous land holdings into private land because customary land was assumed to have zero value (Msisha, 1999). Thus, private landowners (white settlers) had preferential access to land than the indigenous people (Mataya and Tsonga, 2001).

Post-colonial land tenure: After independence, Malawi adopted the English-based Land Act (Chinsinga et al., 2013). The 1965 Land Act maintained the colonial categories of land: private, public, and customary land. The 1965 Land Act recognized customary land as the property of the people of Malawi (Silungwe, 2005). However, the law vested customary land in the president in perpetuity and empowered the traditional leaders to allocate land to families/individuals in their areas. Under the law, the people of Malawi had no enforceable land rights but right of use and occupancy only (Silungwe, 2005). Customary land tenure continued to be assumed insecure. It was perceived as a reserve from which private land was created to further the post-colonial government (under Hastings Kamuzu Banda) development policy of improving agricultural production through estate agriculture (Kanyongolo, 2005). Among the three land categories, customary land dominates in Malawi. Currently, customary land accounts for 80% of the country's total land and supports most of the smallholder farmers. It is on customary land where issues of access are critical.

The 1965 Land Act was the law up until 2016 when a new Land Act was enacted. The 2016 Land Act recognizes two categories of land: private and public land. In the 2016 Land Act, private and public land did not change their meaning. On the contrary, customary land tenure was not recognized in the 2016 Land Act. In order to "secure" the customary land holders through titling and registering customary land, the Act categorizes customary land as either private (customary estate/registered customary land) or public (unallocated/unregistered customary land). The 2016 Customary Land Act disempowers chiefs from allocating land to their community members. The Act places administration of customary land under a village land committee chaired by the chiefs. Similarly, the Act vests all land in Malawi in perpetuity in the Republic (2016 Land Act; section 8) and not in the president. However, on-the-ground titling and registration of customary land has only just started in 2019 in four districts.

Land tenure in the wetlands (dambos): Generally, the land tenure system in Malawi does not distinguish wetlands from other types of land use. No clear distinction is made between upland areas and the wetland (dambo) areas in the land laws. Dambo land is important because it is highly demanded for cultivation. According to Kambewa (2005), Malawi, like many other SADC countries, lacks wetland policy. Even the 2002 land policy made few references to wetlands (dambos), assuming them to be "common access" or "public" land within the overall category of "customary" tenure (Peters and Kambewa, 2007). The 2016 Land Act also does not come out clearly on wetlands. Considering that there are only two categories of land in the 2016 Land Act, wetlands can therefore be assumed to be either private or public depending on whether they are registered or not.

Land tenure and gender: From a gender perspective, land tenure rules and norms are not gender equitable. The Land Acts further obscure the element of gender in land issues. For instance, the 1965 Land Act was gender and class blind. These were entrenched further by the 1967 land reforms. According to Kanyongolo (2013), the 1967 land reforms, instead of facilitating the transformation of class and gender inequalities in land issues, benefited only people who had the purchasing power of leasehold or freehold titles to land. Likewise, the land reforms during the multiparty era do not clearly show gender equality. They give a simplified picture of land rights. For instance, the 2002 land policy simply stated that

> the guarantee of security of tenure shall be "without any gender bias [or] discrimination to all citizens" under section 28 of the Constitution.
>
> *(Government of Malawi, 2001)*

Similarly, the 2016 Land Act, section 22 simply states that

> equality shall be ensured by not favoring certain groups of people and adopting no adverse discriminatory practices towards any person who has applied for a customary estate.
>
> *(Government of Malawi, 2016)*

These assurances do not represent the practical enormous complexity of land rights. A vast range of traditional systems of land ownership and control which overlap with the officially recognized land tenure system (Matchaya, 2009) are left out. The traditional tenure systems often fail to recognize women's rights, particularly in patrilineal societies. Even in matrilineal societies, a WLSA (Women and Land in Southern Africa) study revealed that women's land ownership in matrilineal societies is theoretical, while practically, maternal uncles have power and control over it (WLSA, 2000). This is why under the de facto position, women do not enjoy equal access to/ownership of land due to biases by land administrators though the land law gives both men and women de jure access to and control of land (Mbaya, 2002).

Gender and access to and control of land: Like many African countries, Malawi's major form of agriculture is smallholder. Smallholder agriculture is affected by access to land and labor, the major agricultural inputs. However, to appreciate the implications that gender has on agricultural production in Malawi, one needs to understand the livelihood roles of men and women and how men and women have differential access to land and labor for agriculture.

Most smallholder agriculture in Malawi utilizes customary land, which is titled by the Government of Malawi but administered through traditional authorities to individuals and families. However, customary land is not equally distributed between men and women. Matrilineal and patrilineal cultures determine ownership and access to land. In matrilineal culture, mainly practiced in the central and southern region, women own land through inheritance from their mothers, while males do not own land but only access it through marriage. On the other hand, in patrilineal culture, which is practiced in the northern region where the study was conducted, land is owned by men through inheritance. In this culture, women do not own land but access it through marriage. In cases of divorce, women lose access to land as men are the primary heirs. As a result, rights to cultivation are invested in men. This has a measurable effect on food security given that men decide what type of crops to be grown on their land and they tend to favor cash crops. When these crops are sold, husbands do not always remit income to the household for cooperative decision making. This was observed by Kerr (2005) in Ekwendeni in northern Malawi. In her study, Kerr (2005) also observed that women received less support from kin like land, cash, and food and that they enjoy little or no access to household income.

Gender and agricultural labor: Due to the widespread poverty of smallholder farmers, most use simple tools, such as hand-hoes, to cultivate their farms. Use of hand-hoes and application of fertilizer and harvesting demands a large supply of labor, which is mostly provided by family members. Labor therefore becomes an essential input in agriculture. However, accepted gender roles influence gender division of labor in agricultural production (Karamba and Winters, 2015), especially in societies where men and women perform different tasks or have differential access to household or hired labor (Fisher and Kandiwa, 2014). In cases where men and women perform different tasks, women split their time between home production and agricultural production. Within the household, rural Malawian women usually perform more domestic reproduction tasks than men. They are culturally expected to look after children, the sick, and the elderly and fetch water and firewood and carry out other household chores. Such unshared household tasks hamper women's participation in agricultural production or participation in paid employment (Karamba and Winters, 2015). This may lead to poor harvests as women cannot carry out all their agricultural activities and cannot buy fertilizer as the money is spent on caring for the sick (Kerr, 2005). This becomes worse for women who are in polygamous families. Women in such families are also expected to provide agricultural labor to their husbands or sons, making men have greater access to labor than women.

Apart from gender differences in agricultural labor within the households, studies have revealed that gender differences in agricultural labor also exist between female-headed households and male-headed households (Kassie et al., 2013; Fisher and Kandiwa, 2014). Male-headed households have a higher access to agricultural labor than female-headed households. This is attributed to social norms and access to income. Through social norms, men have command over labor of their wives and female kin; however, female household heads do not have reciprocal command over their male kin for labor. In terms of access to income, male-headed households have greater access to cash than their counterparts. This enables the male household heads to hire labor. Female household heads often have less income due to financial distress after the death of a male head of household or loss of a productive household member (in case the head of the household was a widow) (Kassie et al., 2013). These gender inequalities are not linear, they are linked to other forms of gender inequalities. These inequalities put females/female-headed households at risk of reduced income and food insecurity than male-headed households since few productive agricultural resources are made available to them.

Why gender and wetland agriculture? For more than a decade now, it has come to the attention of development practitioners and policy makers that gender plays an important role in development (Kassie et al., 2013; FAO, 2011; World Bank, 1982). Gender inequalities have been recognized to impede development (Kassie et al., 2013). In search of means to reduce food insecurity and poverty in sub-Saharan Africa, addressing gender inequalities has been argued to be one of the ways that can help to achieve food security and alleviate poverty. Many gender issues have been discussed in agricultural production, but little is known about contribution of gender and wetland agriculture to food security. As discussed earlier, women farmers form an integral part of smallholder agriculture, contributing about 50%–70% of agricultural labor in sub-Saharan Africa, Malawi in particular, but their access to productive agricultural resources like land and labor is unequal to men's access to the same resources (Simtowe, 2010).

Regarding smallholder wetland agriculture, women are usually confined to food crops that are usually grown on small plots due to lack of access to land, labor, and other inputs like fertilizer. Given equal resources as men, women have the potential to increase their income and production by more than 20% (Kiptot et al., 2014). With the increased production, hunger can be reduced in sub-Saharan Africa and Malawi in particular.

Wetland agriculture as adaptation to climate change: With a lot of research on climate, scientists agree that globally climate is changing and will continue to change in the future (Fisher, 2010). Globally, climate is projected to increase by 2°C–4°C by 2100 (IPCC, 2013). At a regional level, predictions indicate that southern Africa is warming at a range of 0.2°C to 0.5°C per decade, accompanied by extensive changes in precipitation (Hulme et al., 2001). For instance, a decrease in rainfall was observed between 1796 and 1996 (Neukom et al., 2014). Similarly, Jenkins et al. (2002) reported that rainfall decreased substantially over the last 60 years. However, other studies have shown that southern Africa experiences more rainfall variability than change in total rainfall (Tadross et al., 2009; Rurinda et al., 2013). This includes longer delays in the onset of the rainy season and more frequent droughts. In general, southern Africa is projected to experience drier and more extreme conditions and an increased frequency of drought in the future (Sheffield and Wood, 2008).

Recent climate trends assessments indicate that between 1960 and 2006, mean annual temperature for Malawi increased by 0.9°C with an average rate of 0.21°C per decade (Nkomwa et al., 2014). During this time, rainfall pattern also changed. According to McSweeney et al. (2008) monthly rainfall slightly decreased by 5.8mm per month but the amount of rain on days with extreme rainfall increased while the number of days with heavy rainfall decreased. Widespread changes in temperature and rainfall pattern resulted in numerous weather-related disasters like droughts, dry spells, and floods (Fisher et al., 2010). Floods and droughts have been recurring, and they are predicted to continue in the future (Chidanti-Malunga, 2009; Peters, 2004). They have increased in intensity and frequency and have been associated with climate change (Msusa, 2011; Chidanti-Malunga, 2009). The situation has put agriculture at risk, adversely affecting the livelihoods of the rural masses who directly depend on rain-fed agriculture for food and income. For instance, flooding in 2001–2002 and drought in 2005 caused chronic food shortages among other problems during those years. The resulting food crisis in these years, as well as in other years, reveals the fragility and vulnerability of rural agricultural livelihoods in Malawi to climate variability.

Climate change related stressors is one of the factors adversely affecting agricultural production, especially of maize, Malawi's staple crop. As a result, Malawi finds herself persistently vulnerable to food insecurity. Being a relative term, food security depends on food preferences of the majority in an area. Malawi's food security depends on adequate production of and access to maize (Kakota et al., 2015). Maize is widely cultivated and consumed across Malawi. In 2006, the Government of Malawi (GoM) pointed out that 80% of cultivated land is maize and accounts for over 50% of calories consumed.

To ensure food security in Malawi, increasing smallholder agricultural production is therefore inevitable. However, erratic rains, recurrent droughts, floods that have affected Malawi since the 1980s, and soil degradation in the uplands among other factors makes it hard for smallholder farmers to increase their production in the uplands. Wetlands being relatively fertile and moist in the dry season have now become alternative sources for crop production. Through the availability of water, nearly exclusively shallow wells and hand-watering are used to water the crops in the dry season (Simon, 2014). Through watering, farmers can grow more than once in a year, offering them opportunity to supplement their meager harvests from the uplands rain-fed crop production (Chinsinga, 2007). In 2005, GoM reported that during the 2003/2004 drought, many farmers across the county turned to wetlands where, using informal irrigation systems, they managed to produce the staple food, maize. During this drought year, the Government of Malawi reported that maize yield under informal irrigation in wetlands was more than that under rain-fed systems in the upland across the entire country. Crop production was higher per unit area in that year in the dimbas than in the uplands. With this scenario, one would therefore argue that when rain fails in Malawi, wetlands become important sources of crop production, contributing significantly to household food security. Studies show that use of dimba as a livelihood adaptation strategy to climate change is spread across sub-Saharan Africa. For instance, in Zimbabwe Nyamadzawo (2014) noted that cultivation of wetlands presents a climate change and variability adaptation option for smallholder farmers. The 2005 Millennium

Ecosystem Assessment noted that of late, wetlands are critical food security safety nets in many parts of sub-Saharan Africa.

In the face of economic shocks, dimbas have also become vital in improving rural household income and providing resilience. Winter cultivation in the dimbas help to produce foodstuffs such as vegetables, sweet potatoes, tomatoes, maize that have higher market price due to generally low supply (Simon, 2014). By selling such foodstuffs, rural households are able to supplement their cash and improve their living standards and consequently alleviate rural poverty (Chinsinga, 2007). In addition, wetland cultivation helps to lower household expenditure for those who do not sell much of their produce (Simon, 2014). Household income is therefore not spent on foodstuffs produced in the wetland gardens.

Methods

Study area

This research was conducted in Nkhata Bay district in northern Malawi (e.g. Figure 24.1). Nkhata Bay district covers a total area of 40,071 km^2, approximately 15% and 4% of the Northern Region and national land area respectively.

Nkhata Bay like the rest of the country is in the warm tropical climate zone. Compared to other parts of the country, it has a fairly moist climate. It experiences both convectional and topographic rainfall. Total average annual rainfall is approximately 1,900 mm. Some places along the Viphya plateau have a rainforest climate. Average temperature ranges between 20°C and 28°C. Low temperature is experienced between June and July and high temperature is experienced between November and December.

With high rainfall but poor soils, cassava is the dominant crop in the area. Maize, sweet potatoes, rice, and bananas are also grown. Besides agriculture, petty trade (i.e. selling local alcoholic drink, fuel wood, charcoal to supplement household incomes), forestry, and fishing (for communities along Lake Malawi) are the main livelihood sources. Fishing is undertaken throughout the year, but peaks from mid-November to March following the rains. In Nkhata Bay district, as in much of the country, agriculture is predominantly rain-fed. Very few households have access to irrigation and non-farm income to supplement their existence. Smallholder agricultural livelihoods in the district are faced with environmental challenges such as droughts and floods, land degradation and declining soil fertility. Socioeconomic and demographic constraints also aggravate the livelihood challenges. Like most parts of the Northern Region, the Nkhata Bay area is largely rural (not urbanized) and commercially isolated with poor road networks. It is one of the country's hard to reach areas and there are limited markets for the agricultural produce (i.e. bitter variety of cassava) grown in the district.

Data collection

Household survey: A household survey was conducted to generate quantitative and qualitative data regarding ownership, access, and use of wetland garden land at the village level and community level. Households' social and demographic characteristics such as household size, households' heads, gender, and other social markers like age, income levels, ethnicity, and marital status were gathered. We conducted 135 interviews with household heads (90 males and 45 females) in Kavuzi, Mpamba, and Sanga. The interviews were conducted in the local language. To ensure adequate representation, all the three study sites were purposefully identified in the study area.

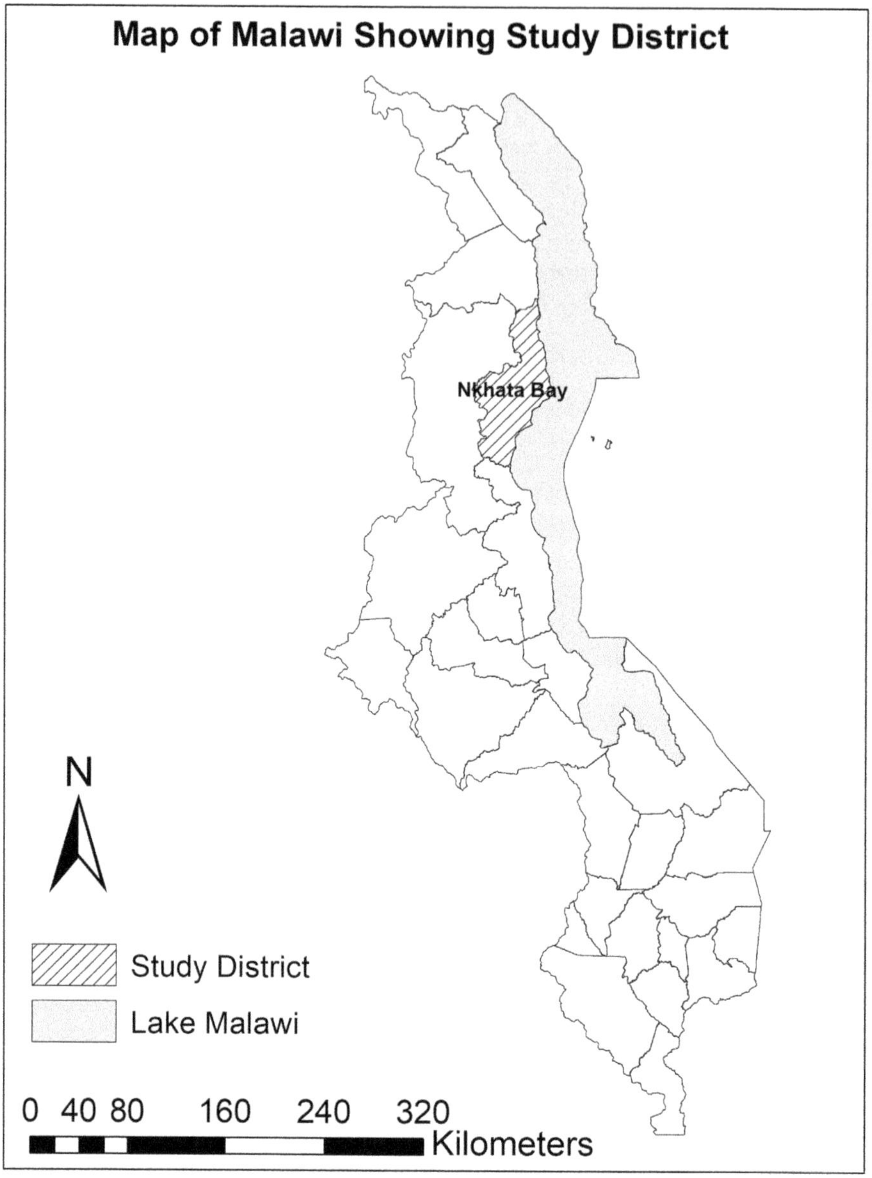

Figure 24.1 Map of Malawi showing study district

Qualitative interviews (FGDs): Focus group discussions (FGD) were conducted to gain an in-depth understanding of ownership, access and use of wetland garden land at the village level and community level, and association with vulnerability and adaptation to climate change in the study site. FGDs targeted specific sub-samples of the population (i.e. female and males separately but simultaneously). In total, 18 FGDs were conducted, six in each study site. FGDs have been argued to be the most used method of gathering qualitative information at a deeper level of understanding and for shared experiences. Thus, through FGD was gained a more balanced view of issues; depth and richness in understanding of ownership, access, and use of wetland gardens; and impact on vulnerability and adaptation to climate change in the study site.

Data analysis

Different but complementary data analysis techniques engaging qualitative and quantitative data were used to analyze collected data.

Qualitative data analysis: Qualitative data analysis involved organizing data into themes. Thematic data analysis was used to structure qualitative data to answer specific research questions. Coding, to sort and structure data, was done manually. Themes generated helped to directly answer the question which explored factors that govern ownership, access and use of wetland garden land at the community level. Analysis drew on multi-scale themes (i.e. household and community) to uncover the underlying structures of ownership, access, and use of wetland gardens. Central to the thematic analysis was gender analysis. Attention was paid to gender dynamics of ownership, access, and use of wetland gardens. Gender was analyzed in relation to other social markers like marital status, age, income levels, ethnicity, and migration status.

Quantitative data analysis: Quantitative data analysis involved use of descriptive plots (e.g. mean, frequency, and boxplots) to explore, describe/summarize social and demographic characteristics such as household size, households' heads, gender, and other social markers such as age, income levels, ethnicity, and marital status. To explore the role/contribution of dimba cultivation to resilience to climate change, comparison was made between households that have access to dimba and households that do not have access to dimba. The study included an analysis test of variance (ANOVA) on household survey data. The test was conducted to evaluate if the number of months the household has adequate food in a year, income from agricultural production, number and types of crops the household grows and yield per year vary between households that have access to dimba and those that do not have access to dimba. A two-way (2-factor) ANOVA included gender and access to a dimba was conducted to further investigate whether or not gender or access to dimba have an effect on agriculture income, number of months a household has adequate food, yield, and the number of crops a household grow. A Kruskal-Wallis test was used because the data were nonparametric.

Results and discussions

Descriptive statistics and exploratory analysis

The results of exploratory analysis indicate no remarkable gender differences in terms of length of residence (Figure 24.2). On the other hand, the results indicate slight gender differences in terms of age of household head, size of active household members, and levels of education. On age of head of household, many male participants were between 42 and 59 years while many of the female participants were between 39 and 45 years (Figure 24.3). In terms of size of active household members, male-headed households have on average seven active household members; female-headed households have about five active household members on average (Figure 24.4). Differences were noted between number of active household members which determines household labor, which affects household production. In terms of levels of education, the majority of the respondents had primary school education. A slight gender difference is seen at the high school (secondary to post-secondary school) level attained mainly by men

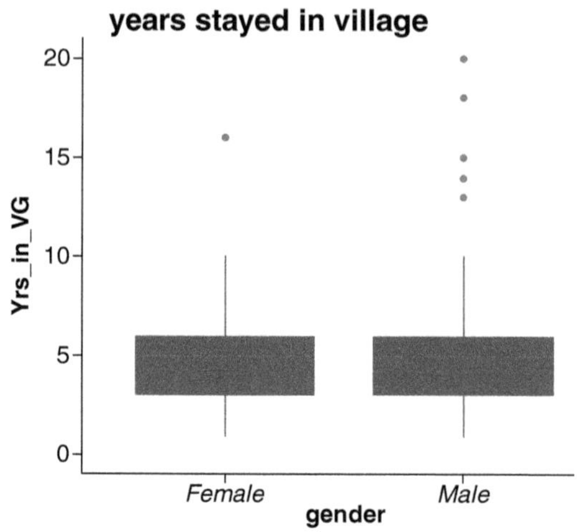

Figure 24.2 Participants' length of stay in the village

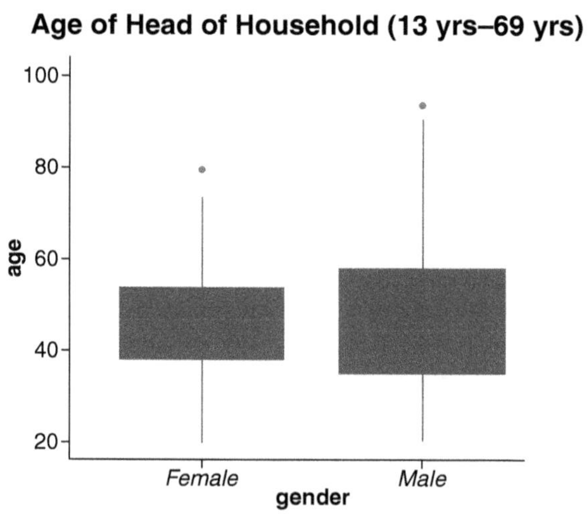

Figure 24.3 Age of head of household (13 yrs–69 yrs)

rather than women (Figure 24.5). Low levels of education affect career choices; as such families are limited to small-scale farming and other small-scale enterprises.

The findings of the household survey show that more than half of the households that participated in the study, 96 (71%) have access to wetland gardens. On the other hand, 39 households (about 29%) had no access to wetland gardens. In terms of gender, there are more male-headed households, and 76 (79%) with access to wetland gardens compared to 20 (about 21%) female-headed households. On the other hand, about 13 (33%) male-headed households indicated that they do not own a wetland garden compared to about 26 (67%) female-headed households that indicated that they don't own a wetland garden. (e.g. Figure 24.6).

Figure 24.4 Number of active household members

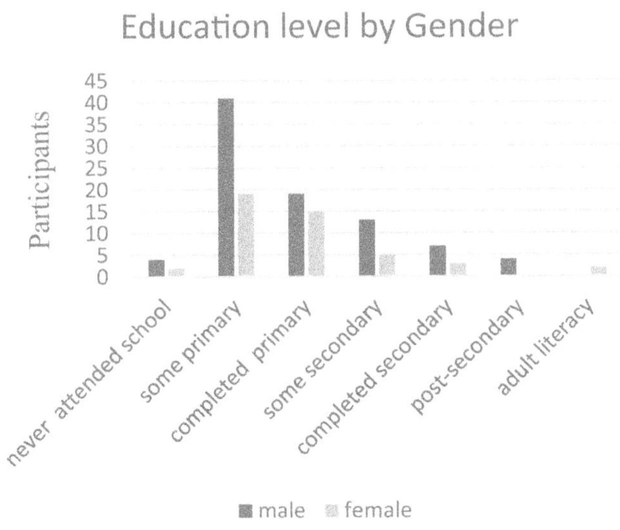

Figure 24.5 Level of education by gender

Exploratory analysis shows that there is a difference in terms of yield, income from agriculture, and the number of months a household has adequate food between households that own wetland gardens and those without wetland gardens (Figures 24.7, 24.8, 24.9, and 24.10). The difference is more pronounced in terms of income from agriculture. There is also difference in terms of gender. The analysis shows that men have a higher average income than women. This may be due to the observation also made in this study in the following paragraphs that many men on average have larger wetland garden land than women. In addition, many men tend to sell their wetland garden crop rather than keeping it for household consumption. This observation may explain the results for Figure 24.8 whereby there is a slight difference in terms

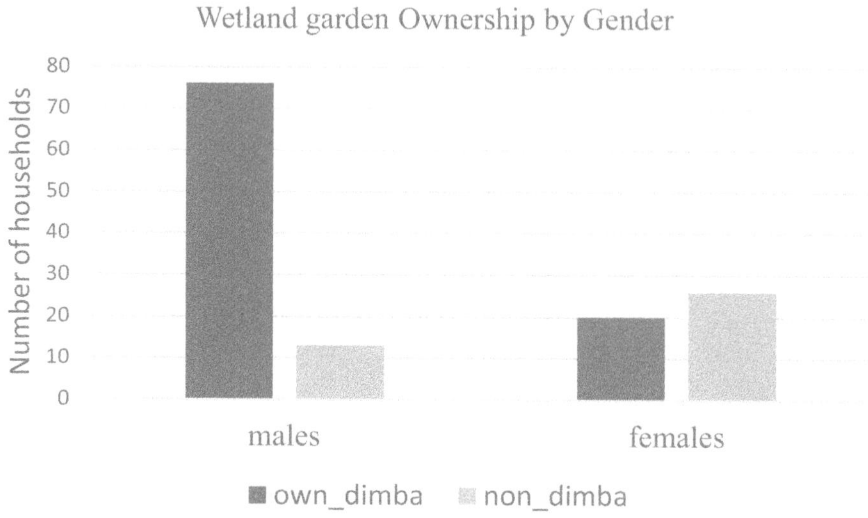

Figure 24.6 Wetland garden ownership by size

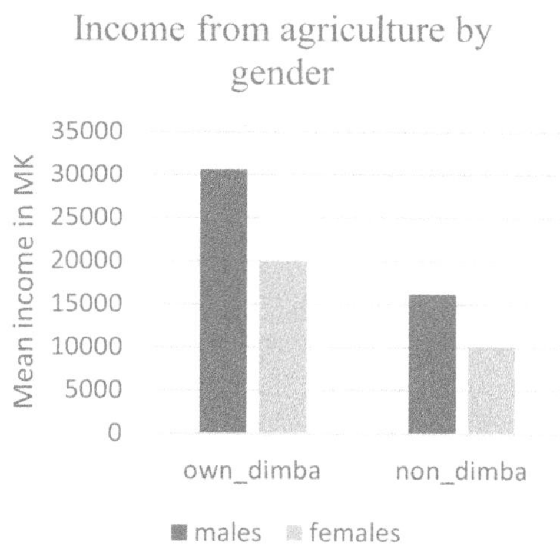

Figure 24.7 Income from agriculture by gender

of yield between men and women although men own larger wetland gardens. However, there is remarkable difference in terms of yield between women who own wetland gardens and those without wetland gardens. On the other hand, the analysis shows that there is no difference in terms of crop diversity between men and women as well those who have wetland gardens and those who do not have wetland gardens.

One-way ANOVA, Kruskal-Wallis test

Statistical tests were conducted to assess if the difference between households with dimba and those without dimba and between men and women in terms of income from agriculture, maize

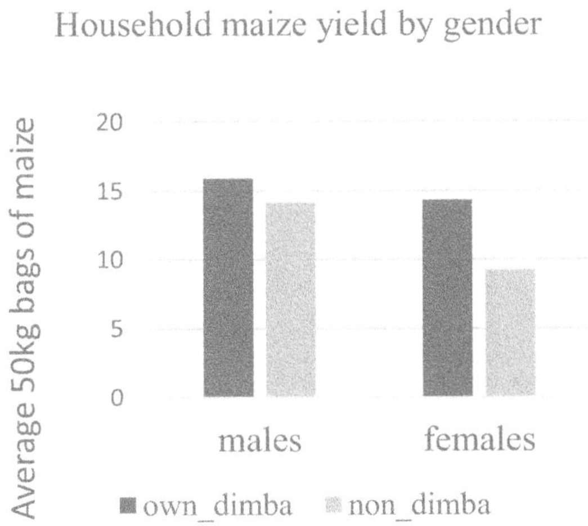

Figure 24.8 Household maize yield by gender

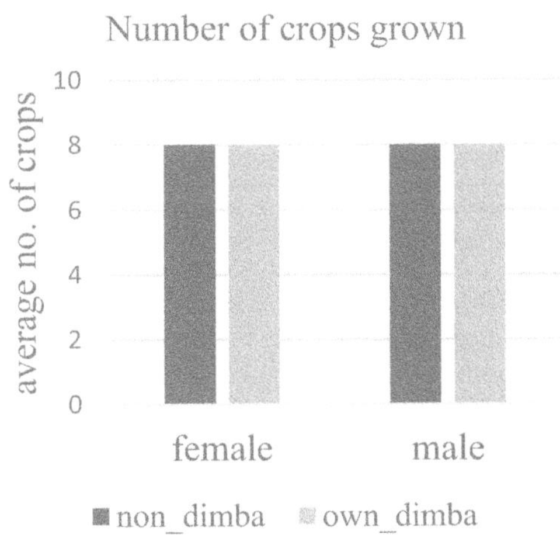

Figure 24.9 Number of crops grown by gender

yield, number of crops grown, and number of months with adequate food are statistically significant. A nonparametric analysis of variance (ANOVA), Kruskal-Wallis test was therefore used to assess the difference for all the four variables. Results show that the variations between the farmers/households that have access to dimba and their counterparts without access to dimba in terms of level of income from agriculture are statistically significant, while variations in terms of yield and number of months a household has adequate food are not statistically significant (e.g. see the detailed Kruskal-Wallis tests in Table 24.1)

Two-way ANOVA, Kruskal-Wallis test

A two-way (2-factor) ANOVA which included gender and access to wetland gardens was conducted to investigate whether or not gender or access to wetland gardens have an effect on

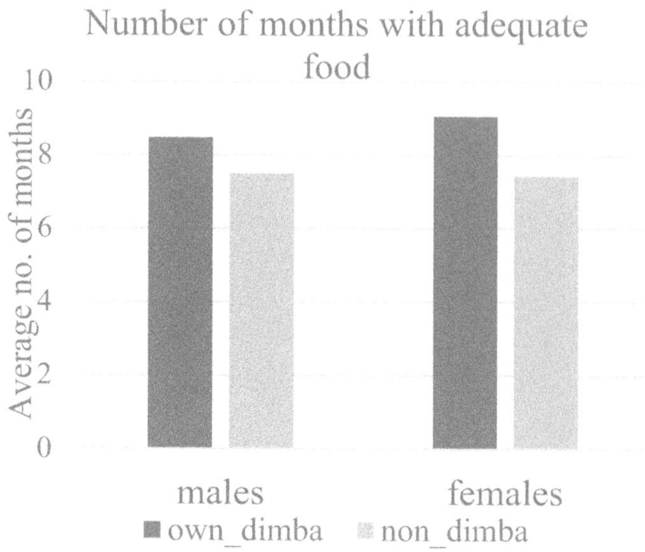

Figure 24.10 Months with adequate food by gender

Table 24.1 One-way ANOVA test output

Variable	*Chi-squared*	*P-value*
Yield (maize)	0.38	0.53
Income from agriculture	3.27	0.01
Months with adequate food	1.48	0.22
Number of crops grown	0.03	0.86

income from agriculture, number of months a household has adequate food, yield, and the number of crops a household grows. Two main effects (gender and access to wetland garden) were analyzed. The two-way ANOVA showed that access to wetland gardens had an impact on a household's income from agriculture (e.g. farmers or households with wetland gardens had on average more income from agriculture compared to their counterparts without wetland gardens, with a p-value of 0.02). The two-way ANOVA also showed a significant impact of gender on the number of months a household has adequate food with a p-value of 0.04 (e.g. see detailed two-way ANOVA tests outputs in Tables 24.2–24.5).

The role of wetland garden cultivation in food security and livelihoods adaptation to climate and economic changes

In this section, I present the findings that address the question of contribution of wetland gardens to household food security and its role in supporting rural livelihoods adapting to climate and economic changes.

Table 24.2 Two-way ANOVA test output for number of crops grown

	Df	*Sum sq*	*Mean sq*	*F value*	*Pr(>F)*
Gender	1	0.3	0.27	0.04	0.85
Own_Wetland garden	1	0.0	0.02	0.00	0.94
Gender: Own_Wetland garden	1	5.9	5.87	0.83	0.36

Table 24.3 Two-way ANOVA test output for income from agriculture

	Df	*Sum sq*	*Mean sq*	*F value*	*Pr(>F)*
Gender	1	1.935e+09	1.935e+09	2.09	0.15
Own_Wetland garden	1	4.793e+09	4.793e+09	5.18	0.02
Gender: Own_Wetland garden	1	3.235e+06	3.235e+06	0.00	0.95

Table 24.4 Two-way ANOVA test output for months with adequate food

	Df	*Sum sq*	*Mean sq*	*F value*	*Pr(>F)*
Gender	1	48.4	48.36	4.15	0.04
Own_Wetland garden	1	7.1	7.11	0.61	0.44
Gender: Own_Wetland garden	1	1.4	0.12	0.83	0.73

Table 24.5 Two-way ANOVA test output for yield – maize

	Df	*Sum sq*	*Mean sq*	*F value*	*Pr(>F)*
Gender	1	525	524.7	1.41	0.24
Own_Wetland garden	1	418	418.4	1.13	0.29
Gender: Own_Wetland garden	1	846	845.9	2.27	0.13

Contribution of wetland cultivation to household food security and income

Participants reported that crops grown in wetland gardens contribute to household food security and income. They explained that though wetland garden crops are mainly grown for subsistence, they usually produce surplus which they sell for income. Participants explained that smallholder farmers who have large wetland gardens and who can afford to buy inputs such as fertilizers and seeds make a considerable income from growing wetland garden crops such as Irish potatoes, early maturing varieties of maize, sweet potatoes, vegetables, and yams (*masimbi*). Farmers further explained that though Irish potatoes are more input intensive they fetch more income than any other wetland garden crop they grow. They reported that a person with a large wetland garden and enough fertilizer can get as much as about MK400,000.00 (about $500) from selling Irish potatoes from the wetland garden.

Further, they reported that second production from wetland gardens usually coincides with the lean season (November–February). Farmers reported to get more income from second wetland garden production as crops tend to fetch very good prices during the lean season. They explained that income from second wetland garden production helps them with household food security in the lean season. They are able to buy maize if they did not have enough or if they did not grow in the wetland garden. Farmers also explained that wetland garden produce also helps to safeguard the little income that households may have as they do not spend money to buy vegetables, sweet potatoes, yams, and other crop produce at the market that they already have in their gardens. Thus, participants reported that farmers with wetland gardens are able to manage food shortage and income problems. Without wetland gardens, participants said, it is difficult for them to find food and income all year round. Participants also said that wetland gardens act as a buffer against reduced yield in the upland production. Those who do not have wetland gardens reported that they spend a lot of money to buy food to beef up reduced yield in the upland fields. Many households, however, reported that they cannot afford to buy food from the market.

Coping and adaptation strategies

In order to understand how wetland gardens help households cope with or adapt to climate and economic change, I asked a series of questions through a household survey to generate a list of coping/adaptation strategies. Results show that there is no remarkable difference regarding strategies employed by men and women. The study also found that strategies employed by smallholder farmers to cope with or adapt to climate and economic shocks are remarkably few (McCusker, 2016). Most common coping/adaptation strategies included wetland gardens, business, ganyu labor (a form of informal casual labor exchange paid in cash or sometimes paid in kind that is prominent among the poor) and Village Savings and Loans (VSL). VSL is a form of microfinance by community members who pool their money together and loan to each other at an agreed interest. Through FGDs participants explained that wetland cultivation is the main strategy that most households in the study area use to cope with and or adapt to climate and economic shocks. Participants explained that wetland garden cultivation helps them to find food and money even when the rains are poor. A woman in Kavuzi pointed out "wetland garden is our main source of livelihood nowadays when there is drought. Without wetland garden we may suffer a lot". With a wetland garden, farmers reported, they can produce twice or more in a year when production has failed in the upland gardens. Participants explained that unlike in the past when they used to have good rains almost all through the rainy season, recently the rains have been unpredictable. Often rains start late and end early and are associated with dry spells. Participants explained that wetland gardens have thus become a buffer against the unpredictable weather patterns.

While ganyu labor and business are also employed as coping/adaptation strategies, they are not as helpful or effective in this area. Participants explained that while doing ganyu they lose time to work in their wetland gardens. As a result, they either plant late or are not able to take good care of their own gardens since much time is spent doing ganyu. In addition, ganyu is not easy to find because very few people can afford to hire ganyu. Since the main source of ganyu in the area is cultivation, ganyu becomes very scarce when the rains are poor as many do not want to spend a lot to cultivate. Sometimes they get their payments very late. Participants had similar views about doing business. They explained they do not have enough money to do business. Small-scale businesses selling mandazi or bananas cannot help them meet their household needs. Other strategies such as fishing were only found in Sanga which is along Lake Malawi. Some women also reported too that they engage in village saving and association, a form of micro-credit that empowers households, particularly women, to be more financially self-sufficient; this helps households become more resilient to climate change through use of and access to a broader set of livelihood strategies like diversifying into micro-businesses. This strategy also is only employed by few women who can afford to make some savings.

Participants further pointed out that even though they rely on wetland gardening as a strategy to cope with or adapt to weather perturbation, climate change is bringing climatic conditions beyond what local farmers have experienced in the past. They explained in the past water was never scarce as rivers/streams would flow all year round. But recently, they said, with poor rains the sources of water for wetland garden irrigation such as streams/rivers or shallow wells dry up. To irrigate crops when the water sources have dried up due to poor rains, men have to dig deep wells to find water. Participants further explained that digging deep wells and irrigating crops from deep wells is painstaking and time consuming.

Female participants explained that due to water scarcity they walk long distances in search of water. This causes them to have less time to work on their wetland gardens as they spend much time walking long distances in search of water for household use. Women also reported that walking long distances to fetch water exposes them to dangers such being raped or attacked by wild animals such as snakes. Some women further explained that water scarcity also threatens their marriages. They explained that some husbands take the two hours' length to fetch water as a lie, "they think we go to meet other men". Women explained that this negatively affects/erodes trust and cause conflicts, thereby affecting productivity like their working together in the wetland garden as a family.

Ownership, access, and use of wetland gardens (dimba): gender and other social markers governing wetland gardens

In this section, I present the findings that address the question of ownership, access, and use of wetland gardens; the effect of gender in relation to other social markers, such as how widowhood and divorce governs ownership, access, and use of wetland gardens. Through qualitative research we conducted 18 FGDs with groups of women separate from men. This was done to establish gendered wetland garden ownership and livelihood patterns.

Ownership and use of wetland gardens

Almost all the land in the study area is under customary tenure. Customary tenure does not necessarily grant land ownership but rather long-term rights of use through communal decision-making authorities (i.e. chiefs). In this study, therefore, land ownership loosely means long-term right of use. As a patrilocal system, the study found that wetland gardens in the study area are owned by men. Women only get to access or use the wetland gardens through their husbands. Participants reported that ownership of wetland gardens are inherited by sons. People without wetland gardens get access to use the gardens through renting. However, participants reported that it is very expensive renting a wetland garden. Women also explained that when one rents a wetland garden and seems to be producing more than the owner, the owners tend to feel jealousy and stop him or her from using their garden.

The plight of women in ownership and accessing wetland gardens is further compounded by other social markers such as marital status (i.e. widowhood, and divorce). Participants explained that most women lose access to wetland gardens when the husband dies. This happens most especially when the widow moves out of the husband's place to somewhere else like her parents'

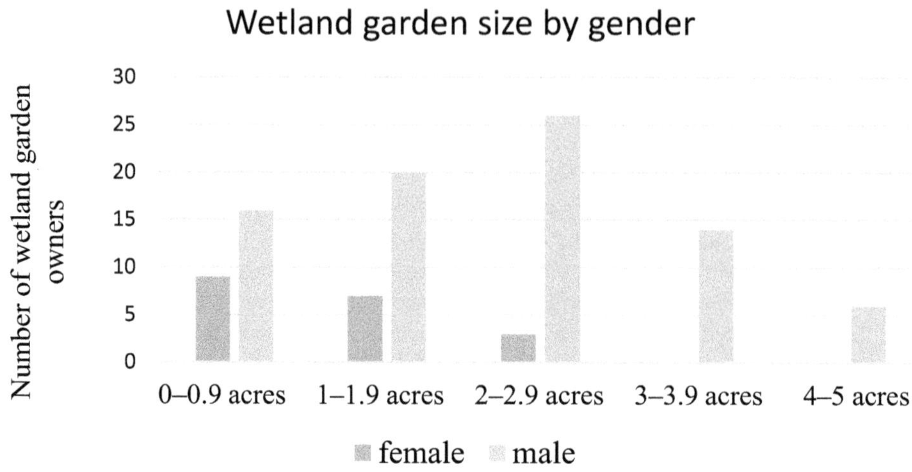

Figure 24.11 Wetland garden size by gender

home. Women participants further pointed out that even for widows who remain at their marital home, their access to wetland gardens after the death of their husbands is not without challenges. Some clan members sharing boundaries with the widow tend to take advantage of her widowhood and push their boundary into the widow's garden, particularly when the widow has no male children or if they are too young to defend her. Participants explained that divorced women lose their access to the wetland garden. When they go to their parents' home, accessing wetland gardens is at the mercy of their brothers who may/may not decide to share the wetland garden with them.

The difference between men and women about access to wetland gardens is also manifested in the size of the wetland gardens. Wetland garden plots are of varying sizes, with some big and some very small. It was reported that on average the smallest size is about half an acre and the biggest size is about 3 acres. Some groups reported the smallest to be a quarter acre and the biggest as 5 acres. In general, more women reported to have smaller sized wetland gardens than men (see Figure 24.11).

Summary and conclusion

Employing a mixed methods approach, this study attempted to address the following questions: (1) How does wetland cultivation contribute to food security and livelihood adaptation to climate change? (2) What governs ownership and use of wetland gardens? The analysis intended to appreciate if wetland gardens contribute to household income and food security, how gender influences access to wetland gardens, and the role of wetland gardens as adaptation to poor rains.

The study has revealed through exploratory analysis that there is difference in terms of food security (measured as number of months a household has adequate food), agricultural income, and yield between households with wetland gardens and those without wetland gardens. Thus, wetland gardens contribute to household's food security and income in Nkhata Bay. This is mainly because wetland gardens enable households to harvest more than once a year. With this, a household can get as much as $500 from wetland garden crops (such as Irish potatoes). Most importantly, the wetland garden harvest coincides with lean season (November–February) and thus it serves as cushion against hunger. Second crops also fetch high market prices, thereby boosting household income. The findings of this study thus strongly support the assertion that wetland gardens play a vital role in rural livelihood security and resilience. Further, the study corroborates with similar research conducted in Simlemba, central Malawi, Cyabayanga in Rwanda, and in Tanzania which indicated that wetland cultivation accounts for more than 40% of domestically consumed food and generates about 50% more household cash income (Msusa, 2011; Mahenya, 2008; Nabahungu and Visser, 2011).

Although wetland cultivation plays a vital role in household food security and income, FGDs revealed that ownership of wetland gardens in Nkhata Bay is affected by gender. Consistent with observations by Simon (2014), Kambewa (2005), and Helliker and Murisa (2011), as a patrilocal system, in Nkhata Bay wetland gardens are owned by men, and women only get access through their husbands. Although women are at the center of wetland cultivation, they have less control over wetland gardens. The difficulty of women in accessing wetland gardens is further compounded by widowhood and divorce. The study has revealed that most women lose access to wetland gardens or face challenges accessing wetland gardens when the husband dies, especially when they do not have older male children.

Analysis of the study took a step further to understand the role of wetland gardens in coping with weather shocks. Generating a list of coping strategies, the study found that smallholders farmers in Nkhata Bay have remarkably few coping strategies. Some of the coping/adaptation strategies included wetland garden cultivation, business, ganyu labor, village loans and savings especially among women, and fishing for those along the lake. Wetland gardening was identified to be the most important and common coping strategy. With unpredictable rainfall, when production fails in the upland gardens, smallholder farmers rely on wetland gardens as a buffer. To echo the arguments raised before, through wetland gardens, smallholder farmers and communities survive long dry seasons and low yields due to poor rains.

References

Chidanti-Malunga, J. F. (2009). *Wetland farming and small-scale informal irrigation in Malawi: The Case of Shire Valley*. PhD Thesis.

Chinsinga, B. (2007). *Hedging food security through winter cultivation: the agronomy of wetland garden cultivation in Malawi*. A Paper read at the Education Development Conference hosted by the National University of Ireland, Galway.

Chinsinga, B., et al. (2013). The political economy of land grabs in Malawi: Investigating the contribution of Limphasa Sugar Corporation to rural development. *Journal of Agricultural and Environmental Ethics*, *26*(6), 1065–1084.

Chirwa, E. W. (2004). *Access to land, growth and poverty reduction in Malawi*. Malawi: University of Malawi, Chancellor College.

Deininger, K., & Xia, F. (2017). *Assessing effects of large-scale land transfers: challenges and opportunities in Malawi's estate sector*. World Bank, Washington DC Research Institute for Global Value Chains, University of International Business & Economics, Beijing.

Dixon, A. B. (2005). Wetland sustainability and the evolution of indigenous knowledge in Ethiopia. *The Geographical Journal*, *174*(4), 306–323.

Fisher, M., Chaudhury, M., & McCusker, B. (2010). Do forests help rural households adapt to climate variability? Evidence from Southern Malawi. *World Development*, *38*(9), 1241–1250.

Fisher, M., & Kandiwa, V. (2014). Can agricultural input subsidies reduce the gender gap in modern maize adoption? Evidence from Malawi. *Food Policy*, *45*, 101–111.

Food and Agriculture Organization (FAO). (2011). *The state of food and agriculture 2010–2011*. Rome: FAO.

Food and Agriculture Organization (FAO). (2012). *The state of food and agriculture in the world. Closing the gender gap. Women in agriculture*. Rome: FAO.

Government of Malawi. (2001). *Malawi National Land Policy*. Lilongwe: Ministry of Lands and Housing.

Government of Malawi. (2016). *Malawi National Land Policy*. Lilongwe: Ministry of Lands and Housing.
Harrigan, J. (2003). U-turns and full circles: two decades of agricultural reform in Malawi 19812000. *World Development, 31*(5), 847–863.
Helliker, K., & Murisa, T (2011). *Land struggles and civil society in Southern Africa*. New Jersey: African World Press.
Hulme, M., et al. (2001). African climate change: 1900–2100. *Climate Research, 17*(2), 145–168.
IPCC. (2013). Summary for policymakers. In T. F. Stocker, D. Qin, G. K. Plattner, M. Tignor, S. K. Allen, J. Boschung, A. Nauels, Y. Xia, V. Bex, & P. M. Midgley (eds.), *Climate change 2013: the physical science basis. Contribution of working group I to the fifth assessment report of the intergovernmental panel on climate change*. Cambridge: Cambridge University Press, 1–27.
Jenkins, G. S., et al. (2002). The challenges of modeling climate variability and change in West Africa. *Climatic Change, 52*(3), 263–286.
Kakota, T., et al. (2015). Determinants of household vulnerability to food insecurity: a case study of semi-arid districts of Malawi. *Journal of International Development, 27*, 73–84.
Kambewa, D. (2005). *Access to and monopoly over wetlands in Malawi*. International Workshop on African water Laws: plural Legislative Frameworks for Rural water management in Africa. Johannesburg, South Africa.
Kanyongolo, F. E. (2005). Land occupations in Malawi: challenging the neoliberal legal order. In S. Moyo & P. Yeros (eds.), *Reclaiming the land: the resurgence of rural movements in Africa, Asia and Latin America*. London; New York: Zed Books, 118–141.
Karamba, R. W., & Winters, P. C. (2015). Gender and agricultural productivity: implications of the farm input subsidy program in Malawi. *Agricultural Economics, 46*, 357–374.
Kassie, M., Ndiritu, S. W., & Stage, J. (2013). What determines gender inequality in household food security in Kenya? Application of exogenous switching treatment regression. *World Development, 56*, 153–171.
Kerr, R. B. (2005). Food security in northern Malawi: gender kinship relations and entitlements in historical context. *Journal of Southern African Studies, 31*(1), 53–74.
Kiptot, E., Franzel, S., & Degrande, A. (2014). Gender, agroforestry and food security in Africa. *Current Opinion in Environmental Sustainability, 6*(1), 104–109.
Kishindo, P. (2004). Customary land tenure and the new land policy in Malawi. *Journal of Contemporary African Studies*, 22(2), 213–225.
Kotze, D. C. (2011). The application of a framework for assessing ecological condition and sustainability of use to three wetlands in Malawi. *Wetlands Ecology and Management, 19*(6), 507–520.
Matchaya, G. (2009). Land ownership security in Malawi. *African Journal of Agricultural Research, 4*(1), 001–013.
Mataya, C. S., & Tsonga, E. W. (2001). Economic aspects of development of agricultural alternatives to tobacco production and export marketing in Malawi. In UNCTAD (ed.), *Analytical Studies on Trade, Environment and Development*. New York and Geneva: United Nations Conference on Trade and Development.
Mbaya, S. (2002, 24th–25th June 2002). *HIV/AIDS and its impact on land issues in Malawi*. FAO/SARPN Workshop on HIV/AIDS and Land, Pretoria.
McCusker, B. (2016). Livelihoods and land use produced together: evidence from rural Malawi. *Human Geography, 9*(3), 63–76.
McSweeney, C., et al. (2008). UNDP climate change country profiles: Ethiopia. Profiles for 52 countries.
Msisha, M. (1999). Land law in Malawi. In Government of Malawi (1999c). *The final report of the presidential commission of inquiry on land policy reform, volume III*. Zomba: Government Printer.
Msusa, H. (2011). Sustainable utilization of wetlands for food security: a case study of the Simulemba Traditional Authority in the Kasungu District of Malawi. *Journal of Developments in Sustainable Agriculture, 6*(1), 86–100.
Mutambikwa, et al. (2000). *Participatory evaluation of vlei utilization and weed problems in communal, resettlement and small scale commercial farming systems of Masvingo and Gutu districts*. Report IDG/00/8. Silsoe Research Institute, Silsoe Bedford.
Nabahungu, N. L., & Visser, S. M. (2011). Contribution of wetland agriculture to farmers' livelihood in Rwanda. *Ecological Economics, 71*(1), 4–12.
Neukom, R., et al. (2014). Multi-proxy summer and winter precipitation reconstruction for southern Africa over the last 200 years. *Climate Dynamics, 42*, 2713–2726.

Nkomwa, E. C., Joshua, M. K., Ngongondo, C., Monjerezi, M., & Chipungu, F. (2014). Assessing indigenous knowledge systems and climate change adaptation strategies in agriculture: a case study of Chagaka Village, Chikhwawa, Southern Malawi. *Physics and Chemistry of the Earth*, *67–69*, 164–172.

Nyamadzawo, G., et al. (2015). Optimizing dambo (seasonal wetland) cultivation for climate change adaptation and sustainable crop production in the smallholder farming areas of Zimbabwe. *International Journal of Agricultural Sustainability*, *13*(1), 23–39.

Peters, P. E. (2004). *Informal irrigation in Malawi's Lake Chilwa Basin: stream-bank and wetland gardens*. BASI CRSP Report. JFK School of Government, Harvard University Press.

Peters, P. E., & Kambewa, D. (2007). CID Working Paper No. 142: whose Security? Deepening Social Conflict over 'Customary' Land in the Shadow of Land Tenure Reform in Malawi by Pauline E. Peters and Daimon Kambewa, Center for International Development at Harvard University, (142).

Rurinda, J., Mapfumo, P., van Wijk, M. T., Mtambanengwe, F., Rufino M. C., Chikowo, R., & Giller, K. E. (2013). Managing soil fertility to adapt to rainfall variability in smallholder cropping systems in Zimbabwe. *Field Crops Research*, *154*, 211–225.

Sheffield, J., Wood, E.F. Projected changes in drought occurrence under future global warming from multi-model, multi-scenario, IPCC AR4 simulations. *Clim Dyn* **31,** 79–105 (2008). https://doi.org/10.1007/s00382-007-0340-z

Silungwe, C. M. (2005). *Customary land tenure reform and development: A critique of customary land tenure reform under Malawi's national land policy*. LLM thesis.UK, University of Warwick.

Simon, S. (2014). Fragile ground, contested soil: dynamic of tenure and policy in the Bamenda wetlands. In M. Ramutsindela, G. Miescher, & M. Boehi (eds.), *The politics of nature and science in southern Africa 2016*. Switzerland: Namibia Resource Centre and Southern Africa Library, 281–304.

Simtowe, F. P. (2010). Livelihoods diversification and gender in Malawi. *African Journal of Agricultural Research*, *5*(3), 204–216.

Tadross, M., Suarez, P., Lotsch, A., et al. (2009). Growing-season rainfall and scenarios of future change in southeast Africa: implications for cultivating maize. *Climate Research*, *40*, 147–161.

WLSA. (2000). *A critical analysis of women's access to land in the WLSA Countries*. Harare: WLSA.

World Bank. (1982). *World development report 1982: Agriculture and economic development*. Washington, DC: World Bank.

25

THE DIALECTIC OF PLACES

Jean-Marie Théodat

For the respect that one has (or lacks) for their roots, it is always intriguing to return to places from childhood to find they have changed. With the passage of time, the eternal things we assume would survive us sometimes disappear. I will mention only a few examples: the Commune building, with its White House aura; the Colombus Quay, with its eloquent statuary at the exit of the Avenue d'Italie; the bronze boar whose snout students touched for good luck the day before exams; the Roxy Bar where Dominican harlots and pimps lingered; the Post Office where you could drop off mail and wait for money transfers from exiled parents; the mausoleum of President Estimé, etc. Every foundation and landmark of my city, which seemed immutable, was in fact very fragile, ephemeral like us, like everything else. I didn't know that.

This exercise, which could have been painful because of an acute nostalgia, becomes a life lesson for the future, giving meaning to our past when handled with the rigor of the profession and a desire to share. Geography is first and foremost sharing the land. Literally and figuratively. Literally, geography is about living together and sharing our little piece of land and only common good: the earth. There is no other place for us in the galaxy. In the figurative sense, geography is about sharing experiences and feelings with others; it is a view of the mind that accumulates to a wealth of knowledge, allowing for a cross-comprehension of problems from different perspectives, but with the same goal: coexistence among cultures, actors and projects. This is the condition for new experiences that respond to collective concerns and correspond to the aspirations of each individual.

Philosophy says a man alone is in bad company because he needs such a long time to understand the world around him that without the experience of others, he would be reduced to reinventing hot water every generation. The same principle applies to the field of geography: working together means sharing with others, students and colleagues, a conglomeration of ideas from around the world that gradually builds knowledge.

The dialectic of places

Places are not fixed. They evolve in their dimension and in their power with time and the projects of those who establish them. They are born and disappear according to a spatial logic resembling a dialectic between diffused tendencies. This dialectic is simultaneously

DOI: 10.4324/9780429450310-29

political, economic, and social, and sometimes leads to real confrontations between those at the limits of these places. As a result of the fights that arise from conflict and the violence that accompanies them, places retain an almost physical memory. People's ubiquitous eyes, ears, and memory keep a diffused recollection, resurfacing on the occasion of adventures that seem distant, yet always remain associated with an initial principle that has conditioned their status. Disorders, in the quasi-psychiatric sense of the term, begin when subsequent arrangements do not take into account this hidden, almost occult dimension, but which remains there, clearly active in determining what could be called only by metaphor, the spirit of the place.

With a group of geography students we made an appointment at Place du Bicentenaire, Place d'Italie, to explore the heart of the capital. In our approach, the city is a mechanism of seven places. At the city's intersection, the proper functioning of the whole is suspended: places of power, sacred places, places for passage, intellectual places, places of mixing, places of pleasure, places of absence. From this perspective, Place du Bicentenaire is a place with multiple functions, both mixing (by the proximity of the port and the Croix-des-Bossales market) pleasure (by the ubiquity of bars, restaurants, and small hotels) and power (with the City Hall, the Parliament), which has become over time a place OF ABSENCE: plagued by the growing impoverishment of the city center and the proximity of the slums of La Saline, City of God, and City of the Eternal. It is also a place where the expression of the social and spatial contradictions of the capital is at its most tense and violent level. Recent events that have marked the political and social scene since the summer of 2018 vividly illustrate this.

Most of my students were born after the 1990s and know nothing about the capital except its traffic jams, the insecurity of surrounding neighborhoods, and the public's general insalubrity. Indeed, the capital city of more than 3 million inhabitants with its high natural population growth (fertility index is 2.9 in the capital) and continuous rural exodus have made the lack of public services and the lack of human resources in the municipal services reduce urbanity to its most ungrateful version. Port-au-Prince is an unhealthy and dirty city where the sanitary conditions of the majority of the population are close to rural conditions: no water, sewerage, electricity, police, or employment in 80% of the urban area. The remaining 20% is accommodated in an upmarket region, inevitably surrounded by slums and precarious neighborhoods. My students, barely in their thirties, find it hard to believe that in the past it was forbidden to walk barefoot in Port-au-Prince or that water games accompanied by classical music once overflowed from the basins of the Place du Bicentenaire. When I evoke for them the atmosphere of the Colombus Quay, with the troubadours who made the people dance on the planks of this deck, the very planks of my childhood through which I admired the sea beneath my feet, they believe me to be a poet. It is so true that this dock, once a place of great pleasure, where crowds from all walks of life flocked, a true symbolic heart where the wealthiest and humblest members of society rubbed shoulders, now exists only in my memory. So, to go back in time, I allowed myself this visit with them, to places where even the Portoprincians only ventured if they were obliged to, by vocation or status: the shopkeepers with their well-stocked stores, and the servants who came to stock up on behalf of their masters. As the place is under permanent tension, we form a single line to avoid dispersing to limit the risk of aggression. An isolated person is quickly in danger in the crowd. This feeling of danger is the primary mark of urbanity when it is not accompanied by an authentic sharing of space. Instead, it is the object of a privative appropriation, which deprives others of the amenities of space for strictly personal purposes. The various neighborhoods that make up the city center are stigmatized by this conflicting dialectic of places.

The *bòdmè*

The waterfront of the capital embodies the memory of the struggles of so many generations that to understand it one must go back to the times of the first inhabitants and the efforts that changed a muddy and mosquito-infested coastline into the capital of the first Negro republic.

In those days, the plain of the Cul de Sac was a den where only the most experienced sailors ventured. Not so much because of the fury of the waves, but because of the reefs between the islets that cluttered the bottom of the bay, and the deleterious nature of the places infested with mosquitoes. Only freebooters and a few buccaneers prospered there. They lived in a synergy of seafarers and hunters, sufficiently advanced to justify a large autonomy from the colonial marshal service, which incessantly controlled this ancient activity for a part of the year to promote the establishment of timid plantations in the middle of these semi-arid lands. The decision, taken by the colonial power to transfer the capital of the colony to this place, was motivated by the double project to rebalance development of the space in response to new plantations in the southern part of the colony, and to better protect the center of power against exactions of pirates.

Le Cap Français was far away from Jeremiah and Les Cayes and travel by horseback or boat could last weeks. The risks of secession were real and made crossing the islands a land or sea odyssey. The location of Le Cap, a bay leaning against a mountainous massif, with no possible strategic retreat, open to the sea breeze, was exposed to incessant pirate attacks. The location of Port-au-Prince was chosen at the junction of the two peninsulas that make up the territory, at the bottom of a bay, offering better potential defense tactics. It was necessary to carry out the hard work of sanitation, landscaping, and backfilling to give rise to this first development from 1743 on a territory measuring barely 78 hectares, Randot's Plantation. Back then, the shoreline reached the level of the Place d'Italie and the mangroves grew sturdily on spongy ground that served as a shelter for the marginalized and for fences of all kinds. The pirates had to get out of the way.

In this model, the *bòdmè*, the waterfront, is the heart of urban activity. It is oriented from north to south, for about 200 meters, from the warehouses to the last store on Bonne Foi street, over a depth of three or four streets going from east to west. The whole formed a quadrilateral of approximately 1 square kilometer where the business of the whole city was done. It was a place of exchange and of life, a place where the lines of force and the colonial colony met. The city had nearly 10,000 inhabitants on the eve of the revolution in 1789. The port was the matrix from which everything else was modeled: the Croix-des-Bossales market, established on a former slave cemetery, the warehouses of the "seaside", and the checkerboard of streets that ensured the fluidity of traffic between the upper town districts (the Cathedral and the Intendancy) and the business center (Rue Bonne Foi, Rue Des Miracles, and Rue des Fronts Forts). Most of the city was built in wood and the roofs covered the whole with an ambiguous layer varying from the gray of the tiles and the ochre of the shingles.

I didn't know that time, but how could I not tell them about the frenzy of the railroad that crossed Rue du Quai, whose vibrations made the facades of wooden residences tremble, threatening to catch fire as the steam locomotives passed by. The stores and boutiques on the waterfront were built of brick, and the doors that closed them were fire and burglar-proof. From their "upper rooms", chanmòt in Creole, the beautiful ladies, confined to the upstairs or backyard where most of the family life took place, looked down on the boys who watched them from the sidewalks. The proximity of businesses and families made social promiscuity more intense. Between Rue des Remparts in the North and Rue Bonne Foi in the South, the Seaside and Rue du Centre were home to a business bourgeoisie and an intellectual elite that contrasted in wealth with the populous suburbs of Bel Air, home to artisans and workers. In some respects, this spatial distribution of wealth reflected another reality, more ethnic and cultural: the seaside was populated by European or Levantine foreign merchants, as often in the ports where this category collaborated in the structuring of wholesale and retail trade. Italian, German, and Danish names could be found alongside other Arab surnames, indicating the presence of this double diaspora, which was destined to play an essential role in national political and economic life. But the lack of urban services and the deterioration of sanitary conditions in the city center led the wealthiest classes to seek refuge outside the initial checkerboard by investing in the first surrounding hills.

The first years of the 20th century gave rise to a first exodus of the elite parties to settle on the hills of Turgeau, Bois Verna, and Bellevue. This is the time when the magnificent gingerbread style residences were built, which are now an invaluable heritage of Haitian architecture. The decorative elements sometimes came directly from France or Italy. The bourgeoisie, whose tastes of luxury were related to the European and more specifically to the French model, cultivated the spirit of a local lifestyle infused with refinement and mystery by the dual French and African origins of its culture. The poor remained confined to Bel Air and La Saline in wooden houses built on a stoop to protect them from floods. The buildings were rarely built on the same level as the street.

This dodge tactic was the way to modernize the city. With suburbs that stood out and detached themselves from the surrounding rurality, the elite sought to avoid contact with the poorest in society while bringing along the accommodation and service personnel who soon set up new informal, unhealthy neighborhoods in the vicinity. The waterfront became the frontier between two categories that were becoming more and more spatially and socially distant in the scale of wealth. And this struggle was already taking place on the sidewalks, under the arcades of the Belle Entrée. As early as the 1920s and '30s, informal traders were already present in front of the stores on the Grand'Rue. They were most often light installations or itinerant merchants who offered fabric, fruit, and food to customers at home. At this stage, the informal market was an itinerant market and the trader was a nomad who went from place to place shouting out his merchandise with a well-defined melody. It is only as the population grew that the needs became clearer, the markets became permanent, and the itinerant traders sedentary. Then began the first conflicts of use on the sidewalks, which became the place of settlement of informal trade pioneers. This trade phagocyted the entire commercial fabric of the capital.

The 1949 renovation and the implementation of an event urbanity

From 1948 on, the space transformed into a place to display new power. President Estimé had embellishment and renovation work carried out that changed the volumes, layout and perspective of the square. For the bicentenary of the capital founding, these major works were supposed to flatter national pride and stimulate the modernization of the capital at an important moment in history, the post-war period. A period marked by renewed optimism and faith in the future and progress, even in a country like Haiti. The result is a vibrant city proud of its Creole, European, and African roots. The statuary was built in classical style: there is a replica of Diana the Huntress, a reproduction of David in Carare marble, in the middle of the Place d'Italie. With its luminous fountains and musical water games, the Bicentenaire offered an opportunity for social mixing where all layers without distinction met on Sundays. Until the 1970s, the Quai Colomb was the meeting place for sunset lovers. The Sunset, a waterfront bar, offered whisky cocktails. Tourism was flourishing, and Sunday family celebrations at the Bicentenaire was a gathering that every good and self-respecting family in the capital enjoyed. The Boulevard Harry Truman offered the prospect of a majestic boulevard on the seafront, lined with bars and hotels with prestigious brands (Carillon, Rond-Point, Tiffany, Beau Rivage) which was said to serve as a model for the one in Santo Domingo.

Fast forward through the years and I find myself with my students in the middle of a field of ruins I do not recognize. Neither the people nor the walls. The earthquake of January 12, 2010, with its 30,000 dead and missing, left the city, the city I knew, only a distant memory. Temporary shelters made of cardboard replaced the bricks of the old stores that are sold by piles on the sidewalk. Where the Place de la Commune used to be, an open-air market has been set up, compact on a pile of garbage. I barely navigate in the middle of the crowd. Once from Place d'Italie, you could see the bell towers of the Cathedral and the dome of the palace which, together with the Town Hall building, formed the golden triangle of the capital. The signs of yesteryear that glorified the Beautiful Entrance have become deposits of pèpè whose piles spill over onto the sidewalks. This is how cities die. The informal market killed off trade, and all good business migrated to Pétion-Ville, in a last dodging pass whereby the bull is still left with its horns – the city is still alive, but for how long?

In this month of November 2019, the Bicentenaire has simply become a lawless zone. Gangs have erected barricades on the main roads leading to it. The Croix-des-Bossales market is as deserted as ever. The sacking of the Senate by the opposition senators at first and its invasion by the crowd to prevent the ratification of the newly appointed Prime Minister marked the crossing of a threshold in the order of the generalized collapse of state structures. After the financial ruin and siphoning of the state coffers by a kleptomaniacal elite, we are witnessing the fall of symbols and the end of all of the illusions that the existence of certain figures, the permanence of certain monuments still helped to safeguard. Nothing remains of what were once tangible landmarks of a common urbanity. There are only ruins and this terrible observation: modern civilizations too are deadly, and sometimes they agonize before our very eyes.

The Bicentenaire has become, like everything else, a place WITHOUT: without water, electricity, public services and legal recognition of the activities that take place there. A no-go area at nightfall, unpleasantness during the day due to the frequent demonstrations of all kinds in front of the Parliament. This place of relaxation and leisure has become the point of convergence of the lines of force of the struggle for power, as well as all of the runoff water that crosses the city. So after the rain, the Bicentenaire is a cesspool. It is here that the deterioration of sanitary conditions is most noticeable and most visible.

We are passing through places where the glances which rest on us, without hostility, are at the very least uneasy and suspicious. Like the bottom of the bay, all the ingredients for a slow and quiet decomposition are there. The toxic fumes from nearby garages and repair shops all end up in the Bois de Chêne and add to this scene a decidedly hostile character to the walk, as it was in the past. It's hard to believe that the Cités de Dieu and the Eternal are built on the alluvial cone of the city's waste! In this unhealthy and filthy environment, a population of half a million inhabitants languishes. We walk around the neighborhood in tight rows, vigilant and studious, hurrying our steps in the hope that the visit will end without any unpleasant surprises.

The brand new National Bank of Haiti building emerges from this urban shipwreck like an ocean liner in the middle of the storm. One wonders if it will announce the end of the disaster or if will sink on the reefs of an erratic urbanity, composed of a social fabric and seismic deconstruction of its environment.

We stop in the cool air conditioning of the only bar in the neighborhood capable of hosting the bank's cafeteria. The best place to appreciate our time together and to analyze our impressions one by one. Geography is sharing land.

Translated by: Elise Miller, Oklahoma Christian University

Further reading

Chaléard, Jean-Louis. Dir. Métropoles au Sud. Le défi des périphéries? Karthala, Paris, 2014.
Conjonction, numéro 223. Le séisme du 12 janvier 2010 en Haïti, Port-au-Prince, 2011.
Corvington, Georges. Port-au-Prince au cours des ans, Éditions Henry Deschamps, 1975.
Lamaute, Emmanuel. Le vieux Port-au-Prince (une tranche de vie haïtienne) suivi de quelques faits et dates. Imprimerie de la Compagnie Lithographique d'Haïti, Port-au-Prince, 1939.
Lucien, Georges Eddy. Une modernisation manquée. Port-au-Prince 1915–1956. Editions de l'Université d'État d'Haïti, 2013.
Prépetit, Claude. La menace sismique en Haïti. Pour que la menace ne soit pas oubliée. Éditions de l'Université d'État d'Haïti, 2011.
Théodat, Jean Marie. Des décombres et des hommes, vie quotidienne à Port-au-Prince après le séisme de 2010. Éditions de l'Université d'État d'Haïti, 2017.
World disasters report. Focus on culture and risk, 2014.

26

FROM *SPECIES LIFE* TO NATURE'S OUTSIDE

New Town "Green City", Kolkata

Ipsita Chatterjee

Lefebvre's claim that capitalism survives through the production of space has been widely accepted in social theory and enthusiastically accepted within geography. In fact, it allows geographers of Marxist bent to understand capitalism's new articulations through one of geography's most coveted concept, space or territory. Although space is a neglected concept in Marx's critique of capitalism, it has been vigorously discussed by Marxist geographers (Harvey, 2018; Lefebvre, 1991). In a parallel line of scholarship, geographers (Marxist and otherwise) have also claimed that capitalism survives through the metabolism of nature (Bakker, 2009; Foster, 2000; Harvey, 1996; McCarthy and Prudham, 2004; Smith, 1984). Unlike space-concept, Marx of course, has written quite a few passages on nature and human consciousness and the production of history. Nature is explicit in Marx's historical materialism where the matter in materialism is both biophysical and biohuman. Notwithstanding Smith's (1996) comment that nature is not much of a Marxist category, I would argue that nature is perhaps more of a Marxist category than space which remains much more implicit in Marx's work, revealed only through excavation by others (Harvey, 2003; Harvey, 1978; Lefebvre, 1991). This chapter accepts the Marxist premise that capitalism exists through the production and metabolism of space and nature. Using the contemporary urban trope of the "Eco-city", and Marx's concept of *species being* and "estrangement" (Marx, 1964; Marx and Engels, 2002), I hope to demonstrate how the *eco* in Eco-city represents an estrangement of nature from itself.

Eco-cities have proliferated in recent times – in 2009 the number was 79 (Joss, 2010); in 2014 it was 174 (Yigitcanlar and Lee, 2014). As an urban policy concept, it represents an attempt to reinvent the post-industrial city and integrate it within a liberal progressive environmental politics of sustainability and climate change. The first noted boost to the concept centered around innovative ways of making "Eco-city Berkeley", thus attempting to provide ecological solutions to the urban-nature contradiction that the city embodies (Register, 1987). Eco-city Berkeley would include transportation, housing, development, public participation policies that would allow for an environmentally and socially just vision of urban planning. The current wave of eco-city proliferation is associated with a general globalization of the concept through the spread of expert knowledge, international conferences, environmental treaties and summits, World Bank reports, and absorption by various national, state, and urban political stakeholders aspiring to keep up the "world class" status of their constituencies. In 2010, the World Bank published the Eco2 city report, where it "resolved" the apparently opposing flows

DOI: 10.4324/9780429450310-30

of economic growth and nature-nurturing through an integrated ecological-economic city model, and described such a city as:

> As the name implies, an Eco2 city builds on the synergy and interdependence of ecological sustainability and economic sustainability and the fundamental ability of these to reinforce and strengthen each other in the urban context. Innovative cities have demonstrated that, supported by the appropriate strategic approach, they are able greatly to enhance their resource efficiency by realizing the same value from a much smaller and renewable resource base, while decreasing harmful pollution and unnecessary waste. By achieving this, they improve the quality of the lives of their citizens, enhance their economic competitiveness and resilience, strengthen their fiscal capacity, provide significant benefits to the poor, and create an enduring culture of sustainability. Urban sustainability of this kind is a powerful and enduring investment that will pay compounding dividends. In a rapidly paced and uncertain global economy, such cities are most likely to survive shocks, attract businesses, manage costs, and prosper.
>
> *(Ecocity, 2019, World Bank: xviii)*

In this excerpt, as well as in the entire report, there is an implicit assumption that "ecological sustainability" and "economic sustainability" can "strengthen each other". It is as if the very ontology of industrial revolution, Fordism and mass production upon which the "modern" industrial city was birthed, could be magically reversed through a discursive sleight of hand that replaces "economic growth" with "economic sustainability". "Sustainability" greenwashes the post-industrial angst associated with "growth", "production", and "development". And somehow this green economy would not work against nature like the industrial economy of the past, and instead, would be fused with ecology in an inseparable tandem: the Eco2 city, where "eco" refers to the coupled "ecology" "economy". "Sustainability" is never actually defined but is taken as an inherently good thing that will allow the post-industrial city to become "resilient". In the earlier excerpt, "urban sustainability" does not possess an inherent essence, it is not valued for itself but for what it can yield, that is, "compounding dividends". Therefore, the Eco2 city meets the need of profit accumulation by being "resilient" enough to absorb shock and crisis, environmental as well as economic. It sustains (sustainability) accumulation by becoming "resilient" to ecological and economic shocks. The rest of the report makes abundantly clear that ecological shocks can be averted and the environment can be sustained through water harvesting, waste management, use of biofuels in meeting power needs, treating waste water for drinking purposes; all of these strategies would open avenues for economic investment. The eco-city model has been embraced with euphoria, making a celebrated journey from Berkeley to Stockholm (Sweden), to Yokohama (Japan), to Curitiba (Brazil), to Gwang Gyo (South Korea), to Dongtan (China), Masdar (UAE), and Sino-Singapore Tianjin (China) (Joss, 2010, 2011; Caprotti et al., 2015). The Eco-city Global Summit was held in Vancouver, Canada in October 2019 and judging by the fanfare on the very interactive website, it was a very global event (ecocity2019.com).

Neil Smith (1996: 40) refers to the local-global assemblage of technocrats, environment ministers, energy ministers, NGOs, social scientists, World Bank, United Nations, reports, conferences, technical assessments as the "environmental policy establishment" that is committed to solving the problems of external nature by reproducing the very conditions of production, consumption that led to those problems in the first place. The environmental policy establishment, needless to say, plays an important role in globalizing "eco-city" as the new urban model. Two things need to be unpacked here: first, "environment", "ecology" (notice nature is not the preferred term in policy discourses like the World Bank's Eco2 city report), "sustainability", "resilience", "economy", are politico-institutional, academic-technocratic concretions of agreed-upon systems of meanings. The agreement has been arrived at through repeated dissemination, circulation, PowerPoint presentations, reports, academic papers, lectures to create a mesh of feedback loops, so much so that it is impossible to trace where and exactly when the "common sense" was produced. The consensual common sense accepts without question that economic growth and environmental sustainability is compatible, although the earth tells us otherwise every day. This common sense takes for granted that profit accumulation within capitalism can be sustained without damaging the environment and in the process ecological and economic justice will be achieved in the urban context. It also accepts that healthy "ecology" means waste management, water harvesting, and emission management, and healthy "economy" means attraction of business, management of costs, compounding dividends, and survival of shocks. It does not matter for instance that the creation of this healthy ecology involves the conversion of nature (farm fields, open meadows, wetlands) into "eco" spaces, and the technocratic solutions to waste and water management are reserved for eco-privatopias, expensive "villages", and satellite cities where private real estate businesses flourish to take advantage of eco-fication. Second, this metabolism of nature through public-private initiatives produces economic spaces of profit accumulation like IT, cyber, finance, technology hubs that are the new industries of the eco-city fortified by green buildings, solar panels, and clean energy. The economy funnels the ecology and vice versa in a loop of capital accumulation that goes back into the pockets of CEOs, software engineers, real estate developers, municipal revenues of elite towns.

Using Marx's ideas of *species being* and estrangement (Marx, 1964; Marx and Engels, 2002), I argue that commodification externalizes nature into environment/ecosystem – a site to be procured/purchased/acquisitioned through creative use of colonial land acquisition acts and zoning laws. Nature as farmlands, wetlands, salt marsh, ponds, and lakes are "legally" territorialized into eco-parks, preserves, green cities – sites available for fixing environmental problems through sustainability approaches and ecosystem management. An eco-city/green city/smart city externalizes nature from itself, creating a "city" and making it available for real estate development, green technology investment, IT and other industrial clustering, and various kinds of private-public partnerships aiding monetization and profit accumulation. An eco-city never fundamentally alters the human-nature relationship (*species being*), it simply creates new forms of human-environment interactions where humans become "anthropogenic agents" – beings that count, record, and fix the environment so that the hyphen in human-environment interaction is preserved for economic growth and capital accumulation.

Using YouTube videos of green projects, speeches of government and non-government officials on green projects, policy and planning documents, websites of government and private stakeholders, land use maps, conference proceedings, I will explore nature and its commodity form, the eco-city/smart city/green city to understand how and why the eco-city represents an estrangement of nature from itself. The purpose behind this conceptual-empirical exercise is to indicate that a critique of nature's commodification is not just about tabling demands for inclusive development, subsidized housing for the poor in expensive green towns, and access of the poor to urban parks, but rather, a critique of commodification must fundamentally re-think the human-nature relationship, it must re-think *species being*. I attempt such a critique through an exploration of New Town-Kolkata, a satellite of the densely populated, old industrial city of Kolkata, India. Although not formally recognized as an "eco-city", New Town represents a recent attempt by the state government (West Bengal) and private interests to produce a "Green City" (NDTV, 2016).

Green is environment, green is economy

Kenneth Jackson (1987) in his *Crabgrass Frontier: Suburbanization of the United States* illustrates how the squalor and congestion of the city, the smog and grayness, the ugly powerlines, and "exotic" immigrants served as the perfect ingredient for enticing the white working class and middle class towards country living in the post-World War II context. The promotion of the "wilderness myth" of pristine sunsets and fresh air, of the manicured American lawn, and picnics in the sun found its way into advertisements for homes and cars. Jackson talks about Henry Ford's auto revolution, the production of the model T car and its avid promotion through advertisements that fueled this suburban wave to the country. Jackson's purpose is to indicate how this desire for the open country was socio-culturally produced and managed through consumption (car, suburban home, lawn mowers); therefore, it was a desire for a piece of nature not in the "hunting gathering style". Nature must be disciplined and wilderness must be tamed through the sanity of the green buffer, the lawn, that would keep the suburban home at a safe distance from anything natural. The lawn itself is deeply "unnatural", infused by non-native varieties of grass, energized by chemicals, and manicured to the dictates of the housing association (Robbins, 2012). The suburban revolution thus created a false dichotomy between city and the suburb, where the city was the site of macabre human production and the suburb was its leafy green antithesis, when in reality, both were produced and deeply commodified.

With the United Nations Earth Summit in 1982 and the resultant sustainable development initiatives (Agenda 21) (Joss, 2011), the need to resolve the false dichotomy was more immediately pursued within the national and the city scales. There is no escape, because the earth is a closed system like a spaceship. Folks can't just escape to the country in their "green havens", leaving the poor in the decaying, polluting city to fend for themselves, because pollution and contamination are not easily contained, so unless the polluting city is re-invented, there is no green future. In terms of urban policy, green, smart, eco meant three possible templates: first, a new city built completely according to green standards of sustainability; second, expansion of pre-existing cities into new areas where nature would be "preserved" as open areas and forested zones within the city limits; and third, a retrofit of existing transport, housing stock, waste and water management infrastructure through green technology and green energy initiatives (Joss, 2010, 2011). Kolkata's New Town is touted as the green satellite of the "non-green" version (Kolkata) and would be an example of the first category.

The smart city/green city/eco-city literature includes a cluster of case studies from Scandinavian countries (Yin et al., 2016), UAE (Cugurullo, 2018; Cugurullo, 2016; Caprotti and Romanowicz, 2013), Singapore (Flynn et al., 2016; Caprotti et al., 2015; Caprotti, 2014), Korea (Hu et al., 2016; Yigitcanlar and Lee, 2014; Shwayri, 2013), India (Datta, 2012), and an overwhelming array of case studies from China (Wang et al., 2017; Hu et al., 2016; Yu et al., 2015; Joss and Molella, 2013). This body of literature provides rich local narratives of sustainable urban strategies put in place in the cities under study, but most tend to be completely uncritical in their stance towards the smart/green/eco-urbanism concept. The mainstream (uncritical) work usually begins with the axiomatic assumption that green urbanism is better than urbanism. Caprotti et al.'s (2015) work on Tianjin eco-city, Caprotti and Romanowicz's (2013) work on Masdar city (UAE), Cugurullo's (2018, 2016) work also on Masdar city, and Datta's (2012) work on Lavasa city (India) are refreshing exceptions and will be discussed later.

The mainstream literature indicates that an eco-city represents not only a happy synthesis of an environment-economy dynamic as portrayed in the World Bank's Eco2 city report, but an aggressive emphasis on digital technology is instrumental in bringing about that synthesis. Yigitcanlar and Lee (2014) note that in case of the Korean cities, greenness is coupled with "ubiquitous computing" (p. 102) to define the new template of the sustainable city. Apart from the ecological angle that emphasizes low carbon emissions, green spaces, renewable energy, green transport, Korea also focuses on smart economy with "ubiquitous computing" where microprocessors pervade all everyday objects, increasing efficiency and safety. Governance is e-governance, commerce is e-commerce, meetings are video conferences. Joss and Molella (2013) indicate similar technological modernization in the Caofeidian International Eco-City, China, with its emphasis on smart grids, GIS maps, digitization of roads and GPS technologies, and applaud the techno-city vision as an inflected template with the eco-city model. Others emphasize the green technology and green infrastructure in the form of renewable and green energy use, water harvesting and treatment, green transportation, waste management as important strategies for a low urban ecological footprint in Chinese cities (Wang et al., 2017; Yu et al., 2016) and in Masdar, Saudi Arabia (Premalatha et al., 2013). Yet others attempt an exploration of the institutional political framework involved in green urbanism. For example, Li and de Jong (2017) indicate the pre-dominance of a top-down government framework in the case of Chinese eco-cities and the importance of firms and businesses in case of Korean cities, whereas German cities seem to be propelled by civil society groups (Hu et al., 2016).

The critical literature that stands in contrast to the mainstream is few, and is emplaced within a broader critique of the previously discussed "ecological modernization thesis", which assumes that any contradiction between environmental preservation and economic growth can be resolved through technological solutions. Cugurullo (2016), based on his work on Masdar city, UAE cautions us that technologies such as smart grids depend on intense extraction of nature in the form of metals, minerals, or plastics that are either mined or derived from petroleum. Caprotti's work on Tianjin eco-city (Caprotti, 2014; Caprotti et al., 2015) argues that the assemblage of advocacy groups, engineers, planners, entrepreneurs within the environmental movement has commodified "climate crisis" and are ubiquitously selling "technological fixes" at the urban scale in the form of green-glossy products like the eco-city and the smart city. A global pattern emerges as batches of consultants, bureaucrats, entrepreneurs travel the world absorbing this green fetish and bringing it home. In UAE, the technological fix is presented in the form of transitioning Abu Dhabi from a traditional oil-based economy into a green R&D cluster. Late capitalism is also green capitalism where the city-nature nexus is crucial for the persistence of accumulation. The critique of green marketing is blended with a critique of green exclusion as Caprotti (2014), Caprotti et al. (2015), and Cugurullo (2016) criticize how green urbanism becomes the prerogative of the cultural and class elites. Cugurullo (2016) for example, points out how the revenue from the Masdar green city flows into the royal family and the Emirati citizens, while the remaining 85% of the population, comprised of expatriates who are the mainstay of all the menial to managerial work, are ghettoized economically and spatially outside the city. Thus, Cugurullo (2018) in his comparative study of Masdar and Hong Kong concludes that eco- and smart cities represent a kind of weird admixture, a "Frankenstein urbanism" that allows for the more profitable components of the green urban experiment to prevail, while phasing out the less lucrative solutions. The "monster" sticks out, awkwardly disconnected from the larger "bioregion"; the broader social and ecological region that surrounds the eco-city does not benefit either environmentally or economically. Datta's (2012) work on planned hill city Lavasa, India critiques bourgeois environmentalism where massive private-public investments go into producing designer "sustainable" cities that routinely displace rural and indigenous populations. Sustainable life-worlds of rural and indigenous people are rudely interrupted, causing forced migration, while green "development" eats up vacated land for elite apartments and condos. Dutta concludes that the solution is probably not creating new eco-cities, but a critical thinking on how existing cities can become eco-friendlier while also catering to the urban poor and building linkages with rural livelihoods.

Estrangement of nature, commodification, and species being

The city and its surroundings (satellites, sprawls, suburbs) is the site of economic growth (neoclassical economics), accumulation of capital (Marxist political economy), and it is in this realization of growth/accumulation that nature/space is commodified. Marxist theory, when simply put, claims that it is in this metabolization of nature/space that nature is commodified for class-specific consumption. And, ecological/environmental theories claim that in the pursuit of growth, the ecology/environment is destroyed through deforestation, carbon emission, water pollution. From this, it follows therefore that urban existence and even rural existence (as in the urban fringe) is always already metabolizing nature and destruction of the environment. Cultivation in the country reproduces nature; creation of farmland is often predicated on cutting down trees, applying fertilizers; cities accelerate this reproduction and ensure that every aspect of nature is completely commodified. If our very existence (urban, rural, forest) is predicated in the transformation of nature, then conceptually speaking, when has nature ceased to exist and where has the "unnatural" started (commodification has begun), and when indeed, in this continuum, have we reverted the unnatural back to "eco", "green", and "smart" again? This seems to be the moot question that needs to be answered if the eco-city/green city model is to be critiqued. Castree (2003) notes that there is very little agreement among human geographers on what constitutes commodification of nature; most do not define it and simply refer to the opening sections of *Capital* Volume I, others take commodification of nature as axiomatic, and yet, most Marxists critique it. Therefore, in order to develop a sharp critique, it is important, argues Castree, that we take stock of this "wide range" that gets labeled as commodification of nature. In answering Castree's challenge, I attempt to dig into some of Marx's work.

For Marx, humans and nature are dialectically conjoined. There is no nature outside humans and all humans are part of nature, and prying them apart is a vacuous conceptual-reduction exercise. In that context, humans are not only natural beings or biophysical components; rather, they are "species being" or "human natural being" (Marx and Engels, 2002, 131), that is, both at once biophysical and social. Nature too is not just a collection of natural stuff or raw material as in bourgeois political economy (classical and neoclassical economic theory), existing outside the human condition waiting to be exploited. No matter how pristine or how untouched, even in its discovery, plotting on a map, image as photography, nature has already been socialized and humanized forming the historical geographical material context of existence. According to Marx and Engels (2002), humans and nature are so inextricably intertwined that nature is human's "inorganic body" (Marx and Engels, 2002, 131). It is through the social act of laboring, this inorganic body evolves, producing history and geography and consciousness as a whole. This dialectic is beautifully explicated by Marx and Engels (2002) in the following quote: it was "in creating a *world of objects* by his practical activity, in his *work upon* inorganic nature", that humans proved herself/himself "a conscious species being". . . . "[T]he universality which makes all nature his inorganic body". Through this production, nature appeared as "*his work*", industry as "the open book of Man's essential powers", and the object of labor as "*the objectification of Man's species-life*". Man and woman would therefore be able to see himself/herself "in a world that he has created". History was the process of Man (and woman) becoming species being. Thus, "history itself is a real part of *natural history* – of nature developing into Man" (Marx and Engels, 2002, 131).

Material existence is natural social existence, and in existing, human labor is objectified or estranged from humans (Marx and Engels, 2002; Marx, 1964). Money objectifies all labor, and humans are alienated not only from the product that their own labor produces, but also from the activity of laboring that is no longer reserved for herself. Her labor belongs to another, and is therefore not free, a process Marx and Engels (2002: 125) calls "self-estrangement". It is here that Marxist theory can locate the transition from nature to "produced nature", or "unnatural", or "commodified nature". The metabolization of nature that Marxist theory critiques happens through the objectification of labor (estrangement), and can only be understood within the context of laboring. Therefore, while all nature is social nature, all nature is humanized, and nature is always human's inorganic body. Marxist critique begins only when this humanization of nature, its socialization, has proceeded through the objectification of labor. If labor is free, then humans labor to satisfy their subsistence needs and, in the process, humanize nature – this is the production of "species life" (Marx, 1964, 114) and history, geography. But, if labor must transform nature into objects to earn a wage (money) and hence subsist, then her labor has been objectified, and in this process of objectification (estrangement), any transformation of nature represents nature's commodification and its unnaturalization. Such a transformation of nature through objectification of labor is no longer [her] "work on inorganic nature" (Marx and Engels, 2002, 131); it is a monetized commodity for satisfying needs external to the species being. Thus, Marx and Engels (2002, 125–126) say:

> Labor was no longer the satisfaction of a need, but "merely a *means* to satisfy needs external to it" – animal needs to maintain physical existence. Thus Man only felt himself "freely active in his animal function". What was animal became human and what was human became animal.

Estrangement of labor from itself means for Marx and Engels that humans are reduced to animals – humans no longer produce "the object of his will and his consciousness" (Marx and Engels, 2002, 125), but produce to fulfill animal needs. This intensifies as private property intensifies until all labor is about *earning* a *living* and therefore, all nature is a means for estranged labor to produce commodities to earn that living. Labor is rendered alien to humans and simultaneously nature is rendered external to species life (Marx, 1964). Species and life get disaggregated so that life becomes acquiring commodities through the destruction of species (nature). Because of the disaggregation of humans from nature and life from species, humans do not realize that they have built their estranged existence on the transformation of nature into raw material, fuel, electricity, energy, biotic and abiotic resources. To summarize, commodification of nature has happened when "human natural being" is rendered into individual-nature interaction through objectified labor working *on* (rather than with) nature to produce estranged existence. Humans are estranged from their own labor, estranged from each other, and estranged from nature, and nature, in turn, is externalized as a means to an end: *earning a living*. Castree (2003) summarizes Marxist geography's various iterations of nature's commodification as privatization (private ownership of nature rather than communal or public use), valuation (using money to represent the worth of nature), alienability (ability to disengage nature from its natural social contexts and make it available for market exchange), individualization (chunking, categorizing, classifying complex nature into discrete products), abstraction (reducing natural diversity through homogenizing and standardizing categories), and displacement (nature made to appear something other than itself thus erasing the exploitation that went into creating this illusive fetishism).

Based on Marx (1964), and Marx and Engels' (2002) theses on species being, species life, estrangement, and objectification, I argue that nature, ecology, and environment are not interchangeable realities, or interchangeable ways of understanding reality. "Nature" becomes "environment" or "ecosystem" as a result of externalization through estrangement. This estrangement I refer to as estrangement of nature from itself. Thus, as objectification of labor and individualization proceeds, and estrangement penetrates all aspects of life, human natural being ceases to exist, the being is separated from her species life to become an individual, and simultaneously, nature is separated from humans becoming the "environment" or "ecosystem". The environment, like objectified labor is a means to earn a living and is therefore reduced to a site rather than a being. The environment becomes a site for analysis, classification, stocktaking, census taking, map making, remote sensing, digitization and satellite imagery, geologic explorations, and resource extraction. All of which involves individuals handling, gazing, measuring, storing the environment as if it is external to the human body (Smith, 1984). The environment becomes a site of opportunity (production of new objects) as well as a site of problem solving (pollution, climate change, drought and flood management). Because nature is estranged from humans, it is rendered external to it, and individuals can now "tap it", or "fix it" without fundamentally altering or revaluating human natural existence. Living becomes extracting from the environment, and when, as a result of this anthropogenic extraction, the environment manifests problems in the form of severe weather, climate change, pollution, and carbon emission, it can be fixed by humans by further tweaking the technology in use, or capping the volumes of emission, or finding "cleaner" ways to go on. In other words, individual-environment interaction is re-evaluated by recalibrating the hyphen, but never disrupting the estrangement that externalizes nature into environment.

The ecosystem approach represents a more focused gaze on the environment whereby estranged nature is abstracted into containers (open and closed) containing species that are intricately dependent on each other through chains and webs of connections. The balance of these systems is delicately maintained through feedback loops of chemicals, heat, energy, and any small change in the quantity or quality of species in a system can upset the system, which must re-adjust for equilibrium to return (Odum, 1983; DeAngelis, 1980; Hannon, 1973). Therefore, individuals/scientists/ecologists must rigorously take stock of species over time and study them to understand how their relative interactions alter the system and if something can be done to keep it at a steady state. In the ecosystem approach, nature is estranged as a system of individualized species externally connected to each other biotically; it is never seen as a species being internally connected to produce human nature. Therefore, deploying an ecosystem perspective would mean ensuring the balance of species, observing the invasive species, curtailing it, and alerting policy makers about system-shocks. But such an analytical-positivist gaze does not ever *see* nature, it merely gazes upon "the system"; it does not conceptualize species life, it only understands "inter-specie interaction"; and it never interprets system-shock as estrangement of humans from nature. Therefore, like the environmental approach, the ecosystem approach also represents estrangement of nature from itself.

The ecological modernization literature (Pearce et al., 1989; Porter, 1995; Lundqvist, 2000; Andersen and Massa, 2000; Hajer, 1995; Pearce et al., 1989; Porter, 1995) addresses the environment, ecosystem, and economic growth symbiosis where an ever-increasing array of green technologies addresses the problems of environmental depletion and ecological destruction while allowing production and growth to proceed. Thus, modern urban societies can go green through clean energy, increased digital communication, cutting down on emissions, recycling and reusing resources, and technologically fixing the environment/ecosystem without having to address human nature estrangement. Indeed, there is no evidence in this literature that estrangement is an issue that has been identified at all. Most recent urban solutions as noted in the "Green is environment, green is economy" section is geared towards reshaping urban environments by adopting the policy-philosophy of ecological modernization. Cities therefore, go "eco" without fundamentally altering class access to urban open areas, or elite-privilege of green technology, or addressing questions of intra-city poverty, and often, actively produce estrangement by usurping farm and indigenous land (Cugurullo, 2018, 2016; Caprotti, 2014; Caprotti et al., 2015; Datta, 2012; Hodson and Marvin, 2010). In that context, to what extent do green city initiatives help stem estrangement of nature from itself? In the next section, I will be exploring this question through the lens of Kolkata's New Town that received a gold rated certificate as a green city (Chakraborty, 2018).

New Town, Kolkata as green city

In 2015, the government of India announced the Smart City Mission, which released a competition-based strategy to create eco-friendly cities in India. As per the rules of the challenge, cities would first compete at the state level, and state-level winners would then compete at the national level (Mission statement and guidelines, Ministry of Urban Development, 2015). And finally, a group of a hundred cities would win at the national level to be officially declared as "smart cities" and get rupees 100 crores per year from the central government under the Smart City Mission between 2015 and 2019. There is no standard template of what this "smart city" should look like, although the mission provides a definition: "the objective is to promote cities that provide core infrastructure and give a decent quality of life to its citizens, a clean and sustainable environment and application of " 'Smart' Solutions" (Ministry of Urban Development, 2015, p. 5). Some core infrastructural elements listed in the same documents include e-governance and citizens participation, waste management, sustainable environment, robust IT connectivity and digitization, inclusive development and housing, open areas, parks, and playgrounds, walkable and bicycle-able localities, energy-efficient buildings, and reduction of congestion, pollution, and resource depletion. In order to compete, cities had to prepare Smart City Proposals (SCPs) containing their vison and strategies. In 2015, the Government of India released a list of 98 nominees, and New Town, a satellite of Kolkata, West Bengal was in the list (Sandilya, 2015).

In 2018, the West Bengal government decided to withdraw from the national government's Smart City Mission, because of the stipulation that state governments must come up with funds matching the national government's funding contribution for each city, each year. The chief minister of the state Mamata Banerjee declared that state investment of rupees[1] 500 crore[2] (100 crore spent each year for five years) for a "smart city" was not very smart; instead, the state should be able to spend that money in any way it likes. Another reason for withdrawal was the stipulation for compulsory imposition of a water tax on smart city residents, which the state government did not approve. While opting out of the national Smart City Mission, the West Bengal government decided to launch its own "Green City" project in Kolkata's eastern suburb called New Town. This new city would be green *and* smart, Bengal's Silicon Valley (The Statesman, 2018), a model that would be imitated in other parts of the state. The objectives are similar to "smart cities", including green buildings, low emission transport, open areas, e-governance (Ghosal, 2016). The West Bengal Housing Infrastructure Development Corporation Ltd. (WBHIDCO), a wholly government owned company was entrusted with the work of planning New Town. The "Land Use and Development Control Plan" prepared by WBHIDCO declares the New Town planning area as an "environment friendly and aesthetically attractive new urban settlement functionally integrated with the future metropolitan structure" (WBHIDCO.com, 2012, p. 11). The "environment" and "aesthetic" assemblage that WBHIDCO proposes, encompasses a vision of an "open space system" (p. 17) however, appropriately "enclosed" from the "encroachment by squatters" (p. 17), including "planned and landscaped" (p. 17) open areas, ecological parks, urban forestry, urban horticulture, parks, playgrounds, and waterbodies for recreation. Canals will be desilted and the excavated earth will be used effectively; this will increase the productivity of canal fisheries and make waterbodies more open and viable (p. 25). The highest percentage of the land use would be residential (29%) with the next highest percentage being open spaces and water bodies (10%) – therefore, it can be inferred that the "environment" (consisting of open spaces and water bodies) would be nurtured to boost real estate sales. Posh homes surrounded by greenery would be the primary draw for class elites (IT professionals, non-resident Indians) looking to find an "outside", a country-club ambience, beyond the congestion and grime of the adjoining metropolis, Kolkata. The green in "Green City" primarily refers to what Housing Infrastructure Development Corporation Ltd. (HIDCO) calls the "open space system", an opposite of and an antithesis to the bursting-to-the-seams old industrial Kolkata model. The "Environmental and Ecological Consideration" (pp. 24–25) section of the plan stresses that existing orchards and water bodies will be integrated, and new ones designed to create "high quality, aesthetically pleasing urban life". "Ecological sustainability" features prominently in the HIDCO plan; for example, as a subtitle on p. 25, however, the main policy thrust seems to be on the connection between green and aesthetic pleasure as a desirable and commodifiable attribute of urban life, rather than any inherent need for an environmentally sustainable city.

In 2017, New Town was conferred the "Green City" title by the Indian Gold Building Council (IGBC), a building certification organization which is a part of Confederation of Indian Industries (CII). According to the IGBC:

> Waste management has been prioritized by the New Town Kolkata Development Authority (NKDA). The NKDA has launched a fleet of 20 waste management vehicles with Global Positioning System (GPS) technology, which enables municipal officials to keep a track of the vehicles when they pick up garbage and transport them. The NKDA officials have set up a monitoring centre [sic], which checks if all the garbage collection vehicles visit their designated areas and collect and transport waste accordingly.
>
> Apart from waste management, NKDA has also focused on improving other civic amenities in New Town, thereby making the city suitable for residential purposes. IGBC has praised the three water ATMs installed in the planned city, which enables people to collect water throughout the day. Electric buses have also been introduced in New Town and cover all areas in the township. Over 20 such buses with 32 seats in each have been rolled out and four charging points have also been set up across New Town.
>
> *(Datta, 2018)*

Similar schemes are approved on a regular basis to keep up New Town's green status; for example, administrative approval was acquired under the Green City Mission for installation of CCTVs, GPS-based monitoring, light and sound dome fountains in different parks, beautification and boundary work for picnic gardens and lakes, installation of waste bins, paddle boats for water-based recreation, creation of roadside pathways, and LED light bulbs for street lighting (Government of West Bengal, 2018).

Numerous YouTube videos produced by WBHIDCO and other public-private organizations list these schemes as inherent features of a "model green city" (HIDCO YouTube Video, 2016a). The technocratic distance created by the new gizmo-based waste management, new transportation, street lighting, and digitization schemes are hastily bridged by nostalgic narratives of nature where lost childhood memories can be revived. It is as if solar panels, LEDs, GPS, CCTVs, e-governance form the technological core of this city, yet it is wrapped up in the warm insulations of nature and nostalgia, a city away from a city, but within ten minutes from the main airport. New Town resolves the contradiction between environment and economy by simultaneously announcing open areas for nature walks, and open avenues for investments. Corporations like IT Accenture, TATA consultancy, Green Pact, Infosys form the resolute "landforms" of this green city. An imitation of the Wall Street financial district's bronze bull statue is strategically placed at a prominent thoroughfare to ground the *eco*nomic dimensions of the *eco*logical town. The YouTube videos extoll the convenience of shopping and entertainment within the malls, multiplexes, movie theatres, and wax museums and the simultaneous ubiquity of educational choices for children ranging from posh play schools to high learning institutions like the Indian Institute of Technology and Delhi Public Schools (HIDCO You tube Video, 2016a). HIDCO's chairman explains that the state and the city government aspire to build a "city in a garden", where the buildings blend with the green in a way that buildings and green become an undistinguishable continuum. Forty thousand new trees will be planted, and competitions would be organized between corporate houses encouraging them to plant trees over the monsoon seasons. The New Town Kolkata Development Authority (NKDA) would check on these trees in a year, codify how many survived, and then, awards would be presented to corporate houses based on best green performance (YouTube video, 2016b) – those that nurture economic growth would be induced to nurture ecological growth. A tree census was carried out in 2017 using GPS tracking as part of the Green City Mission where over 5920 trees over 5ft tall were identified and their latitude, longitude, and Latin name recorded to "quantify New Town's green cover" (The Times of India, 2017). Ecological stocktaking in the form of tree cover, tree census, number of saplings planted, survival rates of species is essential for maintaining "green status quo", upon which the post-industrial new economy of retailing, e-commerce, information technology, and education is predicated.

Eco-park, a 480-acre large green open space is the largest in West Bengal and serves as the crowning jewel of New Town's green image. The HIDCO website under the "Nature tourism tab" claims:

> The vision for the new Eco-Park at New Town embraces a holistic, systems approach to sustainability that seeks to create a balance between ecological regeneration, resource conservation, and optimization with the need to create a recreational urban open space.
>
> This approach offers an additional benefit as compared to traditional models for developing parks and urban open spaces – the added benefit: the chance to create a "living classroom" about nature's service. Here, as the wetlands are filtering water and the plants are filtering air, people can come to relax or rejuvenate, socialize or seek solitude, and learn first-hand how region's flora and fauna are benefitting their city.
>
> *(Parkiti Tirtha,[3] HIDCO)*

Virtual tours available through multiple YouTube videos and HIDCO's websites construct nostalgic nature tours weaving their way through lakeside promenades, award-winning rose gardens, tropical forests, butterfly gardens, bamboo gardens, wildflower meadows, and tea gardens – the ecological, as envisioned through the flora/fauna specific garden spaces is then blended with the socio-cultural with themed spaces like a mask-garden featuring handcrafted masks from different parts of the state of West Bengal, Bengal *haat* (informal market-bazaar) promising local handicrafts for sale, *Adda* zone with benches (area for communal conversation and exchange), artist's cottage, and *Rabi Arayana* (Rabi forest) named after Nobel laureate, Rabindranath Tagore featuring plant species mentioned in his poems and essays (*Prakiti Tirtha*, HIDCO). "The holistic systems approach to sustainability" seems to be an urban garden concept for public consumption (tickets have to be purchased at the entry and for the various recreational activities) where native species of trees from all over the state are housed under poetic names celebrating, displaying, and educating the visitor about *Bengali bhadrolok culture*.[4] "Ecological regeneration" seems to be a nod to the housing and labeling of flora to maintain the integrity of the "eco" in Eco-park. The environmental and ecological mission is about producing and displaying native and non-native species in the middle of New Town where they did not exist, especially showcasing those species that also serve as cultural and literary icons. Eco-Park was not carved out of local nature in its organic glory; therefore, the stated non-traditional objective of Eco-park serving as a "living classroom" is an attempt to "humanize" ecology. The flora and fauna inventoried, landscaped, and organized thematically is not just supposed to be a "botanical garden" with obscure Latin names, the ecology is made accessible through carefully chosen Bengali literary imageries. There is a promise that a Bengali can re-invent oneself as a *species being* as plant species become her inorganic body through culture-nature photosynthesis. Although the Eco-park is visited by people from outside New Town because of its entertainment components like children's playground, boat and kayak rides, zorbing, and other recreational activities, it plays a vitally important role in raising the real estate demands of posh New Town apartments.

"Ecos", a real estate firm, names itself after Eco-park and directs its advertisements at the class elite of Bengal and non-resident Indians. Ecos claims that living in its apartment involves getting up in the mornings to birdcalls, breathing in pure oxygen throughout the day, enjoying "untouched" greens of the 480-acre Eco-park where art and adventure meet. The advertisement asks the viewer "to imagine life with all this and more, to imagine life at Ecos". An apartment purchased at Ecos (priced at over Rs.18,000,000) is to seamlessly blend Eco-park with apartment life with no "barriers or domestic walls" (Ecos_Homes, 2016). And yet, access to nature does not have to be in isolation from modern convenience like gyms, meditation centers, spas, convention centers, banquet halls, playgrounds, swimming pools, and airports. Hundreds of real estate firms like Ecos are trading the promise of nature through expensive apartments sold to upper-class elites. Along with the promise of nature outside is also the promise of green within the homes achieved through energy efficiency, sustainable materials, and green technology used in building the apartments. According to West Bengal's land ceiling policy, developers cannot own more than 25 acres of land in a contiguous parcel, but New Town's policy was recreated to override this limit, allowing developers to own 50 acres as long as they stipulate at least 25% of it for economically weaker section housing (Zee24Ghanta, 2016).

Underlying the Green City, however, lurks difficult stories of land-grab, displacement of local farmers, and theft of livelihood from farmers and sharecroppers. Local farmers claimed that land was forcibly acquired by the government under the Colonial land Acquisition Act of 1894 for the construction of the New Town Green City (About *Jami Bachao* Facebook). Unlike the government's claim that land requisition was peaceful, farmers complained that water supplies to their mature crop-fields were suspended and garbage was dumped on full-grown crops until the farmers were forced to sell their land. Rajarhat *Jami Bachao* Committee (Save Rajarhat's land) and Save Rajarhat Farmers Committee was formed by local farmers resisting the land grab (Dey et al., 2013; Sanhati.com, 2008). The HIDCO office was surrounded by the then leader of the opposition party, and now chief minister, Mamata Banerjee and her party members in protest against forcible land acquisition from farmers (The Indian Express, 2010). *Sanhati*, a promotional organization representing this farmer's committee claims that political parties of all stripes participated in the land-grab model; not only was the ruling CPM (Communist Party Marxist) party goons involved in the land grab at gun point, but the then opposition party's (Trinamool Congress) foot soldiers were equally complicit. Lower than market rates were paid to farmers for prime agricultural land only to be sold by HIDCO to realtors at very high rates. Rajarhat Jami Bachao committee's Facebook page claims:

> The sad irony of it all is the fact that while the farmers were paid only Rs. 6000, the same land was being sold by the government to the businessmen at around Rs. 600 000 a katha,[5] while the promoter and developers in turn were making a profit of around Rs. 15,000,000 to Rs. 20,000,000. This means that the farmers are getting 1% of the profit that the government is making in selling the land to the big business or equivalently the farmers are earning 0.0001 % of what the real estate businesses are earning from the land. It is worth mentioning that in all of Rajarhat municipality the land price is determined as Rs 1500/square foot for multi-storeyed apartments.
>
> *(About Jami Bacaho, Facebook)*

Dey et al. (2013) argue that HIDCO should handsomely compensate people displaced from their life and livelihood and forced into poverty through informal work like rickshaw pulling, hawking, and vending. In its essay documenting the Rajarhat landgrab conflict, *Sanhati* concludes that

> One day Rajarhaat will be completely erased from the agricultural map of the state while the struggle of the helpless farmers will be lost down the memory hole. But their struggle won't be in vain and the ground soaked with their blood will provide the basis for the new struggle for survival.
>
> *(Sanhati.com, 2008)*

Rajarhaat Jami Bachao Committee leader, Nilotpal Dutta, in the organization's Facebook page (About *Jami Bacaho*, Facebook) also claims that New Town is being created and expanded through steady acquisition of fertile land, drainage canals, lakes, ponds, and salt marshes that were home to endangered species of fish. *Dhupir bheel*, a huge natural lake that served as a drainage for two large canals and was a site of rich flora and fauna diversity, and a source of livelihood for families dependent on pisciculture and fishing were filled up as HIDCO declared such natural bodies as "low yielding" and "low return" (Dey et al., 2013, p. 48).

The New Town Green City narrative is consistent with the critical literature discussed earlier where green commodification (of climate crisis, ecosystem management, waste management) and green exclusion (elite access to sustainable life) proceed simultaneously (Datta, 2012; Caprotti, 2014; Caprotti et al., 2015; Cugurullo, 2016). Looking back at that critical literature through the lens of New Town, I agree with the core arguments made there about commodification and exclusion. However, I contend that commodification and class exclusion is not just about zoning nature to produce parks, or the sale of gated apartments to the class rich, or land grab from farmers – these are expressions of estrangement of nature from itself. Only if we conceptualize to the fullest and the deepest extent the context of nature's estrangement can we come to terms with when commodification has happened, and how it must be critiqued. In the next section, I turn to that.

Destroying the environment, disrupting the ecosystem, and reclaiming *Species Life*

The mainstream literature endorsed by the World Bank (Eco2 city report) and reinvigorated by enthusiastic academics reinforce the ecological modernization thesis that technological advancement and good governance can resolve the contradiction between economy and environment and an eco-city will be the site of that resolution. In contrast, the central critique posed by the critical work on the eco-city concept cited in this article claims that eco-cities merely commodify environmental problems and promises to solve them through technological fixes, when in reality, what is actually sustained is economic growth and profit for corporations and class elites. While I agree with the critique proposed, in this article, I pose a larger question: in the face of continuous transformation of nature to make life possible, when has commodification taken place and why is it bad? This question simultaneously led to associated questions like, what does de-commodification look like, and how can it be achieved? In answering these questions, I visited Marx and Engels' concept of *species being* and estrangement, and from these conceptual entry points, I proposed estrangement of nature from itself. I argue that estrangement of nature from itself can reveal what commodification entails, and when it has happened.

A reversal of commodification, therefore, would mean reestablishing species life by disrupting the estrangement of nature from itself. Marx and Engels demonstrated that all nature is social and all humans are natural, hence human natural life is realizing conscious life activity through *species being*. *Species being* is not an early human hunting and gathering in the pure state of nature, but rather, it is humans in relationship with nature prior to objectification of human's labor and hence the objectification of nature. Therefore, while all nature is socialized and humanized and it is human's inorganic body, when this socialization/humanization proceeds through the objectification of labor, then humans have become estranged from their own labor and from other humans (self-estrangement) and therefore, nature becomes commodified. The real estate firm Ecos' promise of waking up to bird calls, "untouched greens", and "pure oxygen" does not represent nature, but rather, represents entrenched commodification through estrangement of nature from itself – every inch of green has been touched by objectification, all birdcalls and oxygen within and outside the gated apartments are objects outside of nature, even when they are emplaced within the environment. Therefore, I argue that when humans are estranged from their labor, their human natural existence becomes human-nature interaction to earn a living. Self-estranged humans exist to fulfill animal functions (not conscious life activity) that are monetized, and these functions are fulfilled by extracting from nature to produce objects that can be sold for a living (untouched greens, birdcalls, and pure oxygen). Estrangement of humans from itself is also estrangement of nature from itself. Nature becomes "external" to human existence through its objectification (estrangement) as raw material, flora, fauna, climate, or land – in this process of estrangement, nature is externalized or commodified from itself as "environment" or "ecosystem". In the estranged human-nature existence, when objectification of labor has proceeded fully, it is impossible to see an "inside" where human nature was once a *species being*, an un-hyphenated natural social existence like the *Dhupir bheel* (natural lake) that once existed filled with birds, fish, and plants, teeming with human life, and bustling with livelihood. On the outside therefore, we must act, and this outside is the unnatural, the external, the environment/ecosystem, and it is here that ecological modernization and its policy variants (eco-city, smart city, green city) emerge. Green City's emergence is predicated on filling up *Dhupir bheel*, acquiring farmland, re-zoning country as city, over-riding land ceiling policy, walling "open spaces" from encroachers, planting poetry through tree species, careful census taking of green cover, and organizing corporate competition for forestation.

Because the ecological modernization thesis does not acknowledge the commodification of nature or is not conscious that it has taken place, or does not even gaze into an "inside" where labor and nature were not estranged, it only understands existence as an "outside" where humans work on the environment "sustainably" to achieve economic growth. Therefore, the solutions it proposes are focused on preserving the environment/ecosystem to make growth feasible. The reality however, is that the environment/ecosystem to be preserved (for growth) is already commodified, as in the case of New Town. In New Town, nature was externalized through land grab, filling lakes, walling and bordering into "preserves", "parks", "open space systems", "tea gardens", mask gardens. Therefore, "a holistic, systems approach to sustainability that seeks to create a balance between ecological regeneration, resource conservation, and optimization" (*Parkiti Tirtha*, HIDCO) happened *after* nature was externalized through colonial land acquisition acts to convert farmland and saltmarsh into Green City. The "holistic systems approach to sustainability" and "balance between ecological regeneration, resource conservation, and optimization" has a purpose, which is, "to create a recreational urban open space", that is, a new "urban environment" that can be "planned and landscaped", enclosed from "encroachment by squatters", and available for horticulture, parks, urban forestry, tree census, ecological gardens (WBHIDCO.com, 2012, p. 11). Why is this labeled as a "systems approach"? Why is it considered holistic? What does it optimize? The New Town, Green City mission calls itself "holistic" because it assumes that by zoning open areas free from development, it can intertwine growth and environment in a non-destructive way. Nature's nurture through smart planning is conceived as (eco) "systems approach", and "optimization" refers to optimization of economic development, growth, and profit that is achieved with judicious attention to conservation and regeneration. The "living classroom" (*Parkiti Tirtha*, HIDCO) is certainly useful, because Eco-park is made accessible to all and sundry who come to enjoy the rides, the poetry inscribed on carefully crafted landscape, the statues, and fountains. But unfortunately, the living classroom, a big chapter (Eco-park) in New Town's green book, does not unmake the "environment" and the "ecosystem" into *species being*.

"Eco" is not nature, "smart" is not human natural existence, because it does not strike at the heart of objectification of labor (farmers getting less than .0001% of what real estate developers earn from the land sale, colonial land acquisition policies used to colonize farmland, and ceiling policies overridden to accommodate developers). Green City-New Town does not disrupt estrangement of humans from themselves and nature from itself; it allows commodification to become a green way of life and recreates it as tea gardens, open spaces, green apartments – an environment/ecosystem externalized. And it is not all good, because any approach to life that is predicated on commodification-estrangement will not comprehensively reify nature from the individual-environment interaction it is trapped in. The eco-city, green city, smart city embodied in New Town is external to nature, an outside where we will continue to tinker with the environment/ecosystem through sustainable approaches to waste management, and ecological approaches to breed and census species. But will we fundamentally disrupt commodification so that environment/ecology becomes nature internalized through human natural existence – a *species being* where the fishermen in the canals, the farmers in their mustard fields, the marshes, wetland, bogs constitute life? Disrupting estrangement of nature from itself is not just about inclusive development through subsidized housing, making eco-parks accessible to the class poor, channelizing company's profit into solar paneling, or going back to pre-urban jungle existence, it requires fundamentally re-thinking human natural existence by re-thinking concepts that create an external, an outside to nature. Concepts that create an external are commons versus private, farm versus city, inhabitants versus squatters, open versus closed spaces, nature versus environment/ecosystem – de-estrangement, de-commodification, return to *species life* require disrupting this conceptual material outside. De-commodification cannot be piecemeal like a clause about economically weaker section housing cursorily attached to land privatization schemes, but rather, de-commodification must mean disrupting the environment and the ecosystem (external to nature) to reclaim *species life*. A trenchant critique towards eco-city approach therefore, is not a call for primitive life, but a life as *species being*.

Notes

1 The average exchange rate between INR (Rupees) and USD for the year 2018 was 1USD = 68.41 Rupees.
2 1 crore = 10 million.
3 Nature tourism.
4 Bengali, as in who resides in the state of West Bengal, and *bhadrolok*, as in the genteel class of Bengal, the cultural elite of the state and even the country, displaying with great pleasure their knowledge about literature, art, politics, and poetry. Bengali *bhadrolok's* gentility is sometimes celebrated, and sometimes ridiculed as Bengali parochialism. West Bengal is known for its literary renaissance (17th to the 19th century), and Bengali *bhadrolok* is a term used in popular parlance to denote the genteel, sophisticated, cultural elite that emerged from that renaissance and continued the cultural revolution/appreciation in periods beyond.
5 1 katha is 720 square feet.

References

About Jami Bacaho, Facebook. *Jami Bachao Committee Rajarhat*. Available at: www.facebook.com/jamibachao.committerajarhat. Accessed on: December 2018.

Andersen, M.S., & Massa, I. 2000. Ecological modernization – origins, dilemmas and future directions. *Journal of Environmental Policy & Planning*, *2*(4), pp. 337–345.

Bakker, K. 2009. Neoliberal nature, ecological fixes, and the pitfalls of comparative research. *Environment and Planning A*, *41*(8), pp. 1781–1787.

Caprotti, F. 2014. Critical research on eco-cities? A walk through the Sino-Singapore Tianjin Eco-City, China. *Cities*, *36*, pp. 10–17.

Caprotti, F., & Romanowicz, J. 2013. Thermal eco-cities: green building and urban thermal metabolism: thermal eco-cities: the case of Masdar, Abu Dhabi. *International Journal of Urban and Regional Research*, *37*(6), pp. 1949–1967.

Caprotti, F., Springer, C., & Harmer, N. 2015. 'Eco' for whom? Envisioning eco-urbanism in the Sino-Singapore Tianjin eco-city, China. *International Journal of Urban and Regional Research*, *39*(3), pp. 495–517.

Castree, N. 2003. Commodifying what nature? *Progress in Human Geography*, *27*(3), pp. 273–297.

Chakraborty, S. 2018. New town gets green city tag. *The Times of India*. Available at: https://timesofindia.indiatimes.com/city/kolkata/new-town-gets-green-city-tag/articleshow/65436740.cms. Accessed on: November 2018.

Cugurullo, F. 2016. Urban eco-modernisation and the policy context of new eco-city projects: where Masdar city fails and why. *Urban Studies*, 53(11), pp. 2417–2433.

Cugurullo, F. 2018. Exposing smart cities and eco-cities: Frankenstein urbanism and the sustainability challenges of the experimental city. *Environment and Planning A: Economy and Space*, *50*(1), pp. 73–92.

Datta, A. 2012. India's ecocity? Environment, urbanisation, and mobility in the making of Lavasa. *Environment and Planning C: Government and Policy*, *30*(6), pp. 982–996.

Datta, S. 2018. *New town gets recognition as Bengal's first green city for effectively building eco-friendly civic amenities*. Available at: https://swachhindia.ndtv.com/west-bengal-waste-management-new-town-green-city-24300/?pfrom=home-campaigns. Accessed on: December 2018.

DeAngelis, D.L. 1980. Energy flow, nutrient cycling, and ecosystem resilience. *Ecology*, *61*(4), pp. 764–771.

Dey, I., Samaddar, R., & Sen, S.K. 2013. *Beyond Kolkata: Rajarhat and the dystopia of urban imagination*. New Delhi, India: Routledge.

Ecocity. 2019. Available at: http://ecocity2019.com. Accessed on: November 9, 2018.

Ecos_Homes. 2016. *Welcome to Ecos! Breathtaking apartments at New town, Rajarhat*. Available at: www.youtube.com/watch?v=fFiEEcGilXc. Accessed on December 2018.

Flynn, A., Yu, L., Feindt, P., & Chen, C. 2016. Eco-cities, governance and sustainable lifestyles: the case of the Sino-Singapore Tianjin Eco-City. *Habitat International, 53*, pp. 78–86.

Foster, J.B. 2000. *Marx's ecology: materialism and nature*. New York: Monthly Review Press.

Ghosal, A. 2016. *West Bengal: Mamata dumps Centre's smart city project, to develop cities on its own*. Available at: https://indianexpress.com/article/india/india-news-india/west-bengal-mamata-dumps-centres-smart-city-project-to-develop-cities-on-its-own-2983994/. Accessed on: December 2018.

Government of West Bengal. 2018. *Nagarayan*. Available at: https://wburbandev.gov.in/home/adm_appv_green_2017_18/YlYzcmZuNkp5d1JhSUFoR3JzZDV1L251SG54N3hxTTNIL3h1M0pySkN-lK2Y2QU1adHFUYXViYSt3b0ZLNVFnVHRGdnowa2NSd001bjRjNlhPOVZFWWc9PQ==. Accessed on: December 2018.

Hajer, M. 1995. *The politics of environmental discourse: ecological modernization and the policy process*. Oxford: Oxford University Press.

Hannon, B. 1973. The structure of ecosystems. *Journal of Theoretical Biology, 41*(3), pp. 535–546.

Harvey, D. 1978. The urban process under capitalism: a framework for analysis. *International Journal of Urban and Regional Research, 2*(1–3), pp. 101–131.

Harvey, D. 1996. *Justice, nature and the geography of difference*. Oxford: Blackwell.

Harvey, D. 2003. *The new imperialism*. London, NewYork, Oxford: OUP.

Harvey, D. 2018. *The limits to capital*. New York: Verso Books.

Hodson, M., & Marvin, S. 2010. Urbanism in the Anthropocene: Ecological urbanism or premium ecological enclaves? *City, 14*(3), pp. 299–313.

Hu, M., Wadin, J.L., Lo, H., Huang, J., Lund University, Institutet för Ekonomisk forskning, Institute of Economic Research & Lunds universitet. 2016. Transformation toward an eco-city: lessons from three Asian cities. *Journal of Cleaner Production, 123*, pp. 77–87.

Jackson, K.T. 1987. *Crabgrass frontier: the suburbanization of the United States*. London, New York: Oxford University Press.

Joss, S. 2010. Eco-cities: a global survey 2009. *WIT Transactions on Ecology and the Environment, 129*(1), pp. 239–250.

Joss, S. 2011. Eco-cities: the mainstreaming of urban sustainability – key characteristics and driving factors. *International Journal of Sustainable Development and Planning, 6*(3), pp. 268–285.

Joss, S., & Molella, A.P. 2013. The eco-city as urban technology: perspectives on Caofeidian International Eco-City (China). *Journal of Urban Technology, 20*(1), pp. 115–137.

Lefebvre, H. 1991. *The production of space*. Cambridge, MA, Blackwell Publishers.

Li, H., & de Jong, M. 2017. Citizen participation in China's eco-city development. Will 'new-type urbanization' generate a breakthrough in realizing it? *Journal of Cleaner Production, 162*, pp. 1085–1094.

Lundqvist, L. 2000. Capacity-building or social construction: explaining Sweden's shift towards ecological modernisation. *Geoforum, 31*, pp. 21–32.

Marx, K. 1964. *The economic & philosophic manuscripts of 1844*. Translated by M. Milligan. Edited by D. J. Struik. New York: International Publishers.

Marx, K., & Engels, F. 2002 (1888). *The communist manifesto*. London: Penguin Classic.

McCarthy, J., & Prudham, S. 2004. Neoliberal nature and the nature of neoliberalism. *Geoforum, 35*(3), pp. 275–283.

Ministry of Urban Development. 2015. *Smart city: mission statement and guidelines*. India: Government of India.

NDTV. 2016. *Not part of smart city, will rather go for green cities: West Bengal*. Available at: www.ndtv.com/india-news/not-part-of-smart-city-will-rather-go-for-green-cities-west-bengal-1445470. Accessed on: November 2018.

Odum, H.T. 1983. *System ecology; an introduction* (No. 04; QH541.15. S5, O3.).

Parkiti Tirtha Salient points of discussion for the inauguration of Eco Park on 29th December 2012 & 27th December 2013. Available at: www.wbhidcoltd.com/org_desc_pjt.php?pg=invprojectfh&typ=ep. Accessed on: December 2018.

Pearce, D., Markandya, A., & Barbier, E.B. 1989. *Blueprint for a green economy*. London: Earthscan Publications.

Porter, M.E., & van der Linde, C. 1995. Toward a new conception of the environment – competitiveness relationship. *Journal of Economic Perspectives, 9*(4), pp. 97–118.
Premalatha, M., Tauseef, S.M., Abbasi, T., & Abbasi, S.A. 2013. The promise and the performance of the world's first two zero carbon eco-cities. *Renewable and Sustainable Energy Reviews, 25*, pp. 660–669.
Register, R. 1987. *Ecocity Berkeley: building cities for a healthy future*. Berkeley, CA: North Atlantic Books.
Robbins, P. 2012. *Lawn people: how grasses, weeds, and chemicals make us who we are*. Philadelphia, PA: Temple University Press.
Sandilya, S. 2015. Government releases list of 98 smart cities, Uttar Pradesh gets maximum share. *Economic Times*. Available at: https://economictimes.indiatimes.com/news/economy/infrastructure/government-releases-list-of-98-smart-cities-uttar-pradesh-gets-maximum-share/articleshow/48704490.cms. Accessed on: December 2018.
Sanhati.com. 2008. *A history of the brutal Rajarhat land acquisition. Bengal's new IT hub*. Available at: http://sanhati.com/about/. Accessed on: December 2018.
Shwayri, S.T. 2013. A model Korean Ubiquitous eco-city? The politics of making Songdo", *Journal of Urban Technology, 20*(1), pp. 39–55.
Smith, N. 1984. *Uneven development*. Oxford: Blackwell.
Smith, N. 1996. The production of nature. In *Future natural* (pp. 47–66). Oxfordshire, UK: Routledge.
The Indian Express. 2010. *Mamata ready to gherao HIDCO office*. Available at: https://indianexpress.com/article/cities/kolkata/mamata-ready-to-gherao-hidco-office/. Accessed on: December 2018.
The Statesman. 2018. *Mamata launches Bengal Silicon Valley Hub*. Available at: www.thestatesman.com/cities/mamata-launches-bengal-silicon-valley-hub-1502673008.html. Accessed on: December 2018.
The Times of India. 2017, March. *Track New town's green cover, one GPS-tagged tree at a time*. Available at: https://timesofindia.indiatimes.com/city/kolkata/track-new-towns-green-cover-one-gps-tagged-tree-at-a-time/articleshow/57741308.cms. Accessed on: December 2018.
Wang, P., Wang, Z., Deng, X., & Chen, J. 2017. Ecological intercorrelation in urban – rural development: an eco-city of China. *Journal of Cleaner Production, 163*, pp. S28–S41.
WBHIDCO. 2012. *Land use and development control plan for New town planning area*. Available at: www.wbhidcoltd.com/uploads/reports/report11.pdf. Accessed on: December 2018.
Yigitcanlar, T., & Lee, S.H. 2014. Korean ubiquitous-eco-city: a smart-sustainable urban form or a branding hoax? *Technological Forecasting and Social Change, 89*, pp. 100–114.
Yin, Y., Rader Olsson, A., & Håkansson, M. 2016. The role of local governance and environmental policy integration in Swedish and Chinese eco-city development. *Journal of Cleaner Production, 134*, pp. 78–86.
YouTube Video. 2016a. *New town: a model green town*. Available at: www.youtube.com/watch?v=. Accessed on: December 2018.
YouTube Video. 2016b. *New town: towards green city*. Available at: www.youtube.com/watch?v=Uh4q5L41P2w. Accessed on: December 2018.
Yu, C., Dijkema, G.P.J., de Jong, M., & Shi, H. 2015. From an eco-industrial park towards an eco-city: a case study in Suzhou, China. *Journal of Cleaner Production, 102*, pp. 264–274.
Zee24Ghanta. 2016. *Smart city or green city for New town?* Available at: www.youtube.com/watch?v=9QaMykwwbaM. Accessed on: December 2018.

PART 4

Policy and governance

27

DIASPORA WITHIN

Territoriality, nationality, and justice for the indigenous community in India

Bikramaditya K. Choudhary

> The reproduction of patronizing ideas and stereotype informs our perceptions of homelessness. It's a powerful, but partial portrayal of how we define notions of home.
>
> *(Michele Lancione, 2019)*

The term "Diaspora" acquired importance in the late 20th century since the term applied principally to Jewish Christians. Of late at least 30 ethnic groups declare that they are a diaspora, or are so deemed by others (Cohen, 1996). Diaspora in the Old Testament refers to "forcible dispersion" or "scattering to other lands" on one or the other pretext. The renewed meaning of diaspora in the post-war period acquired the notion of homelessness which later got extended to "away from home". Homeland is defined as the way we identify home. Steward (2017: 207), while discussing the case of Iran, developed a distinction between "home" and "homeland" and argued that "although home could refer to any place, even a temporary place that can be built or rebuilt, homeland is a mythical place in the mind, carried by memory". It is in this sense that the concept gets linked to idea of "nation"[1] or place of origin. In India the idea of "*desh*" is loosely equated with the notion of nation and later to the concept of nation-state. However, the cultural notion of "*desh*" and "*Videsh*" is different from the legal and sovereign conception of nationality which is broadly dependent on the "color of the passport". The Indian notion of "desh" is more of the communitarian conception, where going away from the usual place of residence to earn livelihood is treated as *videsh*, like in folk songs. With globalization, nationality as an operational concept has been widely debated and Indian nationality has been in the news with electronic media debates like celebrations of "*Pravasi Bhartiya Divas*", voting rights or no voting rights for OCI/PIO (Overseas Citizen of India/Person of Indian Origin). Nevertheless, within the sovereign territory of the nation-state, demands for separate territorial units are not unheard of. I am not going to enter the debate on the merits or demerits or reasons or arguments on these questions. The purpose is to bring the idea of "*desh*" to multiple levels especially at sub-national levels where moving out of the cultural boundary is considered going to "*Videsh*". The popular Bhojpuri genre of "Videshia" is based on the performance comprising songs and dances of women whose husbands/lovers migrate out for employment. In this chapter, I focus on the internally displaced communities. I conceptualize displacement in the

 DOI: 10.4324/9780429450310-32

Indian context and link it to the processes of disempowerment and dispossession. The complexity generated, owing to the process of development-led displacement and the quest for justice by these indigenous communities, is brought in to explain the perils of development processes for these communities.

The development activities across India have resulted in a significant scale of displacement of people, primarily indigenous communities, legally known as Scheduled Tribes, and popularly known as *janjati*. The idea of citizenship and the claims on resources and the entitlements of citizens has been a matter of debate in India for different reasons. The question of loyalty of citizens, especially of the marginalized and more significantly those who question the dominant development processes and the government of the day, is raised by the supporters of the ruling party. Multiple terminologies like that of "anti-national", "(urban)naxals", "traitors" (*gaddar*), and so on have become part of everyday practice to delegitimize those who raise difficult questions on behalf of displaced communities. In recent times with the rise of "nationalist" tendencies, it is not simple to debate the notions, meanings, symbols of a nation. It is becoming increasingly difficult to write about the movements against governments as it is often construed as the movement against the "nation-state", though at times there have been violent attempts to challenge the national authority by different groups. However, it is not absolutely impossible to write about justice of indigenous[2] and marginalized populations at a time when there is an onslaught of state violence, right-wing extremism, financial austerity measures, neoliberal development, racialized policing and surveillance against, among others, the poor, undocumented, minoritized, and displaced – people wrestling with the violence of precarity in its many forms (Nagar and Shirazi, 2019). In India, those who worked and wrote about the perils of indigenous population and argued for their rights even had to face police cases and prolonged legal battles.[3] Seemingly the higher judiciary to some extent protected the rights of these activist-academicians who felt intimidated. Owing to systematic and continuous harassment by the agencies of the state, these activists and academicians often restrained themselves to vociferously arguing for the rights of tribal population and violence perpetuated against them. For the tribal population in India the popular slogan has been "*jal, jangal, jameen*", meaning water, forest, and land. The ownership of these usually has been communitarian and the change in the residence means loss of these vital resources which somehow have become the issue of identity for these indigenous communities.

During fieldwork in the Kendujhar district of Orissa, one of the tribal leaders said that their land does not only give grain but also retains memory for their family and communities, as their ancestors are buried under the same farmland. The memory is one of the most important components that constitute diaspora globally, as memories (in form of their ancestors) are connected with the home and farmland which they have been using for decades or maybe even for centuries. Losing the land and having to relocate because of developmental projects means loss of the home, loss of memory, and loss of identity in one go. The economic loss if intended can be compensated, but compensation for the loss of memory and that of identity is more difficult and it is this characteristic related to place-making and homeland that makes these internally displaced communities a diaspora. Whether to take a "radical-movementalist approach" (Baviskar, 1995; Oliver-Smith, 2009), or a "reformist-managerial approach" (Mahapatra, 1996); or an "institutional approach" (WCD, 1999), the question of risks that the communities will face and the rights which lead to the actual movement or potential for the movement ought to be considered when we develop a framework to study internally displaced people. Ray and Saini (2011), while dealing with the displacement of people in the areas of opencast mining, argued about a possible middle path that can deal with the "appropriate" regulatory framework, livelihood options, meaningful community participation, accountability and commitment of the state and recognize the rights of the displaced. However Ray and Saini did not talk about "cultural genocide", which is not unreal in the case of displacement of communities from their natural abode, especially the tribal communities. The aspect of "cultural uprooting", a concept referred by Padel and Das (2010), is more connected with the conception I am using to place these communities as "diaspora within".

Displacement

In general, availability and unavailability of resources have been associated with the movement of people and the formation, evolution, and vanishing of civilization. The ancient civilizations, including the "Harappa Civilization" which evolved owing to fertile soil and water availability, often vanished mysteriously without having logical or conclusive explanations. Natural calamities including flood and drought, war, disease, and so on have been considered as common causative factors for displacing human settlements from a particular site. Resources, security, and economic prosperity have usually been considered magnetic factors which attract human congregation at certain sites. The Bedouins of the Arab desert located themselves at sites where the most precious resource for human life, "water", was available. They moved from oasis to coast as they found another resource, namely "pearl". The beginning of the use of oil has changed the entire geography of the desert, and the "prosperity" of the oasis faded slowly but significantly (Choudhary and Paul, 2018). The relative significance of resources have created a differentiated geography across the world in which the indigenous population and their lives have become either insignificant or soared in the "process of development". Studies have estimated that about 15 million people in the modern time are forced to leave their homes due to large-scale development projects like dams, irrigation projects, highway projects, mining, for the conservation of nature, and so on (Terminski, 2012; Cernea, 2006).

Internal displacement in India is not new and began far back in the 1950s when the then Prime Minister Nehru considered and propagated dams as "the temple of modern India". The Nehruvian model was based on the notion of "big" and enhanced scientific temper and technological intervention in society. Jawharlal Nehru (1948) while speaking to villagers who were to be displaced by the Hirakud Dam, said "If you are to suffer, you should suffer in the interest of the country". During the last six decades, about 3300 big dams have been commissioned across the length and breadth of the country (Devakumar, 2008). About 60 to 65 million people are said to have been displaced owing to development projects in India as per a report of the *Times of India* (Mukherji, 2012). From 1955 to 2005 about 23 million people were internally displaced due to development-environment activities like mines, dams, industries, wildlife sanctuaries, and so on (Judge, 1977). However, the next ten years witnessed twice the number of internally displaced compared to the previous five decades. The massive increase in the number of internally displaced people can be attributed to the accelerated pace of development in India, which at one point of time was referred to as "Hindu rate of growth".

The Indian economy after 1991 started growing an average of 5–7% annually compared to 2% to 3% in previous decades after independence. Jena (2014), while analyzing the extent of development-induced displacement in Orissa, expressed this sentiment as "with the advent of the New Economic Policy since early 1990s, the Indian State has declared its high-voltage war against poverty through its strategy of rapid industrialization . . . attributed to the establishment of mining based industries". However it also meant that owing to 94 projects, about 6207 hectares of forest land got diverted to non-forest use during 2000–2005. Development-induced displacement is not new to Orissa, and in the garb of "national interest" initially, public sector power projects like Hirakud Dam have displaced tribals, which after the 1990s were replaced by the corporate-state nexus. The movement against POSCO and TATA after 2000 brought notice to the intense resource plunder for profit without sharing the benefits with the local population (Choudhary, 2017). Additionally, enhanced awareness and the rise in activities of socially aware citizens also meant better reporting on those who are displaced. What matters more than the number of displaced is the nature of such displacement, the legislative protection against displacement, the looming dispossession due to displacement, the alienation that comes with the displacement, the resistance put up against such forced displacement, and also the resettlement policy. After displacement, the nature of social and economic classification of the household changes; those in the household become employment seekers ready to sell their labor for wages – this means an owner of a means of livelihood becomes a wage earner. The land ceases to be a household asset and means of livelihood, and the displaced seek unskilled jobs like mine-labor or construction labor in the mining or industrial activities that replace them.

In the process of such displacement all are not equally vulnerable. It is important to analyze the nature of vulnerability of different groups of people, as the rights are distributed unequally. Those who live close to nature have less structured and defined ownership of nature, as they believe in stewardship more than ownership. Unfortunately, these indigenous groups are more vulnerable due to undefined land and other rights and tend to lose more compared to groups who entered the area with defined rights. Interestingly, indigenous groups, usually defined in the Indian constitution as Scheduled Tribes (ST) – also loosely referred as "tribals" (*Janjati*) – constitute a large proportion of the displaced population. Based on the Working Group on Human Rights (WGHR) in India and the UN, Mukherji (2012) claimed that "Of these displaced, over 40 are tribals and another 40% consists of *dalits* (untouchables) and rural poor", as compared to their population percentage of about 7.5% of the total population. Land is one of the pressing issues for the people who are displaced owing to developmental projects, including dams, highways, mines, etc. While doing a social-cost benefit analysis for a coal mining project in the Sarguja district of Madhya Pradesh, the villagers reported that land is the most precious and indispensable resource as it is not only the prime source of livelihood, but also the site where "home" exists(ed). Internally displaced people are primarily engaged in agricultural or livelihood activities and hence land remains the most important resource. A study in Myanmar also suggests that internally displaced people in most cases want to go back and get access to their land, and this remains their most pressing issue (Oxfam, 2018). Along with land, access to the forest is the other crucial issue as it ensures sustenance for those who do not have individual ownership of land and are dependent on non-timber forest products or minor forest products. Hebbar (2005), while discussing the case of Jharkhand, argued that the Forest Bill of 2005 allows multinational and private companies to have a foothold and exploit the forest resource to allow conversion of forest land into revenue-generating villages. The conversion of forest land into non-forest land is predominantly done for mining purposes in states of Jharkhand, Orissa, Madhya Pradesh, and Chhattisgarh, commonly called the central mineral belt of India. Indigenous populations in these areas become doubly dispossessed as they tend to lose rights to access the forest they were acquainted with, and at the site of relocation they do not have either customary rights or legal rights to access the forest. Scholars make distinctions between "land acquisitions" by the state for "public purpose" as opposed to acquisition for private business interests. It is noticed though that the clause "public purpose" gets invoked in favor of private business interests. Some of the not so popular but frequently discussed and debated examples include Singur in West Bengal, POSCCO in Orrisa, Bhatta-Parsaul in Uttar Pradesh. Displacement of people from protected areas is another place-based displacement making millions of tribals "homeless". Implementation of conservation is about placing current and future limits on the rights of people, especially tribal people to use nature (Agrawal and Redford, 2009). It is difficult to argue convincingly about the nature of activity and extent of displacement in India. Mining, especially coal mining, and the construction of dams seem to be the major cause of large-scale displacement. The construction of Hirakud Dam has witnessed protests and movements in Orissa. The displacements owing to these two activities go largely unquestioned as they are "required" to ensure demand of "growing energy" for the "nation" (Baviskar, 1995; Bhengara, 1996; Menon and Saravana, 1996; Ray and Saini, 2011). Displacement is always a contested notion as proponents of development argue that these displacements are inevitable for the development of the nation and if resettlement is taken with sincerity no one should question the displacement. However, the fact is that these displacements not only dispossess tribals of their resources but also make them irrelevant in the new socio-economic context. The technology-driven mining and industrialization that follows the construction and commissioning of multi-purpose dams make the traditional and generational knowledge and livelihood options irrelevant. These people are reduced to mere "urban job-seekers" without adequate "skills". Further, on one hand it denotes physical removal and resettlement, while on the other, which is more real, it means economic, social, cultural, and other forms of loss resulting from the relocation with or without resettlement (Bray and Velazquez, 2009). Another rather silent displacement is taking place in India and is related to conservation. A recent judgment of the Supreme Court of India directed state governments and the union government to ensure that tribals are evicted from the forest land that was dedicated for the purpose of conservation. In the order dated February 13, 2019, the Supreme Court of India directed different state governments to ensure that eviction is completed by the due date and if they failed to ensure the reasons be given on the next date (July 24, 2019). The judgment related to the state Andhra Pradesh, and read as follows:

> The affidavit indicates that the extent of land covered by rejections in respect of STs is 1, 14,400 acres and 66351 claims have been rejected. But the action taken indicates that not even a single order has been complied with. Once the orders of eviction have been passed, the eviction ought to have taken place.

The number of dwellers who ought to be evicted for different states within India is different, and the claims have been rejected by respective state governments. The judgment brought back to debate the issue of the nature and extent of tribal rights and displacement owing to the conservation of forest and wildlife in India. The Scheduled Tribes (Recognition of Forest Rights) Bill 2005 as per section 3(a) ensured the "right to hold and live in the forest land under the individual or common occupation for habitation or for self-cultivation for livelihood by a member or members of a forest dwelling Scheduled Tribe". The Act put a condition of time within the same act, which allowed the state governments to recognize or not recognize the people's existence in the forest prior to the deadline. The Act as in 4(2) reads,

> The recognition and vetting of forest rights under this Act to forest dwelling Scheduled Tribes in relation to any State or Union territory in respect of forest land and their habitat shall be subject to the condition that such Tribes or tribal communities had occupied forest land before the 25th day of October, 1980 or such other date as the Central Government may, by notification in the Official Gazette, specify.

This provision put the onus on respective state government to enlist the tribals living in the forest especially those which have been designated otherwise. The extent of the inhabited forest land unfortunately has never been ascertained. During the colonial period forest and tribal regions were exploited for wood and other forest resources without any accountability for the people of the land. After independence, the constitution did recognize their existence and role in nation-building, allowing them freedom to retain their cultural practice while having access to "national" resources. But this was done in a framework that did not recognize diverse practices and expected population growth of the tribal population. A study of Jharkhand argued that due to migration relative share of tribal population in Jharkhand declined from 36% in 1951 to 27% in 1991; however at the national level data shows a high growth rate of 2.7% annual exponential growth of the tribal population (Maharatna and Chikte, 2004). In this situation a large number of ST population who have been living in these areas are bound to lose the right to live in their own home and are vulnerable to displacement as the February Judgment of the Supreme Court indicates. The Court is implementing the law, but the law is enacted by the democratically elected government and responsibility lies with the elected governments to ensure the disempowerment and dispossession of the indigenous communities from Indian Forests.

Dispossession and disempowerment

The basic purpose of development is said to enlarge people's choice and expand their freedom (HDR, 2010; Sen, 1999). In principle, these choices and extent of freedom can be infinite and are bound to change over time. Development is meant to remove the sources of unfreedom; however when Indian university student leader coined the slogan "*aazadi, aazadi, hum le ke rahenge aazadi*" ("Freedom, freedom, we shall have our freedom, come what may"), it created a national uproar and the "ruling class" presented the slogan as separatist, and a case under IPC section 124 was registered.[4] The argument was constructed as if the youth was seeking separation from the Indian nation-state rather than protesting against continuous and systematic dispossession of tribal people. Over time, more people acting on such tendencies are becoming more real and environmental activists and scholars, including non-governmental organizations like Greenpeace, face the periodic brunt of the state and law-enforcement agencies. Such incidents bring to the forefront the mere nicety of academic argument, like development is meant for freedom. I say so because every nation-state curtails freedom in the name of freedom. The dispassion takes place in the name of development. Those who question development are presented as "someone" questioning the authority of the "state" and thus must be dealt an "iron fist". The democracies allow such actions by the state in the name of ensuring development and freedom. For example, Indian Prime Minister can be lauded for saying "*sabka sath, sabka vikas*" (togetherness and development for all). However, the central question is who constitute "all". It seems that the economy has reached a level where everyone is dealing with the same problem and there is no class existing. Corporations and corporate co-opted states might wish to propagate that the world lives in a classless society – in the line of "the world is flat" – but the question remains, is the world really flat?

After liberalization India chose a path of development which disrespected everything other than "growing the size of the economy" and ensuring "a faster growth rate of GDP", to the extent that Amit Bhaduri (2007) called the model of development "development terrorism". The answer to every question including employment is that if the economy is growing there must be employment generation. This could mean large-scale displacement of indigenous population and dispossession of resources. Mohanty (2009) argued that development-induced displacement in the country has brought severe economic, social, and environmental problems to the displaced people and the magnitude of this displacement and severity of the problem affects tribal women most. What is evident is differentiated levels of vulnerability among displaced ones. Some sections of the population over the last few decades have gained from the development project and therefore the state has been able to present consent of the people for the land acquisition. The Land Acquisition Act in India even today is a remnant of the colonial era with certain modifications. The colonial government could take away land from the land-loads as per the Land Acquisition Act of 1894; the Government of India after independence continued to practice acquisition of land against the owners' wishes for "public purpose" with the amendments in 1967, 1984, and 1999 of the Land Acquisition Act of 1894. The amendment in the Land Acquisition Act proposed in 2007 brought in the component of consent. This continued till the present government repealed the act and changed it to become "development friendly". With the new act, the landowner's consent became mandatory for 60% to 80% of the different purposes for acquiring land, excluding military use, for which land can be acquired without consent. This act induced complications in the tribal areas, especially the areas covered by Schedule V. The social-cost benefit analysis of Sarguja revealed to us the complexity that exists within communities which were to be displaced. Out of the eight villages that we estimated the cost benefit analysis for, two villages showed relatively more resistance to the establishment of the coal field. The people of these villages reported that while they will lose the access to the forest, the environment, the social network they have, they were not getting any compensation. While probing further, we realized that these villages would continue to exist on the margin of the coal field. They had to relocate themselves, as the proposed mining company was not acquiring their village, while the negative externalities in terms of forest and water would not spare the village. The people from the villages which were to be relocated talked about the cost of displacement, including loss of land, loss of forest access, and loss of source of water. The poor faith in the ability of the agencies to ensure resettlement and payment of the compensation amount compounded the expression of misery. The sense of loss in terms of identity and memory was unevenly distributed in the region. While the indigenous communities always spoke about the sense of loss and homelessness, a few households did not have the same sense of loss about their cultural heritage. We tried understanding the sense of hope amongst these people and realized that these households were those who had been relocated from the earlier Rihand Dam site in the neighboring district of Uttar Pradesh. Interestingly, all of them were found to be non-tribal. Tribal population more often than not considers all these development projects as masked "displacement projects" (Padel and Das, 2010).

Padel and Das (2010: 365–372) analyzed impacts of these development-induced displacements and equated such incidents with "cultural genocide", as these displacements not only dispossessed these communities from land and access to forest by converting "common property" into private property, but also destroyed the lived social structure. Destruction of such social structure also means the diminishing of the possibilities of maintained traditional ecological knowledge. It is acceptable almost beyond doubt that "epistemological characteristics of indigenous science include oral transmission; observation over generations; cyclical time orientation; specific cultural/literary style and symbolism; knowledge is context specific to a tribal culture and place; conservation of knowledge through time and generations" (Berkes, 2018). When a tribal population is made to move out of the place, a situation of placelessness becomes real and the context in which the traditional knowledge systems have evolved and grown get lost in the development-induced displacement. It is ironic that countries and societies celebrate the WCED publication "Our Common Future" and refuse to acknowledge in action that "tribal and indigenous people's . . . lifestyle can offer modern societies many lessons in the management of resources in complex forest, mountain and dryland ecosystems" (Burtland, 1987: 12). I say so because displacements across countries and continents in the name of development situate these tribals in new social-political-ecological context where these communities are made to come to terms with the so-called modern world's production system that is distanced from nature at the best. Neil Smith (2010: 49), while invoking Marx, considered that the "idea of the production of nature is indeed paradoxical", and argued that nature is generally seen as precisely that which cannot be produced; it is antithesis of human productive activity. Tribal societies have lived with nature on the basis of their understanding of "use value" rather than "exchange value". The industrialization process not only ensures dispossession of resources of these indigenous communities but also contributes in altering their nature-based value system from "use value" to market-based "exchange value". This means disempowerment for these communities as their so-called celebrated "traditional knowledge systems" become at best "contextual tacit knowledge" and fail to stand against "scientific universal knowledge", which has been the basis of industrial production and global exchange of industrial products. The vast accumulation of these communities' traditional knowledge has ultimately disappeared, owing to the unceremonious displacement of these communities that is often marred with protest and occasional violence.

Quest for justice

The logic for the development-induced displacement lies in the inevitability of the process. Resources are found at certain sites and if these sites are inhabited, there is little chance than to relocate those who are living at those sites. However, the process demands adequate compensation for those who are displaced. The basis of the "cost-benefit analysis" indicate that those who lose land are to be compensated with employment in the modern production system and receive the same or larger quantity of land along with monetary compensation. The study of Malaysia points to the fact that those displaced reported to be in better economic condition compared to pre-displacement condition (Ahsan, 2016). The bargain in such cases could be between losses of identity, loss of culture, loss of place with that of economic gains. The process of resettlement is intrinsically interconnected to the concept of "development-induced displacement", at least since the 1970s. Oliver-Smith (2009: 3) wrote about the harsh and brutal truth about the inadequacy of resettlement processing and said that "unfortunately, resettlement has been so poorly planned, financed, implemented and administered that these projects generally end up being 'development disasters'". Displacement usually affects non-economic aspects of life too, whereas resettlement at the best tries to take care of economic losses, primarily access to resources and ownership of land.

The lack of rehabilitation of those internally displaced ensued discontent against the development model which was adopted after the 1990s. It is argued often that increasing violence in the tribal belt across central India can partly be attributed to the failure of the Indian state to address the discontent which has emerged due to the perpetual neglect of the tribal population who are living in precarious conditions. In its report on "Development Challenges in Extremist Affected Areas", the expert group Planning Commission of India (2008) acknowledged,

> The development paradigm pursued since independence has aggravated the prevailing discontent among the marginalized sections of the society. The development paradigm, as conceived by policy makers, has always imposed on these communities and has caused irreparable damage to these sections. The benefits of this paradigm have been disproportionately cornered by the dominant sections at the expense of the poor, who have borne most of the costs. Development which is insensitive to the needs of these communities have inevitably caused displacement and reduced them to a sub-human existence. In the case of tribes in particular it has ended up in destroying their social organization, cultural identity and resource base which cumulatively makes them increasingly vulnerable to exploitation.

The acceptance of this kind by the expert group of the Planning Commission emphasized the risks and vulnerabilities faced by the tribal communities across the mineral and forest belt in India.

The national policy for resettlement and rehabilitation in India used to view displacement only as an appendix to development (Sah, 1995). Industrial development in the country was seen as the only way forward to tackle the problem of stagnating agriculture and rural deprivation of masses. Development with such "noble" cause has to acquire land "if need be" and the colonial land acquisition policy of 1894 provided the basis for it. The proposed modification in the national policy for resettlement and rehabilitation in the 1990s acknowledged the human misery due to involuntary migration. It noted that "one of the preconditions for the approval of any project should be a strict assessment of the land requirements of the project with a view to keeping the land acquisition to the minimum so that involuntary displacement is minimized" (quoted in Sah, 1995: 3055). The Ministry of Rural Development in 2006 put up a new draft for the National Development, Displacement and Rehabilitation Policy for public discussion. The policy as Shekhar Singh (2006) argued can be assessed on the basis of five criteria, the first being if it can minimize displacement. The other criteria include if it can define the project-affected in a comprehensive manner? How just is the compensation package? Does it prescribe a humane and people-friendly displacement and rehabilitation policy? And, does it contain adequate provision to ensure that it would be implemented in letter and spirit? The outlined criteria by Singh (2006) may seem comprehensive; however, it is good only for the textual analysis as it is not clear how any policy would ensure implementation in letter and spirit. These criteria seems to follow more or less a techno-managerial approach to displacement which can buy peace from the larger section of those affected and minimize the chances of discontent and protest movements. It is interesting to note that the draft put up by the Government of India in 2006 diluted the provisions against those to be affected by the developmental project when it was compared with the draft prepared by the National Advisory Council, which was the guiding force of the UPA Government. To specify, compared to NAC, the GOI draft narrowed the definition of the project-affected to those families which have resided, worked, owned assets, etc. for at least three years prior to the declaration of the affected zone (Singh, 2006). Indian experience with resettlement and rehabilitation has been really poor for decades. Ramaswamy Iyer (2007: 3103), a noted jurist and member of several rehabilitation/resettlement negotiating teams reflected on the act, saying,

> there is a triad of terms: displacement, resettlement and rehabilitation. Displacement may not always be followed by resettlement; and resettlement does not necessarily imply full rehabilitation of displaced persons. In recent years in India, it has come to be accepted that displacement of people must by followed by resettlement and rehabilitation, but this was not always the case.

As many others, Iyer (2007) also considered the Land Acquisition Act 1894 as an instrument of displacement under which private land can be acquired by the state for the "public purpose". In 2011 the Supreme Court emphasized the need to enact a new land acquisition law. A joint bench of Justice Lodha and Justice Khehar in their Judgment remarked that the Land Acquisition Act 1894 (L.A.A 1894) does not adequately protect the interest of owners/persons interested in the land. In this background, the then government drafted Land Acquisition, Rehabilitation and Resettlement Bill 2011, which became an act in 2013 and is known as "Right to Fair Compensation and Transparency in Land Acquisition, Rehabilitation and Resettlement Act 2013". One of the important components of the Act was that for acquisition of land, consent of 80% of the landowners in the case of private acquisition and 70% of the landowner in the case of public acquisition became mandatory (Bharathi, 2012). Though, it did not help the tribal communities, especially where community rights prevail over individual rights.[5] The tribal community still remained out of the purview and the prime reason for that is the Indian ruling class enacted the law to benefit those who have the voice or capability to muscle in their favor.

The tribals on one hand got dispossessed of their assets, including land and their right over forest products and non-timber forest products, and their vast repository of traditional knowledge and on the other they got caught in battle between insurgents and security forces. Their knowledge becomes redundant, they feel alienated in the new economic order, and most likely struggle to find meaning in the new urban-industrial life or relocated rural-jungle life, as the new set-up tends to be "not theirs". These communities fail to understand the role of the state in land acquisition, as in the majority of cases they articulate surprise as to why the state should "fire on them" when they are not protesting against the state but against the owner of the plant, like in the case of POSCO or in the case of TATA. The Supreme Court of India in 2011 on a written petition between Nandini Sundar v. State of Chhattisgarh opined,

> The root cause of the problem, and hence its solution, lies elsewhere. The culture of unrestrained selfishness and greed spawned by modern neo-liberal economic ideology, and the false promise of ever increasing spirals of consumption leading to economic growth that will lift everyone, under-grid this socially, politically and economically unsustainable set of circumstances in vast tracts of India.

The judgment seems favorable, at least acknowledging the plight of tribal people. The provisions of law remain unfavorable to the indigenous community and their rights to resources including land, water, and forest resources. Further, after the Land Acquisition Act of 2013, emphasis was given on cash compensation rather than land, which lays the ground work of destruction of tribal societies, as land is the means of livelihood, while money cannot be the means of livelihood and cannot be a factor of production either unless it is used as capital. If we expect these tribal communities to use the compensation money as capital, we have to bank on our imagination rather than the possibility of justice and rehabilitation of these communities who are uprooted from their own homeland and are forced to live as diaspora in their own country.

Summing up

The hostility against those marginalized and displaced has gained more strength compared to the voice that argues for the rights of those displaced in their own country. The cultural and social uprooting has caused additional burden on the indigenous communities, primarily in the forested areas of the states of Orrisa, Chhattisgarh, and Jharkhand. Construction of dams, the beginning of mining activities, demarcation of sanctuaries, and so on constitute prime reasons for the displacement of people. Statistics are absent on the actual number of families and population displaced, on those who received compensation, and on the revenue or profit generated from the land clearly acquired with the intent to ensure denial of rights to these indigenous communities. Further, accusing these communities of violence and of being untrustworthy to the "nation" adds to the misery of these vulnerable groups. The section of population with relatively lower capabilities to cope with the changed realities is more vulnerable to lose economically as well as culturally. The payment of cash in lieu of their losing their means of production is anything but fair compensation. The long legal battle and diminishing support due to hostile state agencies in recent years contribute to the perils of the tribal communities rather than states protecting these communities.

Notes

1 The idea of nation here is taken differently than the legal or sovereign concept of nation-state only and is taken as the notion argued by Anderson in his book *Nation as Imagined Community*.
2 Tribes/indigenous population/Scheduled Tribes are used interchangeably in this chapter; the Indian Constitution recognizes specific characteristics of indigenous people and provides protection in its Schedule V and Schedule VI; in India they are commonly referred to as Scheduled Tribes (ST); whereas elsewhere they are either referred to as indigenous communities or in some places as tribals or tribes.
3 Archana Prasad and Nandini Sundar, faculty members from Jawaharlal Nehru University and Delhi University, along with journalists and rights groups have been working in the Indian state of Chhattisgarh and have written about the perils of tribal population that is caught between extremists and state agencies. Nandini Sundar has been part of a group that has objected to the practice of "Salwa Judam" (a rather illegitimate practice of raising an informal military group to resist extremists/"right's movements" in the state) in the Apex Court of India and got judgment banning the practice. Archan Prashad and Nandini Sundar have been charged by the state with murder (apparently without any distant connection or proof) of one person in 2016–2017 in a seeming act of vengeance.
4 In 2016, a case of sedition was slapped against JNUSU president for alleged anti-national slogans. One of the slogans, which became popular, concerns "*aazadi*", freedom. The then JNUSU President Kanhaiya Kumar could be seen raising the slogan of *aazadi* against poverty, gender discrimination, caste discrimination, religious fanaticism, and so on. The case under IPC 124 is still pending in the Patiala Court of Delhi.
5 Even this clause was diluted through an Ordinance brought in 2014 by the NDA (National Democratic Alliance) Government in the name of easing the problem of land acquisition.

References

Agrawal, A. and Redford, K. 2009. Conservation and Displacement: An Overview. *Conservation and Society* 7(1).

Ahsan, R. 2016. *Mega Infrastructure Induced Development in East Malaysia: A Study of Social Sustainability*. Working Paper Series, Boston: Massachusetts Intsitute of Technology.

Baviskar, A. 1995. *IN the Belly of the River: Tribal Conflicts Over Development in Narmada Valley*. OUP, New Delhi.

Berkes, F. 2018. *Sacred Ecology*. Routledge, New York.

Bhaduri, A. 2007. Development or Development Terrorism. *Economic and Political Weekly* 42(7).

Bharathi, M. 2012. Tribal Women's Perspective on the Land Acquisition Bill. *Economic and Political Weekly* 48(20).

Bhengara, R. 1996. Coal Mining Displacement. *Economic and Political Weekly* 31(11).

Bray, D. and Velazquez, A. 2009. From Displacement Based Conservation to Place-Based Conservation. *Conservation and Society* 7(1): 11–14.

Burtland, G. H. 1987. *Our Common Future*. WCED, United Nations. New York: Oxford University Press.

Cernea, M. M. 2006. Development-Induced and Conflict-Induced IDPs: Bridging the Research Divide. *Forced Migration Review* 25–27. www.fmreview.org/sites/fmr/files/FMRdownloads/en/brookings/cernea.pdf accessed on 1 May, 2019.

Choudhary, B. K. 2017. Upward Movement vs. Outward Movement: Resource, Development and Indigenous People. *Transactions of Indian Institute of Geographers* 39(1): 35–45.

Choudhary, B. K. and Paul, N. 2018. Transforming Dubai: Oasis to Tourist's Paradise. *Contemporary Review of the Middle East* 5(4): 349–364.

Cohen, R. 1996. Diaspora and the State: From Victims to Challengers. *International Affairs* 72(3): 507–520.

Devakumar, J. 2008. Internal Displacement in Contemporary India: Homeless in Their Own State. *Proceedings of the Indian History Congress* 69: 1211–1225.

Hebbar, R. 2005. Forest Bill 2005 and Tribal Areas: Case of Jharkhand. *Economic and Political Weekly* 41(48): 4952–4955.

Iyer, R. R. 2007. Towards a Just Displacement and Rehabilitation Policy. *Economic and Political Weekly* 42(30): 3103–3107.
Jena, B. K. (2014). Development-Induced Displacement in 21st Century India. *Proceedings of the Indian History Congress* 75: 1183–1191.
Jing, J., 2000. Displacement, resettlement, rehabilitation, reparation and development: China Report. *World Commission on Dams Thematic Review, 1.*
Judge, P. S. 1977. Response to Dams and Displacement in Two Indian States. *Asian Survey* 37(9): 840–851.
Lancione, M. 2019. *On Lessness: Reentering the Politics of Home.* Keynote Address to 7th EUGEO Congress of the Geography of Europe, 15–18 May.
Mahapatra, L. K. 1996. Testing the Risk and Reconstruction Model on India's Resettlement Experience. In M. Cernea (ed.), *The Economics of Involuntary Resettlement: Questions and Challenges.* World Bank, Washington DC, 189–230.
Maharatna, A. and Chikte, R. 2004. Demography of Tribal Population in Jharkhand 1951–1991. *Economic and Political Weekly* 39(46/46): 5053–5062.
Menon, A. and Saravana, V. 1996. Displacement and Rehabilitation Policies: Case of the Kolli Hills Hydroelectricity Project. *Economic and Political Weekly* 31(43).
Mohanty, M. 2009. Development and Tribal Displacement: Reflections on Core Issues. *The Indian Journal of Political Science* 70(2): 345–350.
Mukherji, A. 2012. India Uproots Most People for 'Progress', June 4. *TOI.* https://timesofindia.indiatimes.com/india/India-uproots-most-people-for-progress/articleshow/13792551.cms accessed on 1 May, 2019.
Nagar, R. and Shirazi, R. 2019. Radical Vulnerability. *Antipode* 50.
Oliver-Smith, A. (edited). 2009. *Development and Dispossession: The Crisis of Forced Displacement and Resettlement.* Santa Fe, USA: SAR Press.
Oxfam. 2018. Displaced and Dispossessed. Conflict Affected Communities and Their Land of Origin in Kachin State, Myanmar. https://www-cdn.oxfam.org/s3fs-public/file_attachments/bp-displaced-dispossessed-land-myanmar-210518-en.pdf
Padel, F. and Das, S. 2010. *Out of This Earth: East India Adivasis and the Aluminum Cartel.* Hyderabad: Orient Blackswan.
Planning Commission. 2008. *Expert Group Report on Development Challenges in Extremist Affected Areas.* Government of India, New Delhi. https://niti.gov.in/planningcommission.gov.in/docs/reports/genrep/index.php?repts=b_repgen.htm
Ray, S. and Saini, S. 2011. Development and Displacement: The case of n Opencast Coal Mining Project in Orissa. *Sociological Bulletin 60*(1): 45–64.
Sah, D. C. 1995. Development and Displacement: National Rehabilitation Policy. *Economic and Political Weekly 30*(48): 3055–3058.
Sen, A. 1999. *Development as Freedom.* Oxford: Oxfor University Press.
Singh, S. 2006. Displacement and Rehabilitation: A Comparison of Two Policy Drafts. *Economic and Political Weekly 41*(52).
Smith, N. 2010. *Uneven Development: Nature, Capital and the Production of Space.* Verso, London.
Steward, T. P. 2017. In Between Home and Homeland: Diaspora Identity as a Cultural Hybrid in Mohsen Namjoo's "CielitoLindo". *Popular Communication 15*(3): 207–222.
Terminski, B. 2012. *Environmentally Induced Displacement: Theoretical Framework and Current Challenges.* Liège: University of Liège Press.

28

RATIONALITIES OF GOVERNMENT AND WEBS OF RELATIONS(HIPS) IN THE FUNDING AND IMPLEMENTATION OF SEA DEFENSE SYSTEMS IN THE VOLTA RIVER DELTA OF GHANA

Kwame N. Owusu-Daaku

The pitfall of maladaptation and the potential of governmentality

The Green Climate Fund (GCF) was established by the United Nations Framework Convention on Climate Change (UNFCCC) in 2010 to help majoritarian countries mitigate and adapt to climate change. The fund channels finances to majoritarian countries from various sources (both public and private) that non-majoritarian economies primarily mobilize (Green Climate Fund, n.d.-a). The initial goal of the GCF was to raise \$100 billion per year by 2020 (Green Climate Fund, n.d.-d). This goal has not been realized. However as of 2019, contributors to the GCF had pledged more than \$20.1 billion between its two funding mechanisms (Green Climate Fund, n.d.-d) and funded over 100 projects (Green Climate Fund, n.d.-b). This information demonstrates the magnitude of funding that nations all over the world are directing towards addressing climate change.

As the implementation of formal, donor-funded, and government-supported adaptation projects (Bahadur et al., 2015; USAID, 2012) such as the GCF accelerates (Green Climate Fund, n.d.-c), concerns for maladaptation have moved from the conceptual realm into policy, implementation, and their empirical outcomes. These concerns manifest in practical questions of project design, monitoring and evaluation, and learning. For these reasons, maladaptation is gaining significant attention in the climate change and development communities of practice – particularly with regard to its definitions and development (Atteridge & Remling, 2013; Barnett & O'Neill, 2010; D'Alisa & Kallis, 2016; Jones et al., 2015; Juhola et al., 2016; Klein & Maciver, 1999; Magnan et al., 2016; Scheraga & Grambsch, 1998; Watts, 2015). For example, early definitions of maladaptation were often framed in relation to the initial goals of formal

 DOI: 10.4324/9780429450310-33

adaptation efforts and viewed as "the extent to which adaptation fails" (Klein & Maciver, 1999, p. 193) to deliver on those specific goals. Later definitions of maladaptation incorporated understandings of vulnerability to mean an "action taken ostensibly to avoid or reduce vulnerability to climate change that impacts adversely on, or increases the vulnerability of other systems, sectors or social groups" (Barnett & O'Neill, 2010, p. 211). Other definitions extend maladaptation to the "disorders of adaptive structure", including not just the goals of adaptation but the decision-making pathways and overall governance of various actors involved in adaptation processes (Rappaport in Watts, 2015, p. 27).

This evolution of maladaptation reflects a growing understanding that climate change projects aimed at fostering adaptation and resilience are socio-political processes and not merely techno-managerial ones (Carr, 2019; Chandler, 2016; Chandler & Reid, 2018; Cote & Nightingale, 2012; Eriksen et al., 2015; Grove, 2018; MacKinnon & Driscoll Derickson, 2012; Nightingale, 2009; O'Brien et al., 2004; Ojha et al., 2016; Pronk et al., 2016). At the heart of these socio-political processes are relations of power. Therefore, understanding how maladaptations occur and manifest requires answers to a series of underlying questions: who frames maladaptation as an issue to address?, who benefits from particular framings and/or solutions to maladaptation?, and who is most impacted by processes or outcomes of maladaptation?

The Foucauldian framework of governmentality (Foucault, 1995, 2003; see also Burchell et al., 1991; Dean, 1999) is a useful means of interrogating relations of power and knowledge to demonstrate how good-intentioned efforts to identify and address climate change-related vulnerabilities produce maladaptive outcomes. Governmentality, according to Foucault, has a bit of a complicated origin. Foucault describes governmentality as

> the ensemble formed by institutions, procedures, analyses and reflections, the calculations and tactics that allow the exercise of this very specific albeit complex form of power, which has as its target population, as its principal form of knowledge the political economy, and as its essential technical means apparatuses of security.
>
> *(Burchell et al., 1991, p. 102)*

From Foucault's description, governmentality thus explains the seemingly nonsensical ways in which decisions are made and actions are taken at various scales. As such, these efforts can be productively approached through the lens of governmentality as forms of government – efforts to set the world on a path toward safety and certainty. In the context of governmentality, government is understood as:

> any more or less calculated and rational activity, undertaken by a multiplicity of authorities and agencies, employing a variety of techniques and forms of knowledge, that seeks to shape conduct by working through our desires, aspirations, interests and beliefs, for definite and shifting ends and with a diverse set of relatively unpredictable consequences, effects and outcomes.
>
> *(Dean, 1999, p. 11)*

The diversity and unpredictability of the outcomes of government as defined by Dean (1999) speak directly to the complexities of adaptation, the management of risk, and subsequent maladaptive outcomes. While adaptation policies and projects are efforts to address material challenges in the world, these efforts are also always ways to align the different and sometimes contradictory meanings behind the goal of safety and security from climate change impacts in the context of particular projects and places (Eriksen et al., 2015; Pronk et al., 2016).

Foucauldian analyses of power provide valuable understandings of how disparate people with dissimilar motivations align toward a shared goal (Anderson, 2012; Bettini, 2014; Chandler, 2010; Collier, 2009; Dalby, 2013; Dean, 1999; Dillon, 2007; Dillon & Lobo-Guerrero, 2008; Dillon & Reid, 2001; Foucault, 2004; Grove, 2010, 2014), even when the outcomes of such efforts are maladaptive for some or many in a society, community, or household.

Since such large financial investments are being made all over the word to govern climate change in a Foucauldian sense (Stripple & Bulkeley, 2014), all stakeholders need to give attention to the possibility of maladaptive outcomes from such investments and thus seek to minimize, if not altogether prevent, maladaptation if the most efficient use of these dedicated funds is a concern of all parties involved.

Setting the stage: sea defense systems as a climate change adaptation

An example of a significant financial investment made in the name of adapting to climate change which has visible maladaptive outcomes is coastal protection infrastructure as a response to sea-level rise. These structures, often in the form of seawalls, groins, and revetments, cause further erosion downdrift of where there are no such structures (Angnuureng et al., 2013) and often increase saltwater intrusion into aquifers and surface water (Mensah & Fitzgibbon, 2013).

In Ghana, these systems of groins, revetments, roads, and seawalls are known as sea defense systems (SDS) and the country's Volta River Delta (VRD) contains three of such systems: the Ada Sea Defense System (AdSDS), the Atorkor Sea Defense System (AtSDS), and the Keta Sea Defense System (KSDS) – see Figures 28.1 and 28.2. All three of these systems have

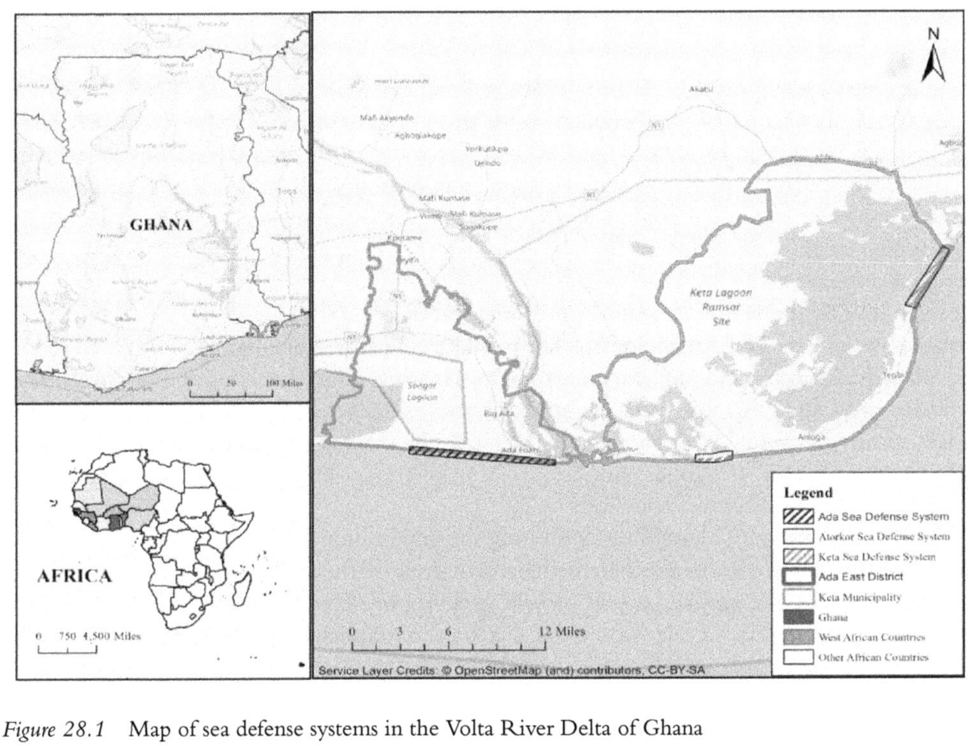

Figure 28.1 Map of sea defense systems in the Volta River Delta of Ghana

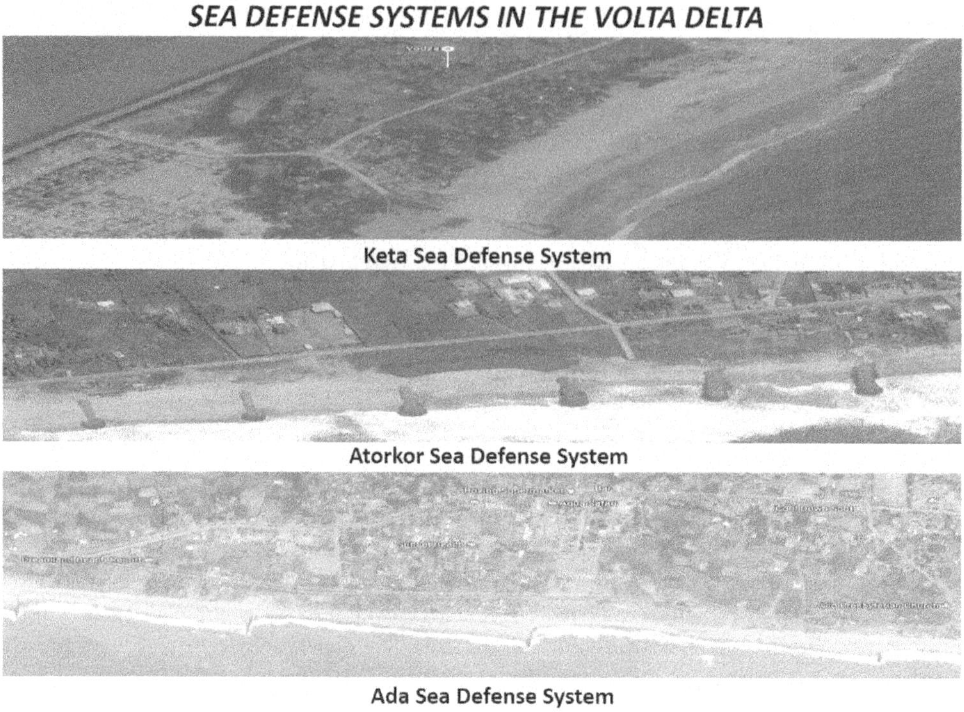

Figure 28.2 Google Earth imagery of sea defense systems in the Volta River Delta of Ghana

also begun to cause downdrift erosion with the Keta and Ada Sea Defense Systems recording evidence of saltwater intrusion (Angnuureng et al., 2013; Mensah & Fitzgibbon, 2013). Governmentality, as applied to SDS in the VRD, can provide insight into the reasons for which a climate change adaptation (CCA) with an apparently obvious maladaptation such as intensified erosion elsewhere would be continually preferred by the implementers of such an adaptation and desired even by individuals who have been affected by such intensified erosion (Owusu-Daaku & Rosko, 2019).

In light of the previous paragraph, a first question to ask is who are the people driving such maladaptive investments? An obvious answer would be the Government of Ghana (GoG) – who are often the initiators and implementers of SDS. However from the definition of government provided by Dean (1999) to encompass "a multiplicity of agencies and authorities", government *in* Ghana for SDS would not only include the GoG but also actors outside of the GoG responsible for the funding, designing, and constructing of these SDS such as international banks, consultants, and contractors respectively.

A second question, which follows outlining the government of SDS in Ghana, is what are the rationalities that inform the statements and actions of these identified actors of government? *Rationalities* are one key means by which things or objects are made governable. Foucault talks about the use of calculations and statistics in rationalizing the need for government intervention in a population. Research that highlights this use of calculations and/or statistics in rationalizing governance of one form or another include Ferguson's *The Anti-Politics Machine*, which illustrates how development institutional framings of Lesotho cast the country as a subsistence economy of farmers, when in reality the majority of the local economy was supported by the wage-labor of men who worked in South African mines (Ferguson, 1994); Scott's *Seeing Like a State*, which argues for the ways in which national governments use surveying and map-making to simplify complexity in order to manage populations (Scott, 1998); and Mitchell's *Rule of Experts*, which demonstrates how even the economy as a taken-for-granted entity is a contrived concept which has been applied to Egypt to demonstrate shortages of food supply, caricature the peasantry, and explain ecological problems in simplistic, unitary ways (Mitchell, 2002). Calculations, definitions, and the attendant documents and statements within which they are contained could be referred to as discourses. *Discourses* can be described as ways of thinking, doing, or being. Thus discourse presents a more collective term to refer to the means of representing rationalities. In other fields, discourses are framed as means to the social construction of reality (Barrett, 1995; Gamson et al., 1992; Phillips & Hardy, 2002; Potter, 1996). The discourses of climate change vulnerability, and the calculations, definitions, statements, and practices of which they are comprised, are critical means to the end of identifying what is to be governed in the context of climate change, and how it should best be governed.

A third and final question of more practical significance would be what lessons can the case of SDS in the VRD of Ghana provide for the governance of CCA? Mees et al. (2012) recognize that the private sector or non-governmental actors also have a part to play in the decision-making process with regard to CCA and hence call for an analysis of responsibilities and capabilities of governmental and non-governmental (both for-profit and non-profit) actors in order to promote efficiency. This call for the involvement of non-governmental actors is another extension of the concept of governmentality beyond the state to understand how non-state actors contribute to projects of rule (Appadurai, 2001; Bryant, 2002; Sending & Neumann, 2006). However this notion of joint or shared responsibility is often not the case in Africa. In many African countries, CCA acts as an idea that travels across multiple sites and through various networks, sometimes propelled by incentives of funding rather than out of direct need in response to a climate impact –whether current or projected (Weisser et al., 2014). As such, in the analysis of pathways and/or frames of decisions and their constitution across scales, it is critical that researchers pay attention to the interactions between governmental officials, and the influence of non-governmental actors such as multi or bilateral development agencies, and funded-research programs where applicable, to determine whether the "politics of a travelling idea" (that is, particular ways of governing or exercising control that seem to circulate within and between policy circles) with regard to CCA (Weisser et al., 2014) are at play within perceptions and whether these identified notions of governing or rule are a major driver for the articulation of these perceptions. This attention to "a travelling idea" reflects identifying particular "rationalities" of government (as defined by Foucault and other governmentality scholars) that condition particular "technologies" of CCA. *Technologies* are thus the means by which populations are ruled or governed. Technologies can be in the form of procedures, infrastructure, tools, programs, or other physical and/or psychosocial objects that serve as the medium through which governing meets certain ends or exerts control. SDS would represent an infrastructural technology of CCA.

State rationalities of climate change

The GoG employs a number of rationalities to employ this technology of SDS as an adaptation to climate change. The first rationality is that of climate change as an economic threat (Owusu-Daaku & Rosko, 2019). In many national planning and climate change policy documents, the

GoG makes this rationality very clear. The rise of climate change in planning in Ghana has resulted in four national documents on climate change – the National Climate Change Policy Framework (NCCPF) and the National Climate Change Adaptation Strategy (NCCAS), which were prepared in 2010; the Ghana National Climate Change Policy (NCCP) published in 2013 (Ministry of Environment Science Technology and Innovation, 2013) and the National Climate Change Policy Action Programme for Implementation (NCCPAPI) published in 2015 (Ministry of Environment Science Technology and Innovation, 2015). In all of these documents, there was a clear recognition by the GoG that Ghana remains vulnerable to climate change. Although the NCCPF provided a broad direction on how climate change issues would be handled, the NCCAS, NCCP, and NCCPAPI provided clear-cut approaches to confront current and anticipated climate change impacts in Ghana. The vision of the NCCP, for instance, is to "ensure a climate-resilient and climate-compatible economy while achieving sustainable development through equitable low-carbon economic growth for Ghana" (Ministry of Environment Science Technology and Innovation, 2013, p. 21) by focusing on adaptation, social development, and mitigation. Again, the GoG is explicit about the priority it gives to framings of impact in terms of the economy. Though it mentions sustainable development as a simultaneous focus and then again mentions low-carbon development, these are mentioned only as such development relates to economic growth. This is the first rationality of the GoG related to climate change: *climate change is a threat to the national economy*.

The NCCPF, NCCAS, and NCCP also emphasized coastal erosion as a challenge. In these documents, sea level rise and coastal erosion and their effects on settlements were identified as a major climate change impact in Ghana. In effect, coastal erosion was part of the climate change discourse in Ghana and policymakers identified coastal erosion as needing attention. This attention reveals the second rationality of the GoG: *coastal erosion is a serious climate change impact that needs attention*.

In the NCCPAPI SDS were clearly emphasized. This emphasis was captured under Focus Area 2 (Build Climate-resilient Infrastructure), Programme 2.7 (Protection of Coastal Resources and Communities), Action 2.7.3 (Construct climate-resilient key coastal infrastructure to protect the communities from storm surges, coastal flooding, sea level rise and ecosystem degradation such as deforestation) (Ministry of Environment Science Technology and Innovation, 2015). Under Action 2.7.3, the GoG specifically outlined the following tasks:

> 2.7.3.1 Develop engineering design and construction standards for appropriate structures such as groynes to stop erosion and sea defence walls for specific littoral segments.
>
> 2.7.3.2 Construct appropriate sea defence structures at Axim, Princess Town and Aketekyi along the west coast and at suitable sites.
>
> 2.7.3.3 Redesign and construct appropriate sea defence structures in areas just east of Keta (that is, from Akplorwotokor to Aflao) to stop the increase in erosion in those areas caused by the sea defence structures at Keta.
>
> *(Ministry of Environment Science Technology and Innovation, 2015, p. 59)*

The irony of task 2.7.3.3 is that the GoG is necessitating more SDS to tackle the erosion created by the KSDS. This unwavering focus on sea defenses, despite their negative effects, presents the third rationality of government: *coastal erosion must be addressed through hard engineering infrastructure such as sea defense systems*.

The protagonist: the Ministry of Water Resources, Works and Housing

Although I have identified the GoG as the initiator of sea defense projects in the VRD, a specific ministry within the GoG has responsibility for the implementation of SDS. This ministry is the Ministry of Water Resources, Works and Housing (MWRWH). Since MWRWH is responsible for the construction of all civil works related to water and housing, the implementation of SDS falls under their prerogative. The MWRWH thus has served as the client in all the SDS projects in the VRD.

Inasmuch as policy documents and briefs serve as official statements and proxies of government viewpoints, it is important to remember that governments (and international agencies) are complex entities made up of individuals with views and perspectives of their own (Mathews, 2005; Robbins, 2000; Carr's contribution in Simon et al., 2011). I therefore found assessing how a representative of the MWRWH viewed SDS as expedient.

In April 2016, the then Minster of the MWRWH Dr. Kwaku Ageymang Mensah stated the following, upon visiting a community in the Keta Municipality which was receiving national attention because the community had recently experienced flooding:

> The NDC [National Democratic Congress] government, through the Ministry of Water Resources, Works and Housing, is implementing a lot of coastal protection works. You can talk about Ada Sea Defense, you can talk about Adjoa, you can talk about Nkontompo, you can talk about New Takoradi. And in the Volta region, we've just completed the Atorkor-Anyanui-Dzita sea defense project at a cost of $60 million. At the moment, we are implementing the Blekusu sea defense project at a cost of $41 million. Now, the President is concerned about what happened two (2) days ago here, and he has instructed that we should take this project as . . . on emergency basis. That is, extend the Atorkor-Anyanui-Dzita project to cover this area. So we've started engaging the contractor. We are also talking to the Ministry of Finance because this one is an emergency and it is Presidential Directive. So we are going to do something for these communities.
>
> *(Gakpo, 2016)*

This statement confirms the third rationality of the GoG regarding addressing coastal erosion along the country's shoreline through the hard engineering infrastructure of SDS. The Minister also articulated how the flooding as a result of sea-level rise required immediate or, as he put it, "emergency" attention. What the Minister's statement did in this situation was to project "the idea of a paternal state that cares and acts immediately and with force when the safety of its people is at stake" (D'Alisa & Kallis, 2016, p. 237). The promise of engaging the contractor to continue an already completed project is not realistic as there is a public procurement process to which every project is subject, which the Minster more than anyone else would know. Thus, the Minster's claim that the Ministry of Finance had been contacted and that this action was a Presidential Directive serves to imply that the order must be obeyed at all cost and with no questions asked, which does not reflect administrative reality in the GoG. While administratively irregular actions do take place, in Ghana such actions are usually carried out under the cover of secrecy and not out in the open in response to such a public statement.

Why is the GoG so keen on the apparatus/technology of infrastructure in implementing its rationalities? For one, research shows that infrastructure tends to buy votes and get persons in government re-elected (D'Alisa & Kallis, 2016, p. 230). As such infrastructural projects are highly visible and can be pointed to as concrete examples of something the government has done. A social protection program or a loan that someone has spent to expand their business and thus increase their productivity is more difficult to trace or take sole responsibility for. The GoG has clearly bought into the project of rule that is SDS as a CCA.

The supporting characters: the capitalists, consultants, and contractors

SDS are technical projects – they involve specialized knowledge and considerable experience in their design, construction, and monitoring and evaluation. This technical knowledge or expertise is often provided by consultants who are often foreign to the country of the project's implementation. Generally, this sourcing of foreign expertise is justified by claims that the country conducting the work does not have the required capacity to carry out such projects on their own. In this section I will trace the different consultants, contractors, and where possible funders involved in these projects. By tracing involvement, I will highlight the total costs of these sea defense systems, and the countries of origin of these different "experts".

KSDS

The Export-Import (EXIM) Bank of the United States funded the KSDS through a loan provided to the GoG (Akyeampong, 2001; RPI staff, 2015). As such, it is unsurprising that all the companies that worked on the KSDS (Great Lakes Dredge and Dock Corporation, Baird and Associates, and Research Planning Incorporated) are American firms, as EXIM Bank exists to support "U.S. exporters to increase their sales abroad and support quality jobs at home" (EXIM Bank, n.d.). The exporter in the case of the KSDS was Great Lakes Dredge and Dock Corporation (GLDDC) and the Bank of America-Chicago, IL applied on behalf of the GoG to EXIM Bank for a loan in the amount of $93,867,006, which EXIM Bank approved to be repaid in ten years (EXIM Bank, 2003). GLDDC was thus awarded a contract from the Ghana Ministry of Works and Housing (now the Ministry of Water Resources, Works and Housing) to protect and stabilize the shoreline from Keta to Hlorve (Akyeampong, 2001, p. 211; GLDD, 2017). GLDDC subcontracted to Baird and Associates (Baird) the design services of the sea defense system (Baird and Associates, 2011), and to Research Planning Incorporated (RPI) the environmental impact assessment of the project (RPI, 2008a), while GLDDC did the actual construction.

I was able to speak to only two staff, from RPI, in person. My attempts for in-person communication with the other members of the KSDS implementation consortium proved futile. These staff emphasized that RPI was responsible for the "science assessment" portion of the project. I find this emphasis unsurprising as on the home page of RPI's website, the company states that it is a "*science*-technology consulting firm that combines the natural *science* disciplines with geospatial and quantitative analysis to address unique challenges in coastal and environmental management" (RPI, 2008b, emphasis mine). The RPI home page also states that the company is "dedicated to bringing *science*-based solutions to our clients in government, industry, and non-profit organizations" (RPI, 2008b). This highlighting of science shows how science is often portrayed as the singular objective basis of decision-making.

AdSDS

The AdSDS was carried out in two (2) phases. The MWRWH estimated the cost of the first phase of the AdSDS at €60 million[1] (Ministry of Water Resources Works and Housing, 2013, p. 7). Another official GoG source stated the second phase as costing €183.4 million (Information Services Department, 2017). Together these two phases total €244.3 million. Thus, the overall cost of the AdSDS has been cited as approximately €240 million (Andoh, 2016; Open Governance, 2017). International Marine and Dredging Consultants (IMDC), Belgium were responsible for the detailed design and engineering of the system (IMDC, 2011). Ecorem – an environmental services company with its headquarters in Belgium – was responsible for the environmental impact assessment (EIA) and the environmental management plan (EMP) (Ecorem, 2009). Ecorem contracted a local Ghanaian non-profit International Centre for Enterprise and Sustainable Development (ICED) to assist with mainly the fieldwork portions of the EIA and EMP (ICED, 2016). The contractors for the AdSDS were Dredging International Services Cyprus Ltd. (DEME, 2014), a subsidiary of Dredging, Environmental, and Marine Engineering (DEME) Group – a Belgian company (Dredging International Services, n.d.). The common factor for all of these companies is that they are all Belgian, and this commonality is not surprising as the funding for the AdSDS (at least the €60 million portion of it) came from BNP Paribas Fortis (Ministry of Water Resources Works and Housing, 2010, p. 22) – an international bank based in Belgium (BNP Paribas Fortis, 2017). Like the KSDS, the funding source of the AdSDS appears to come with terms that require the employment of services from the country of the funding agency.

AtSDS

Between 2008 and 2014, the GoG implemented the construction of a 2.7 km defense system, in addition to the rehabilitation of a road between Srogbe and Kplortorkor in the Keta municipality that had been destroyed by erosion (Andoh, 2014; Blagogee, 2014). The actual groin portion of the AtSDS was about 500 m (Ministry of Finance, 2015). The estimated cost of the AtSDS is debatable because different sources quote different figures. The MWRWH, in its Annual Progress Report (APR) for 2012 on the implementation of the Ghana Shared Growth and Development Agenda (GSGDA) I, stated the total cost of the project as $12.2 million with $8 million as having being disbursed as at the date of reporting (Ministry of Water Resources Works and Housing, 2013). The actual title of the project in the progress report ends with the word "phase" and the comments section on the project states that "there were additional works" (Ministry of Water Resources Works and Housing, 2013, p. 6), so this might refer only to one phase of the project. I hypothesize multiple phases of the project because further news sources quote higher figures ranging from $34 million (GNA, 2011), $41 million (Blagogee, 2014), $48.5 million (Andoh, 2015), and $54 million (Andoh, 2014; Bokpe, 2016). Referring back to the Minister's statement I quoted earlier in this chapter, the cost of the AtSDS was $60 million. I could not find any other government sources that stated the cost of the AtSDS.

The contractors of the AtSDS, Amandi Holdings Limited (Amandi) do not advertise their involvement in the construction of the AtSDS, let alone the cost of the project. Amandi is a member of a larger group of companies known as Super Tech Limited (STL) Ghana (Amandi, 2015), which has had allegations of fraud and a lack of transparency in its capacity building operations related to the 2012 elections for which STL had been contracted to manage the electronic collation of results (GC-staff, n.d.; GhanaWeb Politics, 2016; The New Statesman, 2015; Yusufi & Dowetin, 2015). Amandi is also an Israeli company (Blagogee, 2014; Henya,

2012) but the company's links to Israel or STL is something Amandi is not explicit about. It is news sources and other documents and websites that reveal these ties. Amandi is quite popular with the MWRWH as far as SDS go. According to the Ministry's GSGDA I 2012 APR, Amandi was constructing three sea defense systems: the AtSDS, the Sakumono Sea Defense, and the Ngyiresia Emergency Sea Defense (Ministry of Water Resources Works and Housing, 2013). Outside of the sea defense or coastal protection sector, Amandi had its hands in numerous other projects in Ghana, ranging from the rehabilitation of the Takoradi Port, the expansion of the Tema Port, Information Communication Technology and estate development services, construction of offices for the National Health Insurance Authority, and the rehabilitation and construction of railways lines to revamp the railway sector in Ghana (Amandi, n.d.). With regard to the AtSDS, I practically could not garner any views from the consultants and contractors themselves (despite my efforts beyond trying to glean information from their website).

Mapping money and mentalities

In the following section I briefly illustrate the relationships between the GoG client – the MWRWH, the three sea defense systems in the VRD and their estimated costs, and the international agencies involved in their funding and implementation. In my illustration, I also briefly outline one of the main discourses each of these agencies hold, as well as the three-step rationality of the GoG that I discussed earlier in this chapter, to highlight the similarities between these discourses (as illustrated in Figure 28.3). I garnered the discourses of the international agencies from the website of each organization or firm.

The numbers inside the circles in Figure 28.3 are rounded-off figures of the total estimated costs of each sea defense system. I have the question mark beside the \$60m for the Atorkor sea defense system because the only GoG source for that number is the quote I earlier discussed by the Minster of the MWRWH (Gakpo, 2016). The websites and publicly available documents of each of these actors highlight certain themes. GLDDC, Ecorem, and Dredging International are concerned with climate change and the environment. GLDDC and Dredging International are concerned with coastal erosion and change (particularly the ability of humans to cause that change). Baird, RPI, and IMDC are concerned with engineering or applied science, as well as engineering or applied science as a basis for decision-making. Exim Bank and BNP Paribas Fortis are concerned with finances and protecting (financial) investments. Amandi and Dredging International are concerned with economic growth.

Figure 28.3 shows how an aspect of each of three rationalities of the GoG with regard to climate change, coastal erosion, and sea defense systems is represented to a degree in the discourses of each of the international agencies. "Climate change" on left side of the circles maps onto "environment" on the right side of the circles. "Economy" on the left side of the circles maps onto "economic growth" on the right side of the circles. "Coastal erosion" on the left side of the circles maps on to "shaping" on the right side of the circles. Lastly "sea defense systems" on the left side of the circles maps onto "applied science/engineering" and "science as basis for decision making" on the right side of the circles. The only exceptions to this representation are the discourses of the banks (i.e., "protecting . . . investments" on the right side of the circles does not have a corresponding phrase or word on the left side of the circles), which are logically those of protecting the financial interests of their clients. Though this alignment of GoG rationalities with the rationalities of the implementers of the three sea defense systems may merely be a reflection of the GoG ensuring that it selects companies that will achieve the government's aims, the alignment still illustrates governmentality at work. This implication is that the GoG

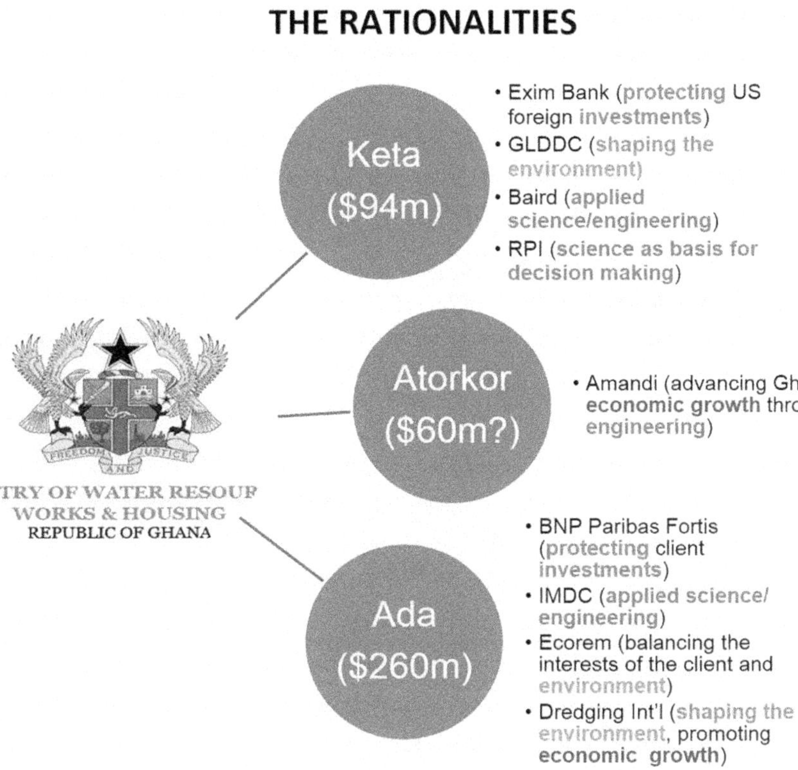

Figure 28.3 Map of national and international actor rationalities on sea defense systems[2]

articulated its rationalities with regard to SDS as CCA well enough that consultants, contractors, and funders who shared some of these rationalities applied to offer their services.

Implications for climate change adaptation governance

Governmentality can help explain how seemingly disparate actors can be aligned to sometimes mutual goals, but governmentality provides no prescription on what can be done in light of such an understanding or analysis. Adaptation decisions are judgment calls that all stakeholders need to make and be explicit about in order to come to some common solutions which may not be ideal for all but would not be extremely inequitable for some (Adger, 2016; de Boer et al., 2010; Sovacool et al., 2015). As such I have the following "judgement call" to suggest.

Overall, the consultants and contractors involved in the implementation of these three SDS were aligned with the GoG discourse on protecting live(lihood)s and property through SDS. However related research has illustrated that people living near these SDS are not as aligned as some of these international actors (Owusu-Daaku & Rosko, 2019).

As such, various sections of society in Ghana need to advocate for the expansion of the scope of CCA planning – from predominantly the national government, international banks, consultants and contractors to include groups such as community residents, the media (Owusu-Daaku & Diko, 2017), and the private sector (Mees et al., 2012). Broadening the range of actors to be mobilized and aligned can change projects of rule and their goals. Sadly in Ghana, community residents have little to no say in GoG adaptation planning that impacts their lives (Sova et al., 2016), an arrangement that becomes even more pronounced when analyzed at an international level (Sova et al., 2015).

Currently the CCA landscape in Ghana is very much influenced by the projects of international donors (Sova et al., 2016) and large international research consortia. However, the study of SDS as CCA through the lens of governmentality makes it clear that such international actors have little to no say in whether the GoG will implement a sea defense system in a particular area as it is the GoG's own rationalities that drive such investments. Hence, the actor to take up expanding the current landscape for CCA decision-making, planning, and implementation is the GoG itself. However, with many other interests competing for the national government's attention it is not likely that such an expanded decision-making landscape will develop without some advocacy and pressure from the other stakeholders – particularly as such an expanded scope ideally implies a devolution of some amount of power on the part of the GoG. Activism as a means to destabilize the current seemingly totalizing governmentality of SDS as a CCA may be the only viable way to ensure that these projects do not continually result in maladaptation.

Notes

1 €1 = $1.17 as at October 18, 2020.

2 Circles represent each sea defense system and bullet points represent the rationalities of various actors associated with a sea defense system.

References

Adger, W. N. (2016). Place, Well-Being, and Fairness Shape Priorities for Adaptation to Climate Change. *Global Environmental Change, 38*, A1–A3. https://doi.org/10.1016/j.gloenvcha.2016.03.009

Akyeampong, E. K. (2001). *Between the Sea and the Lagoon: An Eco-Social History of the Anlo of Southeastern Ghana c. 1850 to Recent Times*. James Currey and Ohio University Press, Athens, OH.

Amandi. (2015). *Amandi Official Website*. www.amandi.com/

Amandi. (n.d.). *Annual Report*. http://dan-go.co.il/wp-content/uploads/2013/04/פרופיל-חברה-מעוצב-אמנדי.pdf

Anderson, B. (2012). Affect and Biopower: Towards a Politics of Life. *Transactions of the Institute of British Geographers, 37*(1), 28–43. http://onlinelibrary.wiley.com/doi/10.1111/j.1475-5661.2011.00441.x/full%0Apapers3://publication/doi/10.1111/j.1475-5661.2011.00441.x

Andoh, D. (2014, December 19). Atorkor Sea Defence Project Completed. *Graphic Online*. www.graphic.com.gh/news/general-news/atorkor-sea-defence-project-completed.html

Andoh, D. (2015, March 12). Keta Sea Defence Project to be Expanded. *Graphic Online*. www.graphic.com.gh/news/general-news/keta-sea-defence-project-to-be-expanded.html

Andoh, D. (2016, February 3). Work on €240 million Ada Coastal Protection Project completed. *Graphic Online*. www.graphic.com.gh/news/general-news/work-on-240-million-ada-coastal-protection-project-completed.html

Angnuureng, D. B., Appeaning Addo, K., & Wiafe, G. (2013). Impact of Sea Defense Structures on Downdrift Coasts: The Case of Keta in Ghana. *Academia Journal of Environmental Sciences, 1*(6), 104–121.

Appadurai, A. (2001). Deep Democracy: Urban Governmentality and the Horizon of Politics. *Environment and Urbanization, 13*(2), 23–43. https://doi.org/10.1177/095624780101300203

Atteridge, A., & Remling, E. (2013). *The Indirect Effects of Adaptation: Pathways for Vulnerability Redistribution in the Colombian Coffee Sector* (No. 2013–10; SEI Working Papers). www.sei-international.org/mediamanager/documents/Publications/Climate/SEI-WP-2013-10-Colombia-coffee-indirect-adaptation-impacts.pdf

Bahadur, A. V, Peters, K., Wilkinson, E., Pichon, F., Gray, K., & Tanner, T. (2015). The 3As: Tracking Resilience Across Braced. In *BRACED Knowledge Manager*. Overseas Development Institute, London, England.

Baird and Associates. (2011). *Keta Coastal Defence*. Project. www.baird.com/what-we-do/project/keta-coastal-defence

Barnett, J., & O'Neill, S. (2010). Maladaptation. *Global Environmental Change*, *20*(2), 211–213. https://doi.org/10.1016/j.gloenvcha.2009.11.004

Barrett, F. J. (1995). The Central Role of Discourse in Large-Scale Change: A Social Construction Perspective. *The Journal of Applied Behavioral Science*, *31*(3), 352–372. https://doi.org/10.1177/0021886395313007

Bettini, G. (2014). Climate Migration as an Adaption Strategy: De-securitizing Climate-Induced Migration or Making the Unruly Governable? *Critical Studies on Security*, *2*(2), 180–195. https://doi.org/10.1080/21624887.2014.909225

Blagogee, E. (2014, December 19). Amandi Completes Sea Defence Projects in Keta, Atorkor, Dzita & Anyanui. *Today Online*. www.todaygh.com/amandi-completes-sea-defence-projects-keta-atorkor-dzita-anyanui/

BNP Paribas Fortis. (2017). *BNP Paribas Fortis*. Official Website. www.bnpparibasfortis.com/home

Bokpe, S. J. (2016, February 6). Opinion Leaders in Dzita Angry with NDC? *Graphic Online*. www.graphic.com.gh/news/politics/opinion-leaders-in-dzita-angry-with-ndc.html

Bryant, R. L. (2002). Non-Governmental Organizations and Governmentality: "Consuming" Biodiversity and Indigenous People in the Philippines. *Political Studies*, *50*(2), 268–292. https://doi.org/10.1111/1467-9248.00370

Burchell, G., Gordon, C., & Miller, P. (1991). *The Foucault Effect: Studies in Governmentality: With Two Lectures by and an Interview with Michel Foucault* (G. Burchell, C. Gordon, & P. Miller (eds.)). The University of Chicago Press. https://doi.org/10.1146/annurev-micro-090110-102957

Carr, E. R. (2019). Properties and Projects: Reconciling Resilience and Transformation for Adaptation and Development. *World Development*, *122*, 70–84. https://doi.org/10.1016/J.WORLDDEV.2019.05.011

Chandler, D. (2010). Review Article: Risk and the Biopolitics of Global Insecurity. *Conflict, Security and Development*, *10*(2), 287–297. https://doi.org/10.1080/14678801003666024

Chandler, D. (2016). How the World Learned to Stop Worrying and Love Failure: Big Data, Resilience and Emergent Causality. *Millennium: Journal of International Studies*, *44*(3), 391–410. https://doi.org/10.1177/0305829816636673

Chandler, D., & Reid, J. (2018). 'Being in Being': Contesting the Ontopolitics of Indigeneity. *The European Legacy: Toward New Paradigms*, *23*(3), 251–268. https://doi.org/10.1080/10848770.2017.1420284

Collier, S. J. (2009). Topologies of Power: Foucault's Analysis of Political Government beyond "Governmentality." *Theory, Culture & Society*, *26*(6), 78–108. https://doi.org/10.1177/0263276409347694

Cote, M., & Nightingale, A. J. (2012). Resilience Thinking Meets Social Theory: Situating Social Change in Socio-Ecological Systems (SES) Research. *Progress in Human Geography*, *36*(4), 475–489. https://doi.org/10.1177/0309132511425708

Dalby, S. (2013). Biopolitics and Climate Security in the Anthropocene. *Geoforum*, *49*, 184–192. https://doi.org/10.1016/j.geoforum.2013.06.013

D'Alisa, G., & Kallis, G. (2016). A Political Ecology of Maladaptation : Insights from a Gramscian Theory of the State. *Global Environmental Change*, *38*, 230–242. https://doi.org/10.1016/j.gloenvcha.2016.03.006

Dean, M. (1999). *Governmentality: Power and Rule in Modern Society*. SAGE, Los Angeles, CA.

de Boer, J., Wardekker, J. A., & van der Sluijs, J. P. (2010). Frame-Based Guide to Situated Decision-Making on Climate Change. *Global Environmental Change*, *20*, 502–510. https://doi.org/10.1016/j.gloenvcha.2010.03.003

DEME. (2014). *Ada Coastal Protection Works*. www.deme-group.com/dredging/references/ada-coastal-protection-works

Dillon, M. (2007). Governing Through Contingency: The Security of Biopolitical Governance. *Political Geography*, *26*(1), 41–47. https://doi.org/10.1016/j.polgeo.2006.08.003

Dillon, M., & Lobo-Guerrero, L. (2008). Biopolitics of Security in the 21st Century: An Introduction. *Review of International Studies*, *34*(2), 265–292. https://doi.org/10.1303/jjaez.2006.68

Dillon, M., & Reid, J. (2001). Global Liberal Governance: Biopolitics, Security and War. *Millennium: Journal of International Studies*, *30*(1), 41–66. https://doi.org/10.1177/03058298010300010501

Dredging International Services. (n.d.). *Dredging International Services (Cyprus)*. www.dredginginternationalservicescyprus.com.cy/

Ecorem. (2009). *Environmental Impact Assessment and Environmental Management Plan for Ada Coastal Protection Works*. Ecorem NV, Aartselaar, Belgium.

Eriksen, S. H., Nightingale, A. J., & Eakin, H. (2015). Reframing Adaptation: The Political Nature of Climate Change Adaptation. *Global Environmental Change*, *35*, 523–533. https://doi.org/10.1016/j.gloenvcha.2015.09.014

EXIM Bank. (2003). *Summary of the Meeting of the Credit Committee-Tuesday, February 9, 1999*. Official Website of the Export-Import Bank of the United States. www.exim.gov/news/minutes/board-meeting-minutes-1999-02-09
EXIM Bank. (n.d.). *Who We Serve*. Official Website of the Export-Import Bank of the United States. Retrieved April 19, 2017, from www.exim.gov/who-we-serve
Ferguson, J. (1994). *The Anti-Politics Machine "Development," Depoliticization, and Bureaucratic Power in Lesotho* (2nd Edition). University of Minnesota Press. www.upress.umn.edu/book-division/books/the-anti-politics-machine
Foucault, M. (1995). *Discipline & Punish: The Birth of the Prison* (A. Sheridan (ed.); Second Vin). Random House, Inc., New York, NY.
Foucault, M. (2003). *"Society Must Be Defended": Lectures at the College de France, 1975–76* (M. Bertani & A. Fontana (eds.); 2nd ed.). Picador, Hampshire, England.
Foucault, M. (2004). *The Birth of Biopolitics: Lectures at the College de France, 1978–79* (M. Senellart (ed.)). Palgrave Macmillan, London, England.
Gakpo, J. O. (2016). *Climate Refugees*. Joy News Hotline Documentary Series. www.youtube.com/watch?v=buZzGVdVfKU
Gamson, W., Croteau, D., Hoynes, W., & Sasson, T. (1992). Media Images and the Social Construction of Reality. *Annual Review of Sociology*, *18*(1), 373–393. https://doi.org/10.1146/annurev.soc.18.1.373
GC-staff. (n.d.). UNDP indicts EC, STL. *Ghana Crusader*. www.ghanacrusader.com/undp-indicts-ec-stl/
GhanaWeb Politics. (2016, April 4). STL's role in elections: EC under pressure. *GhanaWeb*. www.ghanaweb.com/GhanaHomePage/politics/STL-s-role-in-elections-EC-under-pressure-428089
GLDD. (2017). *Keta Sea Defense*. Project Portfolios – International. www.gldd.com/portfolio-item/keta-sea-defense/
GNA. (2011, June). Sea Defence Project Underway in the Volta Region. *Ghana News Agency Website*. http://ghananewsagency.org/science/sea-defence-project-underway-in-the-volta-region-29737
Green Climate Fund. (n.d.-a). *About GCF: Overview*. About Us. www.greenclimate.fund/about
Green Climate Fund. (n.d.-b). *About GCF: Timeline*. About Us. www.greenclimate.fund/about/timeline
Green Climate Fund. (n.d.-c). *Project Portfolio*. Projects and Programmes. www.greenclimate.fund/projects
Green Climate Fund. (n.d.-d). *Resource Mobilisation*. About Us. www.greenclimate.fund/about/resource-mobilisation
Grove, K. (2014). Biopolitics and Adaptation: Governing Socio-Ecological Contingency Through Climate Change and Disaster Studies. *Geography Compass*, *8*(3), 198–210. https://doi.org/10.1111/gec3.12118
Grove, K. (2018). *Resilience (First)*. Routledge, Oxfordshire, England.
Grove, K. J. (2010). Insuring "Our Common Future?" Dangerous Climate Change and the Biopolitics of Environmental Security. *Geopolitics*, *15*(3), 536–563. https://doi.org/10.1080/14650040903501070
Henya, B. E. (2012, March 28). Ghana: Israeli Company Commended – for Completing Sea Defence Project On Time. *AllAfrica*. http://allafrica.com/stories/201203281106.html
ICED. (2016). *ICED Environmental Management*. http://icedghana.com/environmental-management/
IMDC. (2011). *Design of a coastal protection scheme for Ada at the Volta- river mouth* (Coastal Infrastructures).
Information Services Department. (2017). *Ada Sea Defence Project Completed*. Government of Ghana Official Portal. www.ghana.gov.gh/index.php/media-center/news/2377-ada-sea-defence-project-completed
Jones, L., Carabine, E., & Schipper, E. L. F. (2015). *(Re)conceptusalising Maladaption in Policy and Practice: Towards an Evaluative Framework* (June 2015; PRISE Working Papers).
Juhola, S., Glaas, E., Linnér, B.-O., & Neset, T.-S. (2016). Redefining maladaptation. *Environmental Science and Policy*, *55*, 135–140. https://doi.org/10.1016/j.envsci.2015.09.014
Klein, R. J. T., & Maciver, D. C. (1999). Adaptation to Climate Variability and Change: Methodological Issues. *Mitigation and Adaptation Strategies for Global Change*, *4*, 189–198. https://doi.org/10.1023/A:1009690729283
MacKinnon, D., & Derickson, K. D. (2012). From Resilience to Resourcefulness: A Critique of Resilience Policy and Activism. *Progress in Human Geography*, *37*(2), 253–270. https://doi.org/10.1
Magnan, A. K., Schipper, E. L. F., Burkett, M., Bharwani, S., Burton, I., Eriksen, S., Gemenne, F., Schaar, J., & Ziervogel, G. (2016). Addressing the Risk of Maladaptation to Climate Change. *WIREs Climate Change*, 7(5), 646–665. https://doi.org/https://doi.org/10.1002/wcc.409
Mathews, A. S. (2005). Power/Knowledge, Power/Ignorance : Forest Fires and the State in Mexico. *Human Ecology*, *33*(6), 795–820. https://doi.org/10.1007/s10745-005-8211-x

Mees, H. L. P., Driessen, P. P. J., & Runhaar, H. A. C. (2012). Exploring the Scope of Public and Private Responsibilities for Climate Adaptation. *Journal of Environmental Policy & Planning, 14*(3), 305–330. https://doi.org/10.1080/1523908X.2012.707407

Mensah, K. O., & Fitzgibbon, J. (2013). Responsiveness of Ada Sea Defence Project to Salt Water Intrusion Associated with Sea Level Rise. *Journal of Coastal Conservation, 17*, 75–84. https://doi.org/10.1007/s11852-012-0219-y

Ministry of Environment Science Technology and Innovation. (2013). *Ghana National Climate Change Policy*. https://s3.amazonaws.com/ndpc-static/CACHES/NEWS/2015/07/22//Ghana+Climate+Change+Policy.pdf

Ministry of Environment Science Technology and Innovation. (2015). *National Climate Change Policy Action Programme for Implementation*. www.weadapt.org/sites/weadapt.org/files/ghana_national_climate_change_master_plan_2015_2020.pdf

Ministry of Finance. (2015). *Medium Term Expenditure Framework (MTEF) for 2015–2017 Programme Based Budget Estimates for 2015 Ministry of Water Resources, Works and Housing*. Republic of Ghana, Accra, Ghana.

Ministry of Water Resources Works and Housing. (2010). *Monitoring and Evaluation Plan and Budget for 2010–2013* (Issue June). Republic of Ghana, Accra, Ghana.

Ministry of Water Resources Works and Housing. (2013). Annual Progress Report 2012. In *Report on the Status of Implementation of the Ghana Shared Growth and Development Agenda (GSGDA) 2010–2013* (Issue June). Republic of Ghana, Accra, Ghana.

Mitchell, T. (2002). *Rule of Experts: Egypt, Techno-Politics, Modernity*. University of California Press, Berkeley, CA.

Nightingale, A. (2009). Warming Up the Climate Change Debate: A Challenge to Policy Based Adaptation. *Journal of Forest and Livelihood, 8*(1), 85–90.

O'Brien, K., Leichenko, R., Kelkar, U., Venema, H., Aandahl, G., Tompkins, H., Javed, A., Bhadwal, S., Barg, S., Nygaard, L., & West, J. (2004). Mapping Vulnerability to Multiple Stressors: Climate Change and Globalization in India. *Global Environmental Change*. https://doi.org/10.1016/j.gloenvcha.2004.01.001

Ojha, H. R., Ghimire, S., Pain, A., Nightingale, A. J., Khatri, D. B., & Dhungana, H. (2016). Policy without Politics: Technocratic Control of Climate Change Adaptation Policy Making in Nepal. *Climate Policy, 16*(4), 415–433. https://doi.org/10.1080/14693062.2014.1003775

Open Governance. (2017). *Work on Ada Coastal Protection Project completed*. Opengovgh. http://opengov.org.gh/ada-tamale/news/new-blog-post-7/

Owusu-Daaku, K. N., & Diko, S. K. (2017). The Sea Defense Project in the Ada East District and its Implications for Climate Change Policy Implementation in Ghana's Peri-Urban Areas. In Wilson Center Urban Sustainability Laboratory (Ed.), *Edited Volume of the 2016 Annual Reducing Urban Poverty Graduate Paper Competition published by the Wilson Center sponsored by the Cities Alliance, IHC Global, USAID, the Wilson Center, and the World Bank* (2016th ed.). Wilson Center for Scholars.

Owusu-Daaku, K. N., & Rosko, H. (2019). The discursive construction of adaptation subjects via the Ada Sea Defense System in the Volta River Delta of Ghana. *Environment and Planning E: Nature and Space, 2*(3), 617–644. https://doi.org/10.1177/2514848619846087

Phillips, N., & Hardy, C. (2002). Discourse Analysis: Investigating Processes of Social Construction. In *Qualitative research methods v 50*, 50. https://doi.org/10.4135/9781412983921

Potter, J. (1996). Representing Reality: Discourse, Rhetoric and Social Construction. In *Construction*. https://doi.org/10.4135/9781446222119

Pronk, M., Maat, H., & Crane, T. A. (2016). Vulnerability assessments as a political creation: tsunami management in Portugal. *Disasters, 41*(4), 728–747. https://doi.org/10.1111/disa.12223

Robbins, P. (2000). The Practical Politics of Knowing: State Environmental Knowledge and Local Political Economy. *Economic Geography, 76*(2), 126–145.

RPI. (2008a). *Environmental Assessment Keta Sea Defense Works Project (KSDW)*. Projects. www.researchplanning.com/projects/environmental-assessment-keta-sea-defense-works-project-ksdw/

RPI. (2008b). *Home Page*. www.researchplanning.com/

RPI staff. (2015). *Personal Communication*.

Scheraga, J. D., & Grambsch, A. E. (1998). Risks, Opportunities and Adaptation to Climate Change. *Climate Research, 10*(December 10), 85–95. https://doi.org/10.3354/cr011085

Scott, J. C. (1998). *Seeing Like a State: How Certain Schemes to Improve the Human Condition Have Failed* (J. C. Scott (ed.)). Yale ISPS Series, New Haven, CT.

Sending, O. J., & Neumann, I. B. (2006). Governance to governmentality: Analyzing NGOs, states, and power. *International Studies Quarterly*, *50*(3), 651–672. https://doi.org/10.1111/j.1468-2478.2006.00418.x
Simon, D., Sidaway, J. D., Yeboah, I. E. A., O'Reilly, K., & Carr, E. R. (2011). Symposium: Geographers and/in Development. *Environment and Planning A*, *43*(12), 2788–2800. https://doi.org/10.1068/a44572
Sova, C., Vervoort, J., Thornton, T., Helfgott, A., Matthews, D., & Chaudhury, A. (2015). Exploring Farmer Preference Shaping in International Agricultural Climate Change Adaptation Regimes. *Environmental Science and Policy*, *54*, 463–474. https://doi.org/10.1016/j.envsci.2015.08.008
Sova, C. A., Thornton, T. F., Zougmore, R., Helfgott, A., & Chaudhury, A. S. (2016). Power and Influence Mapping in Ghana's Agricultural Adaptation Policy Regime. *Climate and Development*, 1–16. https://doi.org/10.1080/17565529.2016.1154450
Sovacool, B. K., Linnér, B.-O., & Goodsite, M. E. (2015). The political economy of climate adaptation. *Nature Climate Change*, *5*(July), 616–618. https://doi.org/10.1038/nclimate2665
Stripple, J., & Bulkeley, H. (2014). *Governing the Climate: New Approaches to Rationality, Power and Politics* (J. Stripple & H. Bulkeley (eds.)). Cambridge University Press, Cambridge, England.
The New Statesman. (2015, September 23). Voters' Register Saga UNDP REPORT DAMNS STL. *The New Statesman*. www.thestatesmanonline.com/index.php/politics/1333-voters-register-saga-undp-report-damns-stl
USAID. (2012). *Building Resilience to Recurrent Crisis*. United States Agency for International Development, Washington, DC.
Watts, M. J. (2015). Now and Then: The Origins of Political Ecology and the Rebirth of Adaptation as a form of Thought. In T. Perreault, G. Bridge, & J. McCarthy (Eds.), *The Routledge Handbook of Political Ecology* (First, pp. 19–50). Routledge. https://doi.org/10.1016/j.annals.2014.11.004
Weisser, F., Bollig, M., Doevenspeck, M., & Müller-mahn, D. (2014). Translating the 'Adaptation to Climate Change' Paradigm : The Politics of a Travelling Idea in Africa. *The Geograph*, *180*(2), 111–119. https://doi.org/10.1111/geoj.12037
Yusufi, I., & Dowetin, T. (2015). *Conduct of an Institutional Assessment and the Development of a Strategic Plan for the Electoral Commission of Ghana* (Issue August). International Foundation for Electoral Systems, Arlington, VA.

29

POLITICAL ECOLOGY AND POLICY

A case study in engagement[1]

Brent McCusker

This chapter first appeared as McCusker (20150 in *The Routledge Handbook of Political Ecology* (eds.) McCarthy, J., Perreaut, T. and Bridge, G.

Introduction

After several years of being a "practicing" political ecologist, I found myself sitting at the 2010 Annual Meeting of the Association of American Geographers in the Cultural and Political Ecology (CAPE) lecture being delivered by Pier Blaikie (Blaikie, 2012). Well known for both his foundational writings on political ecology and his engagement with policy makers, I was intrigued by his thoughts on whether or not political ecologists could or should engage more closely with that community. As a geographer critical of international development, my first reaction to calls for engagement with policy makers, especially those in the US, was often skepticism. I considered them greatly at fault for the profound unevenness in global development. "What good could come from engagement?" I asked myself. This view was reinforced by often raucous debates on the CAPE listserv that strongly challenged any good that could come out of increased interaction. At about the same time, the Development Geographies Specialty Group had been holding pre-conferences to the main AAG meeting, posing very similar topics and questions. After encouragement from colleagues, I decided to explore closer interaction with policy makers during my sabbatical year via a fellowship from the American Association for the Advancement of Science. In 2011, I applied and was selected to serve as a "science advisor" to the United States Agency for International Development (USAID).

At the onset I decided to use this opportunity simply as a learning experience. Rather than starting from a critique of the agency, I decided to frame my daily work around three questions: first, is greater engagement possible and if so, is it desirable; second, where are there areas of congruity (between this particular development donor and myself) in development epistemologies/approaches; and third, where are there divergences in epistemologies/approaches? Although I have viewed USAID's activities, particularly in southern Africa and South/Central America, as problematic, my goal was to learn why the agency works the way it does. Thus, I did not set out to perform a "participant observation" of the agency as Robertson (2010) had with the Environmental Protection Agency. While I was much more interested at the onset of my tenure at USAID in understanding the inner machinations of the agency than Robertson

 DOI: 10.4324/9780429450310-34

reported being in his piece, on reflection for the preparation of this chapter, we had remarkably similar experiences. I echo his finding that "the idea that such research involves going behind or inside something was eventually replaced by the complexity and internal divisions within the state, and the evident blurring of external and internal critique" (Robertson, 2010: 7). The critiques I brought with me, but suppressed in a misguided notion that I simply needed to "learn", were often voiced by longtime staffers who had intricate knowledge of a particular project or program. They knew there were conceptual (design) or practical (implementation) problems, but their concerns were either never raised to decision makers or were just ignored. The notion that I would simply be a "detached observer" quickly proved naive as my analytical skills were valuable in addressing both conceptual issues with programs/projects but also in helping to demonstrate the value of geography or "spatial thinking" as it was termed. I quickly found myself in the role of an internal advocate for geography and political ecology.

Because I was never a direct employee of the agency, I never had the same level of accountability to it that Robertson had. However, his discussion of the "author-less" text struck an immediate chord with me. In my experience, this was initially frustrating. In academia, the measure of my performance is tied to the words I produce. Echoing Robertson, I gradually learned that the anonymity of authorship often helps push ideas forward that might not otherwise be considered. While I was always treated with respect, I was far too new, and temporary, for the bureaucracy to expect my ideas to be adopted *in toto*. What I did find is that by removing my name, some parts of my ideas could make it forward because they were attributed to my office as a whole.

A final similarity I share with Robertson is the feeling of being completely encompassed by neoliberal ideas while working in "my small part" of the state. My experience was too short to find the same degree of resistance or oppositional thinking that he uncovered; however, I was surprised at how thoroughly neoliberal principles permeated thinking about development policy at USAID. I had hoped at the beginning of my experience to be proven incorrect – a raving radical who finds capitalist conspiracy everywhere. While the halls of the Ronald Reagan Building are indeed *not* filled with capitalist conspirators, their ideas are strongly represented (it is, after all, in a building named after Ronald Reagan!). I did find allies who were critical of the "market-based" solution to any problem, but my tenure was too short to map the network.

I have chosen not to focus on a reflexive assessment of my time at USAID in this chapter, but rather to highlight some of the opportunities for engagement that I observed in my time there. This chapter is qualified by all the normal caveats one might expect: USAID is only one type of policy institution; it is part of the US Government and carries with it a great deal of imperial baggage; these are only my very situated and biased observations (e.g. I *wanted* my time there to be productive); and I was only a very small part of a very large organization and not privy to high-level decision making or decision makers.

In the next section, I summarize the topic of engagement between political ecologists and policy makers in the literature. Given the paucity of critical reflection on the topic in the literature, I draw heavily from a few examples. I then move to my case study at USAID. For purposes of clarity, I distinguish between policy makers and policy implementers and then explore how USAID broadly conceptualizes development and its place in that field. I also problematize my own decision to work for a US government agency. In the penultimate section, I consider several areas of congruence and incongruence between USAID and a political ecological approach to development, noting that there is no singular path for either. My conclusion is that while there are many mechanisms for increased engagement, any relationship between international donors on one hand and political ecologists on the other will remain fraught with challenges and opportunities. My viewpoint, which changed as a result of my experience, is that it is worthwhile for political ecologists to engage with policy implementers.

Where is the policy?

Even a cursory review of major political ecology texts demonstrates a community of practice that is both incredibly varied and deep in analytical richness (Peet and Watts, 2004; Peet et al., 2011; Neumann, 2005; Brannstrom and Vadjunec, 2013; Zimmerer and Bassett, 2003; Bryant and Bailey, 1997; Stott and Sullivan, 2000; Blaikie and Brookfield, 1987; Blaikie, 1985). That same review would find serious consideration paid to the impact of development policies and projects on local people and their environment. It would, however, find less attention paid to policy *making* and policy *implementation*.

One of the few narratives to directly raise this issue was Peter Walker's 2006 article titled "Political Ecology: Where Is the Policy?" But Walker's article did not emerge out of the ether; rather, it was but one materialization of a long-running debate and dialogue between members of the Cultural and Political Ecology Specialty Group of the Association of American Geographers, amongst others. Walker cut straight to the crux of the issue in his paper when he wrote "despite its rich history and professed interest in engaging public debates, the actual engagement of political ecology with fields of research and public debate outside the academy has been limited" (Walker, 2006: 383). To be precise, though, Walker was largely concerned in his text with *public* policy, usually embodied by academics who rise in stature high enough to shift *public* debate through their writing. He discussed the lack of effective counters to such dominant narratives as neo-Malthusianism and environmental determinism.

While he was certainly correct and his intervention long overdue, even it misses an altogether more important point: that political ecologists rarely engage with the people and institutions that create and implement the policies they so directly examine, deconstruct, or challenge (Carr, 2011). A good case is the work of Dick Peet (2007). His book *Geography of Power: The Making of Global Economic Policy* is a trenchant analysis of the ways in which international capitalist development policy is generated and reproduced through regimes of knowledge and power. It deftly describes economic, ideological, and political power circuits in an approachable and persuasive language. I find the book so useful, in fact, that I teach it in my advanced development geography course. What Peet does not do, however, is report on any single interaction with a policy maker or implementer. There is no discussion of policy making and implementing challenges and no discussion of the very heterogeneous composition of such policy makers, their institutions, or their implementing partners. Further, and perhaps most problematically, the book does nothing to identify and build alliances with like-minded policy makers and implementers. Such a book is a wonderful way to stir students' critical thinking skills, but it does little to find and reach out to potential allies inside what are presented as monolithic knowledge and power circuits. Similar strengths and weaknesses extend to a special edition of the journal *Human Geography*, edited by Peet (2013), and indeed to much political ecological work.

Political ecological analysis of policy and policy outcomes has also tended to homogenize policy making and policy implementation (notable exceptions are Bebbington, 2014; Robertson, 2010). In the example that follows, one grounded in experience in Washington, DC, it is important to recognize the differences between policy makers, policy implementers, and implementing partners. Policy *makers* are the legislators who enact policies, usually crafted directly or indirectly by special interest groups and think tanks, at least in the US. They are the members of Congress and their staffs who enact legislation that has increasingly been accompanied by onerous limitations and stringent budget restrictions. Such limitations and restrictions are put in place to limit the flexibility of policy *implementers*, those members of the bureaucracy in the executive branch charged with carrying out (implementing) Congressional and Presidential directives and laws. Policy implementers still do retain some degree of latitude in exactly how they execute such laws and directives, but informal conversations indicate that it is much less than in the past.

In the 1990s, the Clinton administration's efforts at "reinventing government" led to significant reductions-in-force in the civil service. Skilled staff flowed out of executive agencies, replaced by or themselves hired into firms of *implementing partners*. These businesses, pejoratively referred to as "the Beltway Bandits" (or less pejoratively "the Highway Helpers") due to their location around the Washington, DC interstate highway system, perform many of the actual tasks of government, the day-to-day work of *doing*. Most development projects in developing countries are contracted out to either one of these large firms, or, more recently, to local implementing partners. What a political ecologist might see in a developing country is the action of one of these partners. What they might read as a "policy document" is largely written by an implementing partner, albeit with significant guidance from a policy implementer.

A limited engagement

Both Walker (2006) and Blaikie (2012) suggest many reasons for a limited engagement of policy in political ecology. Many political ecologists prefer to "speak to other political ecologists (and perhaps to others in closely related fields)" (Walker, 2006). Blaikie (2012: 234) suggests one possible very good reason for this state of affairs. He wrote that "the content of a political ecology which claims to be emancipating, liberating and politically progressive will be critical of existing knowledges and power relations and maybe of specific actors in policy networks outside the academy altogether". This is a particularly important point. Many political ecologists will be critical of development programs and projects with just cause. However, he goes on to argue that, "deconstruction and other acts of aggressive epistemological attack . . . may not appeal to any audience other than like-minded academics" (Blaikie, 2012: 234).

A second major obstacle in the path of engagement is the structure of reward in academic institutions and increasing workloads. In many departments housing political ecologists, a strict and seemingly unchanging division of labor forces academics into a mix of teaching, research and service, with the former two roughly equal in importance. In those categories, there is often little reward for the creation and publication of policy documents or time spent engaging policy makers or implementers. Simply put, it is impact factor above actual impact in many academic institutions. As university budgets shrink and new sources for funding are pursued, this may well change, but may also result in even more overburdened faculty.

The policy side of this problem is no less inflexible, however. Many funders or policy institutions have very strict limitations on what can be written about them, either from staff, contractors, or consultants working on the inside. Most have publication clearance processes that are cumbersome and time consuming. New forms of communication, such as social media, are looked on with great hesitation – transgressing the boundaries in 140 characters or on a blog post can result in severe penalties. These are not hypothetical scenarios. During my time at USAID, a colleague was subjected to close oversight of all his external communications after a slight misstatement on a blog post. Many actors within these institutions would rather avoid this additional burden.

A common refrain inside some policy institutions is that staff are already so overburdened with work and the institutions themselves so understaffed that engagement with academics is simply an unattainable luxury. While there may be some degree of hyperbole in this claim, and indeed there is great variation from institution to institution, it does evoke conversation about a very real disconnect between policy implementers and academics over timelines. Academic timelines stretch into multiple years to complete analysis, while policy makers and implementers often have much less time, in the range of a few days to a few weeks, to plan assessments of their work. Most academics are also limited in their flexibility by their teaching, with long field research trips confined to summer/winter breaks. The rhythm of the academic year simply does not match that of the quick turnaround needs of the policy world. Given that most political ecologists conduct some type of field research, such time inconsistencies can present a significant obstacle to engagement.

Finally, what analytical tools do political ecologists offer policy makers and implementers? As noted by Blaikie, discourse analysis and dialectics simply won't find an audience within most policy institutions. Some political ecologists have employed geospatial technologies critically in order to demonstrate their utility in answering key questions. I have previously published on the use of such technologies for political ecological research (McCusker and Weiner, 2003) and other political ecologists have used geospatial technologies. Geospatial technologies are becoming increasingly widespread within even development institutions such as USAID. The agency's recently (mid-2011) established GeoCenter promotes spatial thinking and the use of GIS, GPS, and remote sensing amongst USAID staff.

An example: policy implementation at USAID

Rather than remain at levels of abstraction that merely suggest what might be possible, I will discuss in this section an example of a policy implementing body, the US Agency for International Development (USAID) with which I have worked extensively beginning in 2011. USAID is the largest single bi-lateral development donor. It covers a variety of thematic areas related to development including but not limited to health, food security, the environment, science and technology, energy, climate change, gender, democracy, conflict, and humanitarian assistance. Many of these topics are central to political ecology, thus this example seems most appropriate not only for USAID but also other policy implementers that touch on human-environment issues, such as NOAA, NASA, or the EPA (Robertson, 2010).

All of the agency's operating procedures and philosophies of practice, known as the Automated Directives System (ADS), are available to the public via its website (www.usaid.gov/who-we-are/agency-policy; accessed 11/3/2013). The ADS is divided into a series of six collections (Series 100–600), which are further broken down into more specific chapters. Of most direct relevance to this discussion is Series 200, which discusses policy programming, planning, assessment, environmental procedures, and ensuring gender equality.

USAID promotes a set of operating principles, which upon cursory examination, do not seem to be all that divergent from a political ecological approach, including "promote gender equality and female empowerment; apply science, technology and innovation strategically, practice selectivity and focus, measure and evaluate impact, build in sustainability from the start, apply integrated approaches to development, and leverage 'solution-holders' and partner strategically" (USAID, 2012: 14). Each of these is discussed in turn.

The document continues into a section titled "The Program Cycle Overview" that establishes how USAID projects are structured. All project implementation is informed in the first stage of the program cycle by reference to collected strategies and guidance documents. In the second stage of the cycle, country development cooperation strategies (CDCS) are developed in consultation with "host country governments, local civil societies, private sector organizations, the (US) State Department, and the broader USG [US government] interagency and other donors" (USAID, 2012: 23). Project or program goals are structured into a hierarchical set of indicators that "graphically represents the development hypothesis and defines a CDCS goal, development objectives, intermediate results, sub-intermediate results and performance indicators" (USAID, 2012: 24). It is important to note here that while this may seem like an overly structuralist tool, USAID is required to account for each and every dollar spent with indicators of its impact. This leads to data such as "number of women trained" or "number of hectares of land under improved management" that at first glance appear trivial and meaningless but upon further review are essential to reporting and accountability. In my recent work with USAID, I have helped to develop a tool that measures impact more holistically. Called "stocktaking", this method encourages a retrospective, whole of landscape assessment that tries to understand how natural resource utilization/management changes actually occurred against the backdrop of obstacles such as lack of resources or resistance from others. While USAID staff are still required to quantify impact using reductionist indicators, qualitative and less reductionist methods exist and have been well received.

Specific projects are structured to fit into this program cycle and feed into each goal, objective, and result. Each project is required to have a "logical framework" that maps against and contributes directly and clearly to the program. As the project is implemented, the agency stresses attention be paid to "what is learned, adapting project activities, revising work plans; and, if necessary modifying contracts, grants or other implementation modalities" (USAID, 2012: 29).

All projects and programs must contain a rigorous and evidence-based evaluation and monitoring plan. One of the more refreshing changes that has happened in the agency recently is a strong push for decision making to be based on evidence rather than anecdotes or foreign policy goals (although it would be naive to assume the latter have been altogether excised as factors in decision making). As a result of monitoring and evaluation, managers and staff are expected to focus on key learning areas that include:

> facilitating coordination, collaboration and exchange of experimental knowledge internally and with external stakeholders; testing hypotheses, filing critical knowledge gaps and addressing uncertainties in the hypothesis with new research, evaluations or syntheses of existing analyses; ensuring new learning, innovations and performance information gained through monitoring and evaluation inform implementation, policy formulation, and strategy development; and identifying and monitoring game changers or broad conditions that . . . could impede or improve implementation.
>
> *(2012: 31)*

Finally, all of this activity must help the agency demonstrate its budget efficacy. There is a constant struggle for resources, and given that much of the agency's funding comes from Congress as program funding (for instance to promote scientific literacy in country x or reduce deforestation in region y) rather than operation expenses, fiscal accountability is a top priority.

Points of overlap and potential engagement

The brief outline of basic principles is meant only to give the reader an introduction to the program cycle and project design at USAID. In this section, I highlight areas of potential overlap between the work of USAID and political ecologists, and highlight the range of similarities and differences in both philosophy and approach. Sensitive to the fact that such reductionism might overgeneralize political ecology as a set of principles rather than a diverse community of practice and that it takes the ADS at face value and as the single set of principles for a quite diverse agency, I attempt to present a range of possible congruencies and incongruences between the two groups.

USAID and political ecologists share a focus on gender equity in development. The strong overlap includes issues such as women's empowerment, understanding roles and the effect of inequality on natural resource management, and women's role in economic growth. Where political ecologists might part company with USAID is over the implementation of programs designed to promote gender equity. Analysis regarding how USAID programming actually results in change is lacking. There are also deeply Western cultural values that are often attached to gender programming that many political ecologists, especially those in countries that are the target of such programs, might find troubling.

A second area of overlap is the use of science and technology in development. The agency has dramatically increased the presence of all categories of trained scientists, from physicists to social scientists to ecologists, recently. These actors give considerable voice to science within the agency. As with gender equity, there will be a range of responses from political ecologists, from outright rejection of "science" as a label to the use of science as merely a tool of imperialism, to acceptance of its role in development. On this last point, political ecologists can bring a rich heritage of critical assessments of the impact of science and technology in development that is often lacking in a broad sense at USAID, but also in other donors such as the World Bank. The so-called "Green Revolution" is often referred to as an unmitigated success within the development/policy community. Political ecological studies strongly suggest otherwise.

The application of selectivity and focus yields more areas of potential overlap. The ADS refers to selectivity as

> gaining a good understanding of the *conditions on the ground* that are needed to begin movement in a certain development objective and applying clear, measurable, and relevant criteria for selecting countries, sub-national regions, or sectors on the basis of those conditions.
>
> *(USAID, 2012: 15, emphasis added)*

The fact that the agency prioritizes what political ecologists might call "context" is a key area of congruence. How that context is assessed and for what purposes is an altogether different question, however, and one to which political ecologists can contribute.

Following from the understanding of "context", the agency promotes "making assistance investments where there is demonstrable local demand and ownership, and where a broad segment of the community has a stake in ensuring that the activity or service continues after the USAID program or project ends" (USAID, 2012: 16). The participatory and consensual development phase of the 1980s and early 1990s did not pass USAID by. In addition, the agency mandates that projects should build local skills and foster the growth of local governmental and non-governmental institutions and be environmentally sustainable. Again, in principle and in broad terms this appears remarkably similar to much of the political ecological literature on local development.

"Stove-piping" is a common problem in development institutions. This refers to the isolation of programs and projects into their thematic areas (e.g. health, environment, energy) without sufficient recognition of the overlapping nature of development. The agency has recently rethought this issue and has tried to increase integrated approaches to development. This is a particular challenge given the nature of how funding is dispersed by Congress. Political ecology, if anything, is an integrated approach to human-environment relationships and as such could be employed to analyze exactly the interrelationships USAID seeks to better understand.

Throughout ADS, there are general statements that suggest philosophies in congruence with political ecology. As part of explanation of the project cycle, for instance, development is recognized as "not static and rarely linear" (USAID, 2012: 19). Clearly a small bit of post-development has snuck into even the largest development institutions! Host country stakeholder opinions and guidance is referred to repeatedly throughout the document. As mentioned earlier, participation and local ownership is valued. Results are measured based on "evidence (citing specific assessments and evaluations) that illustrates why USAID should reasonably expect a specific investment will produce targeted development impacts" by utilizing "analytic rigor and the best available evidence" that "incorporate[s] continuous learning" (USAID, 2012: 24–25).

On two general points, however, there is serious epistemological divergence. First, USAID promotes "development" in its most mainstream sense. The market and the private sector are continuously referenced in similarly uncritical terms as science and technology. Here, many political ecologists would take issue with USAID. Seeming areas of shared concern, such as focus on local context and integrated approaches described previously, could become areas of disagreement given that the agency situates its solutions in both a neoliberal reading of development and within the context of promoting the interests of the US Government *first*. It should be noted, however, that market limitations are recognized amongst many staff – the agency is simply not a homogeneous block that blindly follows prescribed solutions. The economism in the agency, is, however, pervasive.

In addition, the agency's "results framework" may strike many political ecologists as overly reductionist. The complexity of development, its actors and contexts, are often greatly simplified in order to demonstrate that some action has been taken or some goal achieved. Political ecologists might challenge the narratives that such frameworks normalize and treat without critique or deconstruction.

In reviewing these points of overlap, there is no suggestion that this is a comprehensive list nor will all readers agree in extent or tenor. This dialogue is relatively open for debate exactly because there has been so little engagement.

Conclusion

In this chapter, I have attempted to demonstrate that increased engagement between political ecologists and policy makers and implementers is possible. I have used my own experience at USAID to evoke and discuss areas of both convergence and divergence between the two groups. The areas that separate us may be fewer than we think; however, those that do remain are significant. I tried to avoid presenting this as a binary "either/or" case for or against engagement.

I have described a very specific experience in this chapter. However, a few general insights may apply to other contexts, such as sub-national (state/provincial) or non-governmental institutions. First, policy as written is almost always very different from policy as implemented. A bevy of competing agendas operate behind the public face of any institution. From an NGO perspective, this may result in confusing or contradictory messages. Given the fact that so many different individuals with vastly different agendas shape and reshape policy, largely anonymously, it can be difficult to discern what actually happens "on the inside". Taking time to engage with and understand some of the internal dynamics helps explain what appear to be contradictory or inexplicable policies. Second, building alliances with like-minded individuals within the policy community is critical to understanding how the given agency or department works. This helps identify not only needs, but also pressure points where policy changes can be attempted. Third, there will always be points of divergence between academics and policy makers/implementers over both the philosophies and impacts of policy and implementation, as well as many other areas. These disagreements should not be over-generalized and building on the second point, may be more malleable that outward appearances might suggest. Finally, engaging policy makers and implementers of any type will require flexibility and persistence. Given divergent timelines and respective work pressures, for instance, significant effort and time will be necessary to forge any serious relationship.

Should political ecologists feel that increased engagement is desirable, I would suggest that several issues need further clarification and discussion. First, what should the role of a political ecologist, academic or otherwise, be when working in any type of policy institution? There are clearly a range of possibilities – what are the advantages and disadvantages of each? Second, how can academic political ecologists work together across their respective institutions to start to break down promotion and tenure requirements that either actively discourage such engagement or informally shun it? Third, how can a greater audience for such interaction be cultivated? What are the opportunities to show the value of such work and how is that value translated back into the work of political ecologists?

Of course, there will be many more issues and many more opinions than I can possibly capture here. Further work in this area will demand flexibility and hybridity from both academics and policy practitioners, however, we can at least begin to address the question "should some political ecology be useful?" (Blaikie, 2012).

Note

1 The views set out in this chapter are in no way those of the US Agency for International Development (USAID), its employees, its contractors, or anyone in any way affiliated the agency. As such the chapter has not been cleared by or in any way approved or disapproved by anyone in the agency. As a result, no information about the agency's inner workings are described. I reference no internal documents (or conversations) here, other than those that are made available to the general public.

References

Bebbington, A. 2014. "The Endogenous Scholar: Porous Boundaries and Travelling Ideas in Development." *Third World Quarterly*, 35(3): 520–523.

Blaikie, P. 1985. *The Political Economy of Soil Erosion in Developing Countries*. London: Longman.

Blaikie, P. 2012. "Should Some Political Ecology be Useful? The Inaugural Lecture for the Cultural and Political Ecology Specialty Group, Annual Meeting of the Association of American Geographers, April 2010." *Geoforum*, 43: 231–239.

Blaikie, P. and Brookfield, H. 1987. *Land Degradation and Society*. London: Routledge.

Brannstrom, C. and Vadjunec, J. 2013. *Land Change Science, Political Ecology and Sustainability: Synergies and Divergences*. London: Routledge.

Bryant, R. and Bailey, S. 1997. *Third World Political Ecology*. London: Routledge.

Carr, E. 2011. "If You are Uncomfortable, You are Probably Doing It Right." *Environment and Planning, A*, 43: 2797–2800.

McCusker, B. and Weiner, D. 2003. "GIS Representations of Nature, Political Ecology, and the Study of Land Use and Land Cover Change in South Africa." in K. Zimmerer and T. Bassett (Eds.), *Political Ecology: An Integrative Approach to Geography and Environment-Development Studies*. New York: Guilford, pp. 201–218.

McCusker, B. 2015. "Political Ecology and Policy: A Case Study in Engagement." in J. McCarthy, T. Perreaut, and G. Bridge (Eds.), *The Routledge Handbook of Political Ecology*. New York: Routledge.

Neumann, R. 2005. *Making Political Ecology*. New York: Oxford University Press.

Peet, R. 2007. *Geography of Power: The Making of Global Economic Policy*. London: Zed Books.

Peet, R. 2013. "Comparative Policy Analysis." *Human Geography* 6(2): 1–10.

Peet, R., Robbins, P. and Watts, M. 2011. *Global Political Ecology*. New York, Routledge.
Peet, R. and Watts, M. 2004. *Liberation Ecologies: Environment, Development, Social Movements*. London: Routledge.
Robertson, M. 2010. "Performing Environmental Governance." *Geoforum* 41: 7–10.
Stott, P. and Sullivan, S. 2000. *Political Ecology: Science, Myth and Power*. New York: Oxford University Press.
USAID. 2012. *ADS Chapter 200*. www.usaid.gov/who-we-are/agency-policy; accessed 11/3/2013.
Walker, P. 2006. "Political Ecology: Where is the Policy?" *Progress in Human Geography* 30(3): 382–395.
Zimmerer, K. and Bassett, T. 2003. *Political Ecology: An Integrative Approach to Geography and Environment-Development Studies*. New York: Guilford.

30

SPATIAL POLICYMAKING

Using large, public datasets to illustrate spatial patterns of human vulnerability in Niger

Tim Essam, Brent McCusker, and Jamison Conley

Livelihoods introduction

Introduction

The "Big Data" revolution has not gone unnoticed by policy makers concerned with international development. Traditionally, large datasets pertaining to developing countries have been difficult to obtain or are of such poor quality as to be unusable. Over the past 20 years, this tendency has shifted, however, as the World Bank, CGIAR system, United States Agency for International Development (USAID), Department for International Development (DFID), and many other aid donors have commissioned or supported the collection of large, representative datasets at national or sub-national scale. In the past ten years, these datasets have become increasingly spatial. The location of households or other geographic units have been collected, thus allowing advanced geographic analysis. In this practitioner's report, we describe the process of spatially analyzing a large household dataset to produce maps of human vulnerability to livelihood shocks in the country of Niger in West Africa. This analysis was conducted for the US Agency for International Development (USAID) as part of a larger decision-making process to determine how to better understand the nexus of development and conflict in the Sahel. We show here the value of mapping datasets and how this insight was used in decision-making. In the next section, we review the relevant literature on livelihoods, vulnerability, and poverty mapping. Following that, we present the research problem. This research was somewhat unique in that we were presented with a nebulous research question that continuously changed scope as our stakeholders better understood our approach. In the third section, we describe our methods, which include data collection, data analysis, and presentation of maps. We demonstrate how this research affected policy makers in the next section, and we conclude the chapter with reflections for future research.

Livelihoods, vulnerability, and household data

Vulnerability, or its more positive inversion, *resilience*, is a common focus amongst development agencies as they strive to lift people across the developing world out of poverty. For this research, we adopted Wisner et al.'s (2004) definition of vulnerability as, "the characteristics of a person

 DOI: 10.4324/9780429450310-35

or group in terms of their capacity to anticipate, cope with, resist, and recover from the impact of a natural hazard". We extend this definition to livelihood shocks that include natural hazards, but also economic, social, and political disruptions to normal life. Other definitions of vulnerability link humans to natural systems (Turner et al., 2003), or examine vulnerability as it relates to exposure to coupled economic-environmental forces (O'Brien and Leichenko, 2000). These latter definitions attempt to describe human, economic, and environmental systems and their impact on vulnerability. Wisner et.al. focus on the household and community level, which is more germane to this study.

Nigerien livelihoods are strongly dependent on the natural environment. The majority of the population (81%) is rural and the two main income generating activities are agriculture and pastoralism (LaRovere et al., 2008). Disruptions, or shocks, to these livelihoods increase vulnerability and volatility.

Knowing where the poor live has the distinct advantage of providing policy makers and others with the ability to identify the places where they want to target their interventions – what we call *geographic prioritization*. Until recently, mapping the location of poor and/or vulnerable people was rather rudimentary, either due to a lack of computing power and/or specific data on the distribution of the poor. In its infancy, such mapping often employed small scale (large area) and proxy variables to create maps of the uneven nature of wealth distributions within countries. Over the past decade and a half, as research progressed from simple mapping to more complex correlation and regression analysis, the focus of such analyses also shifted from poverty mapping that simply mapped the location of the poor to more complex livelihood vulnerability mapping and analysis that demonstrated correlation and/or implied causation (Grolle, 2013). These methods contribute to a larger intellectual effort that has emerged from the discipline of geography in recent years referred to as "development geography" (see King and McCusker, 2007). The collection of mapping and analysis efforts we discuss later also had roots in "sustainability science", which "builds toward an understanding of the human-environment condition with the dual objectives of meeting the needs of society while sustaining the life support systems of the planet" (Turner et al., 2003: 8074).

Poverty mapping

Poverty mapping is a relatively straightforward concept that generally involves the "measuring of incidence of poverty and food security in some predetermined area" (Davis, 2003: 1). The two primary uses of poverty mapping, according to Davis, include "spatial identification of the poor" and "to create, as a by-product", explanatory and dependent spatial variables for use in multivariate analysis" (ibid. 2–3). This work is usually undertaken to inform national governments, donor agencies, NGOs, or others engaged in developing and implementing poverty alleviation policies.

Robinson et al. (2007) assert that "maps of poverty are of great use to government and development agencies alike because no matter how they propose to go about reducing poverty, they first need a 'baseline' picture of what it is they are dealing with" (2007: 205). Previous work has demonstrated such utility. For example, Deichmann (1999) demonstrated the heterogeneity of poverty at a national scale in Ecuador, while Alwang et al. (2005) similarly showed the importance of disaggregating poverty spatially for Guatemala, Honduras, and Nicaragua. Burke and Jayne (2010) extended poverty mapping to "spatial poverty trap" mapping, where households are stuck in poverty in or because of their location in particular places. Robinson et al. (2007) analyzed data from the Uganda National Household Survey from 2002/2003 and environmental data to "predict the distribution of poverty at the spatial resolution of the predictor variables" (ibid. 205). Biophysical variables have the strongest relationship to poverty, which itself is concentrated in the northeast.

As part of a special edition of the journal *Food Policy* in 2005, contributors examined the spatial dimensions of poverty in Vietnam (Minot and Baulch, 2005), Malawi (Benson et al., 2005), Ecuador (Farrow et al., 2005), Sri Lanka (Amarasinghe et al., 2005), Bangladesh (Kam et al., 2005), and Mexico (Bellon et al., 2005). One common thread through this collection was the use of the small area estimation approach. This statistical estimate technique uses "detailed household survey data to derive income/expenditure estimators, which are then applied to broad coverage census data to get estimates for a rather large number of households of the target population" (Kam et al., 2005: 552).

For Bangladesh, Kam et al. (2005) chose variables such as landholding size, number of working age household members, and fixed assets used in production to model income for households. The analysis was conducted at the upazila scale, a sub-district geographical unit in Bangladesh. Through the use of a geographical-weighted regression they found "the pockets of high poverty incidence generally coincide with the ecologically poor areas of Bangladesh" in the northeast and northwest of the country, along the major rivers and in the Chittagong Hill tract in the southeast (ibid. 564).

In a similar study in Sri Lanka, Amarasinghe et al. (2005) mapped clusters of poverty at a sub-district scale (the Divisional Secretariat) using cluster analysis and regression modeling. Derived from regression testing, the authors found that rural sub-districts where households are dependent on agriculture tend to be poorer (ibid. 508). Cluster analysis showed that access to irrigation led to clustering of less poor households; however a concentration of many smallholders leads to increased poverty.

Bellon et al. (2005) examined the relationship between the distribution of poverty and biophysical conditions conducive to maize production in Mexico with the intent of better targeting the poor for agricultural extension services. They employed a small area estimation method to produce a map of poverty and regression analysis to determine explanatory variables. This analysis demonstrated a strong correlation between maize production and poverty. In addition, the poor did not benefit from participation in agricultural experiments. Further, the poor in Mexico are clustered in the "mountainous and indigenous areas of central and southern Mexico" and face specific biophysical constraints to maize production, such as slope and soil conditions.

Targeting food poor regions lends itself to mapping, as Farrow et al. (2005) demonstrate in their study of food poverty in Ecuador. Using a panel of experts, they developed a set of variables relating to food poverty including: "social capital, agricultural productivity (climate, soil, management and tenure), labour market structure and access to markets" (ibid. 512). The authors used Moran's I statistics to measure spatial autocorrelation and geographically weighted regressions to "overcome the limitations caused by spatial dependency" (ibid. 517). The results showed a strong clustering of food poverty in the center of the country and that the greater the dependence on agriculture for income/livelihood the higher the level of food poverty.

In a study quite similar to ours, Benson et al. (2005) attempted to uncover the "spatial determinants of poverty" in Malawi (ibid. 532). Using the Integrated Household Survey, a large nationally representative dataset, the authors developed a dependent variable from poverty headcount and regressed that against a set of independent variables that ranged from physical production to social status to livelihood indicators. To map this regression, they used spatial statistics including geographic weighed regression. From this analysis they discovered that "areas with higher maize yields on average will have higher levels of poverty" (ibid. 542) and that higher education levels result in less poverty (ibid. 548). Limited access to services, especially in the isolated far north, is correlated with higher poverty.

Vulnerability and vulnerability analysis

Poverty mapping efforts were initially descriptive, a simple representation of the location of the poor, but recent research, as demonstrated earlier, has begun to include more advanced analytical and geospatial methods in order to uncover underlying correlations to independent variables.

Beginning in the mid to late 1990s, aid donors and host governments began commissioning dataset collection that included livelihood specific questions beyond what was collected in national censuses. Informed by work from the development studies literature, these surveys, such as the World Bank's Living Standards Measurement Survey and the Integrated Household Survey, query households not only on the structure of their livelihoods, but also on recent shocks. Because such data relies on participant recall, it is limited by all the usual issues facing survey research; however, a major advantage is that individuals/households are asked directly whether or not they experienced a range of place-relevant disruptions (shocks) to their livelihoods. Thus, instead of mapping proximate variables that represent only the conditions under which a household might be vulnerable, such surveys provide insight into actual, experienced vulnerability.

Malcomb et al. (2014) modeled vulnerability to climate risks in Malawi. They echoed the limitations of traditional vulnerability approaches mentioned earlier when they noted that in the preparation of their study that only four out of 45 studies "collected some form of primary data regarding socio-economic determinants" (ibid. 20). They consulted a group of experts to determine the variable set to determine climate vulnerability and used the Demographic and Health Survey, a broad based survey of the living conditions and health status of households across Malawi, as a data source for those variables. They then created and mapped an index of vulnerability from the selected variables. Their results showed high levels of climate variability in central Malawi.

Abson et al. (2012) used Principal Components Analysis to determine the key indicators of environmental vulnerability in southern Africa using national datasets; however, most studies rely on some type of regression analysis (Fekete, 2012; Krisjanson et al., 2005) or modeling (Kienberger et al., 2012). Preston et al. (2011) review modeling approaches and distill them into social vulnerability, pressure-and-release, and expanded vulnerability models (ibid. 185–186). Yet other studies compile indices of vulnerability by combining variables using either non-statistical (Dickin et al., 2013; Busby et al., 2014; You and Chamberlin, 2004) or statistical methods (Jakobsen, 2013). Such research provides not only a more nuanced understanding of where people are vulnerable, but importantly for development donors, also insight into the variables related to observed vulnerability.

Research problem and outcome

We drew upon the literature cited earlier to assist the United States Agency for International Development (USAID) to understand the distribution of vulnerability in Niger. This analysis was a first of its kind for USAID. Our research team was provided the question: "In a region plagued by chronic poverty, instability, conflict and rising extremism, what are the strategies and resources that USAID can employ to help the region achieve sustainable, peaceful development?", and asked to provide specific, scientifically justifiable details about (1) the distribution of vulnerability in Niger and (2) the most important socio-economic and demographic factors that correlate to such vulnerability. Our study results were used in a development planning exercise that "gamed" possible outcomes from shifting resources between different sectors (health, environment, security, and agriculture) and places. The game results, alongside our analysis, was presented to the administrator of USAID and the head of the US Special Operations Command (USSOCOM) to inform the development of a collaborative strategy to address the twin problems of poverty and conflict in Sahelian West Africa (USAID, 2014).

USAID's recent push to use open, publicly available data and the unclassified nature of the "development game" led us to use a very large household survey dataset. While we identified two nationally representative, georeferenced datasets, the World Bank's 2011–2012 Living Standards Measurement Study (LSMS) and the 2012 Demographic and Health Survey (DHS), the DHS did not have specific modules covering household shocks or rural income generative activities. The LSMS was conducted in 2011–2012, which was an average year in terms of environmental and economic context and included over 4,000 households. Because the survey included a module on self-reported household shocks, we were able to create five shock vectors of interest to us: agricultural and livestock, weather, food price, financial, and health.

Methodological approach and agency perspective

In formulating our methodological approach we were careful to align testable hypotheses to the major sectors and practices within the agency. For example, in assessing vulnerability to shocks we included relevant demographic (gender, age, ethnicity) and community variables that resonated well with gender and democracy and governance professionals. Accommodating rapidly changing political and fiscal realities of the agency was challenging as well. Initially, we intended to incorporate into our study data mining algorithms focused on links among vulnerability, violent extremism, and terrorism. However, fiscal tightening at DOD and territorial sensitivities severed our ties to our analytical counterparts at USSOCOM. To ensure that our research was open, reproducible, and scalable we meticulously documented all data modifications and analysis in reproducible scripts posted to GitHub, a version control system for facilitating research reproducibility (Ram, 2013).

We used geospatial analysis techniques to highlight the highly varied nature of vulnerability in the study area. Such techniques have the advantage of creating detailed maps of the distribution of selected variables and regression results over space. Given that most USAID members participating in this exercise were unfamiliar with either large datasets and/or geospatial analysis, we employed a step-by-step approach to presenting our results. We first produced simple maps showing the distribution of variables and then we moved into more complex maps of interpolated variables. We concluded by mapping the household vulnerability to each shock with insets of correlated socio-economic and demographic variables.

Our empirical approach consists of exploratory data analysis in the form of interpolated maps, and three regression techniques: (1) logistic regression with geographic fixed-effects, (2) spatial filter models, and (3) localized geographically weighted regressions (GWR). To meet our robustness criteria, we focus on results that are statistically significant across the three approaches. Figure 30.1 depicts interpolated surfaces of the five shocks considered for zones falling within 50 kilometers of sampled households. The five smaller maps show the spatial distribution of each shock while the larger map reflects an aggregate measure of vulnerability. From the exploratory data analysis it is clear agricultural, weather, and food price shocks are the most common across the country, with apparent clustering of these shocks in the west.

To estimate the log odds of a household experiencing one of the five shocks we use a logistic regression model with a linear combination of exogenous household, community, and environmental factors serving as the independent variables. We account for potential spatial

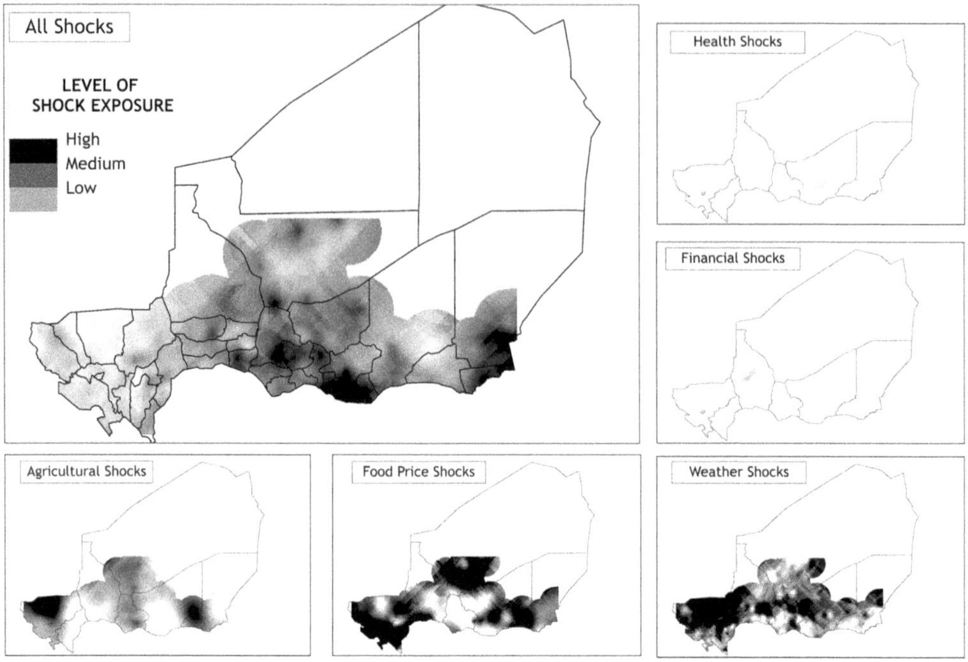

Figure 30.1 Composite shock surface derived from kriging estimates

autocorrelation in the errors of our logistical regression model by estimating a spatial filter model. Spatial filter models have been shown to reduce spatial misspecification error, improve the overall fit of the model, and improve the distributional shape of the model residuals which can help increase the homoscedasticity of model residuals (Thayn and Simanis, 2013). Finally, to investigate spatial heterogeneity we estimate a localized response surface using GWR. This technique differs from the others in that it estimates a local model of the dependent variable by fitting an ordinary linear regression equation to a neighborhood around each feature in the dataset. This gives us a set of regression coefficients, standard errors, and goodness of fit statistics for each household, based solely upon the surrounding area. This can elucidate spatial patterns in the data, such as isolating regions where the relationship between two variables differs from the rest of the country, suggesting the presence of unique local conditions. To estimate each model we used Stata, R, and ArcGIS 10.2 Spatial Statistics Toolbox, respectively.

Data for development: mapping results to agency operations and strategy

In presenting our results to agency practitioners and policy makers we avoided complex numerical tables and statistical jargon and instead focused on simple graphics and maps. For example, the statistical results from our regression models are presented as forest plots for easy intra- and inter-model comparisons, while a thematic map is used to highlight the effect of geography on household vulnerability.

Figures 30.2 and 30.3 summarize the results from the logistic and spatial filter regression models. Certain demographic and community service variables appear to have strong, robust correlations with household shocks. That female-headed households appear to be more

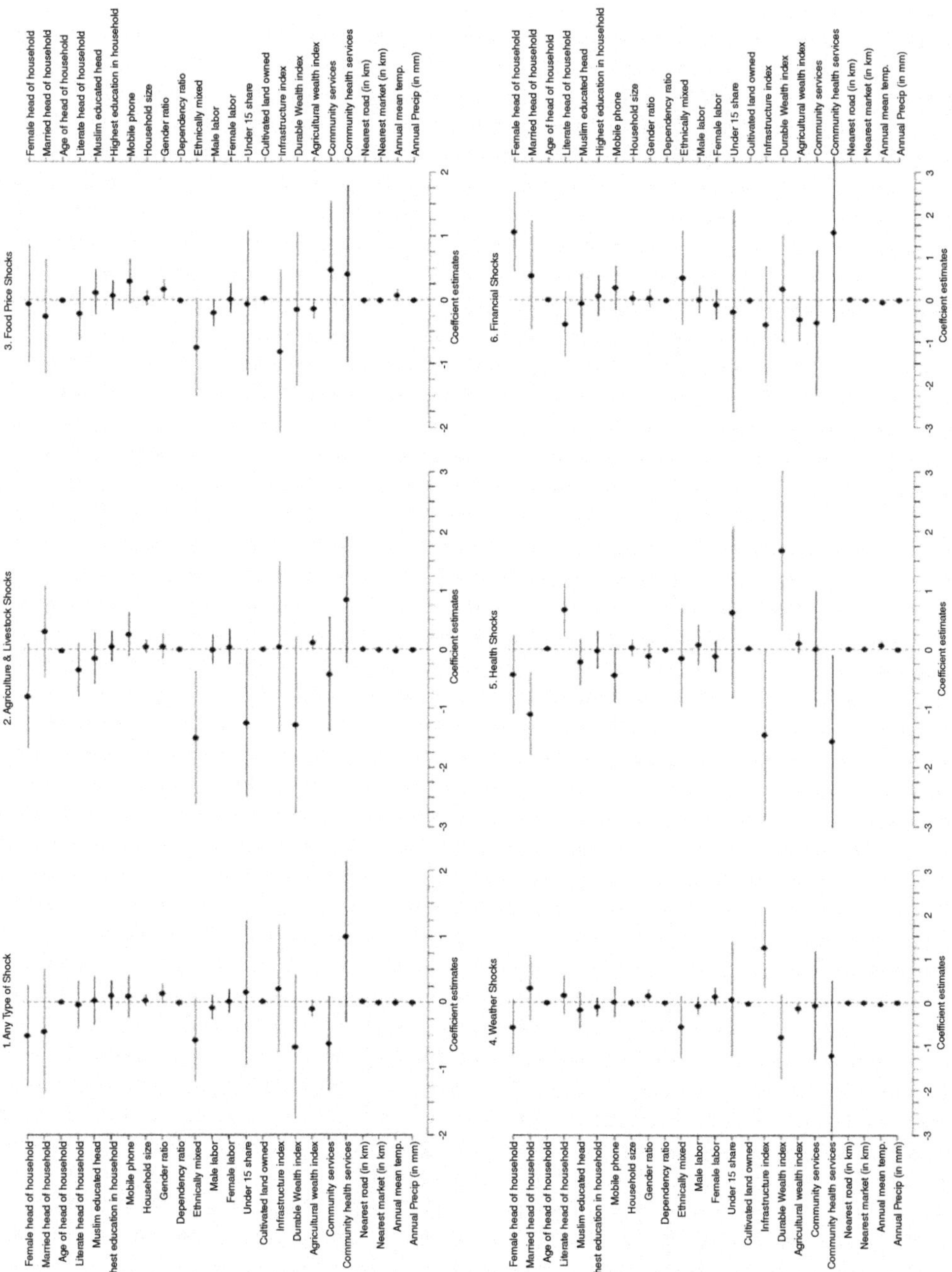

Figure 30.2 Summary of logistic regressions on shocks

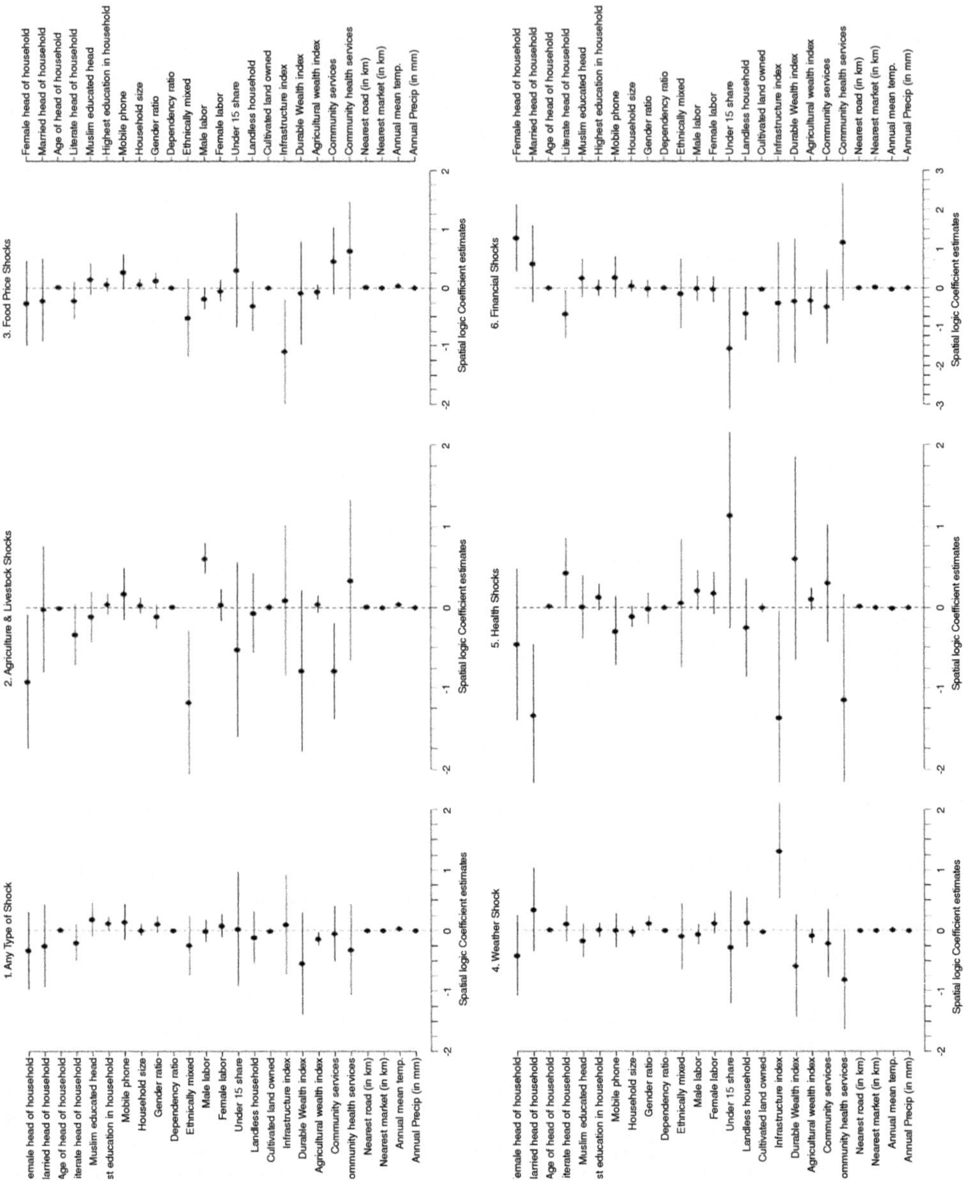

Figure 30.3 Spatial filter logistic regression results

vulnerable to financial shocks and less vulnerable to agricultural shocks was surprising to many within the agency, but is consistent with the livelihoods literature documenting that female-headed households, across countries, are more likely to receive financial transfers (Winters et al., 2009). The strong negative effect of intra-household ethnic heterogeneity on vulnerability may be due to the binary variable capturing unobservable social, political, and financial ties that help insulate ethnically mixed households from the three major shocks. And the effect of being married appears be relevant in protecting against major health shocks.

The GWR results, of which two selected shocks are shown, highlight the spatial heterogeneity of the conditional correlations. For example, female-headed households both east of the capital, Niamey, and near the desert town of Agadez, showed vulnerability to agriculture and livestock shocks. In the east of Niger, near the town of Goure, households with many young people appear to be vulnerable to agriculture and livestock shocks. People living near Tahoua, many of whom reported having poor access to community services, also showed a higher vulnerability to agricultural and livestock shocks. Older-headed households near Agadez showed a greater likelihood to experience food price shocks.

As a final assessment of vulnerability, we also estimate a logistic geographic fixed-effects model based on a composite shock variable and the standard vector of household, community, and environmental factors. Figure 30.4 depicts the geographic fixed-effects results. It clearly shows that, even after controlling for observable factors, households located in specific geographic zones in the eastern part of Niger are more than ten times as likely to experience one of the shocks as similar households in the central part of Niger (Mayahi).

Affecting policy

Our analysis demonstrates that human vulnerability is (1) explicitly spatial and (2) can be explained through correlation to other spatial variables. The use of large datasets and scientifically validated statistical techniques helps legitimatize the outcomes of such research in the minds of policy makers. In our case, the maps we produced were much more palatable for use in the decision-making process than other types of visual representations such as charts, tables, or graphs. Policy makers needed guidance to interpret the output maps, but once explained, the maps became a powerful visual touchstone for discussion. Higher level decision makers were able to frame and reframe key points around the maps.

In the presentation of results, our maps were only intended to demonstrate to spatial nature of vulnerability; however, in the de-briefing to the principles, the power of the maps became evident. Decision makers raised questions about the process through which we generated the maps and queried us on the specific findings we presented. One of our agency stakeholders wrote to us to explain the impact of the analysis and mapping on decision-making in their particular office (or "Bureau" in USAID lingo):

> The Niger mapping project brought leadership attention to the Sahel. . . . DoD believes the tools we [USAID] possess are important to address underlying security challenges in the region before they become "hot zones" that require a military response.
>
> *(pers. comm. 06/2014)*

In addition to this impact, the head of USAID, Rajiv Shah, presented our results in several speeches and recommended that such data-driven exercises are needed in development decision-making.

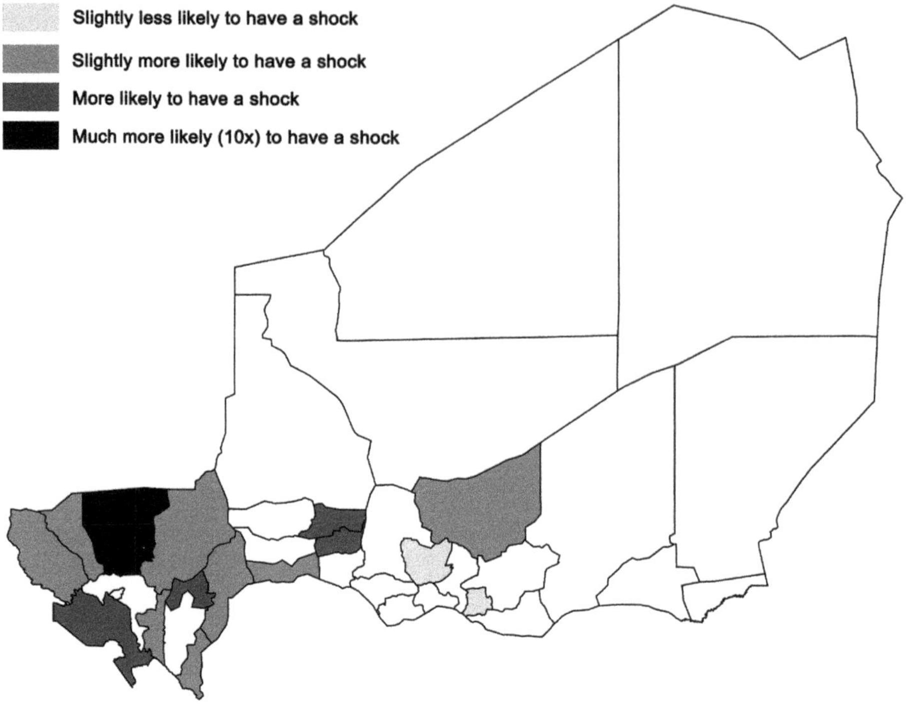

Figure 30.4 Any shock logistic regression results with geographic fixed-effects

In using both large datasets and spatial analysis to assist with development policymaking we learned three key principles. First, the research questions and methods must be relevant. They must actually address a question that decision makers need to answer. Once the question has been posed, then the appropriate data and tools can be selected. All too often, such efforts try to lead policy makers to an outcome by convincing them that a dataset, technique, or communication device (map, website, etc.) is useful without first having identified the actual and real need faced by policy makers. For instance, we presented vulnerability analysis as one way to understand motivating factors for people to join violent extremist organizations. Poverty and vulnerability often lead people to support or join such organizations. The problem was originally framed as an international drugs trafficking one, about which there is a dearth of specific data. By framing our question around vulnerability and by showing that we can understand the conditional correlations of that vulnerability, we were able to convince decision makers of the utility of the research. Another indication of the impact of this particular activity is that the agency has asked similar analysis be conducted for six additional countries.

Second, managing expectations is critical. We often had to explain that correlation does not imply causation and that interpolation was a mathematical estimate of a value where none actually existed. In our project, policy makers often took the fact that our dataset was large and open, that the analysis was statistical and that the maps were clear to imply a level of confidence that we often had to carefully clarify. Finally, we learned the importance of transparency in explaining exactly how we derived our maps and statistical outputs. It was essential that we detailed each step, even if that was not a priority of our working group, because the project was a demonstration of how open analysis and data can be used to inform agency strategy. We were clear about the limitations of the dataset and the analysis. To promote transparency and clarity of explanation, we developed logical, but simple, storylines to guide policy makers through our process of selecting and analyzing data and to explain the printed map results. These storylines removed statistical, geo-statistical and other academic "jargon" and used words such as "related to" in place of "conditional correlation", "related in space" in place of "geographically weighted regression" and pared down complexity into brief explanations.

Conclusion

In this chapter we have summarized our use of a large dataset to address a direct request from policy makers regarding human vulnerability in Niger. We realize that very few practitioners are so lucky as to have a clear question placed in front of them to be addressed. Our method was highly iterative (between we the researchers and the policy makers), which is also unusual. Our use of advanced statistical and geo-statistical techniques was equally unusual for our audience. What we hope we have shown here is that large, open datasets do have significant value for policy and decision makers, when presented in terms they can fully appreciate.

References

Abson, D., Dougill, A., and Stinger, L. 2012. Using principal component analysis for information-rich socio-ecological vulnerability mapping in Southern Africa. *Applied Geography* 35: 515–524.

Alwang, J., Jansen, H., Siegel, P., and Pichon, F. 2005. Geographic space, assets, livelihoods and well-being in rural Central America: Empirical evidence from Guatemala, Honduras and Nicaragua. DSGD Discussion Paper No. 26. International Food Policy Research Institute, Washington.

Amaransinghe, U., Samad, M., and Anputhas, M. 2005. Spatial clustering of rural poverty and food insecurity in Sri Lanka. *Food Policy* 30: 493–509.

Bellon, M., Hodson, D., Bergvinson, D., Beck, D., Martinex-Romero, E., and Montoya, Y. 2005. Targeting agricultural research to benefit poor farmers: Relating poverty mapping to maize environments in Mexico. *Food Policy* 30: 476–492.

Benson, T., Chamberlin, J., and Rhinehart, I. 2005. An investigation of the spatial determinants of the local prevalence of poverty in rural Malawi. *Food Policy* 30: 532–550.

Burke, W., and Jayne, T. 2010. Spatial disadvantages or spatial poverty traps: Household evidence from rural Kenya. ODI Working Paper 327. Overseas Development Institute, London.

Busby, J., Cook, K., Vizy, E., Smith, T., and Bekalo, M. 2014. Identifying hot spots of security vulnerability associated with climate change in Africa. *Climatic Change* 124: 717–731.

Davis, B. 2003. *Choosing a Method for Poverty Mapping*. Rome: FAO.

Deichmann, U. 1999. *Geographic Aspects of Inequality and Poverty*. World Bank Research Paper. www.worldbank.org/poverty/inequal/index.htm

Dickin, S., Schuster-Wallace, C., and Elliott, S. 2013. Developing a vulnerability mapping methodology: Applying the water-associated disease index to dengue in Malaysia. *PLoS One* 8(5): 1–11.

Farrow, A., Larrea, C., Hyman, G., and Lema, G. 2005. Exploring the spatial variation of food poverty in Ecuador. *Food Policy* 30: 510–531.

Fekete, A. 2012. Spatial disaster vulnerability and risk assessments: Challenges in their quality and acceptance. *Natural Hazards* 61: 1161–1178.

Grolle, J. 2013. High-resolution mapping of rural poverty and famine vulnerability in the Sahel: A possible approach for the Republic of Niger. *Population and Environment* 35: 68–97.

Jakobsen, K. 2013. Livelihood asset maps: A multidimensional approach to measuring risk-management capacity and adaptation policy targeting – a case study in Bhutan. *Regional Environmental Change* 13: 219–233.

Kam, S., Hossain, M., Lal Bose, M., and Villano, L. 2005. Spatial patterns of rural poverty and their relationship with welfare-influencing factors in Bangladesh. *Food Policy* 30: 551–567.

Kienberger, S., Blaschke, T., and Zaidi, R. 2013. A framework for spatio-temporal scales and concepts from different disciplines: The vulnerability cube. *Natural Hazards* 68: 1343–1369.
King, B., and McCusker, B. 2007. Environment and development in the former South African bantustans. *The Geographical Journal* 137(1): 6–12.
Krisjanson, P., Radeny, M., Baltenweck, I., Ogutu, J., and Notenbaet, A. 2005. Livelihood mapping and poverty correlates at a meso-level in Kenya. *Food Policy* 30: 568–583.
LaRovere, R., Keulen, H., Hiernaux, P., Szonyi, J., and Schipper, R. 2008. Intensification scenarios in south-western Niger: Implications for revisiting fertilizer policy. *Food Policy* 33: 156–164.
Malcomb, D., Weaver, E., and Krakowka, R. 2014. Vulnerability modeling for sub-Saharan Africa: An operationalized approach in Malawi. *Applied Geography* 48: 17–30.
Minot, N., and Baulch, B. 2005. Spatial patterns of poverty in Vietnam and their implications for policy. *Food Policy* 30: 461–475.
O'Brien, K., and Leichenko, R. 2000. Double exposure: Assessing the impacts of climate change within the context of economic globalization. *Global Environmental Change* 10: 221–232.
Preston, B., Yuen, E., and Westaway, R. 2011. Putting vulnerability to climate change on the map: A review of approaches, benefits, and risks. *Sustainability Science* 6: 177–202.
Ram, K. 2013. Git can facilitate greater reproducibility and increased transparency in science. *Source Code for Biology and Medicine* 8(7): 1–8.
Robinson, T., Emwanu, T., and Rogers, D. 2007. Environmental approaches to poverty mapping: An example from Uganda. *Information Development* 23(2–3): 205–215.
Thayn, J.B., and Simanis, J., 2013. Accounting for spatial autocorrelation in linear regression models using spatial filtering with eigenvectors. *Annals of the Association of American Geographers* 103(1): 47–66.
Turner, B. L., et.al. 2003. A framework for vulnerability analysis in sustainability science. *Proceeding of the National Academy of Sciences* 100(14): 8074–8079.
USAID. 2014. *Development Game Players Handbooks*. Washington, DC: Internal USAID Publication.
Winters, P., Davis, B., Carletto, G., Covarrubias, K., Quinones, E. J., Zezza, A., Azzarri, C., and Stamoulis, K. 2009. Assets, activities, and rural income generation: Evidence from multicountry analysis. *World Development* 37: 1435–1452.
Wisner, B., et al. (2004). *At Risk: Natural Hazards, People's Vulnerability and Disasters*. New York: Routledge.
You, L., and Chamberlin, J. 2004. Spatial analysis of sustainable livelihood enterprises of Uganda cotton production. EPTD Discussion Paper No. 121. International Food Policy Research Institute, Washington, DC.

Appendix

Table A.1 Descriptive statistics for household, community, and weather variables (projected using sampling weights)

	Mean	*Standard error*	*Lower bound*	*Upper bound*
Female-headed household	8%	0.009	0.07	0.10
Married head of household	93%	0.008	0.91	0.94
Age of head of household	45	0.581	43	46
Literate head of household	26%	0.020	0.22	0.30
Muslim-educated head of household	28%	0.021	0.24	0.32
Highest level of education in household	76%	0.060	0.64	0.88
Mobile phone ownership	41%	0.020	0.37	0.45
Household size	6.62	0.128	6.37	6.87
Number of males divided by females in HH	1.23	0.032	1.16	1.29
Dependency ratio	143	3.105	137	149
Ethnically mixed household	6%	0.011	0.04	0.09
Household male labor age>11 & <60	1.44	0.040	1.36	1.52
Household female labor age>11 & <60	1.69	0.044	1.61	1.78
Share of household members under 15	49%	0.007	0.48	0.51
Household is landless	31%	0.032	0.25	0.38
Logged value of cultivated land owned	5.83	0.303	5.23	6.43
Infrastructure index*	0.10	0.010	0.08	0.12
Durable wealth index*	0.18	0.007	0.17	0.20
Agricultural wealth index*	0.03	0.037	-0.05	0.10
Community service index*	0.19	0.023	0.14	0.23
Community health index*	0.03	0.021	-0.01	0.07
HH distance in (KMs) to nearest major road	12.62	1.625	9.40	15.84
HH distance in (KMs) to nearest market	59.89	2.780	54.38	65.40
Annual mean temperature (°C)	28.20	0.045	28.11	28.29
Annual precipitation (mm)	395	6.762	382	409

* All indices are based on factor analysis and final indices were checked for reliability using Cronbach's alpha. N=1,792

Table A.2 Logit analysis of shock variables with geographic fixed-effects

	Any	*Ag/Livestock*	*Food price*	*Weather*	*Health*	*Financial*
Female-headed household	-0.4943	-0.7961*	-0.0590	-0.5553*	-0.4224	1.5981***
	(0.39)	(0.45)	(0.47)	(0.31)	(0.33)	(0.48)
Married head of household	-0.4383	0.3039	-0.2587	0.3420	-1.0880***	0.5843
	(0.48)	(0.40)	(0.46)	(0.38)	(0.36)	(0.65)
Age of head of household	0.0085*	-0.0142***	0.0043	0.0100*	0.0123	0.0131
	(0.00)	(0.01)	(0.01)	(0.01)	(0.01)	(0.01)
Literate head of household	-0.0234	-0.3255	-0.1996	0.1687	0.6610***	-0.5721
	(019)	(0.23)	(0.22)	(0.23)	(0.23)	(0.39)
Muslim educated head of household	0.0314	-0.1621	0.1188	-0.1486	-0.2056	-0.0725
	(0.18)	(0.22)	(0.19)	(0.21)	(0.20)	(0.35)
Highest level of education in household	0.0603	0.0099	0.0365	-0.0376	0.0095	0.0643
	(0.07)	(0.08)	(0.07)	(0.07)	(0.10)	(0.15)
Mobile phone ownership	0.1053	0.2622	0.2993*	0.0145	-0.4388*	0.2967
	(0.16)	(0.19)	(0.17)	(0.18)	(0.24)	(0.26)
Household size	0.0358	0.0509	0.0375	0.0058	0.0266	0.0491
	(0.04)	(0.06)	(0.06)	(0.05)	(0.07)	(0.08)
Number of males divided by females in HH	0.1387*	0.0522	0.1702**	0.1724**	-0.1090	0.0525
	(0.07)	(0.11)	(0.07)	(0.07)	(0.10)	(0.11)
Dependency ratio	0.0003	0.0022	0.0007	-0.0000	-0.0017	0.0019
	(0.00)	(0.00)	(0.00)	(0.00)	(0.00)	(0.00)
Ethnically mixed household	-0.5658*	-1.5001***	-0.7386*	-0.5430	-0.1397	0.5241
	(0.32)	(0.57)	(0.39)	(0.37)	(0.42)	(0.57)
Household male labor age >11 &<60	-0.0688	0.0007	-0.2013*	-0.0691	0.0700	0.0168
	(0.09)	(0.13)	(0.11)	(0.10)	(0.17)	(0.16)
Household female labor age >l1 & <60	0.0204	0.0427	0.0232	0.1502	-0.1163	-0.0973
	(0.09)	(0.15)	(0.12)	(0.10)	(0.13)	(0.18)
Share of household members under 15	0.1614	-1.2622**	-0.0582	0.0956	0.6294	-0.2660

	Any	*Ag/Livestock*	*Food price*	*Weather*	*Health*	*Financial*
	(0.55)	(0.63)	(0.58)	(0.66)	(0.74)	(1.21)
Logged value of cultivated land owned	0.0273*	0.0028	0.0318*	-0.0181	0.0187	0.0020
	(0.02)	(0.02)	(0.02)	(0.01)	(0.02)	(0.03)
Infrastructure index	0.2244	0.0425	-0.8058	1.2482***	-1.4413*	-0.5714
	(0.49)	(0.74)	(0.65)	(0.46)	(0.74)	(0.71)
Durable wealth	-0.6620	-1.2728*	-0.1378	-0.7810	1.6702"	0.2571
	(0.55)	(076)	(0.62)	(0.49)	(0.69)	(0.64)
Agricultural wealth index	-0.0979	0.1128**	-0.1379*	-0.1215*	0.1053	-0.4479
	(0.07)	(0.05)	(0.08)	(0.07)	(0.08)	(0.27)
Community service index	-0.6172*	-0.4312	0.4663	-0.0526	0.0048	-0.5368
	(0.36)	(0.49)	(0.55)	(0.63)	(0.50)	(0.87)
Community health services	1.0083	0.8572	0.4172	-1.2117	-1.567**	1.5818
	(0.67)	(0.55)	(0.71)	(0.87)	(0.75)	(1.07)
HH distance in (KMs) to nearest major road	0.0205***	0.0072	0.0026	0.0013	0.0026	0.0084
	(0.01)	(0.01)	(0.01)	(0.01)	(0.01)	(0.01)
HH distance in (KMs) to nearest market	0.0011	-0.0016	0.0033	0.0055	0.0046	0.0006
	(0.00)	(0.00)	(0.00)	(0.00)	(0.00)	(0.00)
Annual mean temperature (°C * 10)	0.0052	-0.0162	0.0746	-0.0308	0.0678"	-0.0456
	(0.03)	(0.03)	(0.05)	(0.03)	(0.03)	(0.04)
Annual precipitation (mm)	-0.0007	-0.0018	0.0014	-0.0026	-0.0005	0.0034
	-0.4943	-0.7961*	-0.0590	-0.5553*	-0.4224	1.5981***
Geographic fixed-effects	Department	Regional	Regional	Regional	Regional	Regional

Department fixed-effects included in estimation but not reported. Stars indicate level of significance: ***p<0.01, ** p<0.05, * p<0.10. Standard errors, in parentheses, account for sampling design and were estimated using Stata/SE 13.1's survey package (svy) for complex survey data. Sample size equal to 1,752 observations.

Table A.3 Spatial filter logistic regression results

	Any		*Ag/Livestock*		*Food price*		*Weather*		*Health*		*Financial*	
	Est.	*Std Error*	*Est.*	*Std Error*	*Est.*	*Std Error*	*Est.*	*Std Error*	*Est.*	*Std Error*	*Est.*	*Std Error*
Female head of household	-0330	0323	-0.921**	0.418	-0369	0368	-0.408	0334	-0.459	0.473	1367***	0.437
Married head of household	-0251	0344	-0.027	0395	-0321	0362	0342	0347	-1334**	0.441	0.609	0.500
Age of head of household	0.015***	0.005	-0.008	0.005	0.010**	0.005	0.010**	0.004	0.008	0.007	0.006	0.008
Literate head of household	-0.196	0.151	-0342**	0.191	-0323	0.165	0.106	0.152	0.423**	0315	-0.682**	0316
Muslim educated head of household	0.1S1	0.139	-0.119	0.158	0.146	0.139	-0.165	0.138	0.002	0300	0347	0352
Highest level of education in household	0.1 I8**	0.0 58	0.035	0.064	0.052	0.060	0.009	0.055	0.126	0.082	-0.006	0.103
Mobile phone ownership	0.143	0.145	0.162	0.162	0365*	0.149	0.000	0.141	-0395	0315	0371	0363
Household size	0.005	0.047	0.023	0.049	0.051	0.045	-0.014	0.042	-0.114*	0.066	0.041	0.076
Number of males divided by females in HH	0.105	0.067	-0.116	0.075	0.125*	0.065	0.124**	0.060	-0.016	0.095	-0.025	0.113
Dependency ratio	0.001	0.001	0.002	0.001	-0.001	0.001	0.001	0.001	0.000	0.002	0.003**	0.002
Ethnically mixed household	-0346	0347	-1.173***	0.450	-0.516	0341	-0.097	0375	0.053	0.403	-0.155	0.460
Household male lab or age>11 &<60	-0.006	0.092	0.605	0.097	-0.191**	0.090	-0.060	0.083	0304	0.124	-0.019	0.158
Household female labor age>11 &<60	0.0 82	0.096	0.027	0.099	-0.051	0.094	0.123	0.087	0.174	0.130	-0.043	0.161

Share of household members under 15	0.022	0.478	-0.523	0.551	0392	0.499	-0374	0.471	1.137	0.713	-1.560**	0.787
Household is landless	-0.112	0315	-0.073	0349	-0314	0318	0.131	0306	-0351	0305	-0.663*	0358
Logged value of cultivated land owned	-0.009	0.020	0.004	0.023	-0.005	0.021	-0.013	0.019	-0.007	0.029	-0.043	0.032
Infrastructure index	0.096	0.414	0.082	0.472	-1.097**	0.454	1315****	0399	-1363**	0.670	-0394	0.791
Durable wealth	-0.545	0.427	-0.787	0.507	-0.093	0.447	-0.585	0.430	0.599	0.638	-0349	0.811
Agricultural wealth index	-0.133**	0.052	0.035	0.053	-0.071	0.059	-0.085	0.058	0.096	0.069	-0339*	0.183
Community service index	-0.054	0328	-0.787***	0398	0.457	0388	-0311	0384	0395	0368	-0.498	0.482
Community health services	-0321	0383	0334	0.506	0.629	0.425	-0.811*	0.422	-1.140*	0.660	1.160	0.769
HH distance in (KMs) to nearest major road	0.008	0.004	0.004	0.004	0.012***	0.004	0.002	0.004	0.011*	0.006	-0.003	0.007
HH distance in (KMs) to nearest market	0.003	0.002	-0.004	0.002	-0.002	0.003	0.002	0.002	0.000	0.003	0.009**	0.004
Annual Mean Temperature (°C * 10)	0.030****	0.008	0.035***	0.011	0.037***	0.012	0.013	0.009	-0.008	0.012	-0.033**	0.014
Annual Precipitation (mm)	-0.001**	0.001	-0.006****	0.001	0.002*	0.001	-0.001	0.001	0.000	0.001	0.005**	0.002
Eigenvalues retained in order)	1,11,12		10,11,5,	8,2	3,5,2,1,8,	, 12,9,4	10,11,6		8, 10,2,	7	12,3,6,7,	11,1, 10

Spatial filter variables included in estimation but not reported. Stars indicate level of significance: ****p<0.001, ***p<0.01, ** p<0.05, * p<0.10. Sample size equal to 1,751 observations.

INDEX

Page numbers in *italics* indicate figures and in **bold** indicate tables on the corresponding pages.

For Product Safety Concerns and Information please contact our EU
representative GPSR@taylorandfrancis.com
Taylor & Francis Verlag GmbH, Kaufingerstraße 24, 80331 München, Germany

www.ingramcontent.com/pod-product-compliance
Ingram Content Group UK Ltd.
Pitfield, Milton Keynes, MK11 3LW, UK
UKHW050232200926
475702UK00020B/2980
9780367710231